LOGIC, SEMANTICS,
METAMATHEMATICS

LOGIC, SEMANTICS, METAMATHEMATICS

PAPERS FROM 1923 TO 1938

BY

ALFRED TARSKI

TRANSLATED BY

J. H. WOODGER

OXFORD

AT THE CLARENDON PRESS

Oxford University Press, Ely House, London W. 1

GLASGOW NEW YORK TORONTO MELBOURNE WELLINGTON
CAPE TOWN SALISBURY IBADAN NAIROBI LUSAKA ADDIS ABABA
BOMBAY CALCUTTA MADRAS KARACHI LAHORE DACCA
KUALA LUMPUR SINGAPORE HONG KONG TOKYO

FIRST PUBLISHED 1956

REPRINTED LITHOGRAPHICALLY IN GREAT BRITAIN
AT THE UNIVERSITY PRESS, OXFORD
BY VIVIAN RIDLER
PRINTER TO THE UNIVERSITY
1969

TO HIS TEACHER
TADEUSZ KOTARBIŃSKI
THE AUTHOR

TRANSLATOR'S PREFACE

THE setting free of Poland after the First World War was followed by intensive activity in her Universities. In the departments of philosophy and mathematics this took the form, in a number of places, of new and powerful investigations in the fields of mathematical logic, the foundations of mathematics, and the methodology of the sciences. Prominent in this movement was the Warsaw school led by Łukasiewicz, Kotarbiński, and Leśniewski. Under their skilled guidance a younger generation grew up and among these Alfred Tarski quickly distinguished himself.

Ever since I first enjoyed the hospitality of Professor Tarski in Warsaw in 1936, it has seemed to me that the importance and scope of the Polish school of logicians were insufficiently known and appreciated in the English-speaking world. Then came the Second World War, bringing ruin once more to Poland, killing men, destroying laboratories, and burning manuscripts and libraries. After this war it occurred to me that I should be performing a public service, as well as acknowledging in some small measure my debt to my Polish friends, if I prepared a collected edition of some of Professor Tarski's publications in English translation. When he visited England in 1950 to deliver the Sherman Lectures at University College, London, I mentioned my plan to him and received his approval.

This volume contains Tarski's major contributions to logic, semantics, and metamathematics published before the Second World War; their arrangement here corresponds to the chronological order in which they first appeared in print. With the exception of articles II and XI (which are too closely connected with, and too often referred to in, the other articles to be omitted), the volume does not include Tarski's studies in the foundations of special mathematical disciplines—set theory, group theory, etc. Neither does it contain his papers of a pronouncedly mathematical character, dealing with special topics from the domain of set theory, measure theory, abstract algebra, elementary geometry, etc. Also excluded are short notes, abstracts, and preliminary reports which are closely related to some of the articles included in the volume and the contents of which are more fully presented in these articles. A longer paper from the domain of logic and methodology which has been omitted is 'Sur la méthode déductive' in *Travaux du IXe Congrès International de Philosophie* (Paris 1937); this paper is a purely expository one, and its ideas have been fully developed in Chapter VI of Tarski's book *Introduction to logic* (New York, 1941).

In a sense the present work is more than a volume of translations. Naturally an attempt has been made to remove the misprints and errors which occur in the originals. Moreover, the articles have been provided by the author with cross-references to other articles in the volume and with notes referring to later developments and recent literature. Occasionally some new remarks have been added for the purpose of clarifying certain passages in the original text. Articles II and VI contain more serious changes, Tarski having inserted in them several passages which, he hopes, will help to clarify and amplify their contents.

While the work of translation was in progress, the passages which seemed to me doubtful and difficult were noted down and sent to Professor Tarski in Berkeley, California. In this way it was possible to adjust the text of the translations in many points, so as to meet the author's wishes. However, in view of the time limit and the geographical distance between the residences of the author and the translator, it was impracticable to send the whole manuscript to Professor Tarski before it was set in print. Instead he received galley proofs, on which for obvious reasons he could not suggest too extensive changes. It also proved impossible to discuss the suggested changes in detail, and it was left to my decision which changes were actually to be carried through. Thus Professor Tarski is not responsible for the final text or the technical aspect of the book.

Three articles in this volume are joint publications of Tarski and other authors: Professor C. Kuratowski (article VII), Dr. A. Lindenbaum (article XIII), and Professor J. Łukasiewicz (article IV). We are greatly obliged to Professors Kuratowski and Łukasiewicz for their permission to include the translations of the jointly published articles in the volume, Dr. Lindenbaum fell a victim to the Gestapo during the war.

The papers included in the volume originally appeared in the following periodicals and collective works: *Actes du Congrès International de Philosophie Scientifique* (articles XV and XVI), *Comptes Rendus de la Société des Sciences et des Lettres de Varsovie* (articles III and IV), *Ergebnisse eines mathematischen Kolloquiums* (articles XIII and XIV), *Erkenntnis* (article X), *Fundamenta Mathematicae* (articles I, VI, VII, XI, XII, and XVII), *Księga Pamiątkowa Pierwszego Polskiego Zjazdu Matematycznego* (article II), *Monatshefte für Mathematik und Physik* (articles V and IX), *Przegląd Filozoficzny* (articles I, X, XV, and XVI), and *Travaux de la Société des Sciences et des Lettres de Varsovie* (article VIII). Acknowledgements should be made to the publishers and editors of these periodicals for their generosity.

I am obliged to Mr. S. W. P. Steen for kindly reading the proofs of article VIII and throughout the work I have received much help from my son, Michael Woodger. We are also indebted to several colleagues and students of Professor Tarski—Dr. C. C. Chang, Professor A. C. Davis, Professor J. Kalicki, Mr. R. Montague, Professor J. Myhill, Professor D. Rynin, and Mr. D. Scott—for their assistance in revising the original text of the articles and in reading galley proofs.

Finally it is a pleasure to acknowledge the courtesy and help which we have received from the staff of the Clarendon Press at all stages in the production of the book.

J. H. W.

AUTHOR'S ACKNOWLEDGEMENTS

IT is a rare privilege for an author to see a volume of his collected papers published during his lifetime, and especially so if the papers be translated into a single language from originals in a number of other languages. I cannot therefore but be deeply moved by the appearance of this volume and by the thought of the many and great sacrifices which its publication has laid upon my friend, Professor Joseph H. Woodger. For five long years he has devoted to this work an immense amount of effort and time, which otherwise could have been used for fruitful research in his chosen field, theoretical biology and its foundations.

The task of a translator is rarely a gratifying one. Circumstances have made it especially thankless in the present case; let me indicate some of them. The papers whose translations constitute the volume were originally published over a period of fifteen years and in several different languages. They vary considerably in subject-matter, style, and notation. Under these conditions, the task of combining the papers in one book provided with a reasonable degree of terminological consistency and conceptual uniformity presents extreme difficulties. In a few cases (in particular, in the case of the monograph on the concept of truth, which occupies nearly one-third of the volume) the translation had to be based not upon the original, which was published in Polish, but upon the French or German version. This made it even harder for the translator to give a fully adequate rendering of the original intentions and ideas of the author. In addition, due to the factors of space and time, the translator was deprived of the benefit of extensively discussing with the author even the major difficulties encountered in his work, and so achieving a meeting of minds before the text was set up in type. To illustrate this point I may mention that, for various reasons, I have been unable so far to read a considerable part of the present text, and it seems more than likely that I shall not have read it before receiving a copy of the published book. The realization of the difficulties involved makes me feel all the more indebted to him whose initiative, devotion, and labour have brought this volume into existence.

It is needless to say that I fully appreciate the assistance and consideration of all the persons mentioned in the translator's preface. But I should feel unhappy if at this place I did not make special mention of the man who helped me more than anyone else in my part of the job—my younger colleague, the late Jan Kalicki. He spent the last two years of his short life in Berkeley; he generously offered his help on the day of his arrival and continued it untiringly and patiently, with the greatest devotion and conscientiousness, until his last day. He studied the originals of the articles, translated for Professor Woodger various passages from the Polish text, prepared references to recent literature, discussed with me the remarks which I planned to insert in the translations, assumed most of the burden of the extensive correspondence connected with the publication of the book, and read the first batch of galley proofs. His tragic and untimely death (in November 1953) was a cause of considerable delay in the publication of this volume.

A. T.

University of California
Berkeley, August 1955

CONTENTS

I

ON THE PRIMITIVE TERM OF
LOGISTIC†

In this article I propose to establish a theorem belonging to
logistic concerning some connexions, not widely known, which
exist between the terms of this discipline. My reasonings are
based on certain sentences which are generally accepted among
logisticians.. But they do not depend on this or that particular
theory of logical types. Among all the theories of types which
could be constructed[1] there exist those according to which my
arguments in their present form are perfectly legitimate.[2]

The problem of which I here offer a solution is the following:
*is it possible to construct a system of logistic in which the sign of
equivalence is the only primitive sign* (in addition of course to
the quantifiers[3]) ?

This problem seems to me to be interesting for the following
reason. We know that it is possible to construct the system of
logistic by means of a single primitive term, employing for this
purpose either the sign of implication, if we wish to follow the

---

[1] The possibility of constructing different theories of logical types is also
recognized by the inventor of the best known of them. Cf. Whitehead, A. N.,
and Russell, B. (90), vol. 1, p. vii.

[2] Such a theory was developed in 1920 by S. Leśniewski in his course on
the principles of arithmetic in the University of Warsaw; an exposition of the
foundations of a system of logistic based upon this theory of types can be
found in Leśniewski (46), (47), and (47 b).

[3] In the sense of Peirce (see Peirce, C. S. (58 a), p. 197) who gives this name to
the symbols '$\prod$' (universal quantifier) and '$\sum$' (particular or existential
quantifier) representing abbreviations of the expressions: 'for every signification of the terms . . .' and 'for some signification of the terms . . .'.

---

† BIBLIOGRAPHICAL NOTE. This article constitutes the essential part of the
author's doctoral dissertation submitted to the University of Warsaw in 1923.
The paper appeared in print in Polish under the title 'O wyrazie pierwotnym
logistyki' in *Przegląd Filozoficzny*, vol. 26 (1923), pp. 68–89. A somewhat
modified version was published in French in two parts under separate titles:
'Sur le terme primitif de la Logistique', *Fundamenta Mathematicae*, vol. 4
(1923), pp. 196–200, and 'Sur les *truth-functions* au sens de MM. Russell et
Whitehead', ibid., vol. 5 (1924), pp. 59–74. The present English translation
is based partly on the Polish and partly on the French original.

B

example of Russell,[1] or by making use of the idea of Sheffer,[2] who adopts as the primitive term the sign of incompatibility, especially introduced for this purpose. Now, in order really to attain our goal, it is necessary to guard against the entry of any constant special term into the wording of the definitions involved, if this special term is at the same time distinct from the primitive term adopted, from terms previously defined, and from the term to be defined.[3] The sign of equivalence, if we employ it as our primitive term, presents from this standpoint the advantage that it permits us to observe the above rule quite strictly and at the same time to give to our definitions a form as natural as it is convenient, that is to say the form of equivalences.

The theorem which will be proved in § 1 of this article,

$$[p,q] :: p.q \; . \; \equiv \; :.[f] :. p \equiv :[r].p \equiv f(r). \; \equiv \; .[r].q \equiv f(r)^4$$

constitutes a positive answer to the question raised above. In fact it can serve as a definition of the symbol of logical product in terms of the equivalence symbol and the universal quantifier; and as soon as we are able to use the symbol of logical product, the definitions of other terms of logistic do not present any difficulty, as appears, e.g. from the following sentences:

$$[p] :. \sim (p) \equiv : p \equiv .[q].q,$$
$$[p,q] :. p \supset q . \equiv : p \equiv .p.q,$$
$$[p,q] : p \vee q . \equiv . \sim (p) \supset q.^5$$

[1] See Russell, B. (61 d), pp. 16–18.

[2] See Sheffer, H. M. (63 a). See also Russell, B. (61 e).

[3] In this article we regard definitions as sentences belonging to the system of logistic. If therefore we were to use some special symbol in formulating definitions we could hardly claim that only one symbol is accepted in our system as a primitive term. It may be mentioned that, in the work of Whitehead and Russell cited above all the definitions have the form 'a = b Df.' and thus actually contain a special symbol which occurs neither in axioms nor in theorems; it seems, however, that these authors do not treat definitions as sentences belonging to the system.

[4] In this note I adopt the notation of Whitehead and Russell with some slight modifications; in particular, instead of expressions of the form 'ϕx' I write '$\phi(x)$'; also the use of dots differs in some details from that in *Principia Mathematica*.

[5] Some further developments related to this result are contained in the doctoral dissertation of Henry Hiż, 'An economic foundation for arithmetic', Harvard University, 1948; see also Hiż, Henry (31 a).

It will be seen from the discussion in § 2, that the results obtained can be considerably simplified within a system of logistic which contains the following sentence among its axioms or theorems:

$$[p,q,f]:p \equiv q.f(p). \supset f(q).$$

However, this sentence cannot be proved or disproved within any of the systems of logistic which are known from the literature. This gave rise to a further study of the sentence in question, and in particular to a search for other sentences equivalent to it. The results obtained will be presented in §§ 3–6 of this work.

§ 1. Fundamental Theorem

I start by introducing a few definitions, Def. 1–3, which will be used in this and the following sections. Then I give several lemmas, Th. 1–10, and finally prove Th. 11 which is the main result of this work.

The proofs offered are strictly speaking incomplete; they must be regarded as commentaries indicating the course of the reasoning. The structure of these commentaries is in part borrowed from Whitehead and Russell; they do not, I think, require more detailed explanation.

DEF. 1. $[p]:tr(p) \equiv .p \equiv p$

DEF. 2. $[p].as(p) \equiv p$

DEF. 3. $[p]:fl(p) \equiv .p \equiv \overset{.}{\sim}(p)$

For symmetry, one could also introduce the following definition:

DEF. 2'. $[p].ng(p) \equiv \sim(p).$

However, this definition would be quite useless, since the symbol 'ng' would have the same meaning as the negation symbol '$\sim$' already occurring in logic.

TH. 1. $[p].tr(p)$ (Def. 1)

TH. 2. $[p]:[q].p \equiv tr(q). \supset p$

Proof.[1] $[p]:.Hp. \supset :$

(1) $\qquad\qquad\qquad p \equiv .[q].tr(q):$

$\qquad\qquad\qquad\qquad Cn$ $\qquad\qquad\qquad\qquad\qquad$ (1, Th. 1)

TH. 3. $[p,q]:p \supset .p \equiv tr(q)$

Proof. $[p]:Hp. \supset .$

(1) $\qquad\qquad\qquad\qquad tr(q)$ $\qquad\qquad\qquad\qquad$ (Th. 1)

$\qquad\qquad\qquad\qquad Cn$ $\qquad\qquad\qquad\qquad\qquad$ (Hp, 1)

TH. 4. $[p]:[q].p \equiv tr(q). \equiv p$ $\qquad\qquad$ (Th. 2, Th. 3)

TH. 5. $[p,q]::[f]:.p \equiv :[r].p \equiv f(r). \equiv .[r].q \equiv f(r):. \supset q$

Proof. $[p,q]::Hp:. \supset :.$

(1) $\qquad\qquad p \equiv :[r].p \equiv tr(r). \equiv .[r].q \equiv tr(r):.$

(2) $\qquad\qquad [r].p \equiv tr(r). \equiv p:$ $\qquad\qquad$ (Th. 4)

(3) $\qquad\qquad [r].q \equiv tr(r). \equiv q:$ $\qquad\qquad$ (Th. 4)

(4) $\qquad\qquad p \equiv .p \equiv q:$ $\qquad\qquad\qquad$ (1, 2, 3)

(5) $\qquad\qquad p \equiv p. \equiv q:$ $\qquad\qquad\qquad$ (4)[2]

(6) $\qquad\qquad p \equiv p$

$\qquad\qquad Cn$ $\qquad\qquad\qquad\qquad\qquad\qquad$ (5, 6)

TH. 6. $[p]. \sim ([q].p \equiv as(q))$

Proof. $[p].$

(1) $\qquad\qquad\qquad \sim (p \equiv \sim (p)).$

(2) $\qquad\qquad\qquad \sim ([q].p \equiv q).$ $\qquad\qquad\qquad$ (1)

$\qquad\qquad\qquad \sim ([q].p \equiv as(q))$ $\qquad\qquad$ (1, Def. 2)

TH. 7. $[p,q]:[r].p \equiv as(r). \equiv .[r].q \equiv as(r)$

[1] In the proofs of theorems which have the form of a conditional sentence, the symbol 'Hp' stands for the hypothesis of the theorem, and the symbol 'Cn' for its conclusion.

[2] I pass from (4) to (5) by making use of the following theorem which was kindly communicated to me by J. Lukasiewicz:

$$[p, q, r]:.p \equiv .q \equiv r: \equiv :p \equiv q. \equiv r.$$

This theorem, the proof of which is quite easy and is here omitted, expresses an interesting property of equivalence which is analogous to the associative property of logical addition and logical multiplication.

Proof. $[p,q]$:

(1) $\qquad\qquad \sim([r].p \equiv as(r)).$ $\qquad\qquad\qquad$ (Th. 6)

(2) $\qquad\qquad \sim([r].q \equiv as(r))$: $\qquad\qquad\qquad$ (Th. 6)

$\qquad\qquad [r].p \equiv as(r). \equiv .[r].q \equiv as(r)$ $\qquad\qquad$ (1, 2)

TH. 8. $\quad [p,q]::[f]:.p \equiv :[r].p \equiv f(r). \equiv .[r].q \equiv f(r):. \supset p$

Proof. $[p,q]::Hp:. \supset :.$

(1) $\qquad\qquad p \equiv :[r].p \equiv as(r). \equiv .[r].q \equiv as(r):.$

$\qquad\qquad Cn$ $\qquad\qquad\qquad\qquad\qquad\qquad\qquad$ (1, Th. 7)

TH. 9. $\quad [p,q]::[f]:.p \equiv :[r].p \equiv f(r). \equiv .[r].q \equiv f(r):.$

$\qquad\qquad\qquad\qquad\qquad\qquad\qquad \supset .p.q$ $\quad$ (Th. 8, Th. 5)

TH. 10. $\quad [p,q,f]::p.q. \supset :.p \equiv :[r].p \equiv f(r). \equiv .[r].q \equiv f(r)$

Proof. $[p,q,f]::Hp \supset :.$

(1) $\qquad\qquad p$

(2) $\qquad\qquad q$ $\qquad\qquad\qquad\qquad\qquad\qquad\qquad\qquad$ (Hp)

(3) $\qquad\qquad p \equiv q:.$ $\qquad\qquad\qquad\qquad\qquad\qquad\qquad$ (1, 2)

(4) $\qquad\qquad [r]:p \equiv r. \equiv .q \equiv r:.$ $\qquad\qquad\qquad\qquad$ (3)

(5) $\qquad\qquad [r]:p \equiv f(r). \equiv .q \equiv f(r):$ $\qquad\qquad\qquad$ (4)

(6) $\qquad\qquad [r].p \equiv f(r). \equiv .[r].q \equiv f(r):.$ $\qquad\qquad$ (5)

$\qquad\qquad Cn$ $\qquad\qquad\qquad\qquad\qquad\qquad\qquad\qquad$ (1, 6)

TH. 11. $\quad [p,q]::p.q \equiv :.[f]:.p \equiv :[r].p \equiv f(r).$

$\qquad\qquad\qquad\qquad\qquad \equiv .[r].q \equiv f(r)$ $\quad$ (Th. 10, Th. 9)

§ 2. TRUTH-FUNCTIONS AND THE LAW OF SUBSTITUTION

Whitehead and Russell, in their work mentioned above, refer to a function f as a *truth-function*[1] if it takes sentences as argument values and satisfies the condition

(a) $\quad [p,q]:p \equiv q.f(p). \supset f(q).$

The sentence

(A) $\quad [p,q,f]:p \equiv q.f(p). \supset f(q),$

which expresses the fact that every function f (taking sentences as argument values) is a truth-function, will be called the '*law of substitution*'.

---

[1] A. N. Whitehead and B. Russell (90), vol. 1, p. 659 ff.

In Def. 4 I introduce the symbol '$\theta\rho$'; by this definition, the expression '$\theta\rho\{f\}$' will mean the same as 'the function f is a truth-function'. Def. 5 will enable us to replace the law of substitution by a single symbol 'Sb'. Among the theorems proved in this section, Th. 17 is the most important. It shows that, in a system of logistic which contains the law of substitution among its axioms or theorems, the sentence formulated in Th. 11 can be replaced by a simpler sentence,

$$[p,q]:.p.q. \equiv :[f]:p \equiv .f(p) \equiv f(q),$$

which also can serve as a definition of the symbol of logical product in terms of the equivalence symbol and the universal quantifier. From Th. 20 it follows that another closely related sentence

$$[p,q]:.p \vee q. \equiv :[\exists f]:p \equiv .f(p) \equiv f(q)$$

can be used to define the symbol of logical sum.

The remaining theorems of this section are of auxiliary character.

DEF. 4. $[f]:.\theta\rho\{f\} \equiv :[p,q]:p \equiv q.f(p). \supset f(q)$

DEF. 5. $Sb \equiv .[f].\theta\rho\{f\}$

TH. 12. $[f]:.\theta\rho\{f\} \equiv :[p,q]:p \equiv q. \supset .f(p) \equiv f(q)$ (Def. 4)

TH. 13. $Sb \equiv :[p,q,f]:p \equiv q.f(p). \supset f(q)$ (Def. 5, Def. 4)

TH. 14. $Sb \equiv :[p,q,f]:p \equiv q. \supset .f(p) \equiv f(q)$
$$\text{(Def. 5, Th. 12)}$$

TH. 15. $Sb \equiv :.[p,q,f]:p.q. \supset .f(p) \equiv f(q):.$
$$[p,q,f]:\sim (p). \sim (q). \supset .f(p) \equiv f(q)$$
$$\text{(Th. 14)}$$

The theorem just stated shows that the law of substitution is equivalent to the logical product of two sentences, the first of which could be called the *law of substitution for true sentences*, and the second one the *law of substitution for false sentences*. I am unable to solve the problem whether either of these two sentences alone is equivalent to the general law of substitution.

TH. 16. $[p,q,f]:p.q. \supset .f(p) \equiv f(q): \supset :.[p,q]:.p.q. \equiv :$
$$[f]:p \equiv .f(p) \equiv f(q)$$

Proof. $Hp: \supset :.:$

(1)	$[p,q] :: p.q. \supset :.$	
(a)	$p.q :$	
(b)	$[f].f(p) \equiv f(q) :.$	(Hp, a)
(c)	$[f] : p \equiv .f(p) \equiv f(q) :.:$	(a, b)
(2)	$[p,q] :: [f] : p \equiv .f(p) \equiv f(q) : \supset :.$	
(d)	$[f] : p \equiv .f(p) \equiv f(q) :.$	
(e)	$p \equiv .tr(p) \equiv tr(q) :$	(d)
(f)	$tr(p) \equiv tr(q).$	(Th. 1)
(g)	$p :$	(e, f)
(h)	$p \equiv .as(p) \equiv as(q) :$	(d)
(i)	$p \equiv .p \equiv q :$	$(h, \text{Def. 2})$
(j)	$p \equiv p. \equiv q :$	(i)
(k)	$q.$	(j)
(l)	$p.q :.:$	(g, k)
Cn		$(1-c,\ 2-l)$

Th. 17. $Sb \supset :. [p,q] :. p.q. \equiv :[f] : p \equiv .f(p) \equiv f(q)$
$$\text{(Th. 15, Th. 16)}$$

Th. 18. $[p,q,f] : \sim (p). \sim (q). \supset .f(p) \equiv f(q) : \supset :. [p,q] :.$

$\qquad \sim (p). \sim (q). \equiv :[f] : \sim (p) \equiv .f(p) \equiv f(q)$
$$\text{(Th. 1, Def. 2)}$$

I am omitting the proof of Th. 18 which is entirely analogous to that of Th. 16.

Th. 19. $[p,q,f] : \sim (p). \sim (q). \supset .f(p) \equiv f(q) : \supset :. [p,q] :.$

$\qquad p \vee q. \equiv :[\exists f] : p \equiv .f(p) \equiv f(q)$ (Th. 18)

Th. 20. $Sb \supset :. [p,q] :. p \vee q. \equiv :[\exists f] : p \equiv .f(p) \equiv f(q)$
$$\text{(Th. 15, Th. 19)}$$

The converses of Ths. 16 and 19 can easily be established. It is a direct conclusion that the sentence

$$[p,q] :. p.q. \equiv :[f] : p \equiv .f(p) \equiv f(q)$$

not only follows from the law of substitution for true sentences, but is equivalent to it. Similarly the sentence

$$[p, q] : . p \vee q . \equiv : [\exists f] : p \equiv . f(p) \equiv f(q)$$

is equivalent to the law of substitution for false sentences, and the logical product of these two sentences is equivalent to the general law of substitution.

§ 3. Independence of the Law of Substitution of the Axioms of Logistic; the Law of the Number of Functions

To the question whether every function f (having a sentence as argument) is a truth-function, A. N. Whitehead and B. Russell give a negative answer. Their answer is based exclusively on intuitive considerations and does not appear to be convincing.[1]

On the other hand it seems quite clear that the law of substitution can be neither demonstrated nor refuted in any of the systems of logistic hitherto known. Moreover, it is even possible to prove the independence of this sentence of the known systems of axioms of logistic, e.g. the system of Whitehead and Russell, by using the method usually applied in proofs of independence, i.e. by means of a suitable interpretation. Without giving any details we may mention that such an interpretation can be found in a system of logistic based upon the three-valued sentential calculus constructed by Łukasiewicz.[2]

In any case, anyone who regards the sentence (A) as true and wishes to incorporate it in the system of logistic, must either admit it as an axiom or introduce another axiom which, when added to the axioms of the system, implies the sentence (A). The theorems which will be established in the sequel can have the same interest in the construction of such a system of logistic as, for example, theorems concerning the equivalent forms of the axioms of Euclid have for researches on the foundations of geometry.

---

[1] It should be mentioned in this connexion that Leśniewski constructed a general method which makes it possible to eliminate functions not satisfying condition (a) from all known arguments. This method of Leśniewski has not been published.

[2] See article IV, in particular § 3.

Some of the theorems just mentioned are direct corollaries of logically stronger theorems in which conditions are formulated that are necessary and sufficient for a function f to be a truth-function. For instance, as will be seen from Th. 41, the condition

(b) $[p].f(p) \equiv tr(p). \vee .[p].f(p). \equiv as(p). \vee .[p].f(p) \equiv$
$$\sim (p). \vee .[p].f(p) \equiv fl(p)$$

is of this kind. An immediate corollary of Th. 41 is Th. 43 which states that the law of substitution is equivalent to the sentence

(B) $[f]:[p].f(p) \equiv tr(p). \vee .[p].f(p) \equiv as(p). \vee .[p].$
$$f(p) \equiv \sim (p). \vee .[p].f(p) \equiv fl(p).$$

This sentence, owing to its intuitive content, could be called the *law of the number of functions*; it expresses the fact that every function (which has sentences as arguments) is equivalent for every value of the argument, to one of the four functions tr, as, $\sim$, and fl.

DEF. 6.[1] $Tr \equiv .[\exists r].r$

DEF. 7.[1] $Fl \equiv .[r].r$

DEF. 8. $[f,g] := \{f,g\} \equiv .[p].f(p) \equiv g(p)$

TH. 21. Tr (Def. 6)

TH. 22. $[p]:p \equiv .p \equiv Tr$ (Th. 21)

TH. 23. $\sim (Fl)$ (Def. 7)

TH. 24. $[p]: \sim (p) \equiv .p \equiv Fl$ (Th. 23)

TH. 25. $[p]:p \equiv Tr. \vee .p \equiv Fl$

Proof. $[p]:$

(1) $p \vee \sim (p):$

(2) $p \supset .p \equiv Tr$ (Th. 22)

(3) $\sim (p) \supset .p \equiv Fl:$ (Th. 24)

$$p \equiv Tr. \vee .p \equiv Fl$$ (1, 2, 3)

TH. 26. $[p]. \sim (p \equiv Tr.p \equiv Fl)$ (Th. 21, Th. 23)

TH. 27. $[p,f]:\theta\rho\{f\}.f(Tr).f(Fl). \supset f(p)$

---

[1] I write 'Tr' and 'Fl' instead of '1' and '0' which occur, for example, in Schröder, E. (62), vol. 1, p. 188.

Proof. $[p,f] :. Hp. \supset:$

(1)	$p \equiv Tr. \lor .p \equiv Fl:$	(Th. 25)
(2)	$p \equiv Tr. \supset f(p):$	(Def. 4, Hp)
(3)	$p \equiv Fl. \supset f(p):$	(Def. 4, Hp)
	Cn	(1, 2, 3)

Тн. 28. $[f]: \theta\rho\{f\}.f(Tr).f(Fl). \supset = \{f, tr\}$

Proof. $[f] :. Hp. \supset:$

(1)	$[p].f(p):$	(Th. 27)
(2)	$[p].f(p) \equiv tr(p):$	(1, Th. 1)
	Cn	(Def. 8, 7)

Тн. 29. $[f]: \theta\rho\{f\}.f(Tr). \sim (f(Fl)). \supset = \{f, as\}$

Proof. $[f] :: Hp. \supset :.$

(1)	$[p]:$	
(a)	$p \supset.$	
(α)	$p \equiv Tr.$	(Th. 22)
(β)	$f(p):$	(Def. 4, α, Hp)
(b)	$\sim p \supset.$	
(γ)	$p \equiv Fl.$	(Th. 22)
(δ)	$\sim (f(p)):$	(Def. 4, γ, Hp)
(c)	$p \equiv f(p).$	(a—β, b—δ)
(d)	$as(p) \equiv f(p) :.$	(Def. 2, c)
	Cn	(Def. 8, 1—d)

Тн. 30. $[f]: \theta\rho\{f\}. \sim (f(Tr)).f(Fl). \supset = \{f, \sim\}$
$$\text{(Th. 22, Def. 4, Th. 24, Def. 8)}$$

I omit the proof which is analogous to that of Th. 29.

Тн. 31. $[p]. \sim (fl(p))$ (Def. 5)

Тн. 32. $[p,f]: \theta\rho\{f\}. \sim (f(Tr)). \sim (f(Fl)). \supset \sim (f(p))$
$$\text{(Th. 25, Def. 4)}$$

The proof is analogous to that of Th. 27.

Тн. 33. $[f]: \theta\rho\{f\}. \sim (f(Tr)). \sim (f(Fl)). \supset = \{f, fl\}$
$$\text{(Th. 32, Th. 31, Def. 8)}$$

The proof is analogous to that of Th. 28.

TH. 34. $[f]:\theta\rho\{f\}\supset . = \{f,tr\}\vee = \{f,as\}\vee$
$$= \{f,\sim\}\vee = \{f,fl\}$$

Proof.

(1) $[f]:.Hp\supset :$

$f(Tr).f(Fl).\vee .f(Tr). \sim (f(Fl)).\vee$

$\sim (f(Tr)).f(Fl).\vee .\sim (f(Tr)).\sim (f(Fl)):$

Cn   (Hp, 1, Th. 28, Th. 29, Th. 30, Th. 33)

TH. 35. $\theta\rho\{tr\}$

Proof.

(1) $[p,q]:p \equiv q.tr(p).\supset tr(q):.$ (Th. 1)

$\theta\rho\{tr\}$ (Def. 4, 1)

TH. 36. $\theta\rho\{as\}$

Proof.

(1) $[p,q]:$

(a) $p \equiv q.p.\supset q:$

(b) $as(p) \equiv p.$ (Def. 2)

(c) $as(q) \equiv q:$ (Def. 2)

(d) $p \equiv q.as(p).\supset as(q):.$ (a, b, c)

$\theta\rho\{as\}$ (Def. 4, 1—d)

TH. 37. $\theta\rho\{\sim\}$ (Def. 4)

TH. 38. $\theta\rho\{fl\}$ (Th. 31, Def. 4)

The proofs of the last two theorems are analogous to that of Th. 35.

TH. 39. $[f,g]: = \{g,f\}.\theta\rho\{f\}.\supset \theta\rho\{g\}$

Proof. $[f,g]::Hp.\supset :.$

(1) $[p,q]:$

(a) $p \equiv q.f(p).\supset f(q):$ (Def. 4)

(b) $g(p) \equiv f(p).$ (Def. 8, Hp)

(c) $g(q) \equiv f(q):$ (Def. 8, Hp)

(d) $p \equiv q.g(p).\supset g(q):.$ (a, b, c)

Cn   (Def. 4, 1—d)

TH. 40. $[f] :\, = \{f, tr\} \vee \, = \{f, as\} \vee \, = \{f, \sim\} \vee$
$$= \{f, fl\} . \supset \theta\rho\{f\}$$

Proof. $[f] : Hp . \supset .$

(1)	$= \{f, tr\} \supset \theta\rho\{f\}$	(Th. 39, Th. 35)
(2)	$= \{f, as\} \supset \theta\rho\{f\}$	(Th. 39, Th. 36)
(3)	$= \{f, \sim\} \supset \theta\rho\{f\}$	(Th. 39, Th. 37)
(4)	$= \{f, fl\} \supset \theta\rho\{f\}$	(Th. 39, Th. 38)
	Cn	(Hp, 1, 2, 3, 4)

TH. 41. $[f] : \theta\rho\{f\} \equiv . \, = \{f, tr\} \vee \, = \{f, as\} \vee \, = \{f, \sim\} \vee$
$$= \{f, fl\} \quad \text{(Th. 34, Th. 40)}$$

TH. 42. $Sb \equiv . [f] . \, = \{f, tr\} \vee \, = \{f, as\} \vee \, = \{f, \sim\} \vee$
$$= \{f, fl\} \quad \text{(Th. 41)}$$

TH. 43. $Sb \equiv : [f] : [p] . f(p) \equiv tr(p) . \vee . [p] . f(p) \equiv as(p) .$
$$\vee . [p] . f(p) \equiv \sim (p) . \vee . [p] . f(p) \equiv fl(p)$$

§ 4. THE LAW OF DEVELOPMENT

The most important theorem of this section is Th. 50, where I give a new necessary and sufficient condition which a function must satisfy if it is to be a truth-function. According to this condition the function f should possess the following property:

(c) $[p] : . f(p) \equiv : f(Tr) . p . \vee . f(Fl) . \sim (p) .$

Th. 51, which follows from Th. 50, establishes the equivalence of the sentence (A) and the sentence:

(C) $[p, f] : . f(p) \equiv : f(Tr) . p . \vee . f(Fl) . \sim (p);$

in this last sentence it is easy to recognize the *law of development* well known in the algebra of logic.[1]

TH. 44. $[p, f] : . \theta\rho\{f\} . f(p) . \supset : f(Tr) . p . \vee . f(Fl) . \sim p$

Proof. $[p, f] : . Hp . \supset :$

(1)	$p \equiv Tr . \vee . p \equiv Fl :$	(Th. 25)
(2)	$p \equiv Tr . \supset . f(Tr) . p :$	(Def. 4, Hp, Th. 22)
(3)	$p \equiv Fl . \supset . f(Fl) . \sim (p) :$	(Def. 4, Hp, Th. 24)
	Cn	(1, 2, 3)

[1] Cf. L. Couturat (14 a), § 29; Schröder, E. (62), vol. I, p. 409, 44$_+$.

Th. 45. $[p,f]:\theta\rho\{f\}.f(Tr).p. \supset f(p)$ (Th. 22, Def. 4)

Th. 46. $[p,f]:\theta\rho\{f\}.f(Fl). \sim (p). \supset f(p)$ (Th. 24, Def. 4)

Th. 47. $[p,f]:.\theta\rho\{f\}:f(Tr).p. \vee .f(Fl). \sim (p) : \supset f(p)$

(Th. 45, Th. 46)

Th. 48. $[f]::\theta\rho\{f\}\supset :.[p]:.f(p) \equiv :f(Tr).p. \vee .f(Fl). \sim(p)$

(Th. 44, Th. 47)

Th. 49. $[f]::[p]:.f(p)$

$\equiv :f(Tr).p. \vee .f(Fl). \sim (p) :. \supset \theta\rho\{f\}$

Proof. $[f]:.:Hp:. \supset ::$

(1) $[p,q]:.p \equiv q.f(p). \supset :$

(a) $p \equiv q.f(p):$

(b) $f(Tr).p. \vee .f(Fl). \sim (p):$ (Hp, a)

(c) $f(Tr).q. \vee .f(Fl). \sim (q):$ (a, b)

(d) $f(q)::$ (Hp, c)

Cn (Def. 4, 1—d)

Th. 50. $[f]::\theta\rho\{f\} \equiv :.[p]:.f(p)$

$\equiv :f(Tr).p. \vee .f(Fl). \sim (p)$

(Th. 48, Th. 49)

Th. 51. $Sb \equiv :.[p,f]:.f(p)$

$\equiv :f(Tr).p. \vee .f(Fl). \sim (p)$ (Th. 50, Def. 5)

§ 5. The First Theorem on the Bounds of a Function

As well as satisfying the law of development, in virtue of Th. 50, truth-functions also possess other properties which the algebra of logic attributes to its functions. Nevertheless the corresponding theorems are sometimes easier to prove by the use of the specific laws of logistic.

In this section I prove, in particular, Ths. 52–54 according to which all truth-functions satisfy the following conditions:

(d) $[p].f(p). \equiv .f(Tr).f(Fl)$

(e) $[\exists p].f(p). \equiv .f(Tr) \vee f(Fl)$

or, in equivalent terms:

(d') $[p]:f(Tr).f(Fl). \supset .f(p)$,

(e') $[p]:f(p) \supset .f(Tr) \vee f(Fl)$

and also the condition:

(f) $[p]:f(Tr).f(Fl). \supset f(p) \supset .f(Tr) \vee f(Fl)$.

The theorems mentioned only express the conditions which are necessary in order that a function f should be a truth-function. It is impossible to prove that these conditions are sufficient.[1] It is, however, possible to prove that each of the sentences which attributes to each function f (having a sentence as argument) the properties (d), (e), or (f),[2] implies the law of substitution. This has enabled me to establish, in Ths. 58, 62, and 64, the equivalence of this law and the following sentences:

(D) $[f]:[p].f(p). \equiv .f(Tr).f(Fl)$,

(E) $[f]:[\exists p].f(p). \equiv .f(Tr) \vee f(Fl)$,

(F) $[p,f]:f(Tr).f(Fl). \supset f(p) \supset .f(Tr) \vee f(Fl)$.

The sentence (F) is known in the algebra of logic.[3] I call it the *first theorem on the bounds of a function*. Sentences (D) and (E) result from it almost immediately.

Th. 52. $[f]:.\theta\rho\{f\} \supset :[p].f(p). \equiv .f(Tr).f(Fl)$

Proof. $[f]:.Hp \supset :$

(1) $\qquad\qquad [p].f(p). \supset .f(Tr).f(Fl):$

(2) $\qquad\qquad f(Tr).f(Fl). \supset .[p].f(p):$ (Th. 27)

$\qquad\qquad Cn$ (1, 2)

Th. 53. $[f]:.\theta\rho\{f\} \supset :[\exists p].f(p). \equiv .f(Tr) \vee f(Fl)$

---

[1] One can even prove that the sentences expressing the sufficiency of these conditions are independent of the axioms of logistic; as in the case of the law of substitution, the proof is based upon an interpretation in three-valued logic (cf. the remarks at the end of § 2).

[2] Łukasiewicz has already called attention to the fact that in attributing to every function f the properties indicated here as conditions (d) and (e) (and the properties corresponding to these conditions for functions of several arguments) we can bring about notable simplifications in the constructions of logistic. Cf. Łukasiewicz, J. (50 a).

[3] Cf. L. Couturat (14 a), § 28; E. Schröder (62), vol. I, p. 427, 48_+.

Proof. $[f] :. Hp \supset :$

(1) $\sim (f(Tr)) . \sim (f(Fl)) . \supset . [p] . \sim (f(p)) :$ (Th. 32)

(2) $[\exists p] . f(p) . \supset . f(Tr) \vee f(Fl) :$ (1)

(3) $f(Tr) \vee f(Fl) . \supset . [\exists p] . f(p) :$

    Cn (2, 3)

Tʜ. 54. $[f] :. \theta \rho \{f\} \supset : [p] : f(Tr) . f(Fl) . \supset f(p) \supset . f(Tr) \vee f(Fl)$

    (Th. 52, Th. 53)

Tʜ. 55. $[q, r, g] :: [f] : [p] . f(p) . \equiv . f(Tr) . f(Fl) :. g(Tr, Tr) .$

    $. g(Tr, Fl) . g(Fl, Tr) . g(Fl, Fl) :. \supset g(q, r)$

Proof. $[q, r, g] :: Hp :. \supset :$

(1) $[p] . g(Tr, p) . \equiv . g(Tr, Tr) . g(Tr, Fl) :$

(2) $[p] . g(Tr, p) :$ (1, Hp)

(3) $g(Tr, r) :$ (2)

(4) $[p] . g(Fl, p) \equiv . g(Fl, Tr) . g(Fl, Fl) :$ (Hp)

(5) $[p] . g(Fl, p) :$ (4, Hp)

(6) $g(Fl, r) :$ (5)

(7) $[p] . g(p, r) . \equiv . g(Tr, r) . g(Fl, r) :$ (Hp)

(8) $[p] . g(p, r) :$ (7, 3, 6)

    Cn (8)

Tʜ. 56. $[f] : [p] . f(p) . \equiv . f(Tr) . f(Fl) : \supset . [f] . \theta \rho \{f\}$

Proof. $Hp : \supset ::$

(1) $[f] :.$

(a) $Tr \equiv Tr . f(Tr) . \supset f(Tr) :$

(b) $\sim (Tr \equiv Fl) :$ (Th. 21, Th. 23)

(c) $Tr \equiv Fl . f(Tr) . \supset f(Fl) :$ (b)

(d) $Fl \equiv Tr . f(Fl) . \supset f(Tr) :$ (b)

(e) $Fl \equiv Fl . f(Fl) . \supset f(Fl) :.$

(f) $[p, q] : p \equiv q . f(p) . \supset f(q) ::$ (Th. 55, Hp, a, c, d, e)

    Cn (Def. 4, 1—f)

Tʜ. 57. $Sb \equiv : [f] : [p] . f(p) . \equiv . f(Tr) . f(Fl)$

    (Th. 52, Th. 56)

Th. 58. $[p,q,f]:p \equiv q.f(p). \supset f(q): \equiv :[f]:[p].f(p).$
$$\equiv .f(Tr).f(Fl) \quad \text{(Th. 57, Def. 4)}$$

Th. 59. $[f]:[\exists p].f(p). \equiv .f(Tr) \vee f(Fl):\supset:[f]:[p].f(p).$
$$\equiv .f(Tr).f(Fl)$$

Proof. $Hp:\supset:.$

(1) $[f]:[\exists p].\sim(f(p)). \equiv . \sim.(f(Tr)) \vee \sim (f(Fl)):.$

Cn (1)

Th. 60. $[f]:[p].f(p). \equiv .f(Tr).f(Fl): \supset :[f]:[\exists p].f(p).$
$$\equiv .f(Tr) \vee f(Fl)$$

The proof is analogous to that of the preceding theorem.

Th. 61. $[f]:[\exists p].f(p). \equiv .f(Tr) \vee f(Fl): \equiv :[f]:[p].f(p).$
$$\equiv .f(Tr).f(Fl) \quad \text{(Th. 59, Th. 60)}$$

Th. 62. $Sb \equiv :[f]:[\exists p].f(p). \equiv .f(Tr) \vee f(Fl)$
$$\text{(Th. 57, Th. 61)}$$

Th. 63. $Sb \equiv :[f]:[p].f(p). \equiv .f(Tr).f(Fl):[\exists p].f(p).$
$$\equiv .f(Tr) \vee f(Fl) \quad \text{(Th. 57, Th. 62)}$$

Th. 64. $Sb \equiv :[p,f]:f(Tr).f(Fl). \supset f(p) \supset .f(Tr) \vee f(Fl)$
$$\text{(Th. 63)}$$

In addition to the properties of a function f which have been examined in the course of this section, it is possible to study other properties which are stronger from the logical point of view:

(g) $[q]:[p].f(p). \equiv .f(q).f(\sim(q)),$

(h) $[q]:[\exists p].f(p). \equiv .f(q) \vee f(\sim(q)),$

or, in equivalent formulations:

(g') $[p,q]:f(q).f(\sim (q)). \supset f(p),$

(h') $[p,q]:f(p). \supset .f(q) \vee f(\sim(q)),$

and the property:

(i) $[p,q]:f(q).f(\sim (q)). \supset f(p) \supset .f(q) \vee f(\sim (q)).$

By applying analogous methods to those which were used above, the following theorems are easily proved:

Th. 65. $[f]:.\theta\rho\{f\} \supset :[q]:[p].f(p). \equiv .f(q).f(\sim (q))$

Th. 66. $[f]:.\theta\rho\{f\} \supset :[q]:[\exists p].f(p). \equiv .f(q)\lor f(\sim(q))$

Th. 67. $[f]:.\theta\rho\{f\} \supset :[p,q]:f(q).f(\sim(q)). \supset .f(p) \supset .f(q)$
$$\lor f(\sim(q))$$

Th. 68. $Sb \equiv :[q,f]:[p].f(p). \equiv .f(q).f(\sim(q))$

Th. 69. $Sb \equiv :[q,f]:[\exists p].f(p). \equiv .f(q)\lor f(\sim(q))$

Th. 70. $Sb \equiv :[p,q,f]:f(q).f(\sim(q)). \supset f(p) \supset .f(q)\lor f(\sim(q))$

§ 6. The Second Theorem on the Bounds of a Function

I shall now show that every truth-function possesses the following properties (see Ths. 71–73):

(j) $[p].f(p). \equiv f(f(Fl))$,

(k) $[\exists p].f(p). \equiv f(f(Tr))$,

or, in equivalent terms:

(j') $[p].f(f(Fl)) \supset f(p)$,

(k') $[p].f(p) \supset f(f(Tr))$,

as well as the property:

(l) $[p].f(f(Fl)) \supset f(p) \supset f(f(Tr))$.

These conditions, like those of § 5, are only necessary in order that a function f should be a truth-function. Moreover, in contrast to the situation in the preceding section, I am unable to show that one of the following sentences:

(J) $[f]:[p].f(p). \equiv .f(f(Fl))$,

(K) $[f]:[\exists p].f(p). \equiv .f(f(Tr))$,

according to which every function f possesses the properties (j) and (k), implies the law of substitution.

Now we shall see that this law results from the sentence:

(L) $[p,f].f(f(Fl)) \supset f(p) \supset f(f(Tr))$,

which is equivalent to the logical product of the sentences (J) and (K). Thanks to this result I obtain, in Ths. 79 and 80, new formulations equivalent to the sentence which interests me in this work.

We may call the sentence (L), which is also known in the algebra of logic,[1] the *second theorem on the bounds of a function*.

TH. 71. $[f]:.\theta\rho\{f\} \supset :[p].f(p). \equiv f(f(Fl))$

Proof. $[f]::Hp \supset ::$

(1) $[p]:.f(p) \equiv :f(Tr).p \vee .f(Fl). \sim (p)::$ (Th. 48)

(2) $f(f(Fl)) \equiv :f(Tr).f(Fl). \vee .f(Fl). \sim (f(Fl)):.$ (1)

(3) $f(f(Fl)) \equiv .f(Tr).f(Fl):$ (2)

    Cn (Th. 52, Hp, 3)

TH. 72. $[f]:.\theta\rho\{f\} \supset :[\exists p].f(p). \equiv f(f(Tr))$
                               (Th. 48, Th. 53)

The proof is analogous to that of the preceding theorem.

TH. 73. $[f]:.\theta\rho\{f\} \supset .[p].f(f(Fl)) \supset f(p) \supset f(f(Tr))$
                               (Th. 71, Th. 72)

TH. 74. $[f]:[p].f(p). \equiv f(f(Fl)): \supset :[q,f]: \sim (q) \supset .f(q)$
                                            $\equiv f(Fl)$

Proof. $Hp : \supset :::$

(1) $[q,f]:.:$

   (a) $[\exists g]::$

      (α)[2] $[p]:.g(p) \equiv : \sim (p) \supset .f(p) \equiv f(Fl)$

      (β) $g(Fl) \equiv : \sim (Fl) \supset .f(Fl) \equiv f(Fl):.$ (α)

      (γ) $f(Fl) \equiv f(Fl):$

      (δ) $\sim (Fl) \supset .f(Fl) \equiv f(Fl)$ (γ)

      (ϵ) $g(Fl):.$ (β,δ)

      (ζ) $g(g(Fl)) \equiv : \sim (g(Fl)) \supset .f(g(Fl)) \equiv f(Fl):.$ (α)

      (η) $\sim (g(Fl)) \supset .f(g(Fl)) \equiv f(Fl):$ (ϵ)

      (θ) $g(g(Fl)):$ (ζ,η)

      (ι) $[p].g(p):$ (Hp, θ)

      (κ) $g(q):.:$ (ι)

---

[1] Cf. L. Couturat, (14 *a*), § 28 (Remarque).

[2] The auxiliary definition which I introduce at this place and of which I make use in the proof, may seem superfluous. But I have adopted this as a device for making the proof clearer.

(b) $\sim (q) \supset . f(q) \equiv f(Fl) :::$ $\qquad\qquad (a{-}\alpha, \kappa)$

$\qquad Cn$ $\qquad\qquad\qquad\qquad\qquad\qquad\qquad\qquad (1{-}b)$

Th. 75. $[f]:[p].f(p). \equiv f(f(Fl)) : \supset :[p,q,f]:$

$\qquad\qquad\qquad\qquad\qquad\qquad \sim (p). \sim (q).f(p). \supset f(q)$

Proof. $Hp : \supset :.$

(1) $[p,q,f] : \sim (p). \sim (q).f(p). \supset .$

$\quad (a)$ $f(p) \equiv f(Fl).f(q) \equiv f(Fl).f(p)$ $\qquad\qquad$ (Th. 74)

$\quad (b)$ $f(p) \equiv f(q).f(p)$ $\qquad\qquad\qquad\qquad\qquad (a)$

$\quad (c)$ $f(q):.$ $\qquad\qquad\qquad\qquad\qquad\qquad\qquad (b)$

$\qquad Cn$ $\qquad\qquad\qquad\qquad\qquad\qquad\qquad\qquad (1{-}c)$

Th. 76. $[f]:[\exists p].f(p). \equiv f(f(Tr)) : \supset :[q,f]:q \supset .f(q)$

$\qquad\qquad\qquad\qquad\qquad\qquad\qquad\qquad\qquad\qquad \equiv f(Tr)$

Proof. $Hp : \supset :.:$

(1) $[g]:[p]. \sim (g(p)). \equiv . \sim (g(g(Tr)))). :.$

(2) $[q,f] ::$

$\quad (a)$ $[\exists g]:.$

$\qquad (\alpha)^1$ $[p]:g(p) \equiv .p. \sim (f(p) \equiv f(Tr)) :.$

$\qquad (\beta)$ $g(Tr) \equiv .Tr. \sim (f(Tr) \equiv f(Tr)):$ $\qquad\qquad (\alpha)$

$\qquad (\gamma)$ $f(Tr) \equiv f(Tr).$

$\qquad (\delta)$ $\sim (g(Tr)):$ $\qquad\qquad\qquad\qquad\qquad\qquad (\beta, \gamma)$

$\qquad (\epsilon)$ $g(g(Tr)) \equiv .g(Tr). \sim (f(g(Tr)) \equiv f(Tr)):$ $\qquad (\alpha)$

$\qquad (\zeta)$ $\sim (g(g(Tr))):$ $\qquad\qquad\qquad\qquad\qquad\quad (\epsilon, \delta)$

$\qquad (\eta)$ $[p]. \sim (g(p)):$ $\qquad\qquad\qquad\qquad\qquad\quad (1, \zeta)$

$\qquad (\theta)$ $\sim (g(q)) ::$ $\qquad\qquad\qquad\qquad\qquad\qquad (\eta)$

$\quad (b)$ $\sim (q. \sim (f(q) \equiv f(Tr))) ::$ $\qquad\qquad (a{-}\alpha, \beta)$

$\qquad Cn$ $\qquad\qquad\qquad\qquad\qquad\qquad\qquad\qquad (2{-}b)$

Th. 77. $[f]:[\exists p].f(p). \equiv f(f(Tr)):$

$\qquad\qquad\qquad \supset :[p,q,f]:p.q.f(p). \supset f(q)$ $\quad$ (Th. 76)

The proof is analogous to that of Th. 75.

Th. 78. $[f]:[p].f(p). \equiv f(f(Fl)):[\exists p].f(p).$

$\qquad\qquad\qquad\qquad \equiv f(f(Tr)): \supset .[f].\theta\rho\{f\}$

^{1} See the preceding note.

Proof. $Hp : \supset ::$

(1) $\qquad [p,q,f] : . p \equiv q . f(p) . \supset :$

(a) $\qquad p \equiv q . f(p) :$

(b) $\qquad p . q . \vee . \sim (p) . \sim (q) : \qquad\qquad\qquad (a)$

(c) $\qquad p . q . \supset f(q) : \qquad\qquad$ (Th. 77, Hp, a)

(d) $\qquad \sim (p) . \sim (q) . \supset f(q) \qquad$ (Th. 75, Hp, a)

(e) $\qquad f(q) :: \qquad\qquad\qquad\qquad$ (b, c, d)

$\qquad\qquad Cn \qquad\qquad\qquad\qquad$ (Def. 4, 1—e)

Tн. 79. $\quad Sb \equiv : [f] : [p] . f(p) . \equiv f(f(Fl)) : [\exists p] . f(p) .$
$$\equiv f(f(Tr)) \quad \text{(Ths. 71, 72, 78)}$$

Tн. 80. $\quad Sb \equiv . [p,f] . f(f(Fl)) \supset f(p) \supset f(f(Tr)) \qquad$ (Th. 79)

Note I. On Functions of Several Arguments

When we examine functions of several arguments (all these arguments being sentences) we reach results which are analogous to the preceding. Their proofs present no essentially novel features.

For example, if we limit ourselves to functions of two arguments we can establish the following:

In order that such a function f may be a truth-function, i.e. in order that it should satisfy the condition:

(a$_1$) $\quad [p,q,r,s] : p \equiv q . r \equiv s . f(p,r) . \supset . f(q,s)$,

it is necessary that it should possess the following properties:

(b$_1$) $\qquad\qquad [p,q] . f(p,q) . \vee :$
$[p,q] : f(p,q) \equiv . p \vee q : \vee : [p,q] : f(p,q) \equiv . \sim (p) \vee q : \vee :$
$[p,q] : f(p,q) \equiv . p \vee \sim (q) : \vee : [p,q] : f(p,q)$
$$\equiv . \sim (p) \vee \sim (q) : \vee .$$
$[p,q] . f(p,q) \equiv p . \vee . [p,q] . f(p,q) \equiv q . \vee : [p,q] : f(p,q)$
$$\equiv . p \equiv q : \vee :$$
$[p,q] : f(p,q) \equiv . p \equiv \sim (q) : \vee . [p,q] . f(p,q) \equiv$
$$\sim (q) . \vee . [p,q] . f(p,q) \equiv \sim (p) . \vee :$$
$[p,q] : f(p,q) \equiv . p . q : \vee : [p,q] : f(p,q) \equiv . p . \sim (q) : \vee :$
$[p,q] : f(p,q) \equiv . \sim (p) . q : \vee : [p,q] : f(p,q)$
$$\equiv . \sim (p) . \sim (q) : \vee .$$
$$[p,q] . \sim (f(p,q)),$$

(c$_1$) $[p,q]: .f(p,q) \equiv :f(Tr, Tr).p.q. \vee .f(Tr, Fl).p. \sim (q).$

$\qquad \vee f(Fl, Tr). \sim (p).q. \vee .f(Fl, Fl). \sim (p). \sim (q),$

(d$_1$) $[p,q].f(p,q). \equiv .f(Tr, Tr).f(Tr, Fl).f(Fl, Tr).f(Fl, Fl),$

(e$_1$) $[\exists p,q].f(p,q). \equiv .f(Tr, Tr) \vee f(Tr, Fl) \vee f(Fl, Tr) \vee$

$\qquad\qquad\qquad\qquad\qquad\qquad\qquad f(Fl, Fl),$

(f$_1$) $[p,q]:f(Tr, Tr).f(Tr, Fl).f(Fl, Tr).f(Fl, Fl). \supset f(p,q) \supset.$

$\qquad .f(Tr, Tr) \vee f(Tr, Fl) \vee f(Fl, Tr) \vee f(Fl, Fl).$

Conditions (b$_1$) and (c$_1$) are necessary and sufficient in order that a function f should be a truth-function, the others are only necessary.[1]

We can similarly prove the equivalence of the sentences (A$_1$) the *law of substitution*, (B$_1$) the *law of the number of functions*, (C$_1$) the *law of development*, (D$_1$), (E$_1$), and (F$_1$)—the *theorem on the bounds of a function*, which attribute to each function f the properties (a$_1$)–(f$_1$) mentioned above.[2]

It is easy to show that the sentences (A$_1$)–(F$_1$) are equivalent, not only among one another, but also to the sentence (A)—the law of substitution for functions with a single argument. Moreover, without leaving the domain of functions with two arguments, we can again construct a series of sentences equivalent to the sentence (A) and, from the point of view of content, intermediate between the sentences of the preceding sections and those we have just discussed. The following are some examples:

(A$_1'$) $[p,q,r,f]:p \equiv q.f(p,r). \supset f(q,r),$

(D$_1'$) $[p,f]:[q].f(p,q). \equiv .f(p, Tr).f(p, Fl).$

---

[1] I do not cite here the conditions corresponding to conditions (j), (k), and (l) which were studied in the course of § 6, because they will be rather complicated and do not seem to me to be very interesting.

[2] In an unpublished manuscript of Łukasiewicz I have found an argument which may be summarized in these terms: The laws of the number of functions result from the following hypothesis:

$$[p,f]:f(p) \equiv f(Tr). \vee .f(p) \equiv f(Fl)$$

and from analogous hypotheses for functions of several arguments. It is easy to prove that each of these hypotheses is equivalent to the law of substitution (for functions with the same number of arguments).

NOTE II. ON FUNCTIONS OF WHICH THE ARGUMENTS
ARE NOT SENTENCES

Analogous problems present themselves in the study of logistic functions of which the arguments are not sentences but functions.

I shall restrict myself to the consideration of only a single case, to be specific that of a function of one argument which is itself one of the functions studied in §§ 3–6.

By analogy with the terminology of Whitehead and Russell,[1] we call such a function ϕ an *extensional function* if it satisfies the condition:

(a_2) $[f,g] : = \{f,g\} . \phi\{f\} . \supset . \phi\{g\}.$

The sentence

(b_2) $[f,g,\phi] : = \{f,g\} . \phi\{f\} . \supset \phi\{g\},$

which attributes to every function ϕ the property (a_2), is probably independent of the axioms of logistic, even if we add to them the sentence (A). Nevertheless I do not know of a proof of this.

It is possible to formulate a series of theorems (analogous to those of the preceding sections) which express the necessary and sufficient conditions which a function ϕ must satisfy if it is to be extensional or which provide equivalent forms for the sentence (A_2). However, the proofs of these theorems require that the sentence (A) be admitted as hypothesis. Otherwise they are quite like the proofs of §§ 3–6; but the role analogous to that of Th. 25 is here played by the sentence:

$$[f] . = \{f,tr\} \vee . = \{f,as\} \vee . = \{f,\sim\} \vee . = \{f,fl\},$$

which we already know to be equivalent to our hypothesis (see Th. 42 of § 3).

I here give the most characteristic sentences which are equivalent to the sentence (A_2):

(C_2) $[f,\phi] : . \phi\{f\} \equiv : = \{f,tr\} . \phi\{tr\} . \vee . = \{f,as\} . \phi\{as\} . \vee .$
$= \{f,\sim\} . \phi\{\sim\} . \vee . = \{f,fl\} . \phi\{fl\},$

(D_2) $[\phi] : [f] . \phi\{f\} . \equiv . \phi\{tr\} . \phi\{as\} . \phi\{\sim\} . \phi\{fl\},$

---

[1] Whitehead and Russell (90), vol. 1, p. 22.

(E_2) $[\phi] : [\exists f] . \phi\{f\}. \equiv . \phi\{tr\} \vee \phi\{as\} \vee \phi\{\sim\} \vee \phi\{fl\},$

(F_2) $[f, \phi] : \phi\{tr\} . \phi\{as\} . \phi\{\sim\} . \phi\{fl\} . \supset \phi\{f\} \supset . \phi\{tr\} \vee$
$$\phi\{as\} \vee \phi\{\sim\} \vee \phi\{fl\},$$

(M) $[\phi] :. [\exists g] : [f] . \phi\{f\} \equiv g(f(Tr), f(Fl)).$

The sentences (C_2)–(F_2) correspond, strictly speaking, to the sentences (C)–(F) of §§ 2 and 3, or (C_1)–(F_1) of Note I. But the sentence (M) has no correlate in the preceding series of sentences. It may perhaps be added that we can easily derive from (M) the sentence (B_2)—the *law of the number of functions*. For technical reasons I refrain from giving this law here; it would contain sixteen logical summands and its formulation would be very laborious without the help of some auxiliary definitions.

II

FOUNDATIONS OF THE GEOMETRY OF SOLIDS†

SOME years ago Leśniewski suggested the problem of establishing the foundations of a *geometry of solids*, understanding by this term a system of geometry destitute of such geometrical figures as points, lines, and surfaces, and admitting as figures only solids—the intuitive correlates of open (or closed) regular sets of three-dimensional Euclidean geometry.[1] The specific character of such a geometry of solids—in contrast to all point geometries—is shown in particular in the law according to which each figure contains another figure as a proper part. This problem is closely related to the questions discussed in the works of Whitehead and Nicod.[2]

In this résumé I propose to sketch a solution of this problem omitting the question of its philosophical importance.

The deductive theory founded by S. Leśniewski and called by him *mereology*[3] will be essentially involved in our exposition. In this theory the *relation of part to whole* (which is treated

---

[1] By a regular open set, or an open domain, we understand a point set which coincides with the interior of its closure; by a regular closed set, or a closed domain, we understand a point set which coincides with the closure of its interior. These notions have been defined by Kuratowski, C. (42) and (41), pp. 37 f.

[2] See Whitehead, A. N. (88) and (89), and Nicod, J. (56).

[3] The first outline of this theory appeared in the monograph Leśniewski, S. (45 a). See also Leśniewski, S. (47 a). From a formal point of view mereology is closely related to Boolean algebra. In this connexion, cf. XI, p. 333 footnote.

---

† BIBLIOGRAPHICAL NOTE. An address on the foundations of the geometry of solids was given by the author to the First Polish Mathematical Congress, held in Lwów in 1927. A summary of this address appeared in French under the title 'Les fondements de la géométrie des corps' in *Księga Pamiątkowa Pierwszego Polskiego Zjazdu Matematycznego*, supplement to *Annales de la Société Polonaise de Mathématique*, Kraków 1929, pp. 29–33. The present article contains a translation of this summary, with additions designed to clarify certain passages of the French text.

as a relation between individuals[1]) can be taken as the only primitive notion. In terms of this relation various other notions are defined as follows:

DEFINITION I. *An individual X is called a proper part of an individual Y if X is a part of Y and X is not identical with Y.*

DEFINITION II. *An individual X is said to be disjoint from an individual Y if no individual Z is a part of both X and Y.*

DEFINITION III. *An individual X is called a sum of all elements of a class α of individuals[1] if every element of α is a part of X and if no part of X is disjoint from all elements of α.*

The following two statements can be accepted as the only postulates of mereology:[2]

POSTULATE I. *If X is a part of Y and Y is a part of Z, then X is a part of Z.*

POSTULATE II. *For every non-empty class α of individuals there exists exactly one individual X which is a sum of all elements of α.*

The geometry of solids is based upon mereology, in the sense that the relation between part and whole is included in the system of primitive notions of the geometry of solids, and similarly Posts. I and II are included in the postulate system of that theory.

The notion of a *sphere* is accepted as the only *specific* primitive

---

[1] The terms 'individual' and 'class' are used here in the same way as in the well-known work of A. N. Whitehead and B. Russell (90). The terminology adopted in this article differs in several details from that of Leśniewski's mereology. In fact Leśniewski uses the terms 'ingredient', 'part', 'exterior', and 'set' in the sense in which I use 'part', 'proper part', 'disjoint', and 'sum' respectively, and he employs the term 'class' in a sense completely different from ours.

[2] The postulate system for mereology given in this article (Posts. I and II) was obtained by simplifying a postulate system originally formulated by S. Leśniewski in (47 a), *Przegląd Filozoficzny*, vol. 33 (1930), pp. 82 ff. The simplification consisted in eliminating one of Leśniewski's postulates by deriving it from the remaining ones. A postulate system obviously equivalent to the system formed by Posts. I and II can be found in the book of J. H. Woodger, *The Axiomatic Method in Biology*, Cambridge 1937, Appendix E (by A. Tarski), p. 160.

notion of the geometry of solids.[1] In terms of this notion (and of the notions of mereology previously introduced) I shall now successively define a series of further notions which will prove helpful in eventually formulating the specific postulates for the geometry of solids.[2]

DEFINITION 1. *The sphere A is* externally tangent *to the sphere B if (i) the sphere A is disjoint from the sphere B; (ii) given two spheres X and Y containing as a part the sphere A and disjoint from the sphere B, at least one of them is part of the other.*

DEFINITION 2. *The sphere A is* internally tangent *to the sphere B if (i) the sphere A is a proper part of the sphere B; (ii) given two spheres X and Y containing the sphere A as a part and forming part of the sphere B, at least one of them is a part of the other.*

DEFINITION 3. *The spheres A and B are* externally diametrical *to the sphere C if (i) each of the spheres A and B is externally tangent to the sphere C; (ii) given two spheres X and Y disjoint from the sphere C and such that A is a part of X and B a part of Y, the sphere X is disjoint from the sphere Y.*

DEFINITION 4. *The spheres A and B are* internally diametrical *to the sphere C if (i) each of the spheres A and B is internally tangent to the sphere C; (ii) given two spheres X and Y disjoint from the sphere C and such that the sphere A is externally tangent to X and B to Y, the sphere X is disjoint from the sphere Y.*

DEFINITION 5. *The sphere A is* concentric *with the sphere B if one of the following conditions is satisfied: (i) the spheres A and B are identical; (ii) the sphere A is a proper part of the sphere B and besides, given two spheres X and Y externally diametrical to A and internally tangent to B, these spheres are internally diametrical to B; (iii) the sphere B is a proper part of the sphere A and besides,*

---

[1] As regards three-dimensional point geometry, a method of basing it on the notion of *sphere* as the only primitive notion has been developed in Huntington, E. V. (33); but the construction outlined in the present article has very little in common with Huntington's development.

[2] The system of definitions given below embodies some simplifications (in particular in the formulation of Def. 3) for which the author is indebted to B. Knaster. For further simplifications of these definitions see a recent note of Jaśkowski, S. (36 a).

given two spheres X and Y externally diametrical to B and internally tangent to A, these spheres are internally diametrical to A.

DEFINITION 6. *A point is the class of all spheres which are concentric with a given sphere.*[1]

DEFINITION 7. *The points a and b are equidistant from the point c if there exists a sphere X which belongs as element to the point c and which moreover satisfies the following condition: no sphere Y belonging as element to the point a or to the point b is a part of X disjoint from X.*

DEFINITION 8. *A solid is an arbitrary sum of spheres.*

DEFINITION 9. *The point a is interior to the solid B if there exists a sphere A which is at the same time an element of the point a and a part of the solid B.*

It is known that all the concepts of Euclidean geometry can be defined by means of those of *point* and of *equidistance of two points from a third.*[2] Consequently, by regarding the concepts introduced by Defs. 6 and 7 as correlates of their homonyms in ordinary geometry, we are able to define in the geometry of solids the correlates of all the other notions of point geometry. In particular, we are able to establish the meaning of the expression 'the class α of points is a regular open set'.

I turn now to the formulation of the specific postulates which —together with Posts. I and II of mereology—form a postulate system sufficient for the development of the geometry of solids. In the first place I adopt the following:

POSTULATE 1. *The notions of point and of equidistance of two points from a third satisfy all the postulates of ordinary Euclidean geometry of three dimensions.*[3]

In addition to this postulate, which is of fundamental importance, we may admit certain auxiliary postulates which render our system categorical. The postulates which I adopt for this

---

[1] Thus, in the geometry of solids, spheres are treated as individuals, i.e. objects of the first order, while points are classes of spheres and hence objects of the second order. In ordinary point geometry the opposite is true.

[2] Cf. Pieri, M. (59).

[3] A postulate system for ordinary geometry involving only these two notions as primitive can be found in Pieri, M. (59).

purpose establish exact connexions between notions of the geometry of solids—those of a *solid* and of the *relation of a part to a whole*—and the corresponding pair of notions of ordinary point geometry, those of a *regular open set* and of the *relation of inclusion*.

POSTULATE 2. *If A is a solid, the class α of all interior points of A is a non-empty regular open set.*

POSTULATE 3. *If the class α of points is a non-empty regular open set, there exists a solid A such that α is the class of all its interior points.*

POSTULATE 4. *If A and B are solids, and all the interior points of A are at the same time interior to B, then A is a part of B.*

The postulate system given above is far from being simple and elegant; it seems very likely that this postulate system can be essentially simplified by using intrinsic properties of the geometry of solids. For instance, it is not difficult to show that Post. 4 can be replaced by either of the following two postulates:

POSTULATE 4′. *If A is a solid and B a part of A, then B is also a solid.*

POSTULATE 4″. *If A is a sphere and B a part of A, there exists a sphere C which is a part of B.*

It should be noticed that the class of all solids (and not that of all spheres) constitutes what is called the 'universe of discourse' for the geometry of solids. For this reason it may be convenient to adopt the notion of a solid as an additional primitive notion. In this case Def. 8 should, of course, be omitted; in its place the fact that the class of solids coincides with that of arbitrary sums of spheres should be stated in a new postulate.

As regards methodological properties of the geometry of solids, I should like to state the following two results (the proofs of which present no serious difficulties):

THEOREM A. *The postulate system of the geometry of solids is categorical.*

More specifically, given any two models of this postulate system, we can establish a one-to-one correspondence between solids of the first model and those of the second in such a way that if X, Y, Z are any solids of the first model, and X', Y', Z' the corresponding solids of the second model, we have:

(i) X is a sphere if and only if X' is a sphere;

(ii) Y is a part of Z if and only if Y' is a part of Z'.

THEOREM B. *The postulate system of the geometry of solids, with the postulates of mereology included, has a model in ordinary three-dimensional Euclidean geometry. To obtain such a model we interpret spheres as interiors of Euclidean spheres, and the relation of a part to a whole as the inclusion relation restricted to non-empty regular open sets.*[1]

Conversely, the postulate system of three-dimensional Euclidean geometry has a model in the geometry of solids. Such a model is obtained by interpreting points and the relation of equidistance in the way indicated in Defs. 6 and 7.

Consequently, the consistency problems for the two postulate systems are equivalent.

In conclusion, when we compare the results which have been summarized with those of Whitehead and Nicod, the following may be stated: The procedure which has enabled us here to formulate definitions and postulates (especially Defs. 6 and 7 and Post. 1) can be regarded as a special case of the *method of extensive abstraction* developed by Whitehead. Nicod has already drawn attention to the equivalence of the problems of consistency for the two systems of geometry: that of the geometry of solids and that of ordinary point geometry. In my opinion what is to be regarded as a new result is the precise method of establishing the mathematical foundations of the geometry of solids with the help of a categorical system of postulates containing only two primitive notions: the notion of *sphere* and that of *being a part*.

---

[1] This fact has an interesting consequence formulated in terms of Boolean algebra; cf. XI, p. 341, note 2.

III

ON SOME FUNDAMENTAL CONCEPTS
OF METAMATHEMATICS†

OUR object in this communication is to define the meaning,
and to establish the elementary properties, of some important
concepts belonging to the *methodology of the deductive sciences*,
which, following Hilbert, it is customary to call *metamathematics*.[1]

Formalized deductive disciplines form the field of research of
metamathematics roughly in the same sense in which spatial
entities form the field of research in geometry. These disciplines
are regarded, from the standpoint of metamathematics, as sets
of *sentences*. Those sentences which (following a suggestion of
S. Leśniewski) are also called *meaningful sentences*, are themselves regarded as certain inscriptions of a well-defined form.
The set of all sentences is here denoted by the symbol 'S'. From
the sentences of any set X certain other sentences can be obtained
by means of certain operations called *rules of inference*. These
sentences are called the *consequences of the set X*. The set of all
consequences is denoted by the symbol '$Cn(X)$'.

An exact definition of the two concepts, of sentence and of
consequence, can be given only in those branches of metamathematics in which the field of investigation is a concrete
formalized discipline. On account of the generality of the present
considerations, however, these concepts will here be regarded as
primitive and will be characterized by means of a series of

---

[1] Many ideas and results outlined in this report have been presented in
a more detailed way in two later articles of the author, V and XII.

---

† BIBLIOGRAPHICAL NOTE. The main ideas of this article were outlined by
the author in a lecture to the Polish Mathematical Society, Warsaw Section, in
1928. For a summary of this lecture see Tarski, A. (72). The communication
was presented (by J. Łukasiewicz) to the Warsaw Scientific Society on 27
March 1930; it was published under the title 'Über einige fundamentale
Begriffe der Metamathematik' in *Comptes Rendus des séances de la Société des
Sciences et des Lettres de Varsovie*, vol. 23, 1930, cl. iii, pp. 22–29.

axioms. In the customary notation of general set theory these axioms can be formulated in the following way:

AXIOM 1. $\overline{\overline{S}} \leqslant \aleph_0$.

AXIOM 2. *If $X \subseteq S$, then $X \subseteq Cn(X) \subseteq S$.*

AXIOM 3. *If $X \subseteq S$, then $Cn(Cn(X)) = Cn(X)$.*

AXIOM 4. *If $X \subseteq S$, then $Cn(X) = \displaystyle\sum_{Y \subseteq X \text{ and } \overline{\overline{Y}} < \aleph_0} Cn(Y)$.*

AXIOM 5. *There exists a sentence $x \in S$ such that $Cn(\{x\}) = S$.*

With a view to reaching more profound results, other axioms of a more special nature are added to these. In contrast to the first group of axioms those of the second group apply, not to all deductive disciplines, but only to those which presuppose the sentential calculus, in the sense that in considerations relating to these disciplines we may use as premisses all true sentences of the sentential calculus.[1] In the axioms of this second group there occur as new primitive concepts two operations by means of which from simple sentences more complicated ones can be formed, namely the operations of forming implications and forming negations. The implication with the antecedent x and the consequent y is here denoted by the symbol '$c(x,y)$', and the negation of x by the symbol '$n(x)$'. The axioms are as follows:[2]

AXIOM 6*. *If $x \in S$ and $y \in S$, then $c(x,y) \in S$ and $n(x) \in S$.*[3]

---

[1] i.e. all sentences which belong to the ordinary (two-valued) system of the sentential calculus; cf. IV, § 2, Def. 5.

[2] The numbering of the axioms of the second group and of the theorems which follow from them is distinguished from that of the remaining axioms and theorems by the presence of an asterisk '*'.

[3] As already explained, sentences are here regarded as material objects (inscriptions). From this standpoint the content of Ax. 6* does not correspond exactly to the intuitive properties of the concepts occurring in it. It is not always possible to form the implication of two sentences (they may occur in widely separated places). In order to simplify matters we have in formulating this axiom, committed an error; this consists in *identifying equiform sentences* (as S. Leśniewski calls them). This error can be removed by interpreting S as the set of all types of sentences (and not of sentences) and by modifying in an analogous manner the intuitive sense of other primitive concepts. In this connexion by the type of a sentence x we understand the set of all sentences which are equiform with x.

AXIOM 7*. *If* $X \subseteq S$, $y \in S$, $z \in S$ *and* $c(y,z) \in Cn(X)$, *then* $z \in Cn(X+\{y\})$.

AXIOM 8*. *If* $X \subseteq S$, $y \in S$, $z \in S$ *and* $z \in Cn(X+\{y\})$, *then* $c(y,z) \in Cn(X)$.†

AXIOM 9*. *If* $x \in S$, *then* $Cn(\{x, n(x)\}) = S$.

AXIOM 10*. *If* $x \in S$, *then* $Cn(\{x\}) . Cn(\{n(x)\}) = Cn(0)$.

Axioms 8* and 10* are satisfied only in connexion with those formalized disciplines in the sentences of which no free variables occur.[1] Instead of Ax. 8* in its full extent, it suffices to adopt as an axiom the following special case of this sentence:

If $y \in S$, $z \in S$ *and* $z \in Cn(\{y\})$, *then* $c(y,z) \in Cn(0)$.

On the basis of these axioms a series of theorems concerning the concepts involved can be proved, for example:

THEOREM 1. *If* $X \subseteq Y \subseteq S$, *then* $Cn(X) \subseteq Cn(Y)$.

THEOREM 2. *If* $X+Y \subseteq S$, *then*

$$Cn(X+Y) = Cn(X+Cn(Y)) = Cn(Cn(X)+Cn(Y)).$$

This theorem can be generalized to cover an arbitrary (even an infinite) number of summands.

---

[1] This means that expressions (sentential functions) with free variables are not regarded as sentences.

† Ax. 8* is one of the formulations of what is known in the literature as the deduction theorem. This theorem, in its application to the formalism of *Principia Mathematica*, was first established by the author as far back as 1921 (in connexion with a discussion in the monograph of K. Ajdukiewicz (2)); the result was discussed in the author's lecture to the Warsaw Philosophical Institute, Section of Logic, listed by title in *Ruch Filozoficzny*, vol. 6 (1921–2), p. 72 *a*. Subsequently the deduction theorem was often applied in metamathematical discussion. Thus e.g. some theorems stated in the note (85) of A. Lindenbaum and A. Tarski, as well as in the note of A. Tarski (72), were obtained with the essential help of this result. In particular, Th. II in the first of these notes is simply a specialized form of the deduction theorem. A proof of the deduction theorem for a particular formalized theory is outlined in IX, p. 286 (proof of Th. 2 *a*). For reference to other appearances of the deduction theorem in the literature see a review by A. Church (of a book by W. V. Quine) in the *Journal of Symbolic Logic*, vol. 12 (1947), pp. 60–61.

THEOREM 3*. *If $x \in S$, $y \in S$ and $z \in S$, then*

$$c\big(c(x,y), c\big(c(y,z), c(x,z)\big)\big) \in Cn(0), \quad c\big(x, c(n(x), y)\big) \in Cn(0)$$

and $\qquad\qquad c\big(c(n(x), x), x\big) \in Cn(0).$

This theorem asserts that every sentence obtained by substitution from one of the three axioms of the ordinary system of sentential calculus, which are due to J. Łukasiewicz,[1] is a consequence of the null set 0 (and hence is also a consequence of every set X of sentences). By the use of Ax. 7* this theorem can be extended to all substitution instances of true sentences of the sentential calculus.

By means of the concepts S and $Cn(X)$ other important concepts of metamathematics can be defined. For example:

DEFINITION 1. *A set X of sentences is called a* deductive system (*or simply a* system), *in symbols $X \in \mathfrak{S}$, if*

$$Cn(X) = X \subseteq S.$$

The following properties of systems are easily proved:

THEOREM 4. *For every set $X \subseteq S$ there exists the smallest system Y which includes X, and in fact $Y = Cn(X)$.*

In consequence of this theorem the set $Cn(0)$ is the smallest system of all; this set can be called the *system of all logically true sentences*.

THEOREM 5. *If $\mathfrak{R} \subseteq \mathfrak{S}$ and $\mathfrak{R} \neq 0$, then $\prod_{X \in \mathfrak{R}} X \in \mathfrak{S}$ (the intersection of any number of systems is itself a system).*

THEOREM 6. *If for any two systems $X \in \mathfrak{R}$ and $Y \in \mathfrak{R}$ there is a system $Z \in \mathfrak{R}$ with $X \subseteq Z$ and $Y \subseteq Z$, and $\mathfrak{R} \neq 0$, then $\sum_{X \in \mathfrak{R}} X \in \mathfrak{S}$.*

THEOREM 7* (of A. Lindenbaum). *If $\mathfrak{R} \subseteq \mathfrak{S}$, $\overline{\overline{\mathfrak{R}}} < \aleph_0$ and $\sum_{X \in \mathfrak{R}} X \in \mathfrak{S}$, then $\sum_{X \in \mathfrak{R}} X \in \mathfrak{R}$; in other words, no system can be represented as a sum of a finite number of systems distinct from itself.*

THEOREM 8*. *If $\mathfrak{R} \subseteq \mathfrak{S}$, $\overline{\overline{\mathfrak{R}}} < \aleph_0$, $Y \in \mathfrak{S}$, $Y \subseteq \sum_{X \in \mathfrak{R}} X$ and $Y \neq 0$, then there is a system $X \in \mathfrak{R}$ such that $Y \subseteq X$.*

---

[1] Cf. IV, § 2.

We next introduce the notion of *(logical) equivalence*, together with the important notions of *consistency* and *completeness*.

DEFINITION 2. *The sets X and Y of sentences are called* (logically) *equivalent, in symbols X ∼ Y, if X+Y ⊆ S and*

$$Cn(X) = Cn(Y).$$

DEFINITION 3. *The set X of sentences is called* consistent, *in symbols X ∈ 𝔚, if X ⊆ S and if the formula X ∼ S does not hold (i.e. if Cn(X) ≠ S).*

DEFINITION 4. *The set X is said to be* complete, *in symbols X ∈ 𝔙, if X ⊆ S and if every set Y ∈ 𝔚 which includes X satisfies the formula X ∼ Y.*

With the help of the axioms of the second group it can be shown that Defs. 3 and 4 agree with the usual definitions of consistency and completeness:

THEOREM 9*. *X ∈ 𝔚 if and only if X ⊆ S and if for no sentence y ∈ S do we have both y ∈ Cn(X) and n(y) ∈ Cn(X).*

THEOREM 10*. *X ∈ 𝔙 if and only if X ⊆ S and if for every sentence y ∈ S at least one of the formulas y ∈ Cn(X) and n(y) ∈ Cn(X) holds.*

The following theorems are also derivable:

THEOREM 11. *If ℜ ⊆ 𝔚 and if for every finite class 𝔏 ⊆ ℜ a set Y ∈ ℜ exists which satisfies the formula $\sum_{X \in \mathfrak{L}} X \subseteq Y$, then $\sum_{X \in \mathfrak{R}} X \in \mathfrak{W}$.*

THEOREM 12 (of A. Lindenbaum). *If X ∈ 𝔚, then there is a set Y ∈ 𝔖.𝔚.𝔙 such that X ⊆ Y; in other words, every consistent set of sentences can be enlarged to form a consistent and complete system.*

THEOREM 13*. *Let X ⊆ S. In order that y ∈ Cn(X) it is necessary and sufficient that y ∈ S and that the formula X+{n(y)} ∈ 𝔚 should not hold.*

The concept of completeness is often confused with two other concepts which are related to it in content: that of *categoricity* and of *non-ramifiability*.[1] The next notion to be defined is closely related to that of completeness:

DEFINITION 5. *The* degree of completeness *of a set $X \subseteq S$ of sentences, in symbols $\gamma(X)$, is the smallest ordinal number $\alpha \neq 0$ which satisfies the following condition: there exists no increasing sequence of consistent non-equivalent sets X_ξ of sentences of type α which begins with X (i.e. no sequence of sets X_ξ satisfying the formulas: $X_0 = X$, $X_\xi \subseteq X_\eta \subseteq S$ and $Cn(X_\xi) \neq Cn(X_\eta)$ for $\xi < \eta < \alpha$).*

From this definition it follows that:

THEOREM 14. $\gamma(X) = 1$ *if and only if $X \sim S$ (i.e. if $X \subseteq S$ and if the formula $X \in \mathfrak{W}$ does not hold); $\gamma(X) = 2$ if and only if $X \in \mathfrak{W} \cdot \mathfrak{B}$; $\gamma(X) > 2$ if and only if $X \in \mathfrak{W} - \mathfrak{B}$.*

Finally we introduce the following concepts:

DEFINITION 6. *A set X of sentences is called* independent, *in symbols $X \in \mathfrak{U}$, if $X \subseteq S$ and if $Y = X$ always follows from the formulas: $Y \sim X$ and $Y \subseteq X$.*

DEFINITION 7. *A set Y of sentences is called a* basis *of the set X of sentences, in symbols $Y \in \mathfrak{B}(X)$, if $X \sim Y$ and $Y \in \mathfrak{U}$.*

DEFINITION 8. *A set Y of sentences is called a* finite axiom system, *or for short an* axiom system, *of the set X of sentences, in symbols $Y \in \mathfrak{Ax}(X)$, if $X \sim Y$ and $\overline{\overline{Y}} < \aleph_0$.*

DEFINITION 9. *A set X of sentences is said to be* finitely axiomatizable, *or for short* axiomatizable, *in symbols $X \in \mathfrak{A}$, if $\mathfrak{Ax}(X) \neq 0$.*

THEOREM 15*. $X \in \mathfrak{U}$ *if and only if $X \subseteq S$ and for every $y \in X$ the formula $X - \{y\} + \{n(y)\} \in \mathfrak{W}$ holds.*

THEOREM 16. *If $X \subseteq S$ and $\overline{\overline{X}} < \aleph_0$, then there exists a set $Y \subseteq X$ such that $Y \in \mathfrak{B}(X)$; i.e. every finite set of sentences contains a basis as a subset.*

THEOREM 17*. *If $X \subseteq S$, then $\mathfrak{B}(X) \neq 0$; i.e. every set of sentences possesses a basis.*

[1] Cf. the remarks concerning these notions in article V, at the end of § 7.

THEOREM 18. *The following conditions are equivalent*: (1) $X \in \mathfrak{A}$; (2) *there is a set* $Y \subseteq X$, *such that* $Y \in \mathfrak{Ax}(X)$; (3) $\mathfrak{Ax}(X)$. $\mathfrak{B}(X) \neq 0$; (4) *there exists a set* $Y \subseteq X$, *such that* $\overline{\overline{Y}} < \aleph_0$ *and* $Y \in \mathfrak{B}(X)$.

THEOREM 19*. *In order that* $X \in \mathfrak{A}$ *it is necessary and sufficient that* $X \subseteq S$ *and that* X *possess no infinite basis.*

THEOREM 20*. *Let* $X \in \mathfrak{S}$; *in order that* $X \in \mathfrak{A}$ *it is necessary and sufficient that no class* $\mathfrak{R} \subseteq \mathfrak{S}$ *exist which satisfies the following conditions*: $X \overline{\in} \mathfrak{R}$ *and* $X = \sum_{Y \in \mathfrak{R}} Y$ (*i.e. that* X *cannot be represented as a sum of systems distinct from itself*).

THEOREM 21. *We have* $\overline{\overline{\mathfrak{S} \cdot \mathfrak{A}}} \leqslant \aleph_0$ *and* $\overline{\overline{\mathfrak{S} - \mathfrak{A}}} \leqslant \overline{\overline{\mathfrak{S}}} \leqslant 2^{\aleph_0}$; *if an infinite set* $X \in \mathfrak{U}$ *exists, then*

$$\overline{\overline{\mathfrak{S} \cdot \mathfrak{A}}} = \aleph_0 \text{ and } \overline{\overline{\mathfrak{S} - \mathfrak{A}}} = \overline{\overline{\mathfrak{S}}} = 2^{\aleph_0}.$$

It should be noted that in almost all known deductive disciplines an infinite and at the same time independent set of sentences can be constructed, which thus realizes the hypothesis of the second part of the last theorem. In these disciplines there are therefore more non-axiomatizable than axiomatizable systems; systems are, so to speak, only exceptionally axiomatizable.[1]

By some authors the concept of the independence of a set of sentences is sharpened in various directions (*complete independence* by E. H. Moore,[2] *maximal independence* by H. M. Sheffer[3]). These notions will not be discussed here.

Within the conceptual framework of this paper we can carry out metamathematical investigations relating to concrete deductive disciplines. For this purpose in each single case the concepts of sentence and of consequence must first be defined. We then take as a starting-point some set X of sentences in which we are

---

[1] Lindenbaum was the first to draw attention to this fact (in connexion with the sentential calculus), thus indirectly suggesting the notion of axiomatizability; cf. IV, § 3.

[2] *Introduction to a form of general analysis*, New Haven 1910, p. 82.

[3] Sheffer, H. M. (63), p. 32.

interested. We investigate it from the point of view of consistency and axiomatizability; we try to determine its degree of completeness, and possibly to specify all the systems, in particular all the consistent and complete systems, which include X as a subset.†

† For an example of investigations in these directions which deal with the simplest deductive discipline, namely the sentential calculus, the reader is referred to IV. Some results concerning other deductive disciplines are briefly discussed in XII, § 5.

IV

INVESTIGATIONS INTO THE SENTENTIAL CALCULUS†

In the course of the years 1920–30 investigations were carried out in Warsaw belonging to that part of metamathematics—or better metalogic—which has as its field of study the simplest deductive discipline, namely the sentential calculus. These investigations were initiated by Łukasiewicz; the first results originated both with him and with Tarski. In the seminar for mathematical logic which was conducted by Łukasiewicz in the University of Warsaw beginning in 1926, most of the results stated below of Lindenbaum, Sobociński, and Wajsberg were found and discussed. The systematization of all the results and the clarification of the concepts concerned was the work of Tarski.

In the present communication the most important results of these investigations—for the most part not previously published—are collected together.‡

§ 1. General Concepts

It is our intention to refer our considerations to the conceptual apparatus which was developed in the preceding article (see III).

† Bibliographical Note. This joint communication of J. Łukasiewicz and A. Tarski was presented (by Łukasiewicz) to the Warsaw Scientific Society on 27 March 1930; it was published under the title 'Untersuchungen über den Aussagenkalkül' in *Comptes Rendus des séances de la Société des Sciences et des Lettres de Varsovie*, vol. 23, 1930, cl. iii, pp. 30–50.

‡ To avoid misunderstandings it should be stated that the present article does not contain results discovered by both the authors jointly, but is a compilation of theorems and concepts belonging to five different persons. Each theorem and concept is ascribed to its respective originator. Theorem 3, for instance, is not a theorem of Łukasiewicz and Tarski, but a theorem of Lindenbaum. Nevertheless, some scholars mistakenly referred to both authors, Łukasiewicz and Tarski, the many-valued systems of logic ascribed in the article to Łukasiewicz alone. In spite of a correction which appeared in 1933 in the *Journal of Philosophy*, vol. 30, p. 364, this mistake persists till today. It clearly follows from § 3 and notes of this article that the idea of a logic different from the ordinary system called by Łukasiewicz the *two-valued logic*, and the construction of many-valued systems of logic described here, are entirely due to Łukasiewicz alone and should not be referred to Łukasiewicz and Tarski.

For this purpose we wish first to define the notion of a (*meaningful*) *sentence* and that of a *consequence of a set of sentences* with respect to the sentential calculus.

DEFINITION 1. *The set S of all sentences is the intersection of all those sets which contain all sentential variables (elementary sentences) and are closed under the operations of forming implications and negations.*[1]

The concepts of *sentential variable*, of *implication* and of *negation* cannot be explained further; they must rather be regarded as primitive concepts of the metatheory of sentential calculus (i.e. of that part of metamathematics in which sentential calculus is investigated). The fundamental properties of these concepts, which suffice for the construction of the part of metamathematics with which we are here concerned, can be expressed in a series of simple sentences (axioms) which need not now be stated. Usually the letters 'p', 'q', 'r', etc., are used as sentential variables. In order to express in symbols the sentences 'p implies q' (or 'if p, then q') and 'it is not the case that p', Łukasiewicz employs the formulas 'Cpq' and 'Np' respectively.[2] With this notation the use of such technical signs as parentheses, dots, etc., is rendered unnecessary. We shall encounter several examples of sentences written in this symbolism in subsequent sections. In addition to the formation of implications and negations other similar operations are commonly used in the sentential calculus. But as these are all definable by means of the two mentioned above they will not be considered here.

The symbol 'Cpq', which in the sentential calculus *expresses* the implication between 'p' and 'q', is to be clearly distinguished from the metamathematical symbol '$c(x, y)$', which *denotes* the implication with the antecedent x and the consequent y. The expression 'Cpq' is a *sentence* (in the sentential calculus) whilst the expression '$c(x, y)$' is the *name of a sentence* (in the meta-

---

[1] A set—according to the usual terminology of abstract set theory—is said to be *closed under given operations* if as the result of carrying out these operations on elements of the set in question one always obtains elements of this same set.

[2] Cf. Łukasiewicz, J. (51 *a*), p. 610 note, and (51), p. 40.

sentential calculus). An analogous remark applies to the symbolic expressions 'Np' and '$n(x)$'.

The consequences of a set of sentences are formed with the help of two operations, that of *substitution* and that of *detachment* (the modus ponens scheme of inference). The intuitive meaning of the first operation is clear; we shall not, therefore, discuss its character more closely. The second operation consists in obtaining the sentence y from the sentences x and $z = c(x, y)$. We are now in a position to explain the concept of consequence:

DEFINITION 2. *The* set of consequences $Cn(X)$ of the set X of sentences *is the intersection of all those sets which include the set* $X \subseteq S$ *and are closed under the operations of substitution and detachment.*

From this we obtain:

THEOREM 1. *The concepts S and $Cn(X)$ satisfy the axioms 1–5 given in article III.*[1]

We are especially interested in those parts X of the set S which form *deductive systems*, i.e. which satisfy the formula $Cn(X) = X$. Two methods of constructing such systems are available to us. In the first, the so-called *axiomatic method*, an arbitrary, usually finite, set X of sentences—an *axiom system* —is given, and the set $Cn(X)$, i.e. the smallest deductive system over X, is formed. The second method, which can best be called the *matrix method*, depends upon the following definitions of Tarski:[2]

---

[1] See article III, p. 31.

[2] The origin of this method is to be sought in the well-known verification procedure for the usual two-valued sentential calculus (see below Def. 5), which was used by Peirce ('On the algebra of Logic', *Am. Journ. of Math.* vol. 7 (1885), p. 191) and Schröder. This was thoroughly treated in Łukasiewicz, J. (50 a). Łukasiewicz was also the first to define by means of a matrix a system of the sentential calculus different from the usual one, namely his three-valued system (see below, p. 47, note 2). This he did in the year 1920. Many-valued systems, defined by matrices, were also known to Post (see Post, E. L. (60), pp. 180 ff.). The method used by P. Bernays ('Axiomatische Untersuchung des Aussagenkalküls der "Principia Mathematica" ', *Math. Z.*, vol. 25, 1926, pp. 305–20) for the proof of his theorems on independence also rests on matrix formation. The view of matrix formation as a general method of constructing systems is due to Tarski.

DEFINITION 3. *A* (logical) matrix *is an ordered quadruple* $\mathfrak{M} = [A, B, f, g]$ *which consists of two disjoint sets (with elements of any kind whatever) A and B, a function f of two variables and a function g of one variable, where the two functions are defined for all elements of the set* $A + B$ *and take as values elements of* $A + B$ *exclusively.*

The matrix $\mathfrak{M} = [A, B, f, g]$ *is called* normal *if the formulas* $x \in B$ *and* $y \in A$ *always imply* $f(x, y) \in A$.

DEFINITION 4. *The function h is called a* value function *of the matrix* $\mathfrak{M} = [A, B, f, g]$ *if it satisfies the following conditions:* (1) *the function h is defined for every* $x \in S$; (2) *if x is a sentential variable, then* $h(x) \in A + B$; (3) *if* $x \in S$ *and* $y \in S$, *then*

$$h(c(x, y)) = f(h(x), h(y));$$

(4) *if* $x \in S$ *then* $h(n(x)) = g(h(x))$.

The sentence x is satisfied (*or* verified) *by the matrix* $\mathfrak{M} = [A, B, f, g]$, *in symbols* $x \in \mathfrak{E}(\mathfrak{M})$, *if the formula* $h(x) \in B$ *holds for every value function h of this matrix.*

The elements of the set B are, following Bernays,[1] called *designated* elements.

In order to construct a system of the sentential calculus with the help of the matrix method a matrix $\mathfrak{M}$ (usually normal) is set up and the set $\mathfrak{E}(\mathfrak{M})$ of all those sentences which are satisfied by this matrix is considered. This procedure rests upon the following easily provable theorem:

THEOREM 2. *If* $\mathfrak{M}$ *is a normal matrix, then* $\mathfrak{E}(\mathfrak{M}) \in \mathfrak{S}$.

If the set $\mathfrak{E}(\mathfrak{M})$ forms a system (as it always will, according to Th. 2, if the matrix $\mathfrak{M}$ is normal), it is called the *system generated by the matrix* $\mathfrak{M}$.

The following converse of Th. 2, which was proved by Lindenbaum, makes evident the generality of the matrix method described here:

---

[1] See Bernays, P. (5), p. 316.

THEOREM 3. *For every system* $X \in \mathfrak{S}$ *there exists a normal matrix* $\mathfrak{M} = [A, B, f, g]$, *with an at most denumerable set* $A + B$, *which satisfies the formula* $X = \mathfrak{C}(\mathfrak{M})$.†

Each of the two methods has its advantages and disadvantages. Systems constructed by means of the axiomatic method are easier to investigate regarding their axiomatizability, but systems generated by matrices are easier to test for completeness and consistency. In particular the following evident theorem holds:

THEOREM 4. *If* $\mathfrak{M} = [A, B, f, g]$ *is a normal matrix and* $A \neq 0$, *then* $\mathfrak{C}(\mathfrak{M}) \in \mathfrak{W}$.

§ 2. THE ORDINARY (TWO-VALUED) SYSTEM OF THE SENTENTIAL CALCULUS

In the first place we consider the most important of the systems of the sentential calculus, namely the well-known *ordinary system* (also called by Łukasiewicz[1] the two-valued system), which is here denoted by '*L*'.

Using the matrix method, the system *L* may be defined in the following way:

DEFINITION 5. *The ordinary system L of the sentential calculus is the set of all sentences which are satisfied by the matrix* $\mathfrak{M} = [A, B, f, g]$ *where* $A = \{0\}$, $B = \{1\}^2$ *and the functions f and g are defined by the formulas:* $f(0, 0) = f(0, 1) = f(1, 1) = 1$, $f(1, 0) = 0$, $g(0) = 1$, $g(1) = 0$.

---

[1] See (50 *a*).
[2] The set having *a* as its only element is denoted by '$\{a\}$'.

---

† A proof of this theorem has recently been published in J. Łoś, 'O matrycach logicznych' ('On logical matrices', in Polish), *Travaux de la Société des Sciences et des Lettres de Wrocław*, Ser. B, Nr. 19, Wrocław (1949), 42 pp. See also H. Hermes, 'Zur Theorie der aussagenlogischen Matrizen', *Math. Z.*, vol. 53 (1951), pp. 414–18.

From this definition it follows easily that the system L is consistent and complete:

THEOREM 5. $L \in \mathfrak{S}.\mathfrak{W}.\mathfrak{B}.$

The system L can also be defined by means of the axiomatic method. The first axiom system of the sentential calculus was given by C. Frege.[1] Other axiom systems have been given by Whitehead and Russell[2] and by Hilbert.[3] Of the systems at present known the simplest is that of Łukasiewicz; he also has proved in an elementary manner the equivalence of the two definitions of L.[4] His result may be stated thus:

THEOREM 6. *Let X be the set consisting of the three sentences*:

$$\text{'}CCpqCCqrCpr\text{'}, \quad \text{'}CCNppp\text{'}, \quad \text{'}CpCNpq\text{'};$$

then $X \in \mathfrak{Ax}(L)$. Consequently L is axiomatizable, $L \in \mathfrak{A}$.

According to a method for investigating the independence of a set X of sentences developed by Bernays and Łukasiewicz,[5]

---

[1] *Begriffsschrift*, Halle a/S (1879), pp. 25–30. Frege's system is based upon the following six axioms: '$CpCqp$', '$CCpCqrCCpqCpr$', '$CCpCqrCqCpr$', '$CCpqCNqNp$', '$CNNpp$', '$CpNNp$'. Łukasiewicz has shown that in this system the third axiom is superfluous since it can be derived from the preceding two axioms, and that the last three axioms can be replaced by the single sentence '$CCNpNqCqp$'.

[2] *Principia Mathematica*, vol. 1 (1925), p. 91.

[3] See (29 a), p. 153.

[4] Cf. Łukasiewicz, J. (51), pp. 45 and 121 ff. The proof of the equivalence of the two definitions of L amounts to the same thing as proving the completeness of the system L when defined by means of the axiomatic method. The first proof of completeness of this kind is found in Post (60).

[5] Bernays has published, in his article (5), which dates from the year 1926 (but according to the author's statement contains results from his unpublished *Habilitationsschrift* presented in the year 1918), a method based upon matrix formations, which enables us to investigate the independence of given sets of sentences. The method given by Bernays was known before its publication to Łukasiewicz who, independently of Bernays, and following a suggestion of Tarski (cf. I, pp. 8–14), first applied his many-valued systems, defined by means of matrices, to the proofs of independence, and subsequently discovered the general method. On the basis of this method Łukasiewicz had already in 1924 investigated the independence of the axiom systems given by Whitehead and Russell and by Hilbert, and had shown that neither of them is independent. These results (without proof) are contained in the following note by Łukasiewicz: 'Démonstration de la compatibilité des axiomes de la théorie de la déduction', *Ann. Soc. Pol. Math.*, vol. 3 (1925), p. 149.

a normal matrix $\mathfrak{M}_y$ is constructed for every sentence $y \in X$ which verifies all sentences of the set X with the exception of y. With the help of this method Łukasiewicz proved that in contrast to the previously mentioned axiom system the following theorem holds:

THEOREM 7. *The set X of sentences given in Th. 6 is independent; consequently X is a basis of L, $X \in \mathfrak{B}(L)$.*

Tarski developed another *structural* method for the study of independence. Although less general than the method of matrix formation, this can be used successfully in some cases. The following general theorem is due to Tarski:†

THEOREM 8. *The system L, as well as every axiomatizable system of the sentential calculus which contains the sentences: 'CpCqp' and 'CpCqCCpCqrr' (or 'CpCqCCpCqrCsr'), possesses a basis consisting of a single sentence.*[1]

The proof of this theorem enables us in particular to give effectively a basis of the system L which contains a single element.[2]‡ Łukasiewicz has simplified Tarski's proof and, with the help of previous work of B. Sobociński, has established the following:

THEOREM 9. *The set which consists of the single sentence z:*

$$'CCCpCqpCCCNrCsNtCCrCsuCCtsCtuvCwv'$$

is a basis of the system L, i.e. $\{z\} \in \mathfrak{B}(L)$.

---

[1] An analogous, but quite trivial, theorem applies to all axiomatizable systems of those deductive disciplines which already presuppose the sentential calculus and satisfy not only Axs. 1–5, but also Axs. 6*–10* of III.

[2] This result was obtained by Tarski in the year 1925; cf. Leśniewski, S. (46), p. 58. An axiom system of the ordinary sentential calculus consisting of a single axiom was set up by Nicod in the year 1917 (see Nicod, J. (55)). The axiom of Nicod is constructed with the Sheffer disjunction '$p \mid q$' as the only primitive term, and the rule of detachment formulated by Nicod in connexion with this term is stronger than the rule of detachment for implication. This facilitated the solution of the problem.

---

† Compare in this connexion a recent paper of K. Schröter (62 a), in particular pp. 294 ff.

‡ The axiom originally found by Tarski is explicitly formulated in the article by B. Sobociński, 'Z badań nad teorją dedukcji', *Przegląd Filozoficzny*, vol. 35 (1932), pp. 172–93, in particular p. 189. It consists of 53 letters.

THEOREM 15. *In every basis (and in general in every axiom system) of the system L, as well as of every sub-system of L which contains the sentence 'CpCqCrp', at least three distinct sentential variables occur.*[1] *In other words, if X is the set of all those sentences of the system L in which at most two distinct variables occur, then* $L - Cn(X) \neq 0$; *in particular, the sentence 'CpCqCrp' belongs to L but not to* $Cn(X)$.[†]

§ 3. MANY-VALUED SYSTEMS OF THE SENTENTIAL CALCULUS

In addition to the ordinary system of the sentential calculus there are many other systems of this calculus which are worthy of investigation. This was first pointed out by Łukasiewicz who has also singled out a specially important class of such systems.[2] The systems founded by Łukasiewicz are here called *n-valued systems* of the sentential calculus and denoted by the symbol 'L_n' (where either n is a natural number or $n = \aleph_0$). These systems can be defined by means of the matrix method in the following way:

DEFINITION 7. *The n-valued system* L_n *of the sentential calculus (where n is a natural number or* $n = \aleph_0$) *is the set of all*

---

[1] It is not necessary to explain any further the meaning of the expression 'in the sentence x two or three distinct variables occur' since it is intuitively clear. 'Distinct' here means the same as 'not equiform' (cf. III, p. 31, note 3).

[2] What is called the *three-valued* system of the sentential calculus was constructed by Łukasiewicz in the year 1920 and described in a lecture given to the Polish Philosophical Society in Lwów. A report by the author, giving the content of that lecture fairly thoroughly was published in the journal *Ruch Filozoficzny*, vol. 5 (1920), p. 170 (in Polish). A short account of the n-valued systems, the discovery of which belongs to the year 1922, is given in Łukasiewicz, J. (51), pp. 115 ff. The philosophical implications of n-valued systems of sentential calculus are discussed in the article of Łukasiewicz, 'Philosophische Bemerkungen zu mehrwertigen Systemen des Aussagenkalküls', *Comptes Rendus des séances de la Société des Sciences et des Lettres de Varsovie*, vol. 23 (1930), cl. iii, pp. 51–77.   ••

[†] Wajsberg's proof of Th. 15 is given at the end of his paper (87 a). Another proof of the result discussed can be obtained by the use of the method developed in the note of A. H. Diamond and J. C. C. McKinsey, 'Algebras and their subalgebras', *Bulletin of the American Mathematical Society*, vol. 53 (1947), pp. 959–62. For another proof of that part of Th. 15 which concerns the whole system L, see also S. Jaśkowski, 'Trois contributions au calcul des propositions bivalent', *Studia Societatis Scientiarum Torunensis*, section A, vol. 1 (1948), pp. 3–15, in particular pp. 9 f.]

sentences which are satisfied by the matrix $\mathfrak{M} = [A, B, f, g]$ *where, in the case* $n = 1$ *the set* A *is null, in the case* $1 < n < \aleph_0$ A *consists of all fractions of the form* $k/(n-1)$ *for* $0 \leqslant k < n-1$, *and in the case* $n = \aleph_0$ *it consists of all fractions* k/l *for* $0 \leqslant k < l$; *further the set* B *is equal to* $\{1\}$ *and the functions* f *and* g *are defined by the formulas*: $f(x,y) = \min(1, 1-x+y)$, $g(x) = 1-x$.

As Lindenbaum has shown, the system $L_{\aleph_0}$ is not changed if, in the definition of this system, the set A of all proper fractions is replaced by another infinite sub-set of the interval $\langle 0, 1 \rangle$:

THEOREM 16. *Let* $\mathfrak{M} = [A, B, f, g]$ *be a matrix where* $B = \{1\}$, *the functions* f *and* g *satisfy the formulas*

$$f(x,y) = \min(1, 1-x+y), \quad g(x) = 1-x,$$

and A *be an arbitrary infinite set of numbers which satisfies the condition:* $0 \leqslant x < 1$ *for every* $x \in A$, *and is closed under the two operations* f *and* g; *then* $\mathfrak{E}(\mathfrak{M}) = L_{\aleph_0}$.[1]

From Def. 7 the following facts established by Łukasiewicz are easily obtained:

THEOREM 17. (*a*) $L_1 = S$, $L_2 = L$;

(*b*) *if* $2 \leqslant m < \aleph_0$, $2 \leqslant n < \aleph_0$ *and* $n-1$ *is a divisor of* $m-1$, *then* $L_m \subseteq L_n$;

(*c*) $L_{\aleph_0} = \prod\limits_{1 \leqslant n < \aleph_0} L_n$.

THEOREM 18. *All systems* L_n *for* $3 \leqslant n \leqslant \aleph_0$ *are consistent but not complete:* $L_n \in \mathfrak{S} \, . \, \mathfrak{W} - \mathfrak{V}$.

The converse of Th. 17*b* was proved by Lindenbaum:

THEOREM 19. *For* $2 \leqslant m < \aleph_0$ *and* $2 \leqslant n < \aleph_0$ *we have* $L_m \subseteq L_n$ *if and only if* $n-1$ *is a divisor of* $m-1$.

---

[1] Lindenbaum gave a lecture at the first congress of the Polish mathematicians (Lwów, 1927) on mathematical methods of investigating the sentential calculus in which, among other things, he formulated the above-mentioned theorem. Cf. his note 'Méthodes mathématiques dans les recherches sur le système de la théorie de déduction', *Księga Pamiątkowa Pierwszego Polskiego Zjazdu Matematycznego*, Kraków (1929), p. 36.

Th. 17c was improved by Tarski by means of Th. 16:

THEOREM 20. $L_{\aleph_0} = \prod_{1 \leqslant i < \aleph_0} L_{n_i}$ *for every increasing sequence* n_i *of natural numbers.*

Concerning the problem of the degree of completeness for systems L_n the following partial result has been obtained.

THEOREM 21. *If* $n-1$ *is a prime number (in particular if* $n = 3$*), then there are only two systems, namely* S *and* L*, which contain* L_n *as a proper part; in other words, every sentence* $x \in S - L_n$ *satisfies one of the formulas:* $Cn(L_n + \{x\}) = S$ *or* $Cn(L_n + \{x\}) = L; \gamma(L_n) = 3.$

This theorem was proved for $n = 3$ by Lindenbaum; the generalization to all prime numbers given in the theorem is due to Tarski.†

Regarding the axiomatizability of the system L_n we have the following theorem which was first proved by Wajsberg for $n = 3$ and for all n for which $n-1$ is a prime number, and was later extended to all natural numbers by Lindenbaum:

THEOREM 22. *For every* n*,* $1 \leqslant n < \aleph_0$*, we have* $L_n \in \mathfrak{A}.$

The effective proof of Th. 22 enables us to give a basis for every system L_n where $1 \leqslant n < \aleph_0$. In particular Wajsberg has established:

THEOREM 23. *The set* X *consisting of the sentences* '$CpCqp$'*,* '$CCpqCCqrCpr$'*,* '$CCNpNqCqp$'*,* '$CCCpNppp$' *forms a basis of* L_3*, i.e.* $X \in \mathfrak{B}(L_3).$

The following theorem of Wajsberg is one of the generalizations of Th. 22 at present known:

† In May 1930 while the original printing of this article was in progress, Th. 21 was improved and the problem of the degree of completeness was solved for systems L_n with an arbitrary natural n; this was a joint result of members of a proseminar conducted by Łukasiewicz and Tarski in the University of Warsaw. A proof of Th. 21 and its generalizations appeared in print recently; see A. Rose, 'The degree of completeness of m-valued Łukasiewicz propositional calculus', *The Journal of the London Mathematical Society,* vol. 27 (1952), pp. 92–102. The solution of the same problem for $L_{\aleph_0}$ has been given in A. Rose, 'The degree of completeness of the $\aleph_0$-valued Łukasiewicz propositional calculus', *The Journal of the London Mathematical Society,* vol. 28 (1953), pp. 176–84.

THEOREM 24.   *Let* $\mathfrak{M} = [A, B, f, g]$ *be a normal matrix in which the set* $A + B$ *is finite.   If the sentences*

$$'CCpqCCqrCpr', \; 'CCqrCCpqCpr', \; 'CCqrCpp',$$

$$'CCpqCNqNp', \; 'CNqCCpqNp'$$

are satisfied by this matrix, then $\mathfrak{E}(\mathfrak{M}) \in \mathfrak{U}.$†

The Ths. 8, 10, and 15 of § 2 can be applied to the systems L_n. Accordingly we have:

THEOREM 25.   *Every system* L_n, *where* $2 \leqslant n < \aleph_0$, *possesses, for every natural number* m *(and in particular for* $m = 1$), *a basis which has exactly* m *elements.*

THEOREM 26.   *In every basis (and in general in every axiom system) of the system* L_n *at least three distinct sentential variables occur.*

As regards the problem of extending Th. 22 to the case $n = \aleph_0$ Łukasiewicz has formulated the hypothesis that *the system* $L_{\aleph_0}$

† There is a comprehensive literature related to Ths. 22–24 and, more generally, concerning the axiomatizability of various systems of sentential calculus. We list a few papers on this subject in which further bibliographical references can also be found: M. Wajsberg, 'Aksjomatyzacja trójwartościowego rachunku zdań' ('Axiomatization of the three-valued sentential calculus', in Polish), *Comptes Rendus des séances de la Société des Sciences et des Lettres de Varsovie*, vol. 24 (1931), cl. iii, pp. 126–48. A further relevant paper by the same author: 'Beiträge zum Metaaussagenkalkül I', *Monatshefte für Mathematik und Physik*, vol. 42 (1935), pp. 221–42. B. Sobociński, 'Aksjomatyzacja pewnych wielowartościowych systemów teorji dedukcji' ('Axiomatization of certain many-valued systems of the theory of deduction' in Polish), *Roczniki prac naukowych Zrzeszenia Asystentów Uniwersytetu Józefa Pilsudskiego w Warszawie*, vol. 1 (1936), Wydział Matematyczno Przyrodniczy Nr. 1, pp. 399–419. J. Słupecki, 'Dowód aksjomatyzowalności pełnych systemów wielowartościowych rachunku zdań' ('A proof of the axiomatizability of functionally complete systems of many-valued sentential calculus', in Polish), *Comptes Rendus des séances de la Société des Sciences et des Lettres de Varsovie*, cl. iii, vol. 32 (1939), pp. 110–28; and by the same author: 'Pełny trójwartościowy rachunek zdań' ('The full three-valued sentential calculus', in Polish), *Annales Universitatis Mariae Curie-Skłodowska*, vol. 1 (1946), pp. 193–209. J. B. Rosser and A. R. Turquette, 'Axiom schemes for *m*-valued propositional calculi', *Journal of Symbolic Logic*, vol. 10 (1945), pp. 61–82, and 'A note on the deductive completeness of *m*-valued propositional calculi', ibid., vol. 14 (1949), pp. 219–25.

In the first paper of Wajsberg listed above we find a proof of Th. 23, in the second a proof of Th. 24.

is axiomatizable and that *the set consisting of the following five
sentences*

$$\text{'}CpCqp\text{'}, \quad \text{'}CCpqCCqrCpr\text{'}, \quad \text{'}CCCpqqCCqpp\text{'},$$

$$\text{'}CCCpqCqpCqp\text{'}, \quad \text{'}CCNpNqCqp\text{'}$$

forms an axiom system for $L_{\aleph_0}$.†

It must be emphasized that, as defined here, the systems L_n
for $n > 2$ have a fragmentary character, since they are incomplete and are only sub-systems of the ordinary system L. The
problem of supplementing these systems to form complete and
consistent systems which are at the same time distinct from L can
be positively solved, but only in one way, namely by widening
the concept of meaningful sentence of the sentential calculus,
and by introducing, beside the operations of forming implications
and negations, other analogous operations which cannot be
reduced to these two (cf. also § 5).

Finally we may add that the number of all possible systems
of the sentential calculus was determined by Lindenbaum.

THEOREM 27. $\overline{\overline{\mathfrak{S}}} = 2^{\aleph_0}$, *but* $\overline{\overline{\mathfrak{S}.\mathfrak{A}}} = \aleph_0$.‡

This result was improved by Tarski as follows:

THEOREM 28. $\overline{\overline{\mathfrak{S}.\mathfrak{W}.\mathfrak{B}}} = 2^{\aleph_0}$, *but* $\overline{\overline{\mathfrak{S}.\mathfrak{W}.\mathfrak{B}.\mathfrak{A}}} = \aleph_0$.§

---

† This hypothesis has proved to be correct; see M. Wajsberg, 'Beiträge
zum Metaaussagenkalkül I', *Monatshefte für Mathematik und Physik*, vol. 42
(1935), pp. 221–42, in particular p. 240. As far as we know, however, Wajsberg's proof has not appeared in print.

The axiom-system above is not independent: C. A. Meredith has shown that
'$CCCpqCqpCqp$' is deducible from the remaining axioms.

‡ A proof of the first part of Th. 27 (and in fact of a somewhat stronger
result) can be found in the paper of K. Schröter (62 a), pp. 301 ff. The proof
of the second part of Th. 27 is almost obvious.

§ The proof of Th. 28 can be outlined as follows: For any given natural
number $n = 1, 2, 3,...$ let x_n be the sentence which is formed by n symbols 'C'
followed by $n+1$ variables 'p'. Given any set N of natural numbers, let X_N be
the set consisting of all sentences x_{3n} where n belongs to N and of all sentences
x_{3n+1} where n does not belong to N. It can easily be shown that the set $Cn(X_N)$
coincides with the set of all those sentences which can be obtained from sentences of X_N by substitution. Hence the set X_N is consistent and can therefore
be extended to form a complete and consistent system X_N. On the other hand,
if M and N are two different sets of natural numbers, then the sum of X_M and
X_N is clearly inconsistent, and hence the systems X_M^* and X_N^* cannot be
identical. The remaining part of the proof is obvious.

§ 4. THE RESTRICTED SENTENTIAL CALCULUS

In investigations into the sentential calculus attention is some-times restricted to those sentences in which no negation sign occurs. This part of the sentential calculus can be treated as an independent deductive discipline, one which is still simpler than the ordinary sentential calculus and will be called here the *restricted sentential calculus*.

For this purpose we must first of all modify the concept of meaningful sentence by omitting the operation of forming negations from Def. 1. In a corresponding way the concept of substitution is also simplified, and this brings with it a change in the concept of consequence. *After these modifications Th.* 1 *remains valid*.

For the construction of closed systems of the restricted sen-tential calculus both of the methods described in § 1 are used: the axiomatic and the matrix method. But a logical matrix is now defined as an ordered triple $[A, B, f]$ and not as an ordered quadruple (Def. 3); consequently condition (4) in Def. 4 of a value function disappears. *Ths.* 2–4 *remain valid*.

The definition of the ordinary system L^+ of the restricted sentential calculus is completely analogous to Def. 5, with one obvious difference which is called for by the modification in the concept of matrix. This system has been investigated by Tarski. From the definition of the system its consistency and complete-ness are easily derivable; hence *Th.* 5 *holds also in the restricted sentential calculus*. The axiomatizability of the system is estab-lished in the following theorem:

THEOREM 29. *The set* X *consisting of the three sentences* '*CpCqp*', '*CCpqCCqrCpr*', '*CCCpqpp*' *forms a basis of the system* L^+; *consequently* $L^+ \in \mathfrak{A}$.

This theorem originates with Tarski; it contains, however, a simplification communicated to the authors by P. Bernays. In fact the original axiom system of Tarski included, instead of the sentence '*CCCpqpp*', a more complicated sentence,

'$CCCpqrCCprr$'.[1]† The independence of both axiom systems was established by Łukasiewicz.

Ths. 8, 10, 11, 12, 14, *and* 15 *from* § 2 *have been extended to the restricted sentential calculus* by their originators. Tarski, in particular, has succeeded in setting up a basis of the system L^+ consisting of only a single sentence.[2] Two simple examples of such sentences, each containing 25 letters, are given in the next theorem. The first is an organic sentence and was found by Wajsberg, the second is not organic and is due to Łukasiewicz:

THEOREM 30. *The set of sentences consisting either of the single sentence* '$CCCpqCCrstCCuCCrstCCpuCst$' *or of the single sentence* '$CCCpCqpCCCCCrstuCCsuCruvv$' *forms a basis of the system* L^+.‡

Def. 7 of the n-valued system L_n can be applied at once to the restricted sentential calculus provided only that the concept of matrix is suitably modified. *Ths.* 16–22 *as well as* 24–26, which describe the mutual relations among the various systems L_n^+, determine the degree of completeness of the systems and establish their axiomatizability, *have been extended to the restricted sentential calculus* by their originators. (In the case of Th. 21 this was done by Tarski; for Th. 22 by Wajsberg. In Th. 24 the sentences with negation signs are to be omitted.) The problem of the axiomatizability of the system $L_{\aleph_0}^+$ is left open.

Finally, *the number of all possible systems of the sentential*

---

[1] Cf. Leśniewski, S. (46), p. 47, note.   [2] Cf. Leśniewski, S. (46), p. 58.

---

† The original proof of Th. 29 has not been published. But a proof of this result can easily be obtained by means of a method developed in M. Wajsberg, 'Metalogische Beiträge', *Wiadomości Matematyczne*, vol. 18 (1936), pp. 131–68, in particular pp. 154–7; the derivations which are needed for applying Wajsberg's method can be found, for example, in W. V. Quine, *System of Logistic*, Cambridge, Mass. (1934), pp. 60 ff.

‡ More recently Łukasiewicz has shown that the sentence '$CCCpqrCCrpCsp$' can also serve as a single axiom for system L^+ and that there is no shorter sentence with this property. See J. Łukasiewicz, 'The shortest axiom of the implicational calculus of propositions', *Proceedings of the Royal Irish Academy*, section A, vol. 52 (1948), pp. 25–33.

calculus, which was determined by Lindenbaum and Tarski in Ths. 27 and 28, *also remains unchanged in the restricted sentential calculus.*†

§ 5. The Extended Sentential Calculus

By the *extended sentential calculus* we understand a deductive discipline in the sentences of which there occur what are called *universal quantifiers* in addition to sentential variables and the implication sign.[1] For the universal quantifier Łukasiewicz uses the sign '$\prod$' which was introduced by Peirce.[2] With this notation the formula '$\prod pq$' is the symbolic expression of the sentence 'for all p, q (holds)'. The operation which consists in putting the universal quantifier '$\prod$' with a sentential variable x in front of a given sentence y is called *universal quantification of the sentence y with respect to the sentential variable x*, and is

---

[1] In (46) Leśniewski has described the outlines of a deductive system, called by him *Protothetic*, which, compared with the extended sentential calculus, goes still further beyond the ordinary sentential calculus in the respect that, in addition to quantifiers, variable functors are introduced. (In the sentence '*Cpq*' the expression '*C*' is called a *functor*, and '*p*' and '*q*' are called the arguments. The word functor we owe to T. Kotarbiński. In both the ordinary and the extended sentential calculus only constant functors are used.) In addition to this principal distinction, there are yet other differences between the extended sentential calculus and the protothetic as it is described by Leśniewski. In contrast to the extended sentential calculus, in the protothetic only those expressions are regarded as meaningful sentences in which no free, but only bound (apparent) variables occur. Some new operations (rules of inference or directives) are also introduced by means of which consequences are derived from given sentences, such, for example, as the operation of distributing quantifiers, which is superfluous in the extended sentential calculus. Finally it must be emphasized that Leśniewski has formulated with the utmost precision the conditions which a sentence must satisfy if it is to be admitted as a definition in the system of the protothetic, whereas in the present work the problem of definitions has been left untouched. Article I belongs to protothetic. A sketch of the extended sentential calculus is given in Łukasiewicz, J. (51), pp. 154–69; this sketch rests in great part on results of Tarski (cf. loc. cit. Preface, p. vii). The two-valued logic of Łukasiewicz (52 *a*) has many points of contact with the extended sentential calculus. Finally, there are many analogies between the extended sentential calculus and the functional calculus of Hilbert and Ackermann (see Hilbert, D., and Ackermann, W. (30), especially pp. 84–85).

[2] The expression 'quantifier' occurs in the work of Peirce (58 *a*), p. 197, although with a somewhat different meaning.

---

† The footnote concerning Th. 27 on p. 51 applies to the restricted sentential calculus as well.

denoted by '$\pi_x(y)$' in metamathematical discussions. This concept is to be regarded as a primitive concept of the meta-sentential calculus.[1]

DEFINITION 8. The set $S^\times$ of all meaningful sentences (of the extended sentential calculus) *is the intersection of all those sets which contain all sentential variables and are closed under the two operations of forming implications and of universal quantification (with respect to an arbitrary sentential variable).*[2]

The operations of forming negations and of existential quantification (which consists in prefixing to a given sentence y the existential quantifier '$\sum$' with a sentential variable x), are not considered here because, in the system of the extended sentential calculus in which we are interested, they can be defined with the help of the two operations previously mentioned. For example, we can use the formula '$Cp \prod qq$' as definiens for 'Np'.

In deriving consequences from an arbitrary set of sentences Łukasiewicz and Tarski make use of the operations of *insertion* and *deletion of quantifiers*, in addition to those of substitution[3] and detachment. The first of these operations consists in obtaining a sentence $y = c(z, \pi_t(u))$ from a sentence of the form $x = c(z, u)$, where $z \in S^\times$ and $u \in S^\times$, under the assumption that t is a sentential variable which is not free in z.[4] The second operation is the inverse of the first and consists in deriving the sentence $x = c(z, u)$ from the sentence $y = c(z, \pi_t(u))$ (in this case without any restriction concerning the variable t).[5]

---

[1] Cf. the remarks following Def. 1 in § 1.

[2] In contrast to Hilbert, D., and Ackermann, W. (30), p. 52, as well as to the standpoint taken in Łukasiewicz, J. (51), p. 155, the expression $\pi_x(y)$ is also regarded as meaningful when x either occurs as a bound variable in y or does not occur in y at all.

[3] The operation of substitution undergoes certain restrictions in the extended sentential calculus (cf. Łukasiewicz, J. (51), p. 160, and Hilbert, D., and Ackermann, W. (30), p. 54.)

[4] We do not need to discuss the meaning of the expression 'x occurs in the sentence y as a free (or bound) variable' since it is sufficiently clear (cf. Łukasiewicz, J. (51), p. 156, and Hilbert and Ackermann (30), p. 54).

[5] In the restricted functional calculus only the first operation is used. Instead of the second an axiom is set up (cf. Hilbert, D., and Ackermann, W. (30), pp. 53–54). An analogous procedure would not be possible in our calculus; for if we drop the second operation the system $L^\times$ to be discussed below would not have a finite basis.

DEFINITION 9. The set $Cn^\times(X)$ of consequences of the set X of sentences (in the sense of the extended sentential calculus) *is the intersection of all those sets which include the given set* $X \subseteq S^\times$ *and are closed under the operations of substitution and detachment, as well as insertion and deletion of quantifiers.*

With this interpretation of the concepts $S^\times$ and $Cn^\times(X)$ Th. 1 *from* § 1 *remains valid.*

As before, two methods are available for the construction of deductive systems: the axiomatic and the matrix methods. The second method has not yet received a sufficiently clear general formulation, and in fact the problem of a simple and useful definition of the concept of matrix still presents many difficulties. Nevertheless this method has been successfully applied by Łukasiewicz in special cases, namely for the construction of the n-valued systems $L_n^\times$ (for $n < \aleph_0$) and in particular for the construction of the ordinary system $L^\times$ of the extended sentential calculus. The construction of the systems $L_n^\times$ is precisely described in the following.

DEFINITION 10. *First let us introduce the following auxiliary notation:* $b = 'p'$, $g = \pi_b(b)$ (falsehood), $n(x) = c(x, g)$ *for every* $x \in S^\times$ *(the* negation *of the sentence* x*),* $a(x,y) = c(c(x,y), y)$ *and* $k(x, y) = n(a(n(x), n(y)))$ *for every* $x \in S^\times$ *and every* $y \in S^\times$ *(the* disjunction *or rather* alternation, *and the* conjunction *of the sentences* x *and* y*);*[1] *furthermore* $k_{i=1}^m(x_i) = x^1$ *for* $m = 1$ *and* $k_{i=1}^m(x_i) = k(k_{i=1}^{m-1}(x_i), x_m)$ *for every arbitrary natural number* $m > 1$, *where* $x_i \in S^\times$ *for* $1 \leqslant i \leqslant m$ *(the* conjunction *of the sentences* $x_1, x_2,..., x_m$*). Further we put* $b_m = b$ *for* $m = 1$, $b_m = c(n(b), b_{m-1})$ *for every natural number* $m > 1$, *and finally* $a_m = \pi_b(c(b_m, b))$ *for every natural number* m.[2]

---

[1] The logical expressions 'Apq' ('p or q') and 'Kpq' ('p and q') correspond, in the symbolism introduced by Łukasiewicz, to the metalogical expressions '$a(x, y)$' and '$k(x, y)$' respectively. Of the two possible definitions of the alternation, which in the two-valued, but not in the n-valued, system are equivalent: $a(x, y) = c(c(x, y), y)$ and $a(x, y) = c(n(x), y)$, the first was chosen by Łukasiewicz for various, partly intuitive, reasons (cf. Łukasiewicz (50 a), p. 201).

[2] For example,

$$b_1 = 'p', b_2 = 'CCp \prod ppp', b_3 = 'CCp \prod ppCCp \prod ppp',$$

and

$$a_1 = '\prod pCpp', \quad a_2 = '\prod pCCCp \prod pppp',$$
$$a_3 = '\prod pCCCp \prod ppCCp \prod pppp'.$$

Now let n be a definite natural number > 1. We choose n sentences called *basic sentences*, and denote them by the symbols 'g_1', 'g_2',..., 'g_n'; in fact we put $g_1 = g$, $g_2 = a_{n-1}$, and $g_m = c(n(g_2), g_{m-1})$ for every m, $2 < m \leqslant n$.[1] Let G be the smallest set of sentences which contains all sentential variables and basic sentences and is closed with respect to the operation of forming implications.

A function h is called a *value function (of the n-th degree)* if it satisfies the following conditions: (1) the function h is defined for every sentence $x \in G$; (2) if x is a sentential variable, then $h(x)$ is a fraction of the form $(m-1)/(n-1)$, where m is a natural number and $1 \leqslant m \leqslant n$; (3) for every natural number m, $1 \leqslant m \leqslant n$, we have $h(g_m) = (m-1)/(n-1)$; (4) if $x \in G$ and $y \in G$, then $h(c(x,y)) = \min(1, 1-h(x)+h(y))$.[2]

With every sentence $x \in S^\times$ a sentence $f(x) \in G$ is correlated by recursion in the following way: (1) if x is a sentential variable or a basic sentence, then $f(x) = x$; (2) if $x \in S^\times$, $y \in S^\times$, and $c(x, y)$ is not a basic sentence, we put $f(c(x, y)) = c(f(x), f(y))$; (3) if x is a sentential variable which is not free in the sentence $y \in S^\times$, then $f(\pi_x(y)) = f(y)$; (4) but if the sentential variable x is free in the sentence $y \in S^\times$ and $\pi_x(y)$ is not a basic sentence, then we put $f(\pi_x(y)) = k_{i=1}^n(f(y_i))$ where the sentence y_i for every i, $1 \leqslant i \leqslant n$, arises from y by the substitution of the basic sentence g_i for the free variable x.

The *n-valued system $L_n^\times$ of the extended sentential calculus*, where $2 \leqslant n < \aleph_0$, is now defined as the set of all those sentences $x \in S^\times$ which satisfy the formula $h(f(x)) = 1$ for every value function h (of the nth degree); in addition $L^\times$ is set equal to $S^\times$. The system $L_2^\times = L^\times$ is also called the *ordinary system of the extended sentential calculus*.[3]

---

[1] For example, for $n = 3$:
$$g_1 = \text{`} \prod pp \text{'}, \qquad g_2 = \text{`} \prod pCCCp \prod pppp \text{'},$$
$$g_3 = \text{`} CC \prod pCCCp \prod pppp \prod pp \prod pCCCp \prod pppp \text{'}.$$

[2] Cf. Defs. 4 and 7 above.

[3] In the definition adopted by Łukasiewicz, instead of the basic sentences $g_1, g_2,..., g_n$, there occur what are called sentential constants, $c_1, c_2,..., c_n$, i.e. special signs distinct from sentential variables. The concept of meaningful sentence is thereby temporarily extended. The rest of the definition runs quite analogously to the definition in the text. In the final definition of the

From this definition of the systems $L_n^\times$ the following facts easily result (they are partly in opposition to Ths. 18 and 19 of § 3):

THEOREM 31. $L_n^\times \in \mathfrak{S}.\mathfrak{W}.\mathfrak{B}$ for every natural n, $2 \leqslant n < \aleph_0$.[1]

THEOREM 32. For $2 \leqslant m < \aleph_0$ and $2 \leqslant n < \aleph_0$ we have $L_m^\times \subseteq L_n^\times$ if and only if $m = n$ (no system of the sequence $L_n^\times$, where $2 \leqslant n < \aleph_0$, is included in another system of this sequence).

THEOREM 33. The set of all sentences of the system $L_n^\times$ (where $1 \leqslant n < \aleph_0$) in which no bound variables occur is identical with the corresponding system L_n^+ of the restricted sentential calculus.

Regarding the axiomatizability of these systems Tarski has shown that Ths. 8, 10, 22, 29, and 30 also hold in the extended sentential calculus. In this connexion Tarski has also proved the following:

THEOREM 34. Every axiom system of the system $L_2^+ = L^+$ in the restricted sentential calculus is at the same time an axiom system of the system $L_2^\times = L^\times$ in the extended sentential calculus.[1]

On the other hand, not every basis of the system L^+ in the restricted sentential calculus is at the same time a basis in the extended calculus (and not every set of sentences which is independent in the restricted sentential calculus remains independent in the extended calculus).

THEOREM 35. For $3 \leqslant n < \aleph_0$ universal quantifiers and bound variables occur in at least one sentence of every basis (and in general of every axiom system) of the system $L_n^\times$.

systems $L_n^\times$ all expressions which contain sentential constants are eliminated, and the concept of meaningful sentence is reduced to the original expressions. By means of the modification introduced in the text, which is due to Tarski, the definition of the systems $L_n^\times$ certainly takes on a simpler form from the metalogical standpoint, but at the same time it becomes less perspicuous. In order to establish the equivalence of the two definitions it suffices to point out that the expressions chosen as basic sentences satisfy the following condition: for every value function h (in the sense of the original definition of Łukasiewicz),

$$h(f(g_m)) = h(c_m) = \frac{m-1}{n-1}, \quad \text{where} \quad 1 \leqslant m \leqslant n.$$

[1] The completeness and axiomatizability of the system $L_2^\times$ was proved by Tarski in the year 1927. His proof was subsequently simplified by S. Jaśkowski.

It is worthy of note that the proof given by Tarski of Th. 22 in the extended sentential calculus makes it possible to construct effectively an axiom system for every system $L_n^\times$ ($3 \leqslant n < \aleph_0$). Relatively simple axiom systems of this kind were constructed by Wajsberg; in the case $n = 3$ his result is as follows:

THEOREM 36. *Let X be the set consisting of the following sentences:*

$$\text{`}CCCpqCrqCCqpCrp\text{'}, \text{ `}CpCqp\text{'}, \text{ `}CCCpCpqpp\text{'},$$
$$\text{`}C \prod p CCCp \prod pppp C \prod pCCCp \prod pppp \prod pp\text{'},$$
$$\text{`}CC \prod pCCCp \prod pppp \prod pp \prod pCCCp \prod pppp\text{'};$$

then $X \in \mathfrak{Ax}(L_3^\times)$.

An exact definition of the denumerably-valued system $L_{\aleph_0}^\times$ of the extended sentential calculus presents much greater difficulties than that of the finite-valued systems. This system has not yet been investigated.

Ths. 27 *and* 28, which determine the number of all possible systems, *remain correct in the extended sentential calculus.*

In conclusion we should like to add that, as the simplest deductive discipline, the sentential calculus is particularly suitable for metamathematical investigations. It is to be regarded as a laboratory in which metamathematical methods can be discovered and metamathematical concepts constructed which can then be carried over to more complicated mathematical systems.

V

FUNDAMENTAL CONCEPTS OF THE METHODOLOGY OF THE DEDUCTIVE SCIENCES†

INTRODUCTION

THE deductive disciplines constitute the subject-matter of the *methodology of the deductive sciences*, which today, following Hilbert, is usually called *metamathematics*, in much the same sense in which spatial entities constitute the subject-matter of geometry and animals that of zoology. Naturally not all deductive disciplines are presented in a form suitable for objects of scientific investigation. Those, for example, are not suitable which do not rest on a definite logical basis, have no precise rules of inference, and the theorems of which are formulated in the usually ambiguous and inexact terms of colloquial language— in a word those which are not formalized. Metamathematical investigations are confined in consequence to the discussion of formalized deductive disciplines.

Strictly speaking metamathematics is not to be regarded as a single theory. For the purpose of investigating each deductive discipline a special metadiscipline should be constructed. The present studies, however, are of a more general character: their aim is *to make precise the meaning of a series of important metamathematical concepts* which are common to the special metadisciplines, *and to establish the fundamental properties of these concepts.* One result of this approach is that some concepts which can be defined on the basis of special metadisciplines will here be regarded as primitive concepts and characterized by a series of axioms.

An exact proof of the following results naturally requires, besides the above-mentioned axioms, a general logical basis.

† BIBLIOGRAPHICAL NOTE. This article originally appeared under the title 'Fundamentale Begriffe der Methodologie der deduktiven Wissenschaften. I', in *Monatshefte für Mathematik und Physik*, vol. 37 (1930), pp. 361–404. For earlier publications of the author on the same topics see the bibliographical note to III.

This basis should not be thought of as too comprehensive: for example, those chapters of the celebrated work of Whitehead and Russell[1] which comprise the sentential calculus, the theory of apparent variables, the calculus of classes, and the elements of the arithmetic of cardinal and ordinal numbers are quite sufficient for our purposes. The axiom of choice is not used in this discussion, and the axiom of infinity can also be easily eliminated.

As regards notation, we shall for practical reasons formulate these considerations in the terms of everyday language; but we shall also make use of a series of symbols, almost all of which are customary in textbooks and articles from the domain of set theory.

The variables 'x', 'y', 'z'... here denote individuals (objects of lowest type) and in particular sentences; the variables 'A', 'B', 'X', 'Y'..., sets of individuals; '$\Re$', '$\mathfrak{L}$'..., classes of sets (or systems of sets) and finally 'λ', 'μ', 'ξ'..., ordinal numbers. The symbols of the calculus of classes and arithmetic, '$\subseteq$,' '$+$', '$-$', '$.$', '$\sum$', '$\prod$', '$<$', '$\leqslant$', '$\aleph_0$', etc., are used with their usual meanings; 'ω' denotes the smallest ordinal number of the second and 'Ω' of the third number class. The formulas '$x \in A$' or '$x \bar{\in} A$' express as usual that the set A respectively contains or does not contain the element x. '$\bar{\bar{A}}$' denotes the cardinal of the set A. The symbol '$\underset{f(x)}{E}[...]$' denotes the set of all values of the function f corresponding to those values of the argument x which satisfy the condition formulated in the brackets '[...]'. In particular we put $\{x\} = \underset{y}{E}[y = x]$ (the set which contains x as its only element), $\mathfrak{P}(A) = \underset{X}{E}[X \subseteq A]$ (the power set of A), $\mathfrak{Q}(A) = \underset{X}{E}[A \subseteq X]$, $\mathfrak{E} = \underset{X}{E}[\bar{\bar{X}} < \aleph_0]$ (the class of all finite, 'inductive' sets), and $\bar{\xi} = \overline{\underset{\eta}{E}[\eta < \xi]}$ (the cardinal number of the ordinal number ξ).

The results of the present article are applicable to arbitrary deductive disciplines, and in particular to the simplest deductive discipline, the sentential calculus. This article was originally

[1] Whitehead, A. N., and Russell, B. (90).

intended as the first part of a more comprehensive paper. The discussion in the second part dealt with those deductive disciplines which presuppose a certain logical basis, including at least the whole of the sentential calculus (approximately in the sense that all the valid statements of the sentential calculus can be used as premisses in all reasoning within these deductive disciplines). This second part was developed independently of the first part and was based upon a special axiom system. In its original form the second part has never appeared in print.[1]

In conclusion it should be noted that no particular philosophical standpoint regarding the foundations of mathematics is presupposed in the present work. Only incidentally, therefore I may mention that my personal attitude towards this question agrees in principle with that which has found emphatic expression in the writings of S. Leśniewski[2] and which I would call *intuitionistic formalism.*†

§ 1. Meaningful Sentences; Consequences of Sets of Sentences

From the standpoint of metamathematics every deductive discipline is a system of *sentences*, which we shall also call *meaningful sentences*.[3] The sentences are most conveniently regarded as inscriptions, and thus as concrete physical bodies. Naturally, not every inscription is a meaningful sentence of a given discipline: only inscriptions of a well-determined structure are regarded as meaningful. The concept of meaningful sentence has no fixed content, and must be relativized to a

[1] A modified version of the second part is to be found in XII. The discussion in Hertz, P. (27) has some points of contact with the present exposition. As examples of concrete investigations (within special metadisciplines) based upon the conceptual framework of this paper we may mention IV and Presburger, M. (61).

[2] Cf. especially Leśniewski, S. (46), especially p. 78.

[3] Instead of 'meaningful sentence' we could say 'well-formed sentence'. I use the word 'meaningful' to express my agreement with the doctrine of intuitionistic formalism mentioned above.

† This last sentence expresses the views of the author at the time when this article was originally published and does not adequately reflect his present attitude.

concrete formalized discipline. On the basis of the special metadisciplines mentioned in the introduction this concept can be reduced to intuitively simpler concepts;[1] in the present work, however, it must be taken as a primitive concept. The set of all meaningful sentences will be denoted by the symbol 'S'.

Let A be an arbitrary set of sentences of a particular discipline. With the help of certain operations, the so-called *rules of inference*, new sentences are derived from the set A, called the *consequences of the set A*. To establish these rules of inference, and with their help to define exactly the concept of consequence, is again a task of special metadisciplines; in the usual terminology of set theory the schema of such a definition can be formulated as follows:

The set of all *consequences of the set A* is the intersection of all sets which contain the set A and are closed under the given rules of inference.[2]

On account of the intended generality of these investigations we have no alternative but to regard the concept of consequence as primitive. *The two concepts—of sentence and of consequence— are the only primitive concepts which appear in this discussion.* The set of all consequences of the set A of sentences is here denoted by the symbol '$Cn(A)$'.

We will now state four axioms (Ax. 1–4) which express certain elementary properties of the primitive concepts and are satisfied in all known formalized disciplines.

We assume in the first place that the set S is at most denumerable.

AXIOM 1. $\bar{\bar{S}} \leqslant \aleph_0$.

This axiom scarcely requires comment since the contrary hypothesis would be unnatural and would lead to undesired complications in proofs. Moreover, the question arises whether the assumption that the set S is infinite (even denumerably infinite) is consistent with the intuitive view of sentences as con-

---

[1] See IV, p. 39 (Def. 1) and p. 55 (Def. 8).

[2] Cf. IV, p. 40 (Def. 2) and p. 56 (Def. 9). A set is said to be *closed under given operations* if each of these operations, when applied to elements of the set, always yields again an element of the set (cf. Fréchet, M. (17)).

crete inscriptions. Without going deeper into this disputable question, which is irrelevant for our discussion, it may be noted here that I personally regard such an assumption as quite sensible, and that it appears to me even to be useful from a meta-mathematical standpoint to replace the inequality sign by the equality sign in Ax. 1.†

From the definition schema of the set of consequences formulated above it follows that every sentence which belongs to a given set is to be regarded as a consequence of that set, that the set of consequences of a set of sentences consists solely of sentences, and that the consequences of consequences are themselves consequences. These facts are expressed in the next two axioms.

AXIOM 2. *If $A \subseteq S$, then $A \subseteq Cn(A) \subseteq S$.*

AXIOM 3. *If $A \subseteq S$, then $Cn(Cn(A)) = Cn(A)$.*

Finally it should be noted that, in concrete disciplines, the rules of inference with the help of which the consequences of a set of sentences are formed are in practice always operations which can be carried out only on a finite number of sentences (usually even only on one or two sentences). Hence every consequence of a given set of sentences is a consequence of a finite subset of this set and vice versa. This can be shortly expressed by the following

AXIOM 4. *If $A \subseteq S$, then $Cn(A) = \sum\limits_{X \in \mathfrak{P}(A) \cdot \mathfrak{E}} Cn(X)$.*

By way of example we give here some elementary consequences of the above axioms.

THEOREM 1. (a) *If $A \subseteq B \subseteq S$, then $Cn(A) \subseteq Cn(B)$.*

[Ax. 4][1]

(b) *If $A + B \subseteq S$, then the formulas*

$$A \subseteq Cn(B) \quad and \quad Cn(A) \subseteq Cn(B)$$

are equivalent. [Ax. 2, Ax. 3, Th. 1a]

---

[1] In cases where a proof is clear I content myself with a reference to the theorems to be used.

---

† Compare the discussion of related problems in VIII, pp. 174 and 184.

In accordance with Th. 1 a the operation Cn in the domain of sets of sentences is monotonic; this property can be expressed in several equivalent forms, e.g.:

$$\sum_{X \in \Re} Cn(X) \subseteq Cn\Big(\sum_{X \in \Re} X \Big), \quad \text{or} \quad Cn\Big(\prod_{X \in \Re} X \Big) \subseteq \prod_{X \in \Re} Cn(X),$$

for every non-null class $\Re \subseteq \mathfrak{P}(S)$.

THEOREM 2. (a) If $A+B \subseteq S$, then

$$Cn(A+B) = Cn(A+Cn(B)) = Cn(Cn(A)+Cn(B));$$

(b) if, more generally, $\Re$ is an arbitrary class $\subseteq \mathfrak{P}(S)$, then

$$Cn\Big(\sum_{X \in \Re} X \Big) = Cn\Big(\sum_{X \in \Re} Cn(X) \Big).$$

Proof. (a) According to Ax. 2,

$$A+B \subseteq A+Cn(B) \subseteq Cn(A)+Cn(B) \subseteq S,$$

whence by Th. 1 a

(1) $Cn(A+B) \subseteq Cn(A+Cn(B)) \subseteq Cn(Cn(A)+Cn(B)).$

Again according to Th. 1 a and Ax. 2 we have

$$Cn(A) \subseteq Cn(A+B) \subseteq S \quad \text{and} \quad Cn(B) \subseteq Cn(A+B) \subseteq S,$$

thus also $Cn(A)+Cn(B) \subseteq Cn(A+B) \subseteq S$ and therefore

(2) $Cn(Cn(A)+Cn(B)) \subseteq Cn(Cn(A+B)).$

Finally, from Ax. 3 it follows that

(3) $Cn(Cn(A+B)) = Cn(A+B).$

The formulas (1) to (3) give immediately:

$$Cn(A+B) = Cn(A+Cn(B)) = Cn(Cn(A)+Cn(B)), \quad \text{q.e.d.}$$

(b) is proved in an analogous manner.

THEOREM 3. If $A \subseteq S$, $C \in \mathfrak{E}$, and $C \subseteq Cn(A)$, then there exists a set B which satisfies the formulas $B \in \mathfrak{E}$, $B \subseteq A$, and $C \subseteq Cn(B)$.

Proof. By the hypothesis and Ax. 4, a set $G(x)$ can be correlated uniquely with every element $x \in C$, in such a way that the formulas

(1) $x \in Cn(G(x))$, $G(x) \in \mathfrak{E}$, and $G(x) \subseteq A$ for $x \in C$

are satisfied. (Since the set C is finite, the existence of such a correlation can be established without using the axiom of choice).

It follows from (1) that

(2) $$C \subseteq \sum_{x \in C} Cn(G(x)).$$

We put

(3) $$B = \sum_{x \in C} G(x).$$

Since $C \in \mathfrak{E}$, we infer from (1) and (3) that

(4) $$B \in \mathfrak{E} \quad and \quad B \subseteq A.$$

From (3) and (4) we also obtain $G(x) \subseteq B \subseteq A \subseteq S$ for $x \in C$. According to Th. 1a we thus have $Cn(G(x)) \subseteq Cn(B)$ for every element x of C, whence $\sum_{x \in C} Cn(G(x)) \subseteq Cn(B)$; and combining this inclusion with (2) we obtain

(5) $$C \subseteq Cn(B).$$

By (4) and (5) the set B satisfies the conclusion.

THEOREM 4. *Let $\mathfrak{R}$ be a class which satisfies the following condition: (α) for every finite subclass $\mathfrak{L}$ of $\mathfrak{R}$ there exists a set $Y \in \mathfrak{R}$, such that $\sum_{X \in \mathfrak{L}} X \subseteq Y$. If also $\mathfrak{R} \subseteq \mathfrak{P}(S)$, then*

$$Cn\Big(\sum_{X \in \mathfrak{R}} X\Big) = \sum_{X \in \mathfrak{R}} Cn(X).$$

Proof. Let x be a sentence such that

(1) $$x \in Cn\Big(\sum_{X \in \mathfrak{R}} X\Big).$$

Since by the hypothesis $\sum_{X \in \mathfrak{R}} X \subseteq S$, the existence of a set Z which satisfies the formulas

(2) $$Z \in \mathfrak{E}, \quad Z \subseteq \sum_{X \in \mathfrak{R}} X,$$

and

(3) $$x \in Cn(Z)$$

follows from (1) and from Ax. 4. From (2) can further be inferred the existence of a finite subclass $\mathfrak{L}$ of $\mathfrak{R}$, such that $Z \subseteq \sum_{X \in \mathfrak{L}} X$. According to the premiss (α) there corresponds to the class Ω a set $Y \in \Re$ such that $\sum\limits_{X \in \Omega} X \subseteq Y$. Consequently we have

(4) $Y \in \Re \quad and \quad Z \subseteq Y.$

By Th. 1 a, from (4) it follows that $Cn(Z) \subseteq Cn(Y)$, and therefore by (3) $x \in Cn(Y)$; hence with reference to (4) we obtain

(5) $x \in \sum\limits_{X \in \Re} Cn(X).$

Thus we have shown that the formula (1) always implies (5), accordingly the following inclusion holds:

(6) $Cn\Big(\sum\limits_{X \in \Re} X\Big) \subseteq \sum\limits_{X \in \Re} Cn(X).$

On the other hand we have $Y \subseteq \sum\limits_{X \in \Re} X \subseteq S$, whence according to Th. 1 a $Cn(Y) \subseteq Cn\Big(\sum\limits_{X \in \Re} X\Big)$ for every set $Y \in \Re$; consequently

(7) $\sum\limits_{X \in \Re} Cn(X) \subseteq Cn\Big(\sum\limits_{X \in \Re} X\Big).$

The formulas (6) and (7) give

$$Cn\Big(\sum\limits_{X \in \Re} X\Big) = \sum\limits_{X \in \Re} Cn(X), \quad \text{q.e.d.}$$

An immediate consequence of the last theorem is

COROLLARY 5. *Let $\Re$ be a class which satisfies the following condition:* (α) $\Re \neq 0$ *and for any two sets X and Y belonging to $\Re$ either $X \subseteq Y$ or $Y \subseteq X$. If $\Re \subseteq \mathfrak{P}(S)$, then*

$$Cn\Big(\sum\limits_{X \in \Re} X\Big) = \sum\limits_{X \in \Re} Cn(X). \qquad \text{[Th. 4]}$$

This corollary is often applied to the class of all terms of an increasing sequence of sets of sentences.

THEOREM 6. *Let $B+C \subseteq S$, and put*

$$F(X) = C \cdot Cn(X+B)$$

for every set $X \subseteq S$ (and in particular $F(X) = Cn(X+B)$ in case $C = S$). We then have:

(a) $\bar{\bar{C}} \leqslant \aleph_0$;

(b) *if $A \subseteq C$, then $A \subseteq F(A) \subseteq C$;*

(c) *if $A \subseteq C$, then $F(F(A)) = F(A)$;*

(d) *if $A \subseteq C$, then $F(A) = \sum\limits_{X \in \mathfrak{P}(A) \cdot \mathfrak{E}} F(X).$*

(*In other words, Axs.* 1–4 *remain valid if 'S' and 'Cn' are every-where replaced in them by 'C' and 'F' respectively.*)

Proof. First let it be noted that in case $C = S$ we have, by Ax. 2, $Cn(X+B) \subseteq C$ for every $X \subseteq S$, so that the function $F(X) = C.Cn(X+B)$ reduces indeed to

$$F(X) = Cn(X+B).$$

(*a*) follows immediately from Ax. 1.

(*b*) is easily obtained from Ax. 2.

(*c*) By Ax. 2 and Th. 1*a* we have

$$C.Cn(A+B)+B \subseteq Cn(A+B)+B \subseteq S$$

and accordingly

$$Cn(C.Cn(A+B)+B) \subseteq Cn(Cn(A+B)+B);$$

in view of Th. 2*a*

$$Cn(Cn(A+B)+B) = Cn((A+B)+B) = Cn(A+B).$$

From this it follows that

$$C.Cn(C.Cn(A+B)+B) \subseteq C.Cn(A+B);$$

thus by the hypothesis

$$F(F(A)) \subseteq F(A).$$

Since the inverse inclusion $F(A) \subseteq F(F(A))$ results imme-diately from (*b*) (if we replace '*A*' in it by '*F(A)*'), we finally obtain the desired formula $F(F(A)) = F(A)$.

(*d*) By Ax. 4:

(1) $$Cn(A+B) = \sum_{X \in \mathfrak{E}.\,\mathfrak{P}(A+B)} Cn(X).$$

For every set $X \in \mathfrak{P}(A+B)$ we obviously have $X \subseteq X_1+B$, where $X_1 = A.X \in \mathfrak{P}(A)$; from this by Th. 1*a* we have

$$Cn(X) \subseteq Cn(X_1+B).$$

If, moreover, $X \in \mathfrak{E}$, then $X_1 \in \mathfrak{E}$ and accordingly $X_1 \in \mathfrak{E}.\mathfrak{P}(A)$. Thus:

(2) $$\sum_{X \in \mathfrak{E}.\,\mathfrak{P}(A+B)} Cn(X) \subseteq \sum_{X \in \mathfrak{E}.\,\mathfrak{P}(A)} Cn(X+B).$$

On the other hand, if $X \in \mathfrak{P}(A)$, then $X+B \subseteq A+B$, whence $Cn(X+B) \subseteq Cn(A+B)$; therefore

(3) $$\sum_{X \in \mathfrak{E}.\,\mathfrak{P}(A)} Cn(X+B) \subseteq Cn(A+B).$$

The formulas (1)–(3) immediately give

$$Cn(A+B) = \sum_{X\in\mathfrak{E}.\mathfrak{P}(A)} Cn(X+B),$$

and thus

$$C.Cn(A+B) = \sum_{X\in\mathfrak{E}.\mathfrak{P}(A)} C.Cn(X+B).$$

By the hypothesis it follows from this that

$$F(A) = \sum_{X\in\mathfrak{E}.\mathfrak{P}(A)} F(X), \quad \text{q.e.d.}$$

In consequence of the above theorem, the concepts investigated in this discussion can be relativized in two distinct directions: 1. Instead of considering all meaningful sentences we can restrict our attention to the elements of a given set C of sentences—a *sentential domain*. 2. A fixed set of sentences B, a *basic set*, is chosen, and in the formation of the consequences of an arbitrary set X of sentences all the sentences of the basic set are added to the set X, so that the consequences of the set X in the new sense are identical with the consequences of the set $X+B$ in the old sense. According to Th. 6 (above) such a modification in the meaning of the primitive concepts does not affect the validity of the underlying axioms. Therefore all consequences of these axioms also remain valid, and in particular those theorems which will be presented in the following sections; one must only see to it that the non-primitive concepts in these theorems undergo an analogous relativization. On these grounds the theorem in question deserves the name *relativization theorem*.

§ 2. Deductive (Closed) Systems

With the help of the two concepts introduced in the preceding section, those of sentence and consequence, almost all basic concepts of metamathematics can be defined; on the basis of the given axiom system various fundamental properties of these concepts can be established.

In the first place an especially important category of sets of sentences will be singled out, namely the *deductive systems*. Every set of sentences which contains all its consequences is called a *deductive system*, or possibly a *closed system*, or simply a system.[1]

Deductive systems are, so to speak, organic units which form the subject matter of metamathematical investigations. Various important notions, like consistency, completeness, and axiomatizability, which we shall encounter in the sequel, are theoretically applicable to any sets of sentences, but in practice are applied chiefly to systems.

The class of all systems is denoted by the symbol '$\mathfrak{S}$'.

DEFINITION 1. $\mathfrak{S} = \underset{X}{E}[Cn(X) \subseteq X \subseteq S].$

An easy transformation of the above definition yields

THEOREM 7. *In order that $A \in \mathfrak{S}$, it is necessary and sufficient that $Cn(A) = A \subseteq S$.*   [Def. 1, Ax. 2]

Further properties of systems are expressed in the following theorems:

THEOREM 8. *If $B \in \mathfrak{S}$, then the conditions (α) $A \subseteq B$ and (β) $A \subseteq S$ and $Cn(A) \subseteq B$ are equivalent.*   [Th. 1b, 7]

THEOREM 9. *(a) If $A \subseteq S$, then $Cn(A) \in \mathfrak{Q}(A).\mathfrak{S}$ and also $Cn(A) = \prod\limits_{X \in \mathfrak{Q}(A).\mathfrak{S}} X.$*

(b) In particular, $Cn(0) \in \mathfrak{S}$ and $Cn(0) = \prod\limits_{X \in \mathfrak{S}} X.$

Proof. (a) By Axs. 2 and 3,

$$A \subseteq Cn(A) \quad \text{and} \quad Cn(Cn(A)) = Cn(A) \subseteq S,$$

whence by Th. 7 $Cn(A) \in \mathfrak{Q}(A).\mathfrak{S}$. Accordingly

$$\prod\limits_{X \in \mathfrak{Q}(A).\mathfrak{S}} X \subseteq Cn(A);$$

on the other hand from Th. 8 it follows that $Cn(A) \subseteq X$ for every set $X \in \mathfrak{Q}(A).\mathfrak{S}$, thus that $Cn(A) \subseteq \prod\limits_{X \in \mathfrak{Q}(A).\mathfrak{S}} X$. Consequently we have $Cn(A) \in \mathfrak{Q}(A).\mathfrak{S}$ and $Cn(A) = \prod\limits_{X \in \mathfrak{Q}(A).\mathfrak{S}} X,$ q.e.d.

(b) results immediately from (a) if we put $A = 0$.

---

[1] The term 'deductive system' was used in earlier publications of the author; see III and the bibliographical note to that article. In Hertz, P. (27) 'closed system' is used; and in Zermelo, E. (92) we find the phrase 'logically closed system' used in the same sense.

In accordance with the above theorem $Cn(A)$ is the smallest closed system *over* A (i.e. the smallest of all systems which include the set A), and $Cn(0)$ is the smallest system in general. The system $Cn(0)$ can be called the *system of all logically provable or logically valid sentences).*

THEOREM 10.

$$S \in \mathfrak{S} \quad and \quad S = \sum_{X \in \mathfrak{S}} X. \qquad \text{[Def. 1, Ax. 2]}$$

S is thus the largest of all systems.

THEOREM 11. (a) *If* $A \in \mathfrak{S}$ *and* $B \in \mathfrak{S}$, *then* $A \cdot B \in \mathfrak{S}$.

(b) *In general, if* $\mathfrak{R} \subseteq \mathfrak{S}$ *and* $\mathfrak{R} \neq 0$, *then* $\prod_{X \in \mathfrak{R}} X \in \mathfrak{S}$.

Proof. (a) From Def. 1 we obtain $Cn(A) \subseteq A \subseteq S$ and $Cn(B) \subseteq B \subseteq S$. By Th. 1 a we have $Cn(A \cdot B) \subseteq Cn(A)$ and $Cn(A \cdot B) \subseteq Cn(B)$, thus $Cn(A \cdot B) \subseteq Cn(A) \cdot Cn(B)$. Accordingly $Cn(A \cdot B) \subseteq A \cdot B \subseteq S$ whence, by Def. 1, $A \cdot B \in \mathfrak{S}$, q.e.d.

(b) is proved analogously.

In contrast to the preceding theorem, a sum of systems is not always a new system. In this connexion only the following can be proved:

THEOREM 12. *If the class* $\mathfrak{R}$ *satisfies the condition* (α) *of Th.* 4 *and if* $\mathfrak{R} \subseteq \mathfrak{S}$, *then* $\sum_{X \in \mathfrak{R}} X \in \mathfrak{S}$.

Proof. By Def. 1 it follows from the formula $\mathfrak{R} \subseteq \mathfrak{S}$ that $\mathfrak{R} \subseteq \mathfrak{P}(S)$. The premisses of Th. 4 (or of Cor. 5) are thus satisfied, whence $Cn\left(\sum_{X \in \mathfrak{R}} X \right) = \sum_{X \in \mathfrak{R}} Cn(X)$.

From this last formula and the inclusion $\mathfrak{R} \subseteq \mathfrak{S}$ it follows by Th. 7 that $Cn\left(\sum_{X \in \mathfrak{R}} X \right) = \sum_{X \in \mathfrak{R}} X \subseteq S$, and consequently that $\sum_{X \in \mathfrak{R}} X \in \mathfrak{S}$, q.e.d.

The theorem above establishes a sufficient condition for a class of systems to be such that the sum of all systems of this class is itself a system. It can be shown that this condition is not necessary. However, if we restrict ourselves to those deductive disciplines which presuppose sentential calculus, and if we assume the class $\mathfrak{R}$ to be finite, then the converse of Th. 12 holds (a result of Lindenbaum).

§ 3. Logical (Inferential) Equivalence of Two Sets of Sentences

Two sets of sentences are called *logically* or *inferentially equivalent*, or simply *equivalent*, if they have all their consequences in common (i.e. their sets of consequences coincide). The significance of the concept consists in the fact that almost every property to be considered here applies to all sets equivalent to a given set A whenever it applies to A.

The class of all sets which are equivalent to a given set A of sentences is denoted by '$\mathfrak{Aq}(A)$'; the formula '$B \in \mathfrak{Aq}(A)$' accordingly states that the sets A and B are equivalent. But we shall not introduce a special relation symbol for the equivalence of sets.

Definition 2. $\mathfrak{Aq}(A) = \underset{X}{E}[A+X \subseteq S \text{ and } Cn(A) = Cn(X)]$.

Theorem 13. (a) *The formulas* $A \subseteq S$, $A \in \mathfrak{Aq}(A)$ *and* $\mathfrak{Aq}(A) \neq 0$ *are equivalent.*   [Def. 2]

(b) *The formulas* $A \in \mathfrak{Aq}(B)$, $B \in \mathfrak{Aq}(A)$, $\mathfrak{Aq}(A).\mathfrak{Aq}(B) \neq 0$ *and* $\mathfrak{Aq}(A) = \mathfrak{Aq}(B) \neq 0$ *are equivalent.*   [Def. 2]

Theorem 14. *If* $A \subseteq B \subseteq C$ *and* $A \in \mathfrak{Aq}(C)$, *then* $B \in \mathfrak{Aq}(C)$ *and* $\mathfrak{Aq}(A) = \mathfrak{Aq}(B) = \mathfrak{Aq}(C)$.

Proof. By virtue of Def. 2 it follows from the hypothesis that $A \subseteq B \subseteq C \subseteq S$ and $Cn(A) = Cn(C)$; from this by Th. 1*a* we obtain $Cn(A) \subseteq Cn(B) \subseteq Cn(C)$ and $Cn(B) = Cn(C)$. By another application of Def. 2 we then obtain $B \in \mathfrak{Aq}(C)$, which finally on the basis of the hypothesis and of Th. 13*b* yields the second of the required formulas,

$$\mathfrak{Aq}(A) = \mathfrak{Aq}(B) = \mathfrak{Aq}(C).$$

Theorem 15. (a) *If* $A \in \mathfrak{Aq}(B)$ *and* $C \subseteq S$, *then*

$$A+C \in \mathfrak{Aq}(B+C).$$

(b) *If* $A_1 \in \mathfrak{Aq}(B_1)$ *and* $A_2 \in \mathfrak{Aq}(B_2)$, *then*

$$A_1+A_2 \in \mathfrak{Aq}(B_1+B_2).$$

(c) *In general, if to every set* $X \in \mathfrak{R}$ *a set* $Y \in \mathfrak{L}$ *satisfying the formula* $X \in \mathfrak{Aq}(Y)$ *corresponds, and vice versa, then*

$$\sum_{X \in \mathfrak{R}} X \in \mathfrak{Aq}\left(\sum_{Y \in \mathfrak{L}} Y\right).$$

Proof. (*a*) By means of Def. 2 the hypothesis yields

$$(A+C)+(B+C) \subseteq S \quad \text{and} \quad Cn(A) = Cn(B).$$

From this by applying Th. 2*a* we obtain

$$Cn(A+C) = Cn(Cn(A)+C) = Cn(Cn(B)+C) = Cn(B+C).$$

Accordingly by Def. 2 we have $A+C \in \mathfrak{Aq}(B+C)$, q.e.d.

(*b*) and (*c*) follow in an analogous way by means of Th. 2*a, b*.

THEOREM 16. *For every set* $A \subseteq S$ *there is a corresponding sequence of sentences* x_ν *of type* $\pi \leqslant \omega^1$ *which satisfies the formulas*

$$(\alpha) \quad x_\nu \in Cn\Big(\underset{x_\mu}{E}[\mu < \nu] \Big) \text{ for every } \nu < \pi$$

and $\quad (\beta) \quad \underset{x_\nu}{E}[\nu < \pi] \in \mathfrak{Aq}(A).$

Proof. In consequence of Axs. 1 and 2 as well as of the hypothesis of the theorem, the set $Cn(A) \subseteq S$ is at most denumerable; the elements of this set can therefore be ordered in an infinite sequence (with possibly repeating terms) of the type ω, so that

$$(1) \qquad Cn(A) = \underset{y_\lambda}{E}[\lambda < \omega] \subseteq S.$$

Let

(2) λ_0 *be the smallest of the numbers* $\lambda < \omega$ *which satisfy the formula*

$$y_\lambda \in Cn(0);$$

Further, let

(3) *for* $0 < \nu < \omega$, λ_ν *be the smallest of the numbers* $\lambda < \omega$ *which satisfy the formula*

$$y_\lambda \in Cn\Big(\underset{y_{\lambda_\mu}}{E}[\mu < \nu] \Big);$$

Finally, let

(4) π *be the smallest ordinal number with which the conditions* (2) *and* (3) *do not correlate any number* λ_π.

We put

(5) $x_\nu = y_{\lambda_\nu}$ *for* $\nu < \pi$.

---

[1] Where possibly $\pi = 0$ (the empty sequence, which has no terms at all, is regarded as a sequence of type 0).

It follows from (2)–(4) that

$$(6) \qquad \pi \leqslant \omega.$$

From (1)–(5) we easily obtain

$$(7) \qquad x_\nu \bar{\in} Cn\Big(\underset{x_\mu}{E}[\mu < \nu] \Big) \quad \text{for } \nu < \pi$$

and

$$(8) \qquad \underset{x_\nu}{E}[\nu < \pi] \subseteq Cn(A).$$

Let us suppose that

(9) *there exist numbers* $\lambda < \omega$ *such that* $y_\lambda \bar{\in} Cn\Big(\underset{x_\nu}{E}[\nu < \pi] \Big)$.

Accordingly (in view of (5)) let

(10) λ' *be the smallest of the numbers* λ *which satisfy the formulas*
$\lambda < \omega$ *and* $y_\lambda \bar{\in} Cn\Big(\underset{y_{\lambda_\nu}}{E}[\nu < \pi] \Big)$.

If π were less than ω, then, by comparing the statements (2), (3), and (10), we should obtain $\lambda' = \lambda_\pi$, which contradicts the condition (4). Consequently, on the basis of (6),

$$(11) \qquad \pi = \omega.$$

In accordance with Th. 1 a and with reference to (1)–(3) we have $Cn\Big(\underset{y_{\lambda_\mu}}{E}[\mu < \nu] \Big) \subseteq Cn\Big(\underset{y_{\lambda_\mu}}{E}[\mu < \pi] \Big)$ for every $\nu < \pi$. We thus have by (10) $y_{\lambda'} \bar{\in} Cn\Big(\underset{y_{\lambda_\mu}}{E}[\mu < \nu] \Big)$ for $\nu < \pi$ and in particular $y_{\lambda'} \bar{\in} Cn(0)$ for $\nu = 0$, whence, in view of (2) and (3),

$$(12) \qquad \lambda' \geqslant \lambda_\nu \quad \text{for every } \nu < \pi.$$

Finally it is easy to show that

$$(13) \qquad \lambda_{\nu_1} < \lambda_{\nu_2} \quad \text{for } \nu_1 < \nu_2 < \pi.$$

In fact, let $\nu_1 < \nu_2 < \pi$. By Ax. 2 and Th. 1 a, the conditions (1)–(4) imply then the following formulas:

$$\underset{y_{\lambda_\mu}}{E}[\mu < \nu_2] \subseteq Cn\Big(\underset{y_{\lambda_\mu}}{E}[\mu < \nu_2] \Big)$$

and $\qquad Cn\Big(\underset{y_{\lambda_\mu}}{E}[\mu < \nu_1] \Big) \subseteq Cn\Big(\underset{y_{\lambda_\mu}}{E}[\mu < \nu_2] \Big).$

A consequence of the first inclusion is

$$y_{\lambda_{\nu_1}} \in Cn\Big(\underset{y_{\lambda_\mu}}{E}[\mu < \nu_2] \Big);$$

since by (3) (for $\nu = \nu_2$) $y_{\lambda_{\nu_2}} \bar{\in} Cn\Big(\underset{\nu\lambda\mu}{E}[\mu < \nu_2]\Big)$, the numbers λ_{ν_1} and λ_{ν_2} cannot be identical. By combining the last formula with the second inclusion we obtain

$$y_{\lambda_{\nu_2}} \bar{\in} Cn\Big(\underset{\nu\lambda\mu}{E}[\mu < \nu_1]\Big).$$

In other words λ_{ν_2} is one of the numbers λ which satisfy the condition (3) for $\nu = \nu_1$ (or the condition (2) in the case $\nu_1 = 0$); since λ_{ν_1} is by definition the smallest of these numbers, we have $\lambda_{\nu_1} \leqslant \lambda_{\nu_2}$. We thus finally reach the desired formula $\lambda_{\nu_1} < \lambda_{\nu_2}$.

From the formulas (11)–(13) we conclude at once that $\lambda' \geqslant \omega$, which contradicts (10) and hence refutes the assumption (9). It must thus be accepted that

(14) $y_\lambda \in Cn\Big(\underset{x_\nu}{E}[\nu < \pi]\Big)$ for every $\lambda < \omega$.

From (1) and (14) it follows that

$$Cn(A) \subseteq Cn\Big(\underset{x_\nu}{E}[\nu < \pi]\Big);$$

the application of Th. 1 b to the inclusion (8) gives

$$Cn\Big(\underset{x_\nu}{E}[\nu < \pi]\Big) \subseteq Cn(A).$$

Thus we have the identity

$$Cn(A) = Cn\Big(\underset{x_\nu}{E}[\nu < \pi]\Big),$$

which in accordance with Def. 2 leads to the formula

(15) $\underset{x_\nu}{E}[\nu < \pi] \in \mathfrak{Aq}(A).$

The formulas (6), (7), and (15) state that the sequence of sentences x_ν satisfies all conditions of the conclusion.

In the proof above we have used a method of reasoning which is repeatedly employed in set-theoretical considerations. For the idea of applying it to metamathematical problems we are indebted to Lindenbaum, who employed it in the proof of Th. 56 given below.

A sequence of sentences which satisfies the condition (α) of Th. 16 could be called *ordinally independent*; if at the same time the formula (β) is satisfied, then this sequence is to be called an *ordered basis* of the set A of sentences. In this terminology the theorem states that every set of sentences possesses an ordered basis.

THEOREM 17. *If* $A \subseteq S$, *then* $Cn(A) \in \mathfrak{Aq}(A)$ *and*

$$Cn(A) = \sum_{X \in \mathfrak{Aq}(A)} X. \qquad \text{[Def. 2, Axs. 2, 3]}$$

THEOREM 18. (*a*) *In order that* $A \in \mathfrak{Aq}(B)$ *and* $B \in \mathfrak{S}$, *it is necessary and sufficient that* $A \subseteq S$ *and* $B = Cn(A)$.

[Def. 2, Ths. 7, 9 *a*]

(*b*) *If* $A \subseteq S$, *then* $\{Cn(A)\} = \mathfrak{S}.\mathfrak{Aq}(A)$;   [Ths. 13 *a*, 18 *a*]

(*c*) *If* $A \in \mathfrak{S}$, *then* $\{A\} = \mathfrak{S}.\mathfrak{Aq}(A)$.   [Ths. 7, 18 *a*, *b*]

By Ths. 17 and 18 *a*, $Cn(A)$ is both the greatest of all sets equivalent to the set A and the only system equivalent to A; consequently the set $Cn(A)$ can be regarded and used as a representative of the whole class $\mathfrak{Aq}(A)$.

§ 4. AN AXIOM SYSTEM OF A SET OF SENTENCES, AXIOMATIZABLE SETS OF SENTENCES

An *axiom system* of a set of sentences is a finite set which is equivalent to that set; a set of sentences which possesses at least one axiom system is called *axiomatizable*. Lindenbaum has directed our attention to the concept of axiomatizability. In his investigations on the meta-sentential-calculus he has established the interesting fact that, in addition to axiomatizable sets of sentences, non-axiomatizable sets also exist.[1]

Using '$\mathfrak{Ax}(A)$' to denote the class of all axiom systems of a set A of sentences, and '$\mathfrak{A}$' to denote the class of all axiomatizable sets of sentences, we reach

DEFINITION 3. (*a*) $\mathfrak{Ax}(A) = \mathfrak{E}.\mathfrak{Aq}(A)$.

(*b*) $\mathfrak{A} = \underset{X}{E}[\mathfrak{Ax}(X) \neq 0]$.

The following consequences of this definition are easily provable:

THEOREM 19. (*a*) *In order that* $A \in \mathfrak{Ax}(A)$ *it is necessary and sufficient that* $A \in \mathfrak{E}.\mathfrak{P}(S)$.   [Def. 3 *a*, Th. 13 *a*]

(*b*) $\mathfrak{E}.\mathfrak{P}(S) \subseteq \mathfrak{A}$.   [Th. 19 *a*, Def. 3 *b*]

---

[1] Cf. IV, p. 51 (Th. 27).

THEOREM 20. (a) If $\mathfrak{A}\mathfrak{x}(A).\mathfrak{A}\mathfrak{x}(B) \neq 0$, then $A \in \mathfrak{A}\mathfrak{q}(B)$.

[Def. 3 a, Th. 13 b]

(b) If $A \in \mathfrak{A}\mathfrak{q}(B)$, then $\mathfrak{A}\mathfrak{x}(A) = \mathfrak{A}\mathfrak{x}(B)$.   [Th. 13 b, Def. 3 a]

(c) If $A \in \mathfrak{A}$ then $\mathfrak{A}\mathfrak{q}(A) \subseteq \mathfrak{A}$.

THEOREM 21. (a) If $A_1 \in \mathfrak{A}\mathfrak{x}(B_1)$ and $A_2 \in \mathfrak{A}\mathfrak{x}(B_2)$, then

$$A_1 + A_2 \in \mathfrak{A}\mathfrak{x}(B_1 + B_2). \quad \text{[Def. 3 a, Th. 15 b]}$$

(b) If $A \in \mathfrak{A}$ and $B \in \mathfrak{A}$, then $A + B \in \mathfrak{A}$.

[Def. 3 b, Th. 21 a]

THEOREM 22. If $A \in \mathfrak{A}$, then $\mathfrak{P}(A).\mathfrak{A}\mathfrak{x}(A) \neq 0$.

Proof. By Defs. 2 and 3 it follows from the hypothesis that there is a set X which satisfies the formulas

(1) $$X \in \mathfrak{E},$$

(2) $$A + X \subseteq S \quad and \quad Cn(A) = Cn(X).$$

From (2) with the help of Ax. 2 we obtain $A \subseteq S$ and $X \subseteq Cn(A)$; from this, by applying Th. 3 and by reference to (1), we infer that there exists a set Y such that

(3) $$Y \in \mathfrak{E},$$

(4) $$Y \subseteq A \subseteq S \quad and \quad X \subseteq Cn(Y).$$

By Th. 1 a, b the formulas (4) imply

$$Cn(X) \subseteq Cn(Y) \subseteq Cn(A),$$

whence, in view of (2), $Cn(A) = Cn(Y)$; and from this by means of Def. 2 we obtain:

(5) $$Y \in \mathfrak{A}\mathfrak{q}(A).$$

By Def. 3 a the formula $Y \in \mathfrak{A}\mathfrak{x}(A)$ follows from (3) and (5). By combining this formula with (4) we obtain at once

$$\mathfrak{P}(A).\mathfrak{A}\mathfrak{x}(A) \neq 0, \quad \text{q.e.d.}$$

THEOREM 23. *If the class* $\mathfrak{R}$ *satisfies the condition* (α) *of Theorem* 4 *(or of Cor.* 5*) and if* $\sum_{X \in \mathfrak{R}} X \in \mathfrak{A}$, *then*

$$\mathfrak{R}.\mathfrak{A}\mathfrak{q}\Big(\sum_{X \in \mathfrak{R}} X\Big) \neq 0.$$

Proof. By Th. 22 it follows from the hypothesis that

$$\mathfrak{P}\Big(\sum_{X \in \mathfrak{R}} X\Big).\mathfrak{A}\mathfrak{x}\Big(\sum_{X \in \mathfrak{R}} X\Big) \neq 0.$$

Hence by Def. $3a$ we have, for some Y,

$$(1) \qquad\qquad Y \subseteq \sum_{X \in \Re} X,$$

$$(2) \qquad\qquad Y \in \mathfrak{E},$$

and

$$(3) \qquad\qquad Y \in \mathfrak{Aq}\Big(\sum_{X \in \Re} X \Big).$$

From (1) and (2) it is easily inferred that there is a class $\mathfrak{L} \in \mathfrak{E} . \mathfrak{P}(\Re)$ which satisfies the formula $Y \subseteq \sum_{X \in \mathfrak{L}} X$. Since by hypothesis the condition (α) of Th. 4 (or of Cor. 5) is satisfied, it follows that there is a set $Z \in \Re$, which includes the set $\sum_{X \in \mathfrak{L}} X$ and *a fortiori* the set Y. We thus have

$$(4) \qquad\qquad Z \in \Re,$$

$$(5) \qquad\qquad Y \subseteq Z \subseteq \sum_{X \in \Re} X.$$

By Th. 14 the formulas (3) and (5) give at once

$$(6) \qquad\qquad Z \in \mathfrak{Aq}\Big(\sum_{X \in \Re} X \Big).$$

Finally from (4) and (6) we obtain

$$\Re . \mathfrak{Aq}\Big(\sum_{X \in \Re} X \Big) \neq 0, \quad \text{q.e.d.}$$

THEOREM 24. *Let a sequence of sentences x_ν of the type $\pi \leqslant \omega$ be given which satisfies the formula (α) $x_\nu \in S-Cn\Big(\underset{x_\mu}{E}[\mu < \nu] \Big)$ for every $\nu < \pi$. In order that $\underset{x_\nu}{E}[\nu < \pi] \in \mathfrak{A}$, it is necessary and sufficient that $\pi < \omega$.*

Proof. A. Let us suppose that

$$(1) \qquad\qquad \underset{x_\nu}{E}[\nu < \pi] \in \mathfrak{A},$$

and nevertheless assume

$$(2) \qquad\qquad \pi = \omega.$$

By Th. 22 and Def. $3a$ it follows from (1) that there is a set Y which satisfies the formulas

$$(3) \qquad\qquad Y \subseteq \underset{x_\nu}{E}[\nu < \pi], \qquad Y \in \mathfrak{E},$$

and

(4) $$Y \in \mathfrak{Aq}\Big(\underset{x_\nu}{E}[\nu < \pi]\Big).$$

Moreover, from (3) we infer the existence of a number ν such that

(5) $$\nu < \omega \quad and \quad Y \subseteq \underset{x_\mu}{E}[\mu \leqslant \nu].$$

The formulas (2) and (5) give

(6) $$\nu+1 < \pi,$$

whence by the hypothesis of the theorem we have

(7) $$x_{\nu+1} \bar{\in} Cn\Big(\underset{x_\mu}{E}[\mu \leqslant \nu]\Big).$$

Since $\underset{x_\mu}{E}[\mu \leqslant \nu] \subseteq S$, we obtain from (5) with the help of Th. 1a $Cn(Y) \subseteq Cn\Big(\underset{x_\mu}{E}[\mu \leqslant \nu]\Big)$; by comparing this inclusion with (7) we reach the formula

(8) $$x_{\nu+1} \bar{\in} Cn(Y).$$

By Ax. 2 and Def. 3 it results from (4) that

$$\underset{x_\nu}{E}[\nu < \pi] \subseteq Cn\Big(\underset{x_\nu}{E}[\nu < \pi]\Big) = Cn(Y).$$

Thus by virtue of (6) we have $x_{\nu+1} \in Cn(Y)$, which obviously contradicts the formula (8) and refutes the assumption (2).

Hence we conclude that

(9) $$\pi < \omega.$$

B. If on the other hand the inequality (9) is satisfied, then according to the hypothesis we have $\underset{x_\nu}{E}[\nu < \pi] \in \mathfrak{E}.\mathfrak{P}(S)$, from which by Th. 19b the formula (1) follows.

We have thus established the equivalence of the formulas (1) and (9), and this concludes the proof.

THEOREM 25. *The following conditions are equivalent:* (α) $A \in \mathfrak{A}$; (β) *there is no sequence of sentences* x_ν *of type* ω *which satisfies both of the formulas* (α) *and* (β) *of Th.* 16, *and in addition* $A \subseteq S$; (γ) *there is a sequence of sentences* x_ν *of the type* $\pi < \omega$, *which satisfies both of these formulas.*

Proof. If the condition (α) holds, then by Th. 20 c we have $\mathfrak{A}\mathrm{q}(A) \subseteq \mathfrak{A}$ and accordingly $E[\nu < \pi] \in \mathfrak{A}$ for every sequence x_ν of sentences which satisfies the conclusion of Th. 16; from Th. 24 it thus results that the inequality $\pi \geqslant \omega$ does not hold. But since $A \subseteq S$ (by Defs. 2 and 3 a, b), we can assert that (β) follows from (α). On the other hand by Th. 16 there exists for every set $A \subseteq S$ a sequence of sentences either of type ω or of type $\pi < \omega$, which satisfies the conclusion of this theorem; (β) thus implies (γ). Finally, if (γ) is satisfied, then we have $E[\nu < \pi] \in \mathfrak{E}.\,\mathfrak{A}\mathrm{q}(A)$, which on the basis of Def. 3 a, b gives the x_ν condition (α).

Accordingly the conditions (α)—(γ) are equivalent, q.e.d.

Further theorems deal with axiomatizable and unaxiomatizable systems.

THEOREM 26. $\mathfrak{E}.\,\mathfrak{A} = \underset{Cn(X)}{E}\big[X \in \mathfrak{E}.\,\mathfrak{P}(S)\big].$

[Def. 3 a, b, Th. 18 a]

THEOREM 27. *In order that $A \in \mathfrak{E}-\mathfrak{A}$, it is necessary and sufficient that there exists a sequence of sets X_ν of the type ω which satisfies the following formulas:* (α) $X_\nu \in \mathfrak{E}$ *for every* $\nu < \omega$; (β) $X_\mu \subseteq X_\nu$ *and* $X_\mu \neq X_\nu$ *for* $\mu < \nu < \omega$; (γ) $A = \underset{\nu < \omega}{\sum} X_\nu$. *The formula* ($\alpha$) *can be replaced by* ($\alpha'$) $X_\nu \in \mathfrak{E}.\,\mathfrak{A}$.

Proof. A. First let us assume that

(1) $A \in \mathfrak{E}-\mathfrak{A}.$

By applying Ths. 16 and 25 (and taking into account Def. 1) we infer from (1) that a sequence of sentences x_ν of type ω exists which satisfies the formulas

(2) $x_\nu \in Cn\big(\underset{x_\lambda}{E}[\lambda < \nu]\big)$ *for* $\nu < \omega$

and

(3) $\underset{x_\nu}{E}[\nu < \omega] \in \mathfrak{A}\mathrm{q}(A).$

By Th. 18 a from (1) and (3) we obtain

(4) $\underset{x_\nu}{E}[\nu < \omega] \subseteq S$ *and* $Cn\big(\underset{x_\nu}{E}[\nu < \omega]\big) = A.$

We put

(5) $X_\nu = Cn\Big(\underset{x_\lambda}{E}[\lambda < \nu]\Big)$ for $\nu < \omega$.

By Th. 26 it follows from (4) and (5) that

(6) $X_\nu \in \mathfrak{S}$ for $\nu < \omega$

and even that

(6′) $X_\nu \in \mathfrak{S}.\mathfrak{A}$ for $\nu < \omega$.

In accordance with Th. 1 a and by reference to (4) we obtain $Cn\Big(\underset{x_\lambda}{E}[\lambda < \mu]\Big) \subseteq Cn\Big(\underset{x_\lambda}{E}[\lambda < \nu]\Big)$ and from this by (5)

(7) $X_\mu \subseteq X_\nu$ for $\mu < \nu < \omega$.

From (4), (5), and Ax. 2 we obtain $x_\mu \in X_\nu$, but in view of (2) and (5) $x_\mu \bar\in X_\mu$ for $\mu < \nu < \omega$; consequently

(8) $X_\mu \neq X_\nu$ for $\mu < \nu < \omega$.

Finally it is to be noted that the class $\mathfrak{K}'$ of sets of sentences $Y_\nu = \underset{x_\lambda}{E}[\lambda < \nu]$, where $\nu < \omega$, satisfies the hypothesis of Cor. 5, whence $Cn\Big(\underset{Y \in \mathfrak{K}'}{\sum} Y\Big) = \underset{Y \in \mathfrak{K}'}{\sum} Cn(Y)$; on the basis of (4) and (5) we thus have

(9) $A = \underset{\nu < \omega}{\sum} X_\nu$.

From (6)–(9) (as well as (6′)) it results that

(10) *the sequences of sets X_ν of type ω satisfies the formulas* (α)–(γ) *(as well as the formula (α')) of the theorem.*

B. Consider now a sequence of sets X_ν of the kind described in (10) (with the reference to (α') omitted). Put

(11) $\mathfrak{K} = \underset{X_\nu}{E}[\nu < \omega]$.

From (10) and (11) it follows at once that

(12) $A = \underset{X \in \mathfrak{K}}{\sum} X$

and that

(13) *the class $\mathfrak{K}$ satisfies condition (α) of Cor. 5.*

Since further by (10) and (11) $\mathfrak{K} \subseteq \mathfrak{S}$, we infer from (13) and (12) by means of Th. 12 that

(14) $A = \underset{X \in \mathfrak{K}}{\sum} X \in \mathfrak{S}$.

If it were the case that $\Re . \mathfrak{Aq}\left(\sum_{X \in \Re} X\right) \neq 0$, e.g. if

$$Y \in \Re . \mathfrak{Aq}\left(\sum_{X \in \Re} X\right),$$

then by (10) and (14) we should have $Y \in \mathfrak{S} . \mathfrak{Aq}(A)$ and $A \in \mathfrak{S}$, which by Th. 18c gives $Y = A \in \Re$; but the last formula is in contradiction with (10) and (11) (since the sum of all terms of an increasing sequence of sets of type ω contains every term as a *proper* part). Consequently

$$(15) \qquad\qquad \Re . \mathfrak{Aq}\left(\sum_{X \in \Re} X\right) = 0.$$

By the use of Th. 23 we infer from (13)–(15) that

$$A = \sum_{X \in \Re} X \in \mathfrak{A};$$

by combining this formula with (14) we at once obtain (1).

We have thus shown that *the existence of a sequence of sets described in* (10) *forms a necessary and sufficient condition for formula* (1), q.e.d.

From the last two theorems the following corollary is easily obtained:

COROLLARY 28. $\overline{\overline{\mathfrak{S} . \mathfrak{A}}} \leqslant \aleph_0$.

Proof. According to a well-known theorem of set theory Ax. 1 has the consequence that the class $\mathfrak{E} . \mathfrak{P}(S)$ is at most denumerable. By Th. 26 the function Cn establishes a many-to-one correlation between the members of this class and those of the class $\mathfrak{S} . \mathfrak{A}$; accordingly this latter class is also at most denumerable, $\overline{\overline{\mathfrak{S} . \mathfrak{A}}} \leqslant \aleph_0$, q.e.d.

COROLLARY 29. (a) *If* $A \in \mathfrak{S} - \mathfrak{A}$, *then* $\overline{\overline{\mathfrak{P}(A) . \mathfrak{S} . \mathfrak{A}}} = \aleph_0$.

(b) *If* $\mathfrak{S} - \mathfrak{A} \neq 0$ (*or* $\mathfrak{P}(S) - \mathfrak{A} \neq 0$), *then* $\overline{\overline{\mathfrak{S} . \mathfrak{A}}} = \aleph_0$.

Proof. (a) By Th. 27 the hypothesis gives $\overline{\overline{\mathfrak{P}(A) . \mathfrak{S} . \mathfrak{A}}} \geqslant \aleph_0$; since the inverse inequality follows immediately from Cor. 28, we finally have $\overline{\overline{\mathfrak{P}(A) . \mathfrak{S} . \mathfrak{A}}} = \aleph_0$;

(b) results directly from (a) and Cor. 28 (with the help of Ths. 18a and 20c there is no difficulty in showing that the second premiss, $\mathfrak{P}(S) - \mathfrak{A} \neq 0$, implies the first, $\mathfrak{S} - \mathfrak{A} \neq 0$).

THEOREM 30. *Either* $1 \leqslant \overline{\overline{\mathfrak{S}}} \leqslant \aleph_0$ *or* $\aleph_0 \leqslant \overline{\overline{\mathfrak{S}}} \leqslant 2^{\aleph_0}$.

Proof. According to the well-known theorem on the cardinal number of the power set we first infer from Ax. 1 and Def. 1 that $\overline{\overline{\mathfrak{S}}} \leqslant \overline{\overline{\mathfrak{P}(S)}} \leqslant 2^{\aleph_0}$; and by virtue of Th. 10 we also have $\overline{\overline{\mathfrak{S}}} \geqslant 1$. If now $\mathfrak{S} \subseteq \mathfrak{A}$ (or, what amounts to the same thing, $\mathfrak{S} = \mathfrak{S} . \mathfrak{A}$), then by Cor. 28 we have

$$1 \leqslant \overline{\overline{\mathfrak{S}}} \leqslant \aleph_0;$$

if, however, $\mathfrak{S} - \mathfrak{A} \neq 0$, then it follows from Cor. 29*b* that

$$\aleph_0 \leqslant \overline{\overline{\mathfrak{S}}} \leqslant 2^{\aleph_0}, \text{ q.e.d.}$$

It is to be noted that the theorem above can be established without the use of the axiom of choice (otherwise the result would be quite trivial); the same applies also to the generalization of this theorem given below in Cor. 65. If we restrict ourselves to the consideration of deductive disciplines which presuppose sentential calculus, we can improve Th. 30 by showing that the class $\mathfrak{S}$ is either of the power $2^{\overline{\nu}}$ for some $\nu < \omega$ or else of the power $2^{\aleph_0}$.†

§ 5. INDEPENDENT SETS OF SENTENCES; BASIS OF A SET OF SENTENCES

A set of sentences is called *independent* if it is not equivalent to any of its proper subsets. The class of all independent sets of sentences is denoted by '$\mathfrak{U}$':

DEFINITION 4. $\mathfrak{U} = \underset{X}{E}[\mathfrak{P}(X) . \mathfrak{Aq}(X) = \{X\}].$

Some equivalent transformations of the above definition are given in the next two theorems.

THEOREM 31. *The following conditions are equivalent:* (α) $A \in \mathfrak{U}$; (β) $A . Cn(X) \subseteq X \subseteq S$ *for every set* $X \subseteq A$; (γ) *the formulas* $X + Y \subseteq A$ *and* $Cn(X) = Cn(Y)$ *always imply* $X = Y$, *and in addition* $A \subseteq S$.

Proof. A. First we suppose that

(1) $A \in \mathfrak{U}.$

By Defs. 4 and 2 it follows that:

(2) $A \subseteq S.$

† See XII, Ths. 37 and 38 (p. 367).

Let X be any subset of A.

By the use of Th. 2a and with the help of (2) we obtain:

$$Cn((A-Cn(X))+X) = Cn((A-Cn(X))+Cn(X))$$
$$= Cn(A+Cn(X)) = Cn(A+X) = Cn(A),$$

and accordingly by Def. 2, $(A-Cn(X))+X \in \mathfrak{P}(A).\mathfrak{Aq}(A)$. From this, by virtue of Def. 4 and with the help of (1), we infer that $(A-Cn(X))+X = A$ and consequently $A.(Cn(X)-X) = 0$. Thus by (2) we have

(3) $A.Cn(X) \subseteq X \subseteq S$ for every set $X \subseteq A$.

B. We assume next the condition (3). We consider any two sets X and Y which satisfy the formulas $X+Y \subseteq A$ and $Cn(X) = Cn(Y)$. It follows from (3) that

$$A.Cn(Y) = A.Cn(X) \subseteq X;$$

since further (3) gives the inclusion (2), we have by Ax. 2

$$Y \subseteq A.Cn(Y).$$

Consequently $Y \subseteq X$. In an exactly analogous manner we reach the inverse inclusion $X \subseteq Y$, so that finally $X = Y$. We have thus shown that

(4) *the formulas* $X+Y \subseteq A$ *and* $Cn(X) = Cn(Y)$ *always imply* $X = Y$, *and in addition* (2) *holds*.

C. Finally, let (4) be given. According to Def. 2 every set $X \in \mathfrak{P}(A).\mathfrak{Aq}(A)$ satisfies the formula $Cn(X) = Cn(A)$; hence, if in (4) we put $Y = A$, we obtain $X = A$. Thus we have $\mathfrak{P}(A).\mathfrak{Aq}(A) \subseteq \{A\}$; but since by Th. 13$a$ the inclusion

$$\{A\} \subseteq \mathfrak{P}(A).\mathfrak{Aq}(A)$$

also holds, we reach the identity $\mathfrak{P}(A).\mathfrak{Aq}(A) = \{A\}$, which by Def. 4 gives the formula (1).

According to the above argument, (3) follows from (1), (4) from (3), and (1) from (4). *The conditions* (1), (3), *and* (4) *are thus equivalent*, q.e.d.

THEOREM 32. *In order that* $A \in \mathfrak{U}$, *it is necessary and sufficient that* $x \in S-Cn(A-\{x\})$, *for every* $x \in A$.

Proof. A. Let us assume that

(1) $A \in \mathfrak{U},$

and apply Th. 31. According to the condition (γ) of this theorem it follows directly from (1) that

(2) $$A \subseteq S.$$

By putting $X = A - \{x\}$ in condition (β) of the same theorem we further obtain $A \cdot Cn(A - \{x\}) \subseteq A - \{x\}$, whence

$$\{x\} \cdot A \cdot Cn(A - \{x\}) = 0.$$

Under the assumption that $x \in A$ the last formula gives $\{x\} \cdot Cn(A - \{x\}) = 0$ and finally, by virtue of (2),

(3) $$x \in S - Cn(A - \{x\}) \text{ for every } x \in A.$$

B. We now assume formula (3) and note initially that from this the inclusion (2) immediately follows. Let us assume that there exists a set $X \subseteq A$ which satisfies the formula

$$A \cdot Cn(X) - X \neq 0.$$

Let for example $x \in A \cdot Cn(X) - X$ and accordingly $X \subseteq A - \{x\}$, which by Th. 1a gives the inclusion $Cn(X) \subseteq Cn(A - \{x\})$; since then $x \in A \cdot Cn(X)$, we infer at once that $x \in Cn(A - \{x\})$, which contradicts formula (3). Our assumption is thus disproved; consequently, in view of (2), we must assume that

(4) $$A \cdot Cn(X) \subseteq X \subseteq S \text{ for every set } X \subseteq A.$$

By Th. 31 formula (1) follows from (4).

We have thus established the equivalence of the formulas (1) and (3) and thus proved the theorem.

From the last theorem we obtain immediately

COROLLARY 33. (a) $0 \in \mathfrak{U}$;   [Th. 32]

(b) *in order that* $\{x\} \in \mathfrak{U}$, *it is necessary and sufficient that*

$$x \in S - Cn(0). \qquad \text{[Th. 32]}$$

THEOREM 34. *If* $A \in \mathfrak{U}$, *then* $\mathfrak{P}(A) \subseteq \mathfrak{U}$.   [Th. 31]

THEOREM 35. *If* $A \in \mathfrak{U}$, *then* $\mathfrak{P}(A) \cdot \mathfrak{A} = \mathfrak{P}(A) \cdot \mathfrak{E}$.

Proof. According to Th. 32 (or else 31) the hypothesis gives

(1) $$A \subseteq S.$$

Consequently we have $\mathfrak{P}(A).\mathfrak{E} \subseteq \mathfrak{P}(S).\mathfrak{E}$, from which by Th. 19$b$ the inclusion

(2) $$\mathfrak{P}(A).\mathfrak{E} \subseteq \mathfrak{P}(A).\mathfrak{A}$$

follows.

On the other hand let us assume that

(3) $$\mathfrak{P}(A).\mathfrak{A}-\mathfrak{E} \neq 0,$$

and accordingly let

(4) $$X \in \mathfrak{P}(A).\mathfrak{A}-\mathfrak{E}.$$

With the help of Ax. 1 we easily infer from (1)–(4) that the set X is denumerable; accordingly all elements of this set can be ordered in an infinite sequence of the type ω, with all terms distinct, such that

(5) $$X = \underset{x_\nu}{E}[\nu < \omega] \quad \text{where } x_\mu \neq x_\nu \text{ for } \mu < \nu < \omega.$$

According to Th. 34 it follows from (4) and the hypothesis that $X \in \mathfrak{U}$; from this by Th. 32 and with the help of (5) we obtain

(6) $$x_\nu \in S-Cn(X-\{x_\nu\}) \quad \text{for every } \nu < \omega.$$

From (5) and (6) we further obtain

$$\underset{x_\mu}{E}[\mu < \nu] \subseteq X-\{x_\nu\} \subseteq S \quad \text{for } \nu < \omega.$$

By Th. 1a this implies the inclusion

$$Cn\Big(\underset{x_\mu}{E}[\mu < \nu]\Big) \subseteq Cn(X-\{x_\nu\});$$

hence by (6) it follows that

(7) $$x_\nu \in S-Cn\Big(\underset{x_\mu}{E}[\mu < \nu]\Big) \quad \text{for } \nu < \omega.$$

By combining (5) and (7) with Th. 24 we easily obtain

$$X = \underset{x_\nu}{E}[\nu < \omega] \,\bar{\in}\, \mathfrak{A},$$

contrary to formula (4).

With this, assumption (3) is disproved and therefore we have

(8) $$\mathfrak{P}(A).\mathfrak{A} \subseteq \mathfrak{P}(A).\mathfrak{E}.$$

Inclusions (2) and (8) at once give the required identity

$$\mathfrak{P}(A).\mathfrak{A} = \mathfrak{P}(A).\mathfrak{E}, \quad \text{q.e.d.}$$

THEOREM 36. *If $A \in \mathfrak{U} - \mathfrak{E}$, then $\overline{\overline{\mathfrak{P}(Cn(A)).\mathfrak{S}.\mathfrak{A}}} = \aleph_0$, but* $\overline{\overline{\mathfrak{P}(Cn(A)).\mathfrak{S}}} = \overline{\overline{\mathfrak{P}(Cn(A)).\mathfrak{S} - \mathfrak{A}}} = 2^{\aleph_0}$.

Proof. By Ths. 32 and 35 as well as Ax. 1 we infer from the hypothesis that

(1) $$A \subseteq S, \qquad A \bar{\in} \mathfrak{A},$$

and

(2) $$\bar{\bar{A}} = \aleph_0.$$

According to Th. 18a it follows from (1) that $Cn(A) \in \mathfrak{S}$ and $A \in \mathfrak{Aq}(Cn(A))$; from this by Th. 20c we obtain $Cn(A) \in \mathfrak{S} - \mathfrak{A}$. Hence, as a consequence of Cor. 29a, we obtain at once

(3) $$\overline{\overline{\mathfrak{P}(Cn(A)).\mathfrak{S}.\mathfrak{A}}} = \aleph_0.$$

Since by the hypothesis the set A is independent, we can apply Th. 31. Condition (γ) of this theorem asserts that the function Cn maps the class $\mathfrak{P}(A)$ on $\underset{Cn(X)}{E} [X \subseteq A]$ in one-to-one fashion; consequently the two classes have the same cardinal number. By a well-known theorem it follows from (2) that $\overline{\overline{\mathfrak{P}(A)}} = 2^{\aleph_0}$; hence

(4) $$\overline{\overline{\underset{Cn(X)}{E} [X \subseteq A]}} = 2^{\aleph_0}.$$

By means of Ths. 1a and 9a, using (1) we obtain without difficulty that $\underset{Cn(X)}{E} [X \subseteq A] \subseteq \mathfrak{P}(Cn(A)).\mathfrak{S}$; by combining this formula with (4) we get $\overline{\overline{\mathfrak{P}(Cn(A)).\mathfrak{S}}} \geqslant 2^{\aleph_0}$. But since by Th. 30 we also have $\overline{\overline{\mathfrak{P}(Cn(A)).\mathfrak{S}}} \leqslant 2^{\aleph_0}$, we finally obtain

(5) $$\overline{\overline{\mathfrak{P}(Cn(A)).\mathfrak{S}}} = 2^{\aleph_0}.$$

Formulas (3) and (5) at once give

(6) $$\overline{\overline{\mathfrak{P}(Cn(A)).\mathfrak{S} - \mathfrak{A}}} = 2^{\aleph_0} - \aleph_0 = 2^{\aleph_0}.$$

According to (3), (5), and (6) the theorem is completely proved.

COROLLARY 37. *If $\mathfrak{U} - \mathfrak{E} \neq 0$, then $\overline{\overline{\mathfrak{S}.\mathfrak{A}}} = \aleph_0$, but*
$$\overline{\overline{\mathfrak{S}}} = \overline{\overline{\mathfrak{S} - \mathfrak{A}}} = 2^{\aleph_0}.$$

[Th. 36, Cor. 28, Th. 30]

It is to be noted that within almost all deductive disciplines, and in particular within the simplest of them—the sentential calculus—it has been found possible to construct a set of sentences which is both infinite and independent, and thus to realize the hypothesis of the last corollary. Hence, it turns out that in all these disciplines there are more unaxiomatizable than axiomatizable systems; the deductive systems are, so to speak, as a rule unaxiomatizable, although in practice we deal almost exclusively with axiomatizable systems. This paradoxical circumstance was first noticed by Lindenbaum in application to the sentential calculus.[1]

Every independent set of sentences which is equivalent to a given set A is called a *basis of the set* A; the class of all such sets of sentences is here denoted by '$\mathfrak{B}(A)$'.

DEFINITION 5. $\mathfrak{B}(A) = \mathfrak{Aq}(A).\mathfrak{U}.$

The above definition can be expressed otherwise as follows:

THEOREM 38. *In order that* $B \in \mathfrak{B}(A)$, *it is necessary and sufficient that* $\mathfrak{P}(B).\mathfrak{Aq}(A) = \{B\}.$ [Defs. 5, 4, Th. 13 b]

In the usual terminology of set theory a set X is said to be *irreducible with respect to a class* $\mathfrak{R}$ *of sets* when $\mathfrak{P}(X).\mathfrak{R} = \{X\}.$[2] Thus the theorem above asserts that 'a basis of the set A of sentences' means the same as 'a set irreducible with respect to the class $\mathfrak{Aq}(A)$'.

Other properties of this concept are expressed in the following theorems:

THEOREM 39. *In order that* $A \in \mathfrak{B}(A)$, *it is necessary and sufficient that* $A \in \mathfrak{U}.$ [Defs. 5, 4]

THEOREM 40. (a) *If* $\mathfrak{B}(A).\mathfrak{B}(B) \neq 0$, *then* $A \in \mathfrak{Aq}(B)$;   [Def. 5, Th. 13 b]

(b) *if* $A \in \mathfrak{Aq}(B)$, *then* $\mathfrak{B}(A) = \mathfrak{B}(B).$ [Th. 13 b, Def. 5]

THEOREM 41. $\mathfrak{E}.\mathfrak{B}(A) = \mathfrak{U}.\mathfrak{Ax}(A).$ [Defs. 5, 4, 3 a]

THEOREM 42. *If* $A \in \mathfrak{A}$, *then* $\mathfrak{P}(A).\mathfrak{E}.\mathfrak{B}(A) \neq 0.$

---

[1] Cf. IV, Th. 27 (p. 51).
[2] Cf. Tarski, A. (71), p. 48 (Def. 1).

Proof. By Th. 22 and Def. 3 a the hypothesis implies that a set X exists which satisfies the formulas

(1) $$X \subseteq A,$$

(2) $$X \in \mathfrak{E} \quad and \quad X \in \mathfrak{Aq}(A).$$

By Th. 13 a, b it results from (2) that $X \in \mathfrak{Aq}(X)$; consequently the class $\mathfrak{R} = \mathfrak{P}(X) . \mathfrak{Aq}(X)$, which consists of subsets of the finite set X, is distinct from 0. Hence, by a familiar definition of finite sets,[1] we conclude that this class contains among its elements at least one irreducible set Y (i.e. a set with the property $\mathfrak{P}(Y) . \mathfrak{R} = \{Y\}$). By Th. 38 this set Y forms a basis of X, and thus we have

(3) $$Y \subseteq X \quad and \quad Y \in \mathfrak{B}(X).$$

From (1)–(3) we obtain at once

(4) $$Y \subseteq A \quad and \quad Y \in \mathfrak{E}.$$

By Th. 40 b it follows from (2) that $\mathfrak{B}(A) = \mathfrak{B}(X)$, whence by (3)

(5) $$Y \in \mathfrak{B}(A).$$

Formulas (4) and (5) give at once

$$\mathfrak{P}(A) . \mathfrak{E} . \mathfrak{B}(A) \neq 0, \quad \text{q.e.d.}$$

THEOREM 43. *If* $A \in \mathfrak{A}$, *then* $\mathfrak{B}(A) \subseteq \mathfrak{E}$.

Proof. If the conclusion were false, we should have, by Def. 5, $\mathfrak{Aq}(A) . \mathfrak{U} - \mathfrak{E} \neq 0$. Accordingly let $X \in \mathfrak{Aq}(A)$ and $X \in \mathfrak{U} - \mathfrak{E}$. By Th. 35 the second of these formulas yields $X \,\bar{\in}\, \mathfrak{A}$; and hence, with the help of the first formula by using Th. 20 c we infer that $A \,\bar{\in}\, \mathfrak{A}$, contrary to the hypothesis.

We thus have $\mathfrak{B}(A) \subseteq \mathfrak{E}$, and so the theorem is proved.

COROLLARY 44. *The following conditions are equivalent:*

(α) $A \in \mathfrak{A}$; (β) $\mathfrak{B}(A) \neq 0$ *and* $\mathfrak{B}(A) \subseteq \mathfrak{E}$; (γ) $\mathfrak{E} . \mathfrak{B}(A) \neq 0$.

Proof. By Ths. 42 and 43, (β) is at once obtainable from (α); (γ) results immediately from (β); finally, on the basis of Th. 41 and Def. 3 b, (α) follows from (γ). Accordingly the formulas (α)–(γ) are equivalent, q.e.d.

---

[1] Cf. Tarski, A. (71), p. 49 (Def. 3).

By the corollary just established (or by Th. 42) every axiomatizable set of sentences possesses at least one basis. This result cannot be extended to unaxiomatizable sets of sentences on the basis of the axioms underlying this discussion[1] (it can only be shown that every set of sentences possesses an ordered basis; see remarks following Th. 16). On the other hand, for those deductive disciplines which presuppose the sentential calculus it can be proved that every set of sentences has a basis.†

§ 6. CONSISTENT SETS OF SENTENCES

The concepts of consistency and completeness, with which we are concerned in this and the next sections, are among the most important concepts of metamathematics; around these concepts is centred the research which is carried on today within special metadisciplines.

A set of sentences is called consistent if it is not equivalent to the set of all meaningful sentences (or, in other words, if the set of its consequences does not contain as elements all meaningful sentences).

According to the usual definition, a set of sentences is called consistent if there is no sentence which together with its negation belongs to the consequences of this set. Our definition thus diverges from the usual one, and indeed it has a much more general character since knowledge of the concept of negation is not presupposed; in consequence this definition can be applied even to those deductive disciplines in which the negation concept is either entirely lacking or at least does not exhibit the properties usually ascribed to it.[2] However, the two definitions of consistency prove to be equivalent for all those disciplines which are based upon the ordinary system of sentential calculus.

---

[1] For example, in the sentential calculus it is easy to construct sets of sentences without a basis. (A proof of this result appears in Schröter, K. (62 a), pp. 299–301.)

[2] Such a discipline, the so-called restricted sentential calculus is treated in IV, § 4.

---

† Cf. the remarks following Th. 25 in XII (p. 362).

We owe the idea of the definition adopted here to E. Post: in his investigations on the sentential calculus he has made use of a closely related definition.[1]

Denoting the class of all consistent sets of sentences by '$\mathfrak{W}$' we reach

DEFINITION 6. $\mathfrak{W} = \mathfrak{P}(S) - \mathfrak{Aq}(S)$.

The content of this definition is more clearly formulated in the following:

THEOREM 45.   *In order that $A \in \mathfrak{W}$, it is necessary and sufficient that $A \subseteq S$ and $Cn(A) \neq S$.*   [Def. 6, Ths. 10, 18a]

The following elementary properties of consistent sets of sentences deserve consideration:

THEOREM 46.   (a) *If $A \in \mathfrak{W}$, then $\mathfrak{P}(A) \subseteq \mathfrak{W}$;*

[Def. 6, Th. 14]

(b) *if $A \in \mathfrak{W}$, then $\mathfrak{Aq}(A) \subseteq \mathfrak{W}$.*   [Th. 45, Def. 2]

THEOREM 47. $\mathfrak{S} - \mathfrak{W} = \{S\}$.   [Def. 6, Ths. 7, 18c]

THEOREM 48.   *Let $S \in \mathfrak{A}$; in order that $(\alpha) A \in \mathfrak{W}$, it is necessary and sufficient that $(\beta) \mathfrak{P}(A) . \mathfrak{E} \subseteq \mathfrak{W}$.*

Proof. According to Th. 46a the condition (β) is necessary for (α) to hold; it remains only to show that this condition is also sufficient.

Let us assume (β) and suppose that nevertheless

(1) $A \bar{\in} \mathfrak{W}$.

By Def. 6 it follows from (β) that $\mathfrak{P}(A) . \mathfrak{E} \subseteq \mathfrak{P}(S)$; from this we easily obtain:

(2) $A \subseteq S$.

By another application of Def. 6 we obtain from (1) and (2) $A \in \mathfrak{Aq}(S)$. Since by hypothesis $S \in \mathfrak{A}$ holds, we infer by Th. 20c that $A \in \mathfrak{A}$, whence by Th. 22 the formula $\mathfrak{P}(A) . \mathfrak{Ax}(A) \neq 0$ follows. Accordingly by Def. 3 there exists a set X which satisfies the formulas

(3) $X \in \mathfrak{P}(A) . \mathfrak{E}$

---

[1] Post, E. L. (60), p. 177.

and

(4) $X \in \mathfrak{Aq}(A).$

From (β) and (3) it follows at once that $X \in \mathfrak{W}$; according to Th. 13 b, (4) is equivalent to the formula $A \in \mathfrak{Aq}(X)$. Consequently by virtue of Th. 46 b we obtain the formula $A \in \mathfrak{W}$, which coincides with (α) and contradicts the assumption (1).

We have thus shown that (β) is a sufficient condition for (α), and with this the proof is complete.

THEOREM 49. *Let $S \in \mathfrak{A}$; if the class $\mathfrak{R}$ satisfies the condition (α) of Th. 4 (or of Cor. 5) and if at the same time $\mathfrak{R} \subseteq \mathfrak{W}$ holds, then $\sum\limits_{X \in \mathfrak{R}} X \in \mathfrak{W}$.*

Proof. Consider an arbitrary set $Y \in \mathfrak{E} . \mathfrak{P}\Big(\sum\limits_{X \in \mathfrak{R}} X \Big)$. By virtue of condition (α) of Th. 4 we infer without difficulty that there is a set Z which satisfies the formulas $Z \in \mathfrak{R}$ and $Y \subseteq Z$ (cf. the proof of Th. 23). On the basis of the hypothesis the first formula yields $Z \in \mathfrak{W}$; hence by using Th. 46 a and with the help of the second formula we obtain $Y \in \mathfrak{W}$.

It is thus proved that the formula $Y \in \mathfrak{E} . \mathfrak{P}\Big(\sum\limits_{X \in \mathfrak{R}} X \Big)$ always implies $Y \in \mathfrak{W}$. Consequently we have the inclusion

$$\mathfrak{E} . \mathfrak{P}\Big(\sum\limits_{X \in \mathfrak{R}} X \Big) \subseteq \mathfrak{W};$$

in accordance with the hypothesis $S \in \mathfrak{A}$ and with the help of Th. 48 we get from this at once

$$\sum\limits_{X \in \mathfrak{R}} X \in \mathfrak{W}, \quad \text{q.e.d.}$$

The formula $S \in \mathfrak{A}$, which occurs as a premiss in the two preceding theorems, as well as in some later ones (Ths. 51, 56, and 57), cannot be derived from the axioms listed in § 1. It is, however, satisfied for all known formalized disciplines; even the following logically stronger assertion holds: *there is a sentence $x \in S$ such that $Cn(\{x\}) = S$.* Nevertheless it does not seem desirable to include this formula among the axioms, on account of its special and, in a certain sense, accidental character.

§ 7. DECISION DOMAIN OF A SET OF SENTENCES, COMPLETE SETS OF SENTENCES

By the *decision domain* of the set A of sentences we understand the set of all sentences which are either consequences of A or which, when added to A, yield an inconsistent set of sentences. A set of sentences is said to be *complete* or *absolutely complete* if its decision domain contains all meaningful sentences.[1]

With regard to this definition of completeness we can repeat all remarks which were made above in § 6 in connexion with the definition of consistency. This definition is also due to Post, who has used it in his investigations into the sentential calculus.[2] As is easily seen the usual definition of completeness, resting upon the concept of negation, is quite unsuitable for the sentential calculus (as well as for all disciplines which contain sentences with so-called free variables). In fact, by the usual definition, a set of sentences is called complete if, for every sentence, either the sentence itself or its negation belongs to the consequences of this set; and, as is well known, the ordinary system of sentential calculus is complete in the sense that its decision domain contains all meaningful sentences, although it is not complete according to the usual definition. The two definitions prove to be equivalent for all those disciplines which presuppose sentential calculus (and in which no expression containing free variables is regarded as a sentence).

The concept of absolute completeness is of great importance for those metadisciplines in which the objects of investigation are 'poor', elementary disciplines of an uncomplicated logical structure (e.g. the sentential calculus, or disciplines without predicate variables).[3] On the other hand this concept has not yet played an important part in investigations on 'rich', logically more complicated disciplines (e.g. the system of *Principia Mathematica*). The cause of this is perhaps to be sought in the

---

[1] The concept of completeness also occurs with another meaning, discussed, for instance, in Fraenkel, A. (16), § 18.4, pp. 347–54; the meaning used here corresponds to Fraenkel's *Entscheidungsdefinitheit*.

[2] See Post, E. L. (60), p. 177.

[3] Cf. Langford, C. H. (44), p. 115, as well as IV and Presburger, M. (61). For later developments in this domain consult Tarski, A. (84).

widespread, perhaps intuitively plausible, but not always strictly founded, belief in the incompleteness of all systems developed within these disciplines and known at the present day.[1] It is nevertheless to be expected that the concept of completeness will some day attain a greater importance even for the latter disciplines. For, although all known consistent systems of this kind may be incomplete, yet Th. 56 to be established below offers at least a theoretical possibility of extending every such system to one which is both consistent and complete. The question now arises how this extension is to be carried out so as to be 'effective', as natural as possible, and at the same time agreeing with some philosophical viewpoint.[2]

Moreover, it is to be noted that some more special problems which are closely connected with the concept of completeness have already been successfully investigated even with respect to those 'rich' disciplines. The investigations referred to have aimed at and succeeded in proving that all meaningful sentences of a particular logical form (e.g. all sentences without function variables) belong to the decision domain of some given system. It might be useful for such investigations to introduce the concept of relative completeness: a set A of sentences is said to be *relatively complete with respect to the set B of sentences* if the decision domain of A includes the set B; this concept will not be discussed further here.[3]

The decision domain of the set A of sentences will be denoted by '$\mathfrak{Ent}(A)$' and the class of all complete sets of sentences by '$\mathfrak{V}$'.

DEFINITION 7. (a) $\mathfrak{Ent}(A) = Cn(A) + S . \underset{x}{E}[A + \{x\} \bar{\in} \mathfrak{V}]$.

(b) $\mathfrak{V} = \mathfrak{P}(S) . \underset{X}{E}[\mathfrak{Ent}(X) = S]$.

---

[1] See Fraenkel, A. (16), pp. 347–54.

[2] Since this article first appeared in print, several important contributions concerned with the concept of completeness have been published, throwing much light on the questions discussed in the last paragraph. See in the first place Gödel, K. (22).

[3] The investigations on absolute completeness in a 'poor' discipline are as a rule equivalent to those on relative completeness in a more extensive discipline. For this reason the works cited on p. 93, note 3 can be regarded as examples of investigations on relative completeness, if they are related to an extended discipline (e.g. to *Principia Mathematica*).

The following elementary properties of decision domains should be noted:

THEOREM 50. *(a) If $A \subseteq S$, then $A \subseteq \mathfrak{Ent}(A) \subseteq S$.*

[Ax. 2, Def. 7 a]

(b) If $A \subseteq B \subseteq S$, then $\mathfrak{Ent}\,(A) \subseteq \mathfrak{Ent}(B)$.

[Ths. 1 a, 46 a, Def. 7 a]

(c) If $A \in \mathfrak{Aq}(B)$, then $\mathfrak{Ent}(A) = \mathfrak{Ent}(B)$.

[Def. 2, Ths. 13 b, 15 a, 46 b, Def. 7 a]

THEOREM 51. *Let $S \in \mathfrak{A}$; if $A \subseteq S$, then*

$$\mathfrak{Ent}(A) = \sum_{X \in \mathfrak{P}(A).\,\mathfrak{E}} \mathfrak{Ent}(X).$$

Proof. We consider a sentence $y \in S$ such that $A + \{y\} \bar{\in} \mathfrak{W}$. Hence, by Th. 48 and with the help of the hypothesis, we conclude that $\mathfrak{P}(A + \{y\}).\,\mathfrak{E} - \mathfrak{W} \neq 0$. Thus there is a set Y which satisfies the formulas $Y \in \mathfrak{P}(A + \{y\}).\,\mathfrak{E}$ and $Y \bar{\in} \mathfrak{W}$. The first formula gives us $Y - \{y\} \in \mathfrak{P}(A).\,\mathfrak{E}$, and with the help of Th. 46 a obtain from the second formula $Y + \{y\} = (Y - \{y\}) + \{y\} \bar{\in} \mathfrak{W}$. Consequently there exists a set X such that $X \in \mathfrak{P}(A).\,\mathfrak{E}$ and $X + \{y\} \bar{\in} \mathfrak{W}$ (and in fact $X = Y - \{y\}$); thus we have

$$y \in \sum_{X \in \mathfrak{P}(A).\,\mathfrak{E}} \left(\underset{x}{E}[X + \{x\} \bar{\in} \mathfrak{W}] \right).$$

From this argument it follows immediately that

(1) $S.\underset{x}{E}[A + \{x\} \bar{\in} \mathfrak{W}] \subseteq \sum_{X \in \mathfrak{P}(A).\,\mathfrak{E}} \left(S.\underset{x}{E}[X + \{x\} \bar{\in} \mathfrak{W}] \right).$

According to Ax. 4 we have further

(2) $Cn(A) = \sum_{X \in \mathfrak{P}(A).\,\mathfrak{E}} Cn(X).$

Formulas (1) and (2) give at once

$Cn(A) + S.\underset{x}{E}[A + \{x\} \bar{\in} \mathfrak{W}] \subseteq \sum_{X \in \mathfrak{P}(A).\,\mathfrak{E}} \left(Cn(X) + S.\underset{x}{E}[X + \{x\} \bar{\in} \mathfrak{W}] \right),$

whence by Def. 7 a

(3) $\mathfrak{Ent}(A) \subseteq \sum_{X \in \mathfrak{P}(A).\,\mathfrak{E}} \mathfrak{Ent}(X).$

On the other hand it follows from Th. 50 b that

$$\mathfrak{Ent}(X) \subseteq \mathfrak{Ent}(A)$$

for every set $X \in \mathfrak{P}(A) . \mathfrak{E}$, whence

(4) $$\sum_{X \in \mathfrak{P}(A) . \mathfrak{E}} \mathfrak{Ent}(X) \subseteq \mathfrak{Ent}(A).$$

Finally, from (3) and (4) we obtain

$$\mathfrak{Ent}(A) = \sum_{X \in \mathfrak{P}(A) . \mathfrak{E}} \mathfrak{Ent}(X), \quad \text{q.e.d.}$$

Def. 7 b can be modified in various ways:

THEOREM 52. *The following conditions are equivalent:*

(α) $A \in \mathfrak{B}$; (β) $A \subseteq S$ and $\mathfrak{Q}(A) . \mathfrak{W} \subseteq \mathfrak{Aq}(A)$; ($\gamma$) $A \subseteq S$ and $\mathfrak{Q}(A) . \mathfrak{S} . \mathfrak{W} \subseteq \{Cn(A)\}$; ($\delta$) $\mathfrak{Q}(A) . \mathfrak{S} = \{S, Cn(A)\}$.

Proof. A. Let us assume formula (α) and consider any set $X \in \mathfrak{Q}(A) . \mathfrak{W}$. Thus according to Def. 6 $A \subseteq X \subseteq S$; hence by Def. 7 a, b and with the help of (α) we infer that $X \subseteq \mathfrak{Ent}(A)$ and that consequently for every $x \in X$ either $x \in Cn(A)$ or $A + \{x\} \overline{\in} \mathfrak{W}$; but since $A + \{x\} \subseteq X$, it follows from Th. 46 a that $A + \{x\} \in \mathfrak{W}$ and therefore $x \in Cn(A)$ for every $x \in X$. From this it follows that $X \subseteq Cn(A)$ and further that

$$A \subseteq X \subseteq Cn(A);$$

on the basis of Th. 1 a, b this last formula gives

$$Cn(A) \subseteq Cn(X) \subseteq Cn(A), \quad \text{i.e.} \quad Cn(A) = Cn(X),$$

whence by Def. 2 the formula $X \in \mathfrak{Aq}(A)$ follows.

From these considerations we conclude at once that

$$\mathfrak{Q}(A) . \mathfrak{W} \subseteq \mathfrak{Aq}(A);$$

by Def. 7 b, (α) gives also the inclusion $A \subseteq S$. It is thus proved that

(1) (β) *follows from* (α).

B. Further by Th. 18 b we have

(2) (β) *implies* (γ).

C. It is likewise easy by Ths. 47 and 9 a to establish that

(3) (δ) *follows from* (γ).

D. Now we assume that (δ) holds; from this it results immediately that $A \subseteq S$. Let further an arbitrary sentence $x \in S$ be given. By virtue of Th. 9 a we have

$$Cn(A+\{x\}) \in \mathfrak{Q}(A+\{x\}). \mathfrak{S}.$$

Thus *a fortiori* $Cn(A+\{x\}) \in \mathfrak{Q}(A). \mathfrak{S}$. Hence by ($\delta$) we obtain

$$Cn(A+\{x\}) \in \{S, Cn(A)\}.$$

If $Cn(A+\{x\}) = S$, then it follows from Th. 45 that $A+\{x\} \bar{\in} \mathfrak{W}$; but if

$$Cn(A+\{x\}) = Cn(A),$$

then, by Ax. 2, we have $x \in Cn(A+\{x\})$ and consequently $x \in Cn(A)$. Thus by Def. 7 a it follows in either case that

$$x \in \mathfrak{Ent}(A).$$

By means of this argument we reach the inclusion $S \subseteq \mathfrak{Ent}(A)$, which by Th. 50 a leads to the identity $\mathfrak{Ent}(A) = S$; and, in accordance with Def. 7 b, (α) is a consequence of this identity. Thus

(4)   (α) *follows from* (δ).

By combining (1)–(4) we infer at once that *the conditions* (α)–(δ) *are equivalent*, q.e.d.

THEOREM 53. *The following conditions are equivalent:* (α) $A \in \mathfrak{S}. \mathfrak{W}. \mathfrak{V}$; ($\beta$) $A \in \mathfrak{W}$ *and for every* $x \in S$ *either* $x \in A$ *or* $A+\{x\} \bar{\in} \mathfrak{W}$; ($\gamma$) $\mathfrak{Q}(A). \mathfrak{W} = \{A\}$.

Proof. A. By comparing Def. 7 a, b and Th. 7 we easily see that

(1)   (β) *follows from* (α).

B. Let us assume (β). We consider an arbitrary set

$$X \in \mathfrak{Q}(A). \mathfrak{W},$$

which thus, by Def. 6, satisfies the formula $A \subseteq X \subseteq S$. For every $x \in X$ we have $A+\{x\} \subseteq X$ and consequently, by Th. 46 a $A+\{x\} \in \mathfrak{W}$, which by virtue of (β) gives the formula $x \in A$; consequently we have $X \subseteq A$ and in fact $X = A$. Hence we conclude that $\mathfrak{Q}(A). \mathfrak{W} \subseteq \{A\}$; since the inverse inclusion follows directly from (β) we finally reach the formula (γ), i.e.

$$\mathfrak{Q}(A). \mathfrak{W} = \{A\}.$$

It is thus proved that

(2)   (γ) *follows from* (β).

C. Next we assume that (γ) holds. From this it results immediately that $A \in \mathfrak{W}$ and therefore $A \subseteq S$. By Ths. 9a and 18b we thus have $Cn(A) \in \mathfrak{Q}(A) . \mathfrak{S} . \mathfrak{Aq}(A)$; hence, with the help of Th. 46b, we obtain $Cn(A) \in \mathfrak{Q}(A) . \mathfrak{W}$. The comparison of the last two formulas with (γ) gives first

$$Cn(A) = A \in \mathfrak{S}$$

and secondly $\mathfrak{Q}(A) . \mathfrak{W} \subseteq \mathfrak{Aq}(A)$.

Now, since the condition (β) of Th. 52 is satisfied, we have also $A \in \mathfrak{B}$, so that finally the whole formula (α), $A \in \mathfrak{S} . \mathfrak{W} . \mathfrak{B}$, is derived. Accordingly we have

(3)   (γ) *implies* (α).

From (1)–(3) it follows at once that *the conditions* (α)–(γ) *are equivalent*. This completes the proof.

THEOREM 54. (a) *If* $A \in \mathfrak{B}$, *then* $\mathfrak{Q}(A) . \mathfrak{P}(S) \subseteq \mathfrak{B}$;

[Def. 7b, Th. 50a,b]

(b) *if* $A \in \mathfrak{B}$, *then* $\mathfrak{Aq}(A) \subseteq \mathfrak{B}$.   [Defs. 2, 7b, Th. 50c]

THEOREM 55. (a) $\mathfrak{P}(S) = \mathfrak{W} + \mathfrak{B}$;   [Th. 45, Defs. 6, 7a,b]

(b) $S \in \mathfrak{B}$.   [Ths. 47, 55a]

The following interesting theorem, due to Lindenbaum, asserts that (provided the set S is axiomatizable) every consistent set of sentences can be extended to a consistent and complete system.

THEOREM 56. *Let* $S \in \mathfrak{A}$; *if* $A \in \mathfrak{W}$, *then*

$$\mathfrak{Q}(A) . \mathfrak{S} . \mathfrak{W} . \mathfrak{B} \neq 0.$$

Proof. According to Def. 6 and Th. 47 the hypothesis gives $A \subseteq S$ and $A \neq S$, whence $S \neq 0$. Thus in view of Ax. 1 all sentences can be ordered in an infinite sequence of type ω (not necessarily with all terms distinct), such that

(1) $S = \underset{x_\lambda}{E}[\lambda < \omega]$.

Let

(2) λ_0 *be the smallest of the numbers* $\lambda < \omega$ *which satisfy the formula* $A + \{x_\lambda\} \in \mathfrak{W}$.

Further, let

(3) λ_ν (where $0 < \nu < \omega$) be the smallest of the numbers $\lambda < \omega$ which satisfy the conditions

$$A + \underset{x_{\lambda_\mu}}{E}[\mu < \nu] + \{x_\lambda\} \in \mathfrak{W} \quad and \quad \lambda > \lambda_\mu \text{ for every } \mu < \nu.$$

Finally, let

(4) π be the smallest ordinal number with which the conditions (2) and (3) correlate no number λ_π.

Now we put

(5) $$X_0 = A, \quad X_{\nu+1} = A + \underset{x_{\lambda_\mu}}{E}[\mu \leqslant \nu] \quad for \quad \nu < \pi$$

and

(6) $$X = X_0 + \sum_{\nu < \pi} X_{\nu+1}.$$

The conditions (2)–(4) give at once

(7) $$\pi \leqslant \omega.$$

From (4)–(6) we obtain

(8) $$A \subseteq X,$$

and in general

(9) $$X_0 \subseteq X_{\mu+1} \subseteq X_{\nu+1} \subseteq X \quad for \quad \mu < \nu < \pi.$$

From (2)–(5) we infer without difficulty that

(10) $$X_0 \in \mathfrak{W} \quad and \quad X_{\nu+1} \in \mathfrak{W} \quad for \quad \nu < \pi.$$

Let $\mathfrak{K} = \{X_0\} + \underset{X_{\nu+1}}{E}[\nu < \pi]$. By (9) and (7) the class $\mathfrak{K}$ satisfies the condition (α) of Cor. 5; by (10) the inclusion $\mathfrak{K} \subseteq \mathfrak{W}$ holds. Taking into account the formula $S \in \mathfrak{A}$ assumed in the hypothesis we see that all premisses of Th. 49 are satisfied. Accordingly we have $\sum_{Y \in \mathfrak{K}} Y \in \mathfrak{W}$, or, by (6),

(11) $$X \in \mathfrak{W}.$$

From (4)–(6) we obtain the formula

(12) $$X = A + \underset{x_{\lambda_\mu}}{E}[\mu < \pi].$$

By an indirect argument (analogous to the one used in refuting the assumption (9) in the proof of Th. 16) we easily infer from

(2)–(5) that for every number $\lambda < \omega$ which is distinct from all numbers λ_μ with $\mu < \pi$ either $X_0 + \{x_\nu\} \bar\in \mathfrak{W}$ holds or there is a number $\nu < \pi$ which satisfies the formula

$$X_{\nu+1} + \{x_\lambda\} \bar\in \mathfrak{W}.$$

By virtue of (9) and Th. 46 a we have *a fortiori* $X + \{x_\nu\} \bar\in \mathfrak{W}$. Hence by (1) and (12) we obtain

(13)   *for every $x \in S$ either $x \in X$ or $X + \{x\} \bar\in \mathfrak{W}$.*

According to (11) and (13) the set X satisfies the condition (β) of Th. 53 (for $A = X$); consequently,

(14) $X \in \mathfrak{S} . \mathfrak{W} . \mathfrak{V}.$

The formulas (8) and (14) at once give

$$\mathfrak{Q}(A) . \mathfrak{S} . \mathfrak{W} . \mathfrak{V} \neq 0, \quad \text{q.e.d.}$$

Besides the concept of completeness, and perhaps more often than this concept (especially with regard to logically more complicated and comprehensive disciplines), two other concepts are treated in metamathematics which are related in content to completeness, although they are logically weaker and thus more general, viz. *non-ramifiability* and *categoricity* (monomorphy).[1] These concepts are not reducible to those of sentence and consequence.† The definition and the establishment of the fundamental properties and of the mutual connexions of these concepts, as well as the clarification of their relation to the concept of completeness, must be left to special metadisciplines.

§ 8. CARDINAL AND ORDINAL DEGREE OF COMPLETENESS

In order to obtain a classification and characterization of incomplete sets of sentences, we introduce here the concept of the degree of completeness of a set of sentences, and we do this in two ways, namely by correlating a cardinal number and an ordinal number with every set of sentences. The *cardinal degree*

---

[1] Information about these concepts is given in Fraenkel, A. (16), pp. 347–54, where the literatute on the subject is also listed. For the definitions of these notions see also X, pp. 310–314 ff., and XIII, p. 390.

† Nor can they be defined in terms of those notions which are discussed in XII.

of completeness of the set A of sentences, symbolically $g(A)$, is the number of all systems which include the set A; the *ordinal degree of completeness*, in symbols $\gamma(A)$, is identical with the smallest ordinal number π for which there is no strictly increasing sequence of type π of consistent systems which include the set A. In formulas we thus have

DEFINITION 8. (a) $g(A) = \overline{\overline{\mathfrak{Q}(A).\mathfrak{S}}}$;

(b) $\gamma(A)$ *is the smallest ordinal number* π *for which there is no sequence of type* π *of sets* X_ν *which satisfy the formulas:*

(α) $X_\nu \in \mathfrak{Q}(A).\mathfrak{S}.\mathfrak{W}$ *for* $\nu < \pi$,   *where* $A \subseteq S$,

and

(β) $X_\mu \subseteq X_\nu$ *and* $X_\mu \neq X_\nu$ *for* $\mu < \nu < \pi$.

Def. 8 b may be transformed as follows:

THEOREM 57. $\gamma(A)$ *is identical with the smallest ordinal number* π *for which there is no sequence of type* π *of sentences* x_ν *which satisfy the formula* (α) $x_\nu \in S - Cn\left(A + \underset{x_\mu}{E}[\mu < \nu]\right)$ *for* $\nu < \pi$, *where* $A \subseteq S$.

Proof. Let a number ξ be given such that

(1) $\xi < \gamma(A)$.

By Def. 8 b there is a sequence of sets X_ν of type ξ which satisfy the formulas:

(2) $X_\nu \in \mathfrak{Q}(A).\mathfrak{S}.\mathfrak{W}$ *for* $\nu < \xi$,   *where* $A \subseteq S$;

(3) $X_\mu \subseteq X_\nu$ and $X_\mu \neq X_\nu$ for $\mu < \nu < \xi$.

We put further

(4) $X_\xi = S$ (*in case* ξ *is not a limit number*).

By Ths. 45 and 47 the formulas (2) and (4) give

(5) $X_\nu \subseteq S = X_\xi$ and $X_\nu \neq S = X_\xi$ for $\nu < \xi$.

From (3) and (5) we infer at once that the sets $X_{\nu+1} - X_\nu$, where $\nu < \xi$, are distinct from 0; hence with every such set one of its elements x_ν can be correlated in such a way that

(6) $x_\nu \in X_{\nu+1} - X_\nu$ *for* $\nu < \xi$.

(This correlation does not require the axiom of choice since by

(5) and Ax. 1 all sets $X_{\nu+1}-X_\nu$ with $\nu < \xi$ are subsets of the at most denumerable set S.)

It follows from (2), (3), and (6) that $A+\underset{x_\mu}{E}[\mu < \nu] \subseteq X_\nu$ for every $\nu < \xi$; hence with the help of (2) and Th. 8 we obtain $Cn\big(A+\underset{x_\mu}{E}[\mu < \nu]\big) \subseteq X_\nu$. But since by (5) and (6) $x_\nu \in S-X_\nu$, we finally have

(7) $\quad x_\nu \in S-Cn\big(A+\underset{x_\mu}{E}[\mu < \nu]\big)$ for $\nu < \xi$ (where $A \subseteq S$).

With this we have proved that (1) always implies (7): for every number $\xi < \gamma(A)$ a sequence of sentences of type ξ can be constructed which satisfies formula (α) of the theorem here discussed. But since by hypothesis there is no such sequence for the number π, the inequality $\pi < \gamma(A)$ cannot hold. Consequently,

(8) $\qquad\qquad\qquad\qquad \pi \geqslant \gamma(A)$.

By an analogous argument the inverse inequality

(9) $\qquad\qquad\qquad\qquad \pi \leqslant \gamma(A)$

can be proved. For let us consider an arbitrary number $\xi < \pi$. According to the hypothesis there is a sequence of sentences x_ν of type ξ which satisfies the formula (7). By putting

$$X_\nu = Cn\big(A+\underset{x_\mu}{E}[\mu < \nu]\big) \text{ for } \nu < \xi,$$

we easily conclude from Ths. 9 a, 47, and 1 a that the sequence of sets X_ν satisfies the formulas (2) and (3). But by Def. 8 b no such sequence of sets can be constructed for the number $\gamma(A)$. Consequently the inequality $\gamma(A) < \pi$ cannot hold and we obtain the formula (9).

Formulas (8) and (9) at once give the required identity

$$\gamma(A) = \pi, \quad \text{q.e.d.}$$

THEOREM 58. (a) If $A \subseteq B$, then

$$\mathfrak{g}(A) \geqslant \mathfrak{g}(B) \quad and \quad \gamma(A) \geqslant \gamma(B); \qquad [\text{Def. } 8\,a, b]$$

(b) if $A \in \mathfrak{Aq}(B)$, then $\mathfrak{g}(A) = \mathfrak{g}(B)$ and $\gamma(A) = \gamma(B)$.

Proof of (b). By Def. 2 the hypothesis gives $Cn(A) = Cn(B)$. From this by Th. 8 we infer that for every set $X \in \mathfrak{S}$ the inclusions $A \subseteq X$ and $B \subseteq X$ are equivalent, whence

$$\mathfrak{Q}(A).\,\mathfrak{S} = \mathfrak{Q}(B).\,\mathfrak{S}.$$

By virtue of Def. 8 a, b we obtain at once from the last formula the two required identities:

$$g(A) = g(B) \quad \text{and} \quad \gamma(A) = \gamma(B).$$

It is left to the reader to establish the connexions between the numbers $g(A)$ and $g(B)$, or $\gamma(A)$ and $\gamma(B)$, under more special assumptions (e.g. $A \in \mathfrak{P}(B) - \mathfrak{Aq}(B)$).

In the following theorem a characterization is given of the most important categories of sets of sentences, in fact of consistent and complete sets, in terms of the notions just defined.

THEOREM 59. (a) *The formulas* $A \subseteq S$, $g(A) \geqslant 1$ *and* $\gamma(A) \geqslant 1$ *are equivalent;*   [Def. 8 a, b, Th. 10][1]

(b) *the formulas* $A \in \mathfrak{B}$, $1 \leqslant g(A) \leqslant 2$, *and* $1 \leqslant \gamma(A) \leqslant 2$ *are equivalent;*   [Def. 8 a, b, Ths. 52, 10, 47, 9 a, 46 a]

(c) *the formulas* $A \in \mathfrak{W}$, $g(A) \geqslant 2$, *and* $\gamma(A) \geqslant 2$ *are equivalent.*
                                          [Def. 8 a, b, Ths. 9 a, 10, 45, 47, 46 a]

The proof is quite elementary.

COROLLARY 60. (a) *The formulas*

$$A \in \mathfrak{Aq}(S) \ (or \ A \in \mathfrak{P}(S) - \mathfrak{W}), \ g(A) = 1, \ and \ \gamma(A) = 1$$

are equivalent;   [Th. 59 a, c, Defs. 2, 6]

(b) *the formulas* $A \in \mathfrak{W}.\mathfrak{B}$, $g(A) = 2$, *and* $\gamma(A) = 2$ *are equivalent;*   [Th. 59 b, c]

(c) *the formulas* $A \in \mathfrak{W} - \mathfrak{B}$, $g(A) \geqslant 3$, *and* $\gamma(A) \geqslant 3$ *are equivalent.*   [Th. 59 b, c]

THEOREM 61. *For every set* A *we have*

$$g(A) \leqslant 2^{\aleph_0} \cdot and \quad \gamma(A) \leqslant \Omega.$$

Proof. By Def. 1 we have $\mathfrak{S} \subseteq \mathfrak{P}(S)$; hence by Def. 8 a it follows that $g(A) \leqslant \overline{\overline{\mathfrak{P}(S)}}$. On the other hand (by a well-known theorem on the cardinal number of the power set) Ax. 1 implies the formula $\overline{\overline{\mathfrak{P}(S)}} \leqslant 2^{\aleph_0}$. Thus we finally reach the required inequality $g(A) \leqslant 2^{\aleph_0}$.

If $\gamma(A) > \Omega$ were the case, then by Defs. 8 b and 6 there would exist an increasing sequence (without repeating terms) of the type Ω of subsets of the set S; but this contradicts Ax. 1

[1] It is assumed that in the case of the hypothesis $A \subseteq S$ at least the empty sequence of type 0 satisfies (vacuously) the conditions (α) and (β) of Def. 8 b.

according to which the set S is at most denumerable. Consequently $\gamma(A) \leqslant \Omega$, and with that the proof is complete.

THEOREM 62. *If $A \subseteq S$ and if there is a set X such that $X \bar{\in} \mathfrak{C}$ and $x \in S - Cn\big(A + (X - \{x\})\big)$ for every $x \in X$, then*

$$\mathfrak{g}(A) = 2^{\aleph_0} \quad and \quad \gamma(A) = \Omega.$$

Proof. Applying Ths. 7 and 32, the following is obtained from Cor. 37:

(1) *if there is a set X such that $X \bar{\in} \mathfrak{C}$ and $x \in S - Cn(X - \{x\})$ for every $x \in X$, then*

$$\overline{\overline{\underset{Y}{E}[Cn(Y) = Y \subseteq S]}} = 2^{\aleph_0}.$$

We put

(2) $\qquad F(Y) = Cn(A + Y) \quad$ *for every set $Y \subseteq S$.*

By the relativization theorem, Th. 6, all the axioms postulated in § 1 remain valid if everywhere in them 'Cn' is replaced by the symbol 'F' just defined (although the variables in the axioms must also be renamed). Therefore all consequences of these axioms also remain valid (cf. the remarks after Th. 6). This applies in particular to (1) which, by the transformation decribed above, and in view of (2), becomes

(3) *if there is a set X such that $X \bar{\in} \mathfrak{C}$ and $x \in S - Cn\big(A + (X - \{x\})\big)$ for every $x \in X$, then*

$$\overline{\overline{\underset{Y}{E}[Cn(A + Y) = Y \subseteq S]}} = 2^{\aleph_0}.$$

From (3) with the help of the hypothesis we obtain

(4) $\qquad \overline{\overline{\underset{Y}{E}[Cn(A + Y) = Y \subseteq S]}} = 2^{\aleph_0}.$

By means of Ths. 7 and 9a it is now easy to show that the formulas $Cn(A + Y) = Y \subseteq S$ and $Y \in \mathfrak{Q}(A) . \mathfrak{S}$ are equivalent. Accordingly it follows from (4) that $\overline{\overline{\mathfrak{Q}(A) . \mathfrak{S}}} = 2^{\aleph_0}$; and hence by means of Def. 8a we obtain immediately the required formula

(5) $\qquad\qquad\qquad \mathfrak{g}(A) = 2^{\aleph_0}.$

We now consider an arbitrary number ξ such that

(6) $\qquad\qquad\qquad\qquad \xi < \Omega,$

thus $\bar{\xi} \leqslant \aleph_0$. By Ax. 1 the set $X \subseteq S$ involved in the hypothesis is countable. Hence a sequence of type ξ can be constructed consisting entirely of distinct sentences x_ν of the set X and thus satisfying the formulas:

(7) $\quad \underset{x_\nu}{E}[\nu < \xi] \subseteq X \subseteq S \quad and \quad x_\mu \neq x_\nu \quad for \quad \mu < \nu < \xi.$

Let ν be an arbitrary number $< \xi$. From (7) we infer that $A + \underset{x_\mu}{E}[\mu < \nu] \subseteq A + (X - \{x_\nu\}) \subseteq S$, whence by Th. $1a$ the inclusion

$$Cn\Big(A + \underset{x_\mu}{E}[\mu < \nu]\Big) \subseteq Cn(A + (X - \{x_\nu\}))$$

follows. But since by virtue of the hypothesis

$$x_\nu \in S - Cn(A + (X - \{x_\nu\})),$$

we have further

(8) $\qquad x_\nu \in S - Cn\Big(A + \underset{x_\mu}{E}[\mu < \nu]\Big) \quad for\ every\ \nu < \xi.$

Thus (6) implies (8): for every number $\xi < \Omega$ there is a sequence of sentences of type ξ which satisfies the formula (8). But according to Th. 57 there exists no such sequence for the number $\gamma(A)$. Consequently $\gamma(A) \geqslant \Omega$, and comparing this formula with Th. 61 we obtain at once

(9) $\qquad\qquad\qquad \gamma(A) = \Omega.$

The formulas (5) and (9) form the conclusion of the theorem.

The theorem and the next corollary immediately following were obtained jointly by Lindenbaum and the author.

THEOREM 63. (a) If $\gamma(A) \leqslant \omega$, then $g(A) \leqslant \aleph_0$;
$\qquad\qquad$ (b) if $\gamma(A) \geqslant \omega$, then $g(A) \geqslant \aleph_0$.

Proof. (a) If A is not a subset of S, then by Th. $59a$ $g(A) = 0$, and so $g(A) \leqslant \aleph_0$. Hence, we restrict ourselves to the case where

(1) $\qquad\qquad\qquad A \subseteq S.$

We put

(2) $\qquad F(X) = Cn(A + X) \quad for\ every\ set\ X \subseteq S.$

By an easy argument (which is completely analogous with the derivation of (3) in the previous proof) we infer from Th. 6 with the help of (1) and (2) that Axs. 1–4 as well as all their consequences remain valid if 'Cn' is everywhere replaced by 'F'. In particular, by Def. 2, Th. 16 can be transformed in the following way:

(3) *for every set $X \subseteq S$ there is a corresponding sequence of sentences x_ν of type $\pi \leqslant \omega$ which satisfies the formulas*

$$(\alpha)\ x_\nu \in S - F\Big(\underset{x_\mu}{E}[\mu < \nu]\Big)\ \ for\ \nu < \pi$$

and $$(\beta)\ F\Big(\underset{x_\nu}{E}[\nu < \pi]\Big) = F(X).$$

By Th. 57 the type π of every sequence of sentences which satisfies the formula (α) of the condition (3) is less than $\gamma(A)$, thus according to the hypothesis $< \omega$. Further by Th. 7 it follows from (2) that for every set $X \in \mathfrak{Q}(A) . \mathfrak{S}$ the formula $F(X) = X$ holds. In view of this we obtain from (3)

(4) *for every set $X \in \mathfrak{Q}(A) . \mathfrak{S}$ there is a corresponding sequence of sentences $x_\nu \in S$ of type $\pi < \omega$ which satisfies the formula*

$$F\Big(\underset{x_\nu}{E}[\nu < \pi]\Big) = X.$$

According to Ax. 1 the set S is at most denumerable; hence, as is well known, every set of finite sequences consisting only of elements of S is likewise at most denumerable.

Now (4) shows that the function F maps such a set of sequences onto the class $\mathfrak{Q}(A) . \mathfrak{S}$; consequently this class is also at most denumerable. From this by Def. 8 a it follows immediately that

$$\mathfrak{g}(A) \leqslant \aleph_0, \quad \text{q.e.d.}$$

(b) By Th. 57 the hypothesis implies that

(5) $$A \subseteq S,$$

and that, with every number $\pi < \omega$ (thus $< \gamma(A)$), a sequence of sentences $x_\nu^{(\pi)}$ of type π can be correlated so that

(6) $$x_\nu^{(\pi)} \in S - Cn\Big(A + \underset{x_\mu^{(\pi)}}{E}[\mu < \nu]\Big)\ \ for\ \ \nu < \pi < \omega.$$

(The axiom of choice is not used in this step, for, as we have already mentioned, the set of all finite sequences of sentences is at most denumerable.)

We now put

(7) $X_\nu^{(\pi)} = Cn\Big(A + \underset{x_\mu^{(\pi)}}{E}[\mu < \nu]\Big)$ *for* $\nu < \pi < \omega$,

and let

(8) $\mathfrak{R} = \underset{X_\nu^{(\pi)}}{E}[\nu < \pi < \omega]$.

By the use of Th. 9 a it easily results from (4)–(8) that

(9) $\mathfrak{R} \subseteq \mathfrak{Q}(A) . \mathfrak{S}$.

From (8) it follows at once that $\overline{\overline{\mathfrak{R}}} \leqslant \aleph_0$. But on the other hand the class $\mathfrak{R}$ is infinite. For by Ax. 2 we infer without difficulty from (5)–(7) that $x_\mu^{(\pi)} \in X_\nu^{(\pi)} - X_\mu^{(\pi)}$ and therefore $X_\mu^{(\pi)} \neq X_\nu^{(\pi)}$ for $\mu < \nu < \pi < \omega$; hence by (8) we obtain $\overline{\overline{\mathfrak{R}}} \geqslant \bar{\pi}$ for every $\pi < \omega$. Accordingly

(10) $\overline{\overline{\mathfrak{R}}} = \aleph_0$.

In view of Def. 8 a the formulas (9) and (10) give the required inequality

$$\mathfrak{g}(A) \geqslant \aleph_0.$$

COROLLARY 64. (a) *If* $\gamma(A) = \omega$, *then* $\mathfrak{g}(A) = \aleph_0$;

[Th. 63 a, b]

(b) *if* $\mathfrak{g}(A) < \aleph_0$, *then* $\gamma(A) < \omega$;   [Th. 63 b]

(c) *if* $\mathfrak{g}(A) > \aleph_0$, *then* $\gamma(A) > \omega$.   [Th. 63 a]

COROLLARY 65. *For every set* $A \subseteq S$ *we have*

either $1 \leqslant \mathfrak{g}(A) \leqslant \aleph_0$ *or* $\aleph_0 \leqslant \mathfrak{g}(A) \leqslant 2^{\aleph_0}$.

[Ths. 59 a, 61, 63 a, b]

From this corollary Th. 30 is at once derivable as a special case (for $A = 0$).

THEOREM 66. *If* $\gamma(A) \neq \Omega$, *then* $\overline{\gamma(A)} \leqslant \mathfrak{g}(A)$.

Proof. In the case $\gamma(A) = 0$ the conclusion is obvious. If $0 < \gamma(A) < \omega$, then the class $\mathfrak{Q}(A) . \mathfrak{S}$, by virtue of Def. 8 a, contains at least $\overline{\gamma(A) - 1}$ consistent systems and moreover, by Th. 47, one contradictory system, namely S. Thus we have $\overline{\overline{\mathfrak{Q}(A) . \mathfrak{S}}} \geqslant \overline{\gamma(A) - 1} + 1 = \overline{\gamma(A)}$, which by Def. 8 b gives the formula $\overline{\gamma(A)} \leqslant \mathfrak{g}(A)$. Finally, if we have $\omega \leqslant \gamma(A) < \Omega$, then $\overline{\gamma(A)} = \aleph_0$ and, by Th. 63 b, $\mathfrak{g}(A) \geqslant \aleph_0$; from this again the conclusion follows: $\overline{\gamma(A)} \leqslant g(A)$. Since by Th. 61 and the hypothesis $\gamma(A) < \Omega$, all possible cases have been dealt with and so the theorem is proved.

It remains undecided whether this theorem can be extended to the case $\gamma(A) = \Omega$.

The following theorem is of a more special nature.

THEOREM 67. *Let $S \in \mathfrak{A}$; then, if π is a limit number, we have $\gamma(A) \neq \pi+1$ for every set A.*

Proof. Let us suppose, in contradiction with the conclusion, that for a certain set A

$$(1) \qquad\qquad \gamma(A) = \pi+1.$$

Hence, by Def. 8 b, for the number π (as well as for every number $\xi < \pi+1$) there exists a sequence of sets X_ν of type π which satisfies the formulas:

$$(2) \qquad\qquad X_\nu \in \mathfrak{Q}(A).\mathfrak{S}.\mathfrak{W} \quad for \quad \nu < \pi,$$

$$(3) \qquad X_\mu \subseteq X_\nu \quad and \quad X_\mu \neq X_\nu \quad for \quad \mu < \nu < \pi.$$

Let us put

$$(4) \qquad\qquad X_\pi = \sum_{\nu < \pi} X_\nu.$$

Let $\mathfrak{R} = \underset{X_\nu}{E}[\nu < \pi]$. Since $\pi \neq 0$, we also have $\mathfrak{R} \neq 0$; from this by (3) we conclude that the class $\mathfrak{R}$ satisfies condition (α) of Cor. 5. Thus, by (2), $\mathfrak{R} \subseteq \mathfrak{S}.\mathfrak{W}$ holds. Consequently we can apply Ths. 12 and 49 to this class; thus by the formula $S \in \mathfrak{A}$ assumed in the hypothesis we obtain $\sum_{X \in \mathfrak{R}} X \in \mathfrak{S}.\mathfrak{W}$. But since, by (2) and (4), $A \subseteq X_\pi = \sum_{X \in \mathfrak{R}} X$, we have $X_\pi \in \mathfrak{Q}(A).\mathfrak{S}.\mathfrak{W}$, and accordingly (2) can be generalized in the following way:

$$(5) \qquad\qquad X_\nu \in \mathfrak{Q}(A).\mathfrak{S}.\mathfrak{W} \quad for \quad \nu < \pi+1.$$

Since by hypothesis π is a limit number, (4) provides an analogous generalization of (3):

$$(6) \qquad X_\mu \subseteq X_\nu \quad and \quad X_\mu \neq X_\nu \quad for \quad \mu < \nu < \pi+1.$$

In consequence of Def. 8 b the existence of a sequence of

sets (of type $\pi+1$) satisfying the formulas (5) and (6) contradicts the assumption (1). We must have therefore

$$\gamma(A) \neq \pi+1, \quad \text{q.e.d.}$$

The results obtained here which concern the notions $g(A)$ and $\gamma(A)$ are rather fragmentary. Only by restricting ourselves to those deductive disciplines which presuppose the sentential calculus are we able to obtain more complete results. In particular we can then show that, for every set A of sentences, $g(A)$ is either finite or equals $2^{\aleph_0}$, and similarly $\gamma(A)$ is either finite or equals Ω.

In general, when developing the metamathematics of deductive disciplines which presuppose the sentential calculus, we chiefly concentrate upon the same notions which have been discussed in the present paper. Since, however, the development is based upon logically stronger (though more special) assumptions, various results obtained here can be supplemented and improved.

VI

ON DEFINABLE SETS OF REAL NUMBERS†

MATHEMATICIANS, in general, do not like to operate with the notion of definability; their attitude towards this notion is one of distrust and reserve. The reasons for this aversion are quite clear and understandable. First, the meaning of this notion is not at all precise: a given object may or may not be definable with respect to the deductive system in which it is studied, the rules of definition which are adopted, and the terms that are taken as primitive. It is thus permissible to use the notion of definability only in a relative sense. This fact has often been neglected in mathematical considerations and has been the source of numerous contradictions, of which the classical example is furnished by the well-known antinomy of Richard.[1] The distrust of mathematicians towards the notion in question is reinforced by the current opinion that this notion is outside the proper limits of mathematics altogether. The problems of making its meaning more precise, of removing the confusions and misunderstandings connected with it, and of establishing its fundamental properties belong to another branch of science —metamathematics.

In this article I shall try to convince the reader that the opinion just mentioned is not altogether correct. Without doubt

[1] Cf. Fraenkel, A. (16) where precise bibliographical references will be found.

† BIBLIOGRAPHICAL NOTE. The main ideas of this article originated in 1929 and were developed by the author in his talk to the Polish Mathematical Society, Section of Lwów, in 1930; see a summary of this talk under the title 'Über definierbare Mengen reeller Zahlen', *Annales de la Société Polonaise de Mathématique*, vol. 9 (1930), pp. 206–7. The article itself first appeared under the title 'Sur les ensembles définissables de nombres réels. I.' in *Fundamenta Mathematicae*, vol. 17 (1931), pp. 210–39. The present English edition has been supplemented by some new passages which contain elaborations of the original text. The most important addition has been made in § 2, after Th. 6; see in this connexion footnote † on p. 120.

the notion of definability as usually conceived is of a meta-mathematical nature. I believe, however, that I have found a general method which, with a reservation to be discussed at the end of § 1, allows us to reconstruct this notion in the domain of mathematics. This method is also applicable to certain other notions of a metamathematical nature. The reconstructed concepts do not differ at all from other mathematical notions and need arouse neither fears nor doubts; their study remains entirely within the domain of normal mathematical reasoning. Finally, it seems to me that this method allows us to derive certain results which could not be obtained by operating only with the metamathematical conception of the notions studied.

A description of the method in question which is quite general and abstract would involve certain technical difficulties, and, if given at the outset, would lack that clarity which I should like it to have. For this reason I prefer in this article to restrict consideration to a special case, one which is particularly important from the point of view of the questions which interest mathematicians at the present time. I shall in particular analyse the notion of definability of a single category of objects, namely, sets of real numbers. Moreover, my considerations will be in the nature of a sketch; I shall content myself simply with constructing some precise definitions, either omitting the consequences which follow from them, or presenting them without demonstration.[1]

§ 1. The Concept of Definable Sets of Real Numbers from the Metamathematical Standpoint

The problem set in this article belongs in principle to the type of problems which frequently occur in the course of mathematical investigations. Our interest is directed towards a term of which we can give an account that is more or less precise in its intuitive content, but the significance of which has not at present been rigorously established, at least in mathematics.

---

[1] In this connexion it should be noted that an analogous method can be profitably applied for the definition of many concepts in the field of meta-mathematics, e.g. that of *true sentence* or of a *universally valid sentential function*. Cf. VIII.

We then seek to construct a definition of this term which, while satisfying the requirements of methodological rigour, will also render adequately and precisely the actual meaning of the term. It was just such problems that the geometers solved when they established the meaning of the terms 'movement', 'line', 'surface', or 'dimension' for the first time. Here I present an analogous problem concerning the term 'definable set of real numbers'.

Strictly speaking this analogy should not be carried too far. In geometry it was a question of making precise the spatial intuitions acquired empirically in everyday life, intuitions which are vague and confused by their very nature. Here we have to deal with intuitions more clear and conscious, those of a logical nature relating to another domain of science, metamathematics. To the geometers the necessity presented itself of choosing one of several incompatible meanings, but here arbitrariness in establishing the content of the term in question is reduced almost to zero.

I shall begin then by presenting to the reader the content of this term, especially as it is now understood in metamathematics. The remarks I am about to make are not at all necessary for the considerations that will follow—any more than empirical knowledge of lines and surfaces is necessary for a mathematical theory of geometry. These remarks will allow us to grasp more easily the constructions explained in the following section and, above all, to judge whether or not they convey the actual meaning of the term. I shall confine myself to sketching this matter briefly without attempting too much precision and rigour in its formulation.

As already mentioned, the notion of definability should always be relativized to the deductive system in which the investigation is carried out. Now in our case it is quite immaterial which of the possible systems of the arithmetic of real numbers is chosen for discussion. It would be possible, for example, to regard arithmetic as a certain chapter of mathematical logic, without separate axioms and primitive terms. But it will be more advantageous here to treat arithmetic as an independent deductive science, forming as it were a superstructure of logic.

The construction of this science may be thought of more or less in the following way: as a basis we admit some system of mathematical logic, without altering its rules of inference and of definition; we then enlarge the system of primitive terms and axioms by the addition of those which are specific to arithmetic. As a suitable basis we can use, for example, the system developed in *Principia Mathematica* by Whitehead and Russell.[1] In order to avoid needless complications it is, however, convenient to subject this system first to certain simplifications, especially to simplify the language, taking care of course not to diminish its capacity for expressing all the ideas which could be formulated in the original system. These modifications concern the following points: the ramified theory of types is replaced by the simplified theory, so that the axiom of reducibility is rejected;[2] also the axiom of extensionality is adopted. All constant terms, except the primitive terms, are eliminated. Finally, we shall suppress sentential variables and variables representing two- and many-termed relations.[3]

In view of this last point we shall only have variables in the system which represent (1) individuals, that is to say objects of the first order (and in particular real numbers, which are to be regarded as individuals); (2) sets (classes) of individuals, i.e. objects of the second order; (3) sets (families) of these sets, i.e. objects of third order, etc. It is desirable, further, to fix exactly the form of the signs which are to serve as variables, and especially to distinguish them according to the order of the objects which they represent. Thus we can agree to employ the signs $'x_l'$, $'x_l''$,..., $'x_l^{(k)}'$,... as variables of order 1; $'x_n''$, $'x_n''$,..., $'x_n^{(k)}'$,... as variables of order 2; and, in general, the signs

[1] Whitehead, A. N., and Russell, B. A. W. (90).

[2] Cf. Chwistek, L. (14) and Carnap, R. (8).

[3] An arbitrary two-termed relation R can be replaced in all discussions by the set of ordered couples $[x, y]$ satisfying the formula xRy; ordered couples can also be interpreted as certain sets (cf. Kuratowski, C. (38); Chwistek, L. (13)). In an analogous fashion many-termed relations can be eliminated. As regards sentential variables they are completely avoided in Neumann, J. v. (54).

'$x'_{(l)}$' '$x''_{(l)}$',..., '$x^{(k)}_{(l)}$',... as variables of order l, taking as values objects of the same order.[1]

As primitive terms of the system of logic it is convenient to adopt the sign of *negation* '$\overline{}$'; the signs of *logical sum* '$+$' and of *logical product* '$.$'; the quantifiers: *universal* '$\prod$' and *existential* '$\sum$'; and finally the sign '$\in$' of *set membership*. The meaning of these signs does not require further remark.[2] It now remains to add to this system the specific primitive terms of the arithmetic of real numbers. The three signs 'v', 'μ', and 'σ' are known to suffice for the definition of all the concepts of this science. The sentential functions '$v(x^{(k)}_r)$', '$\mu(x^{(k)}_r, x^{(l)}_r)$' and '$\sigma(x^{(k)}_r, x^{(l)}_r, x^{(m)}_r)$' assert respectively that $x^{(k)}_r = 1$; $x^{(k)}_r$ and $x^{(l)}_r$ are real numbers such that $x^{(k)}_r \leqslant x^{(l)}_r$; and $x^{(k)}_r, x^{(l)}_r$, and $x^{(m)}_r$ are real numbers such that $x^{(k)}_r = x^{(l)}_r + x^{(m)}_r$.

In addition to variables and constants it is necessary to have some technical signs in the system, namely *left* '(' and *right* ')' *parentheses*. From these three kinds of sign a great variety of expressions can be formed. *Sentential functions* constitute an especially important category.[3] In order to make this notion precise we distinguish first the *primitive sentential functions*, namely expressions of the type '$(x^{(k)}_{(m)} \in x^{(l)}_{(m+1)})$', '$v(x^{(k)}_r)$', '$\mu(x^{(k)}_r, x^{(l)}_r)$', and '$\sigma(x^{(k)}_r, x^{(l)}_r, x^{(m)}_r)$'. Then we consider certain operations on expressions called the *fundamental operations*. These are the operation of negation which when applied to a given expression 'p' yields its *negation* '$\bar{p}$', the operations of logical addition and multiplication, which form from two expressions 'p' and 'q', their *logical sum* '$(p+q)$' and *logical*

---

[1] I use the symbols '$x^{(k)}_r$',..., '$x''_{(l)}$',..., '$x^{(k)}_{(l)}$',... only as schemata: in every particular case, instead of the indices '(k)' and '(l)' there is an appropriate number of small strokes.

[2] The signs '$\prod$' (or 'Σ') and '$.$' (or '$+$') are not indispensable, for they can be defined by means of other signs. In principle the sign '$\in$' is also super-fluous, for in the place of such expressions as '$x \in X$' ('x is an element of the set X') we can use equivalent expressions of the form '$X(x)$' ('x has the property X', or 'x satisfies the condition X'); I prefer here, however, to conform to the current language of mathematical works.

[3] I especially emphasize that I constantly use the term 'sentential function' as a metamathematical term, denoting expressions of one category (for this term is sometimes interpreted in a logical sense, a meaning being given to it like that of the terms 'property', 'condition', or 'class').

product '$(p.q)$'. Finally, the operations of universal and of existential quantification which from the expression 'p' and the variable '$x_{(l)}^{(k)}$' yield the *universal quantification* '$\prod\limits_{x_{(l)}^{(k)}} p$' and the *existential quantification* '$\sum\limits_{x_{(l)}^{(k)}} p$' of '$p$' with respect to the variable '$x_{(l)}^{(k)}$'.

The *sentential functions* are those expressions which can be obtained from the primitive functions by applying to them, in any order and any (finite) number of times, the five fundamental operations. In other words the set of all sentential functions is the smallest set of expressions which contains as elements all the primitive functions and is closed with respect to the above operations. The following are examples of sentential functions:

$$\sum_{x_i''} \sigma(x_i'', x_i', x_i'), \quad \prod_{x_i'} ((x_i' \in x_n') + (x_i' \in x_n')), \quad \sum_{x_{i''}'} \prod_{x_{i'}'} (x_n' \in x_m').$$

Within this style of notation (which clearly departs from the symbolism of *Principia Mathematica*[1]), we introduce in practice many simplifications. Thus we shall use as variables of the three lowest orders the signs '$x^{(k)}$', '$X^{(k)}$', and '$\mathscr{X}^{(k)}$' respectively. In the place of 'x''', 'x'''', 'x''''', 'x'''''' we write 'x', 'y', 'z', 'u', and similarly for the two following orders. As negations of the primitive sentential functions we write

'$(x_{(m)}^{(k)} \bar{\in} x_{(m+1)}^{(l)})$', '$\bar{v}(x^{(k)})$', '$\bar{\mu}(x^{(k)}, x^{(l)})$', and '$\bar{\sigma}(x^{(k)}, x^{(l)}, x^{(m)})$'.

Finally, setting aside the mechanical way of using parentheses exemplified above in the construction of sentential functions,[2] we omit parentheses everywhere so long as no misunderstanding will arise.

We can classify sentential functions according to the order of the variables which occur in them. The function which contains at least one variable of order n, but no variable of higher order, is called a sentential function of order n; thus in the examples given above we have functions of orders 1, 2, and 3.

---

[1] On the other hand our notation agrees with that of Schröder, E. (62).

[2] This method is due to Lewis (cf. Lewis, C. I. (48), p. 357). By using the idea of Łukasiewicz it is possible to avoid the use of parentheses completely in this system: for this purpose it suffices to use the expressions '$\in xX$', 'vx', 'μxy', 'σxyz', '$+pq$', and '$.pq$', instead of '$(x \in X)$', '$v(x)$', '$\mu(x, y)$', '$\sigma(x, y, z)$', '$(p+q)$', and '$(p.q)$' (cf. IV).

Among the variables which enter into the composition of a given sentential function we can distinguish *free* (or *real*) and *bound* (or *apparent*) *variables*. The precise distinction between these two kinds presents no difficulty.[1] Thus in the function '$\prod_x ((x \bar{\in} Y) + (x \in Z))$' the variables '$Y$' and '$Z$' are free and '$x$' is bound. Sentential functions which are devoid of free variables, like the expression '$\sum_X \prod_x (x \bar{\in} X)$', are called *sentences*.

In order to carry out the construction of the formal system of arithmetic which I shall sketch here, it would be necessary to formulate explicitly those sentences which are to be regarded as *axioms* (both the general logical ones and those which are specifically arithmetical), and then to formulate the *rules of inference* (*rules of proof*) with the help of which it is possible to derive from the axioms other sentences called *theorems* of the system. The solution of these problems occasions no great difficulty. If I omit their analysis here, it is because they are not of much importance for what is to follow.[2]

Let us now consider the situation metamathematically. For each deductive system it is possible to construct a particular science, namely the 'metasystem', in which the given system is subjected to investigation. Hence, to the domain of the metasystem belong all such terms as 'variable of the nth order', 'sentential function', 'free variable of a sentential function', 'sentence', etc., that is to say, terms which denote the individual expressions of the system under consideration, sets of these expressions, and relations between them. On the other hand nothing forbids our introducing into the metasystem arithmetical notions, in particular real numbers, sets of real numbers, etc. By operating with these two categories of terms (and with general logical terms) we can try to define the sense of the following phrase: '*A finite sequence of objects satisfies a given sentential function.*' The successful accomplishment of this task raises difficulties which are greater than would

---

[1] Cf. for example, Hilbert, D., and Ackermann, W. (30), pp. 52–54; cf. also article VIII, § 2, Def. 11, p. 178.

[2] As regards the logical axioms and the rules of deduction see article VIII, § 2.

appear at first sight.† However, in whatever form and to whatever degree we do succeed in solving this problem, the intuitive
meaning of the above phrase seems clear and unambiguous.
It will therefore suffice to illustrate it here by some particular
examples. Thus the function '$X \in \mathscr{X}$' is satisfied by those,
and only those, sequences of two terms of which the first term
is a set X of individuals and the second term is a family $\mathscr{X}$ of
those sets which contains X as an element. The functions
'$\sigma(x, y, z)$' and '$v(x) \cdot v(y) \cdot \mu(y, z)$' are satisfied by all sequences
composed of three real numbers x, y, z, where $x = y+z$ and
$x = 1 = y \leqslant z$ respectively. The function

$$'\sum_{z} \sum_{u} \big(\sigma(x, y, z) \cdot \mu(u, z) \cdot v(u) \big)'$$

is satisfied by sequences of two real numbers x and y where
$x \geqslant y+1$. The function '$\sigma(x, y, y) + \bar{\sigma}(x, y, y)$' is satisfied by
any sequence formed by two individuals, and the function
'$(x \in X) \cdot (x \in X)$' is satisfied by no sequence. Finally, a function
without free variables, i.e. a sentence, is satisfied either by the
empty sequence or by no sequence, according as the sentence is true or false. As we see from these examples, in all
situations where we say that a given finite sequence satisfies
a sentential function, it is possible to establish a one-one
correspondence between the terms of the sequence and the
free variables which occur in the function. Also the order
of every object which constitutes a term of the sequence is
equal to that of the corresponding variable. The special case
of the notion in question where the sentential function contains
only a single free variable is of particular importance for us.
The sequences which satisfy such a function are also composed
of a single term. Instead, therefore, of sequences we may speak
quite simply of *objects* (namely the unique terms of the corresponding sequences) and say that they *satisfy the given function*.
Thus the function '$\sum_{X} (x \in X)$' is satisfied by all individuals,
the function '$\sum_{x} (x \in X)$' by any set of individuals except the

† For an exhaustive discussion of the subject, see VIII.

null set. The function '$\sigma(x, x, x)$' is satisfied by only one individual, the real number 0, and the function '$\sum_{y} (\mu(y, x) . \sigma(y, y, y))$' by every non-negative real number and by no other individual.

With every sentential function with m variables we can now correlate uniquely the set of all m-termed sequences which satisfy this function; in case $m = 1$, we again replace the set of one-termed sequences by the set of objects which are the unique terms of these sequences. Consequently, a function which contains a variable of order 1 as its only free variable determines a certain set of individuals, which, in particular, may be a certain set of real numbers. The sets thus determined by sentential functions are precisely the *definable sets in the arithmetical system considered*. Among them we can distinguish sets of the first, second, etc., orders, corresponding to the orders of the functions which determine them.

The metamathematical definition of definable sets, based on the notion of satisfaction (of sentential functions by objects), then takes the following form:

A set X is a *definable set* (or a *definable set of order n*) if there is a sentential function (or a sentential function of order n at most[1]) which contains some variable of order 1 as its only free variable, and which satisfies the condition that, for every real number x, $x \in X$ if and only if x satisfies this function.

The following will serve as examples of definable sets in the sense of the foregoing definition—and in particular of those of order 1: the set composed of the single number 0, the set of all positive numbers, the set of all numbers x such that $0 \leqslant x \leqslant 1$, and many others. The functions which determine these sets are, respectively: '$\sigma(x, x, x)$', '$\bar{\sigma}(x, x, x) . \sum_{y} (\mu(y, x) . \sigma(y, y, y))$', '$\sum_{y} \sum_{z} (\mu(y, x) . \sigma(y, y, y) . \mu(x, z) . v(z))$', etc. A relatively simple example of a definable set of order 2, of which it can be shown that it is not of order 1, is furnished by the set of all natural numbers (including 0), determined by the sentential

---

[1] It is easy to show that the words 'at most' can be omitted without modifying the meaning of the notion defined.

function

$$'\prod_{X}\Bigl(\sum_{y}\,(\sigma(y,y,y).(y\,\bar{\in}\,X))+$$

$$+\sum_{y}\sum_{z}\sum_{t}\,(\sigma(z,y,t).v(t).(y\in X).(z\,\bar{\in}\,X))+(x\in X)\Bigr)'.$$

Such examples can be multiplied indefinitely. Each particular set of numbers with which we are concerned in mathematics is a definable set, inasmuch as we have no other means of introducing any set *individually* into mathematics than by constructing the sentential function which determines it, and this construction is itself the proof of the definability of the set. Moreover it is not difficult to show that the family of all definable sets (as well as that of the functions which determine them) is only denumerable, while the family of *all* sets of numbers is not denumerable. The existence of undefinable sets follows immediately. Also, the definable sets can be arranged in an ordinary infinite sequence; by applying the diagonal procedure it is possible *to define in the metasystem a concrete set which would not be definable in the system itself*. In this there is clearly no trace of any antinomy, and this fact will not appear at all paradoxical if we take proper note of the relative character of the notion of definability.

In the next section I shall attempt a partial reconstruction of the notion just defined within mathematics itself. It is evident *a priori* that there can be no question of a total mathematical reconstruction of the notion of definability. In fact, if this were possible, we could, by applying the diagonal procedure, define within the system of arithmetic a set of numbers provably not definable in this system, and we should this time find ourselves involved in the original antinomy of Richard. For the same reason, it is impossible to define in mathematics the general notion of a definable set of order n. In other words, it is impossible to construct a function which correlates with every natural number n the family of all definable sets of order n. I shall show, however, that it is possible to reconstruct by purely mathematical methods the notions of definable sets of orders

1, 2,..., i.e. the reconstruction can be carried out for each natural number separately.

The idea of the reconstruction is quite simple in principle. We notice that every sentential function determines the set of all finite sequences which satisfy it. Consequently, in the place of the metamathematical notion of a sentential function, we can make use of its mathematical analogue, the concept of a set of sequences. I shall therefore introduce first those sets of sequences which are determined by the primitive sentential functions. Then I shall define certain operations on sets of sequences which correspond to the five fundamental operations on expressions. Finally, in imitation of the definition of sentential function, I shall define the concept of definable set of sequences of order n. This notion will lead us easily to that of definable sets of individuals of order n.

The whole construction will be carried out in detail for the case $n = 1$ and in its main features for $n = 2$. I believe that after this explanation the method of construction for higher values of n will be clear.†

§ 2. DEFINABLE SETS OF ORDER 1 FROM THE STANDPOINT OF MATHEMATICS

As the field of our considerations in this section, we can make use of any of the known systems of mathematical logic including, in particular, the calculus of classes, the logic of relations, and the arithmetic of real numbers, that is to say a system of logic with a system of arithmetic axiomatically constructed upon it. For example, we can use for this purpose the system of logic of *Principia Mathematica*. It would, however, be necessary to replace the ramified theory of types by the simplified theory.[1]

---

[1] I omit completely the other simplifications of the system mentioned in § 1, since here they would be very inconvenient.

---

† This article was conceived as the first part of a more comprehensive paper. The second half was intended to include a discussion of definable sets of order $n \geqslant 2$. Since the second half has never appeared in print, we have included in the present translation some remarks concerning definable sets of higher order, and in particular a precise mathematical definition of definable sets of order 2 (see remarks in § 2, following Th. 6). In this connexion compare also a recent article by the author 'A problem concerning the notion of definability', *Journal of Symbolic Logic*, vol. 13 (1948), pp. 107–11.

Most of the symbols to be used are commonly found in works on set theory. In particular, the symbols '0' and '$\{x, y, ..., z\}$' denote respectively the null set and the finite set composed of the elements x, y, ..., z. The symbol '$E\phi(x)$' denotes
$$\quad x$$
the set of all objects x which satisfy the condition ϕ. The symbol 'Nt' denotes the set of all natural numbers (zero included) and the symbol 'Rl' that of all real numbers. Further, when I have need to refer to the considerations of the preceding section, I shall employ the logical and arithmetical signs which have there been explained.[1]

The principal instrument of these investigations will be formed by finite sequences of real numbers and sets of such sequences. In order to avoid all possible misunderstanding the notion of finite sequence will first be considered more closely.

Let r be any binary relation. By the *domain of the relation r*, in symbols $D(r)$, I understand the set E (*there exists a y such*
$$\quad x$$
that xry). The *counter domain* (or *range*) $C(r)$ is defined by the formula $C(r) = E$ (*there exists an x such that xry*). The relation r
$$\quad v$$
is called a *one-many relation* (or *function*) if for all x, y, z the formulas xrz and yrz always imply the equality $x = y$.†

Every function s, whose counter domain is a finite subset of Nt I call a *finite sequence*. The set of all finite sequences will be denoted by 'Sf'. The unique x which satisfies the formula xsk for the given sequence s and natural number k, will be called the *k-th term of the sequence s*, or the *term with the index k of the sequence s*, and denoted by 's_k'. It is not at all necessary (contrary to the customary interpretation of the notion of a finite sequence) that the set $C(s)$ should be a

---

[1] Thus the symbols '+' and '.' are used in several different senses: namely as logical signs, as signs of the theory of sets, and as signs of arithmetic. I do not believe that this will occasion any error within the limits of this article.

---

† Contrary to the notation adopted in this paper, functions are usually identified not with one-many but with many-one relations, i.e. with relations r such that for arbitrary x, y, z, the formulas xry and xrz imply $y = z$. Consequently, what is called in this paper the domain of a function (or, in particular, of a sequence) is usually referred to as the range or counter domain of this function, and conversely.

segment of the set Nt. A sequence s may, for example, have a second term without a first. When we have $s \in Sf$ and $D(s) \subseteq Rl$, the sequence s is called a *finite sequence of real numbers*. In accordance with these conventions, the null relation, i.e. the relation r such that $D(r) = 0 = \mathcal{C}(r)$, also belongs to the set Sf.

Two relations r and s are *identical*, in symbols $r = s$, when the formulas xry and xsy are equivalent for every x and y. In particular, the identity of the sequences r and s is characterized by the conditions $\mathcal{C}(r) = \mathcal{C}(s)$ and $r_k = s_k$ for every $k \in \mathcal{C}(r)$. I denote by the symbol 'r/X' (where r is any relation and X any set) the relation t which satisfies the condition: for any x and y we have xty if and only if we have xry and $y \in X$. We assume it can easily be proved that $s \in Sf$ always implies $s/X \in Sf$. s/X is called a *subsequence of the sequence* s, i.e. it is composed of all the terms of the sequence s whose indices belong to the set X. Conversely, the sequence s may be called a *prolongation of the sequence* s/X.

I shall be concerned for the present with sets of finite sequences. I introduce in the first place the notions of the *domain* and *counter domain of a set of sequences* (or *of arbitrary relations*) S, in symbols $D(S)$ and $\mathcal{C}(S)$. I define them by the formulas

DEFINITION 1.   (a) $D(S) = \sum_{s \in S} D(s)$,

             (b) $\mathcal{C}(S) = \sum_{s \in S} \mathcal{C}(s)$.

The counter domain of a set of finite sequences is evidently the union of the finite sets of natural numbers.

I shall distinguish here a category of sets of sequences which is especially important for us, namely, the *homogeneous sets*.

DEFINITION 2. S is a *homogeneous set* of sequences (or of *arbitrary relations*), if $\mathcal{C}(s) = \mathcal{C}(S)$ for all $s \in S$.

All the sets of sequences with which we shall deal in the sequel will be homogeneous sets. This is explained by the fact that I operate here with sets of sequences as the mathematical analogues of sentential functions. Now all sets of sequences determined by sentential functions are composed of sequences having the same counter domain; they are thus homogeneous in the sense of Def. 2 (cf. § 1 in this connexion).

The largest homogeneous set of sequences with a given counter domain N is the set R_N defined by the formula

DEFINITION 3. $R_N = Sf . \underset{s}{E} (\mathcal{C}(s) = N).$

By putting $N = \mathcal{C}(S)$ we can associate a finite set $N \subseteq Nt$ such that $S \subseteq R_N$ with every homogeneous set S of sequences. Except in the case $S = 0$, this correspondence is unique. Conversely, when $S \subseteq R_N$, S is a homogeneous set and we have (again except when $S = 0$) $\mathcal{C}(S) = N$.

I next define certain special sets of sequences U_k, $M_{k,l}$ and $S_{k,l,m}$ which I call *primitive sets*:

DEFINITION 4. (a) $U_k = R_{\{k\}} . \underset{s}{E} (s_k = 1),$

(b) $M_{k,l} = R_{\{k,l\}} . \underset{s}{E} (s_k \leqslant s_l),$

(c) $S_{k,l,m} = R_{\{k,l,m\}} . \underset{s}{E}(s_k = s_l + s_m).$

Thus the primitive sets are the sets of sequences determined by the primitive sentential functions of order 1: '$v(x^{(k)})$', '$\mu(x^{(k)}, x^{(l)})$', and '$\sigma(x^{(k)}, x^{(l)}, x^{(m)})$'.

I next introduce five fundamental operations on sets of sequences: *complementation*, *addition*, and *multiplication*, as well as *summation* and *multiplication with respect to the k-th terms*. The results of these operations will be denoted respectively by the symbols '$\overset{\circ}{S}$', '$S \overset{\circ}{+} T$', '$S \overset{\circ}{\cdot} T$', '$\overset{\circ}{\underset{k}{\sum}} S$', and '$\overset{\circ}{\underset{k}{\prod}} S$'. The results of these operations are always homogeneous sets of sequences.

DEFINITION 5. $\overset{\circ}{S} = R_{\mathcal{C}(S)} - S.$

DEFINITION 6.

(a) $S \overset{\circ}{+} T = R_{\mathcal{C}(S) + \mathcal{C}(T)} . \underset{u}{E} (u/\mathcal{C}(S) \in S \ or \ u/\mathcal{C}(T) \in T),$

(b) $S \overset{\circ}{\cdot} T = R_{\mathcal{C}(S) + \mathcal{C}(T)} . \underset{u}{E} (u/\mathcal{C}(S) \in S \ and \ u/\mathcal{C}(T) \in T).$

According to these definitions the set $\overset{\circ}{S}$ consists of all sequences whose counter domains equal $\mathcal{C}(S)$ and which do not belong to the set S. The set $S \overset{\circ}{+} T$ is the set of all sequences of real numbers with the counter domain $\mathcal{C}(S) + \mathcal{C}(T)$ which contain among their subsequences either a sequence of the set S or a sequence of the set T (or both); similarly the set $S \overset{\circ}{\cdot} T$ is the set of all sequences with the counter domain $\mathcal{C}(S) + \mathcal{C}(T)$ which contain among their subsequences both a sequence of the set S and a sequence of the set T. Thus $\overset{\circ}{U}_1 \overset{\circ}{+} M_{2,3}$ is the set of all sequences s composed of three real numbers s_1, s_2, and s_3 such that we have $s_1 \neq 1$ or $s_2 \leqslant s_3$, and $M_{1,2} \overset{\circ}{\cdot} M_{2,3} \overset{\circ}{\cdot} S_{3,4,4}$ is the set of all sequences s with four terms s_1, s_2, s_3, and s_4 satisfying the formula $s_1 \leqslant s_2 \leqslant s_3 = 2 . s_4$.

The operations of complementation, addition, and multiplication just introduced approximate to the corresponding operations of the algebra of sets and offer many analogous formal properties, especially in the domain of homogeneous sets. For example, they satisfy the *commutative* and *associative laws*, the two *distributive laws*, the *law of De Morgan*, $\overline{S \overset{\circ}{+} T} = \overset{\circ}{S} \overset{\circ}{\cdot} \overset{\circ}{T}$, the formulas

$$S \overset{\circ}{+} 0 = S, \quad S \overset{\circ}{\cdot} 0 = 0, \quad S \overset{\circ}{\cdot} \overset{\circ}{S} = 0,$$

and many others.

This correspondence is impaired only in a single detail, namely, there does not exist any set having exactly the same formal properties as the universal set 1 of the algebra of sets, and there exist many different sets S such that $\overset{\circ}{S} = 0$ (among homogeneous sets all sets of the type R_N have this property), while $\overset{\circ}{0}$ is a special set, namely R_0, composed of a single element, the null relation. Consequently the formula

$$S \overset{\circ}{+} \overset{\circ}{S} = T \overset{\circ}{+} \overset{\circ}{T}$$

is not in general satisfied; the *law of double complementation* $\overset{\circ}{\overset{\circ}{S}} = S$ fails in the case where $\overset{\circ}{S} = 0$; the *second law of De Morgan*,

$$\overline{S \overset{\circ}{\cdot} T} = \overset{\circ}{S} \overset{\circ}{+} \overset{\circ}{T},$$

requires the hypothesis $S \overset{\circ}{\cdot} T \neq 0$; etc. From the practical point of view this fact does not involve much inconvenience;

in the place of the set 1 we can usually employ $\overset{\circ}{0}$, which possesses a series of analogous properties, e.g.

$$S\overset{\circ}{+}\overset{\circ}{0} = S\overset{\circ}{+}\overset{\circ}{S}, \quad S\overset{\circ}{?}\overset{\circ}{0} = S, \quad \text{etc.}$$

If we restrict ourselves to the consideration of subsets of any fixed set R_N, by letting $R_N = 1$ and changing Def. 5 in only one point, namely by putting $\overset{\circ}{0} = 1$, the operations $\overset{\circ}{S}$, $S\overset{\circ}{+}T$, and $S\overset{\circ}{?}T$ coincide completely, as is easily seen, with the corresponding operations in the algebra of sets.

The operations of addition and of multiplication can easily be extended to an arbitrary finite number (and even to an infinite number) of sets of sequences; the results of these generalized operations are denoted respectively by the symbols

$$\text{`}S_1\overset{\circ}{+}S_2\overset{\circ}{+}...\overset{\circ}{+}S_n\text{'} \quad \text{and} \quad \text{`}S_1\overset{\circ}{?}S_2\overset{\circ}{?}...\overset{\circ}{?}S_n\text{'}.$$

DEFINITION 7. (a) $\overset{\circ}{\underset{k}{\sum}} S = R_{\mathcal{C}(S)-\{k\}} . \underset{s}{E}$ (there exists a sequence $t \in R_{\mathcal{C}(S)}$ satisfying the formulas $t/\mathcal{C}(s) = s$ and $t \in S$),

(b) $\overset{\circ}{\underset{k}{\prod}} S = R_{\mathcal{C}(S)-\{k\}} . \underset{s}{E}$ (for all sequences $t \in R_{\mathcal{C}(S)}$ the formula $t/\mathcal{C}(s) = s$ implies the formula $t \in S$).

Let S be any homogeneous set of sequences. If $k \overline{\in} \mathcal{C}(S)$, the sets $\overset{\circ}{\underset{k}{\sum}} S$ and $\overset{\circ}{\underset{k}{\prod}} S$ coincide, according to the above definition, with the set S. If, however, $k \in \mathcal{C}(S)$, then $\overset{\circ}{\underset{k}{\sum}} S$ is the set of all the sequences which are obtainable from those of the set S by the omission of the kth terms, and the set $\overset{\circ}{\underset{k}{\prod}} S$ is composed of all sequences not containing the kth terms and whose every prolongation by the addition of some kth term belongs to the set S. For example, we verify immediately that the set $\overset{\circ}{\underset{2}{\sum}} (M_{2,1}\overset{\circ}{?}S_{2,2,2})$ is formed of all sequences composed of a single term s_1 which is a non-negative real number; the set $\overset{\circ}{\underset{3}{\sum}} (S_{1,2,3}\overset{\circ}{?}S_{2,1,3})$ is composed of all sequences with two identical terms $s_1 = s_2$; $\overset{\circ}{\underset{2}{\prod}} M_{1,2}$ is the empty set, and $\overset{\circ}{\underset{3}{\prod}} (\overline{\overset{\circ}{U}}_3\overset{\circ}{+}S_{1,2,3})$ is the set of all sequences s with two terms where $s_1 = s_2+1$.

The laws which express the formal properties of the operations $\overset{\circ}{\underset{k}{\sum}} S$ and $\overset{\circ}{\underset{k}{\prod}} S$ and the relations between these operations and those examined above are completely analogous, especially where homogeneous sets are concerned, with the laws of logic for the quantifiers '$\sum$' and '$\prod$', for the signs of negation, logical sum, and logical product. As examples, I mention the formulas

$$\overset{\circ}{\underset{k}{\sum}} \overset{\circ}{\underset{l}{\sum}} S = \overset{\circ}{\underset{l}{\sum}} \overset{\circ}{\underset{k}{\sum}} S, \quad \overset{\circ}{\underset{k}{\prod}} \overset{\circ}{\underset{l}{\prod}} S = \overset{\circ}{\underset{l}{\prod}} \overset{\circ}{\underset{k}{\prod}} S, \quad \overset{\circ}{\underset{k}{\sum}} \overset{\circ}{\underset{l}{\prod}} S \subseteq \overset{\circ}{\underset{l}{\prod}} \overset{\circ}{\underset{k}{\sum}} S,$$

also the distributive laws

$$\overset{\circ}{\underset{k}{\sum}} (S \dotplus T) = \overset{\circ}{\underset{k}{\sum}} S \dotplus \overset{\circ}{\underset{k}{\sum}} T \quad \text{and} \quad \overset{\circ}{\underset{k}{\prod}} (S \overset{\circ}{.} T) = \overset{\circ}{\underset{k}{\prod}} S \overset{\circ}{.} \overset{\circ}{\underset{k}{\prod}} T,$$

and finally the laws of De Morgan

$$\overline{\overset{\circ}{\underset{k}{\sum}} S} = \overset{\circ}{\underset{k}{\prod}} \bar{S} \quad \text{and} \quad \overline{\overset{\circ}{\underset{k}{\prod}} S} = \overset{\circ}{\underset{k}{\sum}} \bar{S}$$

(this last formula fails for the case where $\overset{\circ}{\underset{k}{\prod}} S = 0$).[1]

The analogy between the laws just discussed and those of logic is not at all a matter of chance. It is easy to show that it is possible to set up a very strict correspondence between the fundamental operations on sentential functions studied in § 1, and those on sets of sequences introduced in Defs. 5–7. This can be done in the following way: let $f_1, f_2, \dots$ be any sentential functions (of order 1) and $S_1, S_2, \dots$ the sets of sequences determined by them; every function obtained by one of the fundamental operations when applied to the given functions determines the set of sequences which can be obtained with the help of the appropriate operation when applied to the sets corresponding to these functions. A certain departure from this general phenomenon only appears in the case where the function considered is obtained by the negation of a function which determines the null set. Thus, for example, to the function '$\overline{\sigma(x', x'', x''')} . \bar{\sigma}(x', x'', x''')$' corresponds, to be sure, the set

---

[1] For the reasons discussed above, p. 124.

$R_{\{1,2,3\}} = S_{1,2,3} \dotplus \overset{\circ}{S}_{1,2,3}$ and not the set $R_0 = \overline{S_{1,2,3} \overset{\circ}{\circ} \overset{\circ}{S}_{1,2,3}}$ (cf. the remarks on the subject of Defs. 5–6).

Only one more operation on sets of sequences remains to be mentioned, namely the replacement of the index k by l in the sequences of a given set. The result of this operation will be denoted by the symbol $\dfrac{{}^{\backprime}k\,{}^{\prime}}{l}\,S$.

DEFINITION 8. $\dfrac{k}{l}\,S = Sf\underset{t}{.\,E}$ (there exists a sequence $s \in S$ such that we have either $k \in \mathcal{C}(s)$, $\mathcal{C}(t) = \mathcal{C}(s) - \{k\} + \{l\}$, $s_k = t_l$, and $s/(\mathcal{C}(s) - \{k\}) = t/(\mathcal{C}(S) - \{k\})$ or $k \,\overline{\in}\, \mathcal{C}(s)$ and $s = t$).

Thus, when S is a homogeneous set of sequences, $l \in Nt$ and $k \in \mathcal{C}(S)$, the set $\dfrac{k}{l}\,S$ is formed of all the sequences which can be obtained from certain sequences of the set S by replacing in the kth terms the index k by l, without altering any other terms. For example,

$$\tfrac{2}{3}M_{1,2} = M_{1,3}, \qquad \tfrac{3}{2}S_{1,2,3} = S_{1,2,2},$$

$$\tfrac{1}{2}\Big(\overset{\circ}{M}_{1,2} \overset{\circ}{\circ} \overset{\circ}{\textstyle\sum_1} S_{1,2,3}\Big) = \overset{\circ}{M}_{2,2} \overset{\circ}{\circ} \overset{\circ}{\textstyle\sum_1} S_{1,2,3}, \quad \text{etc.}$$

If, however, k does not belong to the counter domain of the set S of sequences, we have $\dfrac{k}{l}\,S = S$; e.g.

$$\tfrac{3}{2}U_1 = U_1, \qquad \tfrac{2}{3}\overset{\circ}{\textstyle\sum_2} S_{2,1,2} = \overset{\circ}{\textstyle\sum_2} S_{2,1,2}, \quad \text{etc.}$$

The operation $\dfrac{k}{l}\,S$, when applied to a homogeneous set of sequences, always yields a homogeneous set of sequences; if, in particular, $S \subseteq R_N$, where $N \subseteq Nt$ and $l \in Nt$, we have $\dfrac{k}{l}\,S \subseteq R_{N-\{k\}+\{l\}}$ or $\dfrac{k}{l}\,S \subseteq R_N$, according to whether $k \in N$ or $k \,\overline{\in}\, N$. For homogeneous sets, the definition of this operation can be simplified and formulated in the following way:

$$\frac{k}{k}\,S = S; \quad \frac{k}{l}\,S = \overset{\circ}{\textstyle\sum_k} (M_{k,l} \overset{\circ}{\circ} M_{l,k} \overset{\circ}{\circ} S) \quad \textit{if} \quad k \neq l.$$

In the same sense in which the fundamental operations on sentential functions correspond to those on sets of sequences, a certain operation, which I have not had occasion to mention in § 1, corresponds to the operation $\frac{k}{l} S$. This is the operation of *substituting the free variable* '$x^{(l)}$' *for the free variable* '$x^{(k)}$' in a given sentential function. The intuitive significance of this operation is, incidentally, quite clear. For example, by substituting in the sentential function

$$'\mu(x, y) . \sum_x \sigma(x, y, z)'$$

the free variable 'y' for the free variable 'x' we obtain

$$'\mu(y, y) . \sum_x \sigma(x, y, z)'.$$

These functions determine the sets of sequences

$$\mathring{\bar{M}}_{1,2} \overset{\circ}{:} \overset{\circ}{\sum_1} S_{1,2,3} \quad \text{and} \quad \mathring{\bar{M}}_{2,2} \overset{\circ}{:} \overset{\circ}{\sum_1} S_{1,2,3}$$

respectively.

Now we are ready to define the notion of *definability of order* 1, first for finite sets of sequences of real numbers, and then for sets of real numbers themselves. The family of definable sets of sequences will be denoted by '$\mathscr{D}f$', and that of definable sets of individuals by '$\mathscr{D}$'. Instead of saying *definable of order* 1 we shall say *elementarily definable* or *arithmetically definable*.

DEFINITION 9. $\mathscr{D}f$ is the product (intersection) of all the families of sets $\mathscr{X}$ which satisfy the following conditions: (α) $U_k \in \mathscr{X}$, $M_{k,l} \in \mathscr{X}$, and $S_{k,l,m} \in \mathscr{X}$ for any natural numbers k, l, and m; (β) if $S \in \mathscr{X}$, then $\mathring{\bar{S}} \in \mathscr{X}$; ($\gamma$) if $S \in \mathscr{X}$ and $T \in \mathscr{X}$, then $S \mathring{+} T \in \mathscr{X}$ and $S \overset{\circ}{:} T \in \mathscr{X}$; ($\delta$) if $k \in Nt$ and $S \in \mathscr{X}$, then $\overset{\circ}{\underset{k}{\sum}} S \in \mathscr{X}$ and $\overset{\circ}{\underset{k}{\prod}} S \in \mathscr{X}$.

DEFINITION 10. $\mathscr{D} = \underset{X}{E}$ (there exists a set $S \in \mathscr{D}f$ such that $D(S) = X$ and $\mathit{Q}(S)$ is a set composed of only one element).

Now the question arises whether *the definitions just constructed* (the formal rigour of which raises no objection) *are also adequate materially*; in other words *do they in fact grasp the current meaning of the notion as it is known intuitively*? Properly understood, this question contains no problem of a purely mathematical nature, but it is nevertheless of capital importance for our considerations.

In order to put this question into a precise form, let us suppose that the material adequacy of the metamathematical definition of definable set of order n, which we reached in the preceding section, is beyond doubt. The question proposed then reverts to a problem which is quite concrete, namely, whether Def. 10 is equivalent to a certain particular case of this metamathematical definition, namely the case where $n = 1$. This last problem obviously belongs to the domain of metamathematics and is easily solved in the affirmative.

We have in fact already established that there is a strict correspondence between primitive sets of sequences and the fundamental operations on these sets on the one hand, and the primitive sentential functions of order 1 and the fundamental operations on these expressions on the other. We have even rigorously defined this correspondence (pp. 122 and 127). From these facts, Def. 9, and the definitions of § 1 of a sentential function and its order, we can show without difficulty by induction that the family $\mathscr{D}f$ is exactly the family of sets of sequences which are determined by sentential functions of order 1.[1] It follows almost immediately that the family $\mathscr{D}$ coincides with that of the definable sets of order 1 in the sense of § 1.

If we wish to convince ourselves of the material adequacy of Def. 10 and of its conformity with intuition without going beyond the domain of strictly mathematical considerations, we must have recourse to the empirical method. In fact, by examining various special sets which have been arithmetically defined (in the intuitive sense of this term), we can show that all of them belong to the family $\mathscr{D}$; conversely, for every particular set belonging to this family we are able to construct an elementary definition. Moreover, we easily notice that the same

---

[1] The breakdown mentioned on p. 126 in the correspondence between the negation of sentential functions and complementation of sets of sequences only complicates the proof a little.

completely mechanical method of reasoning can be applied in all the cases concerned.

The following shows what this method involves: Let A be any set of numbers elementarily defined with the help of the primitive notions '1', '$\leqslant$', and '$+$'. The definition of the set A may be put in the form:

$$(1) \qquad\qquad A = \underset{x^{(k)}}{E}\, \phi(x^{(k)}).$$

The symbol '$\phi(x^{(k)})$' here represents a certain sentential function containing the variable '$x^{(k)}$' of order 1 as its sole free variable. It may also contain a series of bound variables '$x^{(l)}$', '$x^{(m)}$', etc., provided that all these variables are also of order 1 (since if this were not the case the definition could not be called elementary). As we know from logic, the function '$\phi(x^{(k)})$' can be constructed in such a way that it contains no logical constants other than the signs of negation, of logical sum and product as well as the universal and existential quantifiers. In the same way we can eliminate all the arithmetical constants except the signs 'v', 'μ', and 'σ', corresponding to the three primitive notions.

It is not difficult to show in each particular case that the formula (1) can be transformed in the following way: the sentential functions '$v(x^{(k)})$', '$\mu(x^{(k)}, x^{(l)})$', and '$\sigma(x^{(k)}, x^{(l)}, x^{(m)})$' (or '$x^{(k)} = 1$', '$x^{(k)} \leqslant x^{(l)}$', and '$x^{(k)} = x^{(l)} + x^{(m)}$') are replaced by symbols of the form 'U_k', '$M_{k,l}$', and '$S_{k,l,m}$', respectively; the signs of logical operations by those of the corresponding operations on sets of sequences, introduced in Defs. 5–7; and finally the symbol '$\underset{x^{(k)}}{E}$' by 'D'.[1] By this transformation the formula (1) takes the form

$$(2) \qquad\qquad A = D(S),$$

where in the place of 'S' there is a composite symbol, the structure of which shows at once that it denotes a set of sequences of the family $\mathscr{D}f$ with counter domain consisting of a single

---

[1] In the course of these transformations use may be made of certain formulas which are established in VII.

element. By applying Def. 10 we conclude at once that the set A belongs to the family $\mathscr{D}$.

I shall now give some concrete examples:

1. Let A be a set consisting of the single number 0. We see at once that $A = E(x = x+x)$, i.e. that $A = E\,\sigma(x,x,x)$. By transforming this formula in the way just described we obtain $A = D(S_{1,1,1})$, so that $A \in \mathscr{D}$.

2. Let A be the set of all positive numbers. As is easily established, we have $A = E\,(x \neq x+x$ *and there exists a y such that* $y \leqslant x$ *and* $y = y+y)$, i.e.

$$A = E_x \Big(\bar{\sigma}(x,x,x) \cdot \sum_y \big(\mu(y,x) \cdot \sigma(y,y,y) \big) \Big),$$

whence $A = D\Big(\overset{\circ}{S}_{1,1,1} \overset{\circ}{\,\circ\,} \sum_2 (M_{2,1} \overset{\circ}{\circ} S_{2,2,2}) \Big)$, and again we have $A \in \mathscr{D}$.

3. Finally let $A = E_x (0 \leqslant x \leqslant 1)$. We transform this formula successively as follows:

$$A = E_x \, (\textit{there exist y and z such that } y \leqslant x,\ y = y+y,$$
$$x \leqslant z, \textit{ and } z = 1)$$
$$= E_x \sum_y \sum_z \big(\mu(y,x) \cdot \sigma(y,y,y) \cdot \mu(x,z) \cdot \upsilon(z) \big)$$
$$= D\Big(\sum_2 \sum_3 (M_{2,1} \overset{\circ}{\circ} S_{2,2,2} \overset{\circ}{\circ} M_{1,3} \overset{\circ}{\circ} U_3) \Big).$$

We thus have in this case also $A \in \mathscr{D}$.

The same method can be applied in the opposite direction. A single example will suffice to illustrate this:

4. Let A be a particular element of the family $\mathscr{D}$, for example let $A = D\Big(\sum_2 \sum_3 (S_{2,1,3} \overset{\circ}{\circ} S_{2,2,2} \overset{\circ}{\circ} U_3) \Big)$. We easily obtain the following transformation of this formula

$$A = E_x \sum_y \sum_z \big(\sigma(y,x,z) \cdot \sigma(y,y,y) \cdot \upsilon(z) \big) = E_x \, (\textit{there exist y}$$
$$\textit{and z such that } y = x+z,\ y = y+y, \textit{ and } z = 1).$$

We see that the set A can be arithmetically defined. It is, in fact, composed of a single number, -1.

As a consequence of these considerations *the intuitive adequacy of Def.* 10 *seems to be indisputable.*

The sets of finite sequences and especially the arithmetically definable sets of numbers in the sense established above are (from the point of view of analytic geometry) objects of a very elementary structure. In order to characterize them in terms as familiar as possible, I shall use the well-known concept of a *linear polynomial with integer coefficients.* Polynomials are here understood as functions which correlate real numbers with finite sequences of real numbers (in other words as one-many relations whose domains contain only real numbers and counter domains only finite sequences of such numbers). More precisely, the linear polynomial p with integer coefficients is determined by a finite set N of natural numbers, by a sequence a of whole numbers with counter domain N, and by a number b, and it correlates the number

$$p(s) = \sum_{k \in N} a_k \cdot s_k + b$$

with every sequence $s \in R_N$. With the help of this concept I shall define a special category of sets of finite sequences, the *elementary linear sets.*

DEFINITION 11. S is called an *elementary linear set* if there exists a linear polynomial p with integer coefficients such that either

$$S = \mathop{E}_{s}(p(s) = 0 \quad \text{and} \quad D(s) \subseteq Rl),$$

or

$$S = \mathop{E}_{s}(p(s) > 0 \quad \text{and} \quad D(s) \subseteq Rl).$$

An elementary linear set is simply the set of all sequences s of real numbers which are also solutions either of a linear equation of the type '$p(s) = 0$', or of an inequality of the type '$p(s) > 0$'.

From Def. 11 and the interpretation of a linear polynomial adopted here, it follows that every elementary linear set is homogeneous.

We are now in a position to formulate the following theorem characterizing the family of sets $\mathscr{D}f$.

THEOREM 1. *In order that a set S of sequences of real numbers should belong to the family $\mathscr{D}f$, it is necessary and sufficient that S should be a finite sum of finite products of elementary linear sets (with a common counter domain).*[1]

Proof. That the condition is sufficient is easily proved from Def. 9 and an easy lemma stating that every elementary linear set belongs to the family $\mathscr{D}f$. To show that this condition is necessary we use an argument by induction: we show that the family of all sets which satisfy the given condition contains the sets U_k, $M_{k,l}$ and $S_{k,l,m}$ (which is clear), and then we show that it is closed with respect to the five operations introduced in Defs. 5–7. Hence, we conclude at once, with the help of Def. 9, that every set $S \in \mathscr{D}f$ satisfies this condition. A certain difficulty arises with the operation $\overset{\circ}{\underset{k}{\sum}}$ (to which by the De Morgan laws the operation $\overset{\circ}{\underset{k}{\prod}}$ is reducible). The question reduces to proving a lemma of algebra showing that a necessary and sufficient condition for a system of linear equations and inequalities in several unknowns to have a solution with respect to one of the unknowns, is that it be representable as a logical sum of systems in which only the other unknowns appear.

We see in this demonstration the intimate connexion between the operations $\overset{\circ}{\underset{k}{\sum}}$ and $\overset{\circ}{\underset{k}{\prod}}$ and the elimination of an unknown from a system of equations and inequalities familiar in algebra.

Sets of numbers of the following types: $\underset{x}{E}(x = a)$, $\underset{x}{E}(x > a)$, $\underset{x}{E}(x < a)$, and $\underset{x}{E}(a < x < b)$,[2] where a and b are two arbitrary rational numbers, I shall call *intervals with rational end points*. With this convention we can easily deduce from Th. 1 the following corollary concerning the family $\mathscr{D}$.

---

[1] Sum and product may here be understood rather in the sense of Def. 6 (and the condition of the common counter domain is then superfluous) than in the ordinary sense of the algebra of sets.

[2] And, if necessary, such other analogous types as $\underset{x}{E}(x \leqslant a)$, $\underset{x}{E}(a \leqslant x \leqslant b)$, etc., but this is not indispensable here.

THEOREM 2. *In order that a set X of real numbers should belong to the family $\mathscr{D}$, it is necessary and sufficient that X is the sum of a finite number of intervals with rational end points.*

We can easily formulate (and prove) a metamathematical theorem which is an exact analogue of Th. 1. This metamathematical result leads us to a conclusion that, in the system of arithmetic described in § 1, every sentence of order 1 can be proved or disproved. Moreover, by analysing the proof of this result, we see that there is a mechanical method which enables us to decide in each particular case whether a given sentence (of order 1) is provable or disprovable.

It should be emphasized that the two preceding theorems have only an accidental character and have only a slight bearing upon the leading idea of this work. These theorems are due to the fact that we have chosen the sets U_k, $M_{k,l}$, and $S_{k,l,m}$ as the primitive sets of sequences, i.e. restricted ourselves to a system of arithmetic which contains as the only primitive notions the number 1, the relation $\leqslant$ and the operation $+$. Now it is well known that arithmetic can also be based on many other systems of primitive concepts, and these systems may even contain superfluous concepts which can be defined by means of other notions. Nothing compels us, then, to make the definitions of the families $\mathscr{D}f$ and $\mathscr{D}$ depend just on the particular sets U_k, $M_{k,l}$, and $S_{k,l,m}$. On the contrary, these sets may be replaced by others, and the extension and the structure of the two families may thus undergo quite essential modifications. It is easy to see, for example, that the arithmetic of real numbers can be based on the two primitive notions of sum and of product.[1] We can then replace in Def. 9 the sets U_k and $M_{k\,l}$ by the sets

$$P_{k,l,m} = R_{\{k,l,m\}} \cdot \underset{s}{E} (s_k \in Rl,\ s_l \in Rl,\ s_m \in Rl,\ \text{and}\ s_k = s_l \cdot s_m),$$

while retaining the sets $S_{k,l,m}$; Def. 10 remains unchanged. To obtain for the new notion of definability the results analogous to Ths. 1 and 2, we make the following changes:

(1) In Def. 11 and Th. 1 we delete the term *'linear'* everywhere (i.e. instead of linear polynomials, we consider poly-

[1] Cf. Veblen, O. (87).

nomials of arbitrary degrees); (2) in Th. 2 we replace the term
'*rational*' by '*algebraic*'. Ths. 1 and 2 in the modified formulations then still hold; their proof, however, becomes more involved and difficult.† Thus in this case the family $\mathscr{D}$ has
undergone an extension. We have here, in fact, a system of
primitive concepts which is stronger in the logical sense. In
many cases this extension is still greater, especially when the
notion of natural number is taken as a primitive notion, and
the sets $N_k = R_{\{k\}} \cdot \underset{s}{E}(s_k \in Nt)$ are primitive sets. The family $\mathscr{D}$
may then contain sets which are quite complicated from the
point of view of analytic geometry, e.g. the so-called projective
sets of arbitrarily high classes.[1]

In order to deprive the notion of elementary definability (of
order 1) of its accidental character, it is necessary to relativize
it to an arbitrary system of primitive concepts or—more precisely—to an arbitrary family of primitive sets of sequences.
In this relativization we no longer have in mind the primitive
concepts of a certain special science, e.g. of the arithmetic of
real numbers. The set Rl is now replaced by an arbitrary set
V (the so-called universe of discourse or universal set), and
the symbol Sf is assumed to denote the set of all finite sequences
s such that the domain of s is included in V; the primitive sets
of sequences are certain subsets of Sf. We can even abstract
from the type of objects which constitute the set V, and we
can treat the terms occurring in the definition which we are to
construct as 'systematically ambiguous'[2] terms, intended to

---

[1] Cf. VII. I would point out that it seems profitable from the practical point
of view to base arithmetic on 'strong' systems of primitive notions, for it is
convenient to have a sufficiently large category of arithmetical concepts which
do not require definitions of order greater than 1.

[2] Cf. (90), vol. 1, pp. 39–41.

---

† By analysing this proof we obtain (just as in the case of the original
Ths. 1 and 2) some conclusions of a metamathematical nature; we see that,
in the modified system of arithmetic, every sentence of order 1 is still provable
or disprovable, and that there is a mechanical method which permits us in
each particular case to decide whether or not such a sentence is provable.
For a detailed proof of all these results see the monograph A. Tarski, *A Decision
Method for Elementary Algebra and Geometry*, 2nd ed., Berkeley, California,
1951.

denote simultaneously objects of different types. In this way, while considerably exceeding the limits of the task we set ourselves at first, we reach the general notion of an *arithmetically definable set of finite sequences* (with all the terms of a given type) *with respect to the family $\mathscr{F}$ of primitive finite sequences* (with all the terms of the same type). The family of all these definable sets will be denoted by the symbol '$\mathscr{D}f(\mathscr{F})$'.

DEFINITION 12. $\mathscr{D}f(\mathscr{F})$ is the product of all the families $\mathscr{X}$ which satisfy the condition (α) $\mathscr{F} \subseteq \mathscr{X}$, as well as the conditions (β)–(δ) of Def. 9.

In imitation of Def. 10, we shall make use of the family $\mathscr{D}f(\mathscr{F})$ in order to define the family $\mathscr{D}(\mathscr{F})$ of all *arithmetically definable sets* (composed of objects of a given type) *with respect to the family $\mathscr{F}$ of primitive sets of sequences* (with terms of the same type):

DEFINITION 13. $\mathscr{D}(\mathscr{F}) = \underset{X}{E}$ (there exists a set $S \in \mathscr{D}f(\mathscr{F})$ such that $D(S) = X$ and $\mathcal{C}(S)$ is a set composed of a single element).

In a completely analogous way we introduce the notion of an *arithmetically definable two-, three-, and n-termed relation* (between objects of a given type) *with respect to the family $\mathscr{F}$*.

From Defs. 12 and 13 it is easy to deduce various elementary properties of the families $\mathscr{D}f(\mathscr{F})$ and $\mathscr{D}(\mathscr{F})$, such, for example, as the following:

THEOREM 3. *(a) $\mathscr{D}f(\mathscr{F})$ is a family which satisfies the conditions (α)–(δ) of Def. 12; it is in fact the smallest of these families;*
(b) $\mathscr{D}f(\mathscr{F}) = \sum_{\mathscr{G} \in \mathfrak{G}} \mathscr{D}f(\mathscr{G})$ and $\mathscr{D}(\mathscr{F}) = \sum_{\mathscr{G} \in \mathfrak{G}} \mathscr{D}(\mathscr{G})$, where $\mathfrak{G}$ is the set of all finite subfamilies of the family $\mathscr{F}$;
(c) if $\mathscr{F} \subseteq \mathscr{G}$, we have $\mathscr{D}f(\mathscr{F}) \subseteq \mathscr{D}f(\mathscr{G})$ and $\mathscr{D}(\mathscr{F}) \subseteq \mathscr{D}(\mathscr{G})$;
(d) $\mathscr{D}f(\mathscr{D}f(\mathscr{F})) = \mathscr{D}f(\mathscr{F})$ and $\mathscr{D}(\mathscr{D}f(\mathscr{F})) = \mathscr{D}(\mathscr{F})$;
(e) if the family $\mathscr{F}$ is at most denumerable, then the families $\mathscr{D}f(\mathscr{F})$ and $\mathscr{D}(\mathscr{F})$ are also at most denumerable.

It is to be noted that in the proof of the above theorem we need not make use of the specific properties of the operations which figure in the conditions (β)–(δ) of Def. 12. It can be based exclusively on the fact that the family $\mathscr{D}f(\mathscr{F})$ is defined as the product of all the families $\mathscr{X}$ containing $\mathscr{F}$ and closed with respect to certain operations applicable to a finite number of sets of families $\mathscr{X}$ (this last detail occurs in the proof of (b) and (e)).

We are able—and this is an important point—*to prove Th. 3 (e) in an effective fashion*; we are able to define a (well-determined) function which correlates an infinite sequence S^* satisfying the formula

$$\mathscr{D}f(\mathscr{F}) = D(S^*) \quad \text{(i.e. } \mathscr{D}f(\mathscr{F}) = \{S_1^*, S_2^*, ..., S_n^*, ...\}\text{)}$$

with every infinite sequence S of sets (of type ω) such that $\mathscr{F} = D(S)$ (i.e. such that $\mathscr{F} = \{S_1, S_2, ..., S_n, ...\}$).

In all the applications of Defs. 12 and 13 to concrete families we never have to deal with families $\mathscr{F}$ of a completely arbitrary nature; they always satisfy certain conditions, defining what we shall call *regular families*.

DEFINITION 14. $\mathscr{F}$ is a *regular family*, if (α) $\mathscr{F}$ is a non-null family which is at most denumerable; (β) every set $S \in \mathscr{F}$ is a homogeneous set of sequences; (γ) if $k \in Nt$, $l \in Nt$, and $S \in \mathscr{F}$, we have $\dfrac{k}{l} S \in \mathscr{F}$.

An example of a regular family is the family composed of all the sets U_k, $M_{k,l}$, and $S_{k,l,m}$, where k, l, and m are any natural numbers.

The postulate of regularity, imposed on the families $\mathscr{F}$ to which we apply Defs. 12 and 13, appears quite natural when we recall that we are operating with primitive sets as the mathematical representatives of the primitive terms of some deductive science. Now, as is well known, it is possible to choose for each deductive science a system of primitive terms consisting exclusively of so-called *predicates*, i.e. of signs such as 'v', 'μ', and 'σ' which, when accompanied by a certain number of variables, the *arguments*, form sentential functions. It is precisely these sentential functions which we call primitive functions when we introduce the sets of sequences determined by them into the family $\mathscr{F}$. All the expressions, in particular therefore all the primitive functions, in a given deductive system are at

most denumerable in number—it is for this reason that we
have condition (α) of Def. 14; the sets of sequences determined
by sentential functions are always homogeneous—hence we
have condition (β). Finally, every time we consider a sentential function as one of the primitive functions, we consider also every function which can be obtained from it by
substituting one variable for another—therefore the family $\mathscr{F}$
should be closed with respect to the operation $\dfrac{k}{l}S$ which is
the mathematical analogue of the operation of substitution.[1]
It is precisely this which explains condition (γ) of Def. 14.

It is to be noted that the condition (γ) may be omitted in
Def. 14 provided we introduce into Def. 12 an analogous assumption on the families $\mathscr{X}$. Such a procedure would conform to
that metamathematical attitude towards the method of construction of deductive systems which admits, as primitive
expressions of the system, not predicates, e.g. 'v', 'μ', and 'σ',
but concrete sentential functions, e.g. '$v(x)$', '$\mu(x,y)$', and
'$\sigma(x,y,z)$', which are identified with the primitive functions;
then the sentential functions are obtained from primitive
expressions not only by using the five fundamental operations,
but also with the help of the operation of substitution. Such
an attitude, which incidentally is not common in practice, is in
principle possible and correct, although it would involve some
inessential complications.

However, according to the point of view we have adopted,
the inclusion of this assumption in Def. 12 (and in Def. 9) is
superfluous, as the following shows:

THEOREM 4. *If $\mathscr{F}$ is a regular family, then $\mathscr{D}f(\mathscr{F})$ is also
regular.*

This is easily obtained from Defs. 5–8 and 12 and Th. 3 e.

The other properties of the family $\mathscr{D}f(\mathscr{F})$ under the hypothesis
that $\mathscr{F}$ is regular can be obtained by analogy from theorems
of metamathematics, especially from those which express relations between arbitrary sentential functions and the primitive

---

[1] Cf. the remarks on p. 127 following Def. 8.

functions. As an example I give here the following theorem, which corresponds to the well-known theorem on the reduction of every sentential function to the so-called normal form,[1] and which greatly facilitates reasoning in this domain:

THEOREM 5. *For any regular family* $\mathscr{F}$, *a necessary and sufficient condition that* $S \in \mathscr{D}f(\mathscr{F})$ *is that* $S = \overset{\circ}{0}$ *or that* S *is of the form*

$$\overset{\circ}{\underset{k_{1,1}}{\sum}}\ \overset{\circ}{\underset{k_{1,2}}{\sum}}\ \cdots\ \overset{\circ}{\underset{k_{1,l_1}}{\sum}}\ \overset{\circ}{\underset{k_{2,1}}{\prod}}\ \overset{\circ}{\underset{k_{2,2}}{\prod}}\ \cdots\ \overset{\circ}{\underset{k_{2,l_2}}{\prod}}\ \cdots\ \overset{\circ}{\underset{k_{n-1,1}}{\sum}}\ \overset{\circ}{\underset{k_{n-1,2}}{\sum}}\ \cdots\ \overset{\circ}{\underset{k_{n-1,l_{n-1}}}{\sum}}\ \overset{\circ}{\underset{k_{n,1}}{\prod}}\ \overset{\circ}{\underset{k_{n,2}}{\prod}}\ \cdots\ \overset{\circ}{\underset{k_{n,l_n}}{\prod}}\ T,$$

where T *is a finite sum of finite products of sets* $X \in \mathscr{F}$ *and their complements* $\overset{\circ}{\overline{X}}$,[2] *where* n *is an even natural number and where* l_i (*for* $1 \leqslant i \leqslant n$) *and* $k_{i,j}$ (*for* $1 \leqslant i \leqslant n$ *and* $1 \leqslant j \leqslant l_i$) *are arbitrary natural numbers.*

Proof. If follows immediately from Th. 3 a that the condition in question is sufficient. To prove that it is necessary we proceed by induction: from Def. 14 and the formal properties of the fundamental operations on sets of sequences, we show that the family $\mathscr{X}$ of all the sets which satisfy the condition of the theorem has all the properties (α)–(δ) of Def. 12. Consequently every set $S = \mathscr{D}f(\mathscr{F})$ belongs to this family.[3]

As an example I also give an easy consequence of Defs. 12–14 concerning the family $\mathscr{D}(\mathscr{F})$.

THEOREM 6. *If* $\mathscr{F}$ *is a regular family, and* $X \in \mathscr{D}(\mathscr{F})$ *and* $Y \in \mathscr{D}(\mathscr{F})$, *then we have*

$$\overline{X} \in \mathscr{D}(\mathscr{F}),\text{[4]}\quad X + Y \in \mathscr{D}(\mathscr{F}),\quad and\quad X.Y \in \mathscr{D}(\mathscr{F}).$$

The relativized notions $\mathscr{D}f(\mathscr{F})$ and $\mathscr{D}(\mathscr{F})$ are of fundamental importance for the whole mathematical theory of definability. With their help we can, for example, easily define the notions of definable sets of orders higher than 1. In fact, for any given natural number n, let us agree to denote by '$\mathscr{D}f_n$' the family of all

---

[1] Cf. Hilbert, D., and Ackermann, W. (30), pp. 63–64.

[2] Cf. p. 133, note 1.

[3] This is analogous to the corresponding proof in metamathematics; cf. note 1, above.

[4] Here $\overline{X}$ is the complement of the set X (with respect to the universe of discourse V).

definable sets of order n of finite sequences of real numbers, and by '$\mathscr{D}_n$' the corresponding family of all definable sets of order n of real numbers themselves. Thus the families $\mathscr{D}f_1$ and $\mathscr{D}_1$ coincide with $\mathscr{D}f$ and $\mathscr{D}$ respectively. We now restrict ourselves to the case $n = 2$ and construct mathematical definitions of $\mathscr{D}f_2$ and $\mathscr{D}_2$ in the following way:

For the universe of discourse we take the family of all sets of real numbers (i.e. of all sub-sets of Rl); thus Sf is now the set of all finite sequences whose terms are sets of real numbers, and the meaning of the symbol 'R_N' (see Def. 3) changes correspondingly. X being any set of finite sequences of real numbers, we replace, in every sequence $s \in X$, each term s_k by the set $\{s_k\}$, and we denote the set of all finite sequences thus obtained by X^*. In particular we have (cf. Def. 4):

(a) $U_k^* = R_{\{k\}} \cdot \underset{s}{E}(s_k = \{1\})$;

(b) $M_{k,l}^* = R_{\{k,l\}} \cdot \underset{s}{E}\big($for some x and y, $s_k = \{x\}$, $s_l = \{y\}$, and $x \leqslant y\big)$;

(c) $S_{k,l,m}^* = R_{k,l,m} \cdot \underset{s}{E}\big($for some x, y and z, $s_k = \{x\}$, $s_l = \{y\}$, $s_m = \{z\}$, and $x = y + z\big)$.

We also put

(d) $E_{k\,l} = R_{\{k,l\}} \cdot \underset{s}{E}\big($for some x, $s_k = \{x\}$ and $x \in s_l\big)$.

The sets U_k^*, $M_{k,l}^*$, $S_{k,l,m}^*$, and $E_{k,l}$ are regarded as primitive sets of sequences, and the family of all these primitive sets is denoted by '$\mathscr{F}_0$'.

We now define $\mathscr{D}f_2$ as the family of all sets X such that $X^* \in \mathscr{D}f(\mathscr{F}_0)$. The definition of $\mathscr{D}_2$ is obtained from Def. 10 by providing the symbols '$\mathscr{D}$' and '$\mathscr{D}f$' with subscripts '2'; we could also define $\mathscr{D}_2$ as the family of all sets X such that $\{X\} \in \mathscr{D}(\mathscr{F}_0)$. To convince ourselves that the definition just outlined is materially correct we proceed as in the case of $\mathscr{D}$ (cf. the remarks following Def. 10).

In a similar way we define the notions $\mathscr{D}f_n$ and $\mathscr{D}_n$ as well as the relativized notions $\mathscr{D}f_n(\mathscr{F})$ and $\mathscr{D}_n(\mathscr{F})$ for every natural number n. We can show that each of the families $\mathscr{D}_n$ (or $\mathscr{D}f_n$), is contained as a part in $\mathscr{D}_{n+1}$ (or $\mathscr{D}f_{n+1}$); moreover, by means of a diagonal procedure, we can always construct a set which belongs to $\mathscr{D}_{n+1}$ without belonging to $\mathscr{D}_n$.

Professor Kuratowski has drawn my attention to a geometrical interpretation of the concepts introduced here. In particular, a finite sequence with counter domain $\{k, l, ..., p\}$ can be interpreted as a point in a space of a finite number of dimensions and of which X_k, X_l,..., X_p are the coordinate axes. The set $R_{\{k,l,...,p\}}$ is then the entire space; the homogeneous sets are the sets contained in one such space. The set U_k is composed of a single point with coordinate 1 situated on the axis X_k. The sets $M_{k,l}$ and $S_{k,l,m}$ become known elementary figures of analytic geometry, namely half-planes and planes in certain special positions. The geometrical interpretation of the operations on sets introduced in Defs. 5–8 is not at all difficult. In particular, the operation $\overset{\circ}{\underset{k}{\sum}}$ is easily recognized as that of projection parallel to the X_k axis. In this interpretation it is not necessary to distinguish between a term and the sequence composed of this term; Defs. 10 and 13 become superfluous. Ths. 1 and 2 take the form of certain theorems of analytic geometry.

From the abstract character of these considerations we see that the hypothesis that the axes X_k, X_l,..., X_p are Euclidean straight lines is no longer necessary. They may, on the contrary, be abstract spaces of absolutely any kind; the space $R_{\{k,l,...,p\}}$ is then their combinatorial or Cartesian product.

The joint note by Kuratowski and myself, which immediately follows this article, as well as further investigations by Kuratowski,[1] seem to testify to the great heuristic importance of the geometrical interpretation of the constructions sketched in this article. In particular, as we shall show in the next article, the following theorem is a consequence of Def. 12:

If $\mathscr{F}$ is a family of projective sets, so also is $\mathscr{D}f(\mathscr{F})$.

In conclusion we may point out that, instead of finite sequences, we can operate in all the preceding constructions,

---

[1] See Kuratowski, C. (39), and also (41), pp. 168 ff. and 243 ff.

with ordinary infinite sequences (with counter domain identical with the set of natural numbers) or, in other words, with points of space with an infinite number of dimensions (Hilbert space). We could even use exclusively operations on sets of sequences which never lead out of this space. Moreover the operations of complementation, addition, and multiplication would then coincide with ordinary operations of the algebra of sets. Thus, the modification suggested would certainly simplify the development and enhance its intrinsic elegance. On the other hand, the modified construction would have some defects from the standpoint of applications, and its heuristic value might possibly be diminished.

VII

LOGICAL OPERATIONS AND PROJECTIVE SETS†

A SENTENTIAL function $\phi(x_1,...,x_n)$ of n real variables will be called *projective* if the set $\underset{x_1...x_n}{E}\ \phi(x_1,...,x_n)$,[1] i.e. the set of points of n-dimensional space satisfying the function $\phi(x_1,...,x_n)$, is projective.[2]

We propose to show that *the five logical operations* (see § 1) *when applied to projective functions* (a finite number of times) *always yield projective functions.* That is to say *every set of points definable by means of sentential functions is projective.*[3]

In the definitions which we usually encounter, especially in classical mathematics, the functions $\phi(x_1,...,x_n)$ are projective functions of a particularly simple type: the sets $\underset{x_1...x_n}{E}\ \phi(x_1,...,x_n)$ which correspond to them are curves, surfaces, etc., which are elementary in the sense of analysis (such as the plane $z = x+y$,

---

[1] This notation is due to Lebesgue. If, for example, $\phi(x, y, z)$ stands for the equation $z = x+y$, the set $\underset{xyz}{E}\ \phi(x, y, z)$ is the plane given by this equation; if

$$\phi(x, y) \equiv (x < y),$$

the set $\underset{xy}{E}\ \phi(x, y)$ is the upper part of the XY plane determined by the diagonal $x = y$.

[2] A set of points is called *projective*, according to Lusin, when it is obtained from a closed set by applying, any finite number of times, two operations: (1) orthogonal projection, (2) passage to the complement.

[3] It is a question of *explicit* (cf. § 4) and *arithmetical* (or *elementary*) definitions in the sense established in VI, Dfs. 9 and 12, pp. 128 and 136. Cf. here the theorem on p. 141 of the latter article, which is formulated without using the concept of sentential function.

---

† BIBLIOGRAPHICAL NOTE. The main results of this article were presented by the authors to the Polish Mathematical Society, Section of Lwów, in two talks on 12 June 1930 and 15 December 1930. The summaries of these talks are given in Kuratowski, C.´(40) and Tarski, A. (74). The article itself appeared under the title 'Les opérations logiques et les ensembles projectifs', *Fundamenta Mathematicae*, vol. 17 (1931), pp. 240–8.

the hyperbolic paraboloid $z = xy$, the surface $z = x^y$, the set of positive integers, etc.). All these definitions define projective sets. It is, on the other hand, remarkable that it is possible to specify by means of these elementary functions sets non-measurable B, functions which cannot be represented analytically, and projective sets of an arbitrary class.

In order to specify non-projective sets (in an explicit fashion) it is indispensable to use, apart from real variables, variables of a higher type which admit sets of points as values.[1]

We thus see that the concept of the projective set of Lusin suggests itself in a natural way as soon as we wish to study sets (or functions) which can be specified effectively.

§ 1. LOGICAL NOTATION. TRANSFORMATION FORMULAS

If α represents a sentence, then α' represents the negation of this sentence; if α and β represent two sentences, then $\alpha + \beta$ (read 'α or β') represents the logical sum, and $\alpha . \beta$ (read 'α and β') represents the logical product of these sentences.

If $\phi(x)$ represents a sentential function, then $\sum_x \phi(x)$ is read 'there exists an x such that $\phi(x)$', $\prod_x \phi(x)$ is read 'for every x we have $\phi(x)$'.[2]

If A and B represent two sets (of real numbers), and A_x represents a set dependent on a parameter x, then the following equivalences hold:

$$(1) \qquad (x \in A)' \equiv (x \in A'),$$

$$(2) \qquad (x \in A) + (x \in B) \equiv x \in (A + B),$$

$$(3) \qquad (x \in A) . (x \in B) \equiv x \in A . B,$$

$$(4) \qquad \sum_x (t \in A_x) \equiv t \in \sum_x A_x,$$

$$(5) \qquad \prod_x (t \in A_x) \equiv t \in \prod_x A_x.$$

---

[1] The definable non-projective sets are then of the order $n \geqslant 2$ in the sense of VI.

[2] For example, $\prod_x \sum_y (x < y)$ states that for every x there exists a y such that $x < y$. The condition for a function $f(x)$ to be bounded is expressed thus:
$$\sum_y \prod_x |f(x)| < y.$$

The symbols "'", '$+$', '$.$', '$\sum$', and '$\prod$' on the left-hand sides of these equivalences denote the five logical operations and on the right-hand sides the corresponding set-theoretical operations (complementation, set addition, and set multiplication).

Remembering the formula

$$(6) \qquad t \in E_x \phi(x) \equiv \phi(t),$$

which defines the operation E, we can easily deduce, from the preceding formulas, the following identities:

$$(7) \qquad E_x(\phi(x))' = \Big(E_x \phi(x)\Big)',$$

$$(8) \qquad E_x(\phi(x)+\psi(x)) = E_x \phi(x)+E_x \psi(x),$$

$$(9) \qquad E_x(\phi(x).\psi(x)) = E_x \phi(x).E_x \psi(x),$$

$$(10) \qquad E_x \sum_y \phi(x,y) = \sum_y E_x \phi(x,y),$$

$$(11) \qquad E_x \prod_y \phi(x,y) = \prod_y E_x \phi(x,y).$$

By way of example, we prove the formula (10).

By (6), $t \in E_x \sum_y \phi(x,y) \equiv \sum_y \phi(t,y)$. Denoting by A_y the set $E_x \phi(x,y)$ we have, according to (6), $\phi(t,y) \equiv t \in E_x \phi(x,y) \equiv t \in A_y$, whence, by (4),

$$\sum_y \phi(t,y) \equiv \sum_y t \in A_y \equiv t \in \sum_y A_y \equiv t \in \sum_y E_x \phi(x,y).$$

§ 2. LOGICO-MATHEMATICAL DUALITY

The formulas (7)–(11) show that the operation E may be permuted with each of the five logical operations, each of which changes the logical sense into the mathematical sense (or vice versa). We thus see that if the sentential function $\alpha(x)$ is obtainable from the functions $\phi_1,..., \phi_k$ (with different variables or not) by applying the five logical operations, then the set $E_x \alpha(x)$ is obtained *in the same way* from the sets $E_x \phi_1,..., E_x \phi_k$.[1]

---

[1] This justifies the use of the same symbols in two senses, logical and mathematical, without danger of misunderstanding. We thus see that the passage from the sentential function $\alpha(x)$ to the set $E_x \alpha(x)$ is only *another way of reading* the expression which defines $\alpha(x)$.

This also remains true if α is a function of *several* variables, for the foregoing formulas can be generalized immediately by replacing the variable x by a system of n variables.[1]

In the sequel this logico-mathematical duality will be expressed in another form by taking into account the following *geometrical interpretation* of the logical operation $\sum$:

(12) *The set* $E \sum_{x} \sum_{y} \phi(x,y)$ *is the projection of the set* $E_{xy} \phi(x,y)$ *parallel to the Y axis.*

From (6) we see

$$x_0 \in E_x \sum_y \phi(x,y) \equiv \sum_y \phi(x_0,y) \equiv \sum_y \left[(x_0,y) \in E_{xy}\phi(x,y) \right]$$

and this last sentence asserts that there exists a point having the abscissa x_0 which belongs to the set $E_{xy}\phi(x,y)$; in other words that x_0 belongs to the projection of this set onto the X axis.

Examples

1. Let A be the set of all integers. Let $\mathscr{E}(x)$ be the largest integer which does not exceed x; in symbols

$$[y = \mathscr{E}(x)] \equiv (y \in A).(y \leqslant x).(x < y+1).$$

Hence, by formula (9),

$$E_{xy}[y = \mathscr{E}(x)] = E_{xy}(y \in A).E_{xy}(y \leqslant x).E_{xy}(x < y+1).$$

We thus see that the graph of the equation $y = \mathscr{E}(x)$ is the common part of the three sets: (1) the set of horizontal straight lines with integer ordinates, (2) the half plane $y \leqslant x$, (3) the half plane $x < y+1$.

2. Evidently we have $(x \geqslant 0) \equiv \sum_y (x = y^2)$. Consequently the set of non-negative numbers $= E(x \geqslant 0) = E_x \sum_y (x = y^2)$ $=$ the projection parallel to the Y axis of the parabola $E_{xy}(x = y^2)$.

3. Let us express in symbols the fact that, if x is a rational

---

[1] The case $n = 0$ is included; in this case $\phi(x)$ becomes a sentence and, denoting by 0 and 1 its logical value (i.e. 'false' or 'true') and employing the same symbols to denote the empty set and the space of real numbers respectively, we have

$$(x \in 0) \equiv 0, \quad (x \in 1) \equiv 1, \quad E_x 0 = 0, \quad E_x 1 = 1.$$

number $(x \in R)$, it is of the form y/z where y and z are integers of which the second is not zero:

$$R = E \sum_x \sum_y \sum_z (y \in A).(z \in A).(xz = y).(z \neq 0).$$

The set $\underset{xyz}{E} (y \in A).(z \in A).(xz = y).(z \neq 0)$ is, according to (9), the common part of four sets: of the set formed by the planes $\underset{xyz}{E} (y \in A)$, of the set formed by the planes $\underset{xyz}{E} (z \in A)$, of the hyperbolic paraboloid $y = xz$, of the whole space minus the plane $z = 0$. By projecting this set onto the XY plane, and then onto the X axis, we obtain, in conformity with (12), the set R.

REMARKS

(1) In all the foregoing the hypothesis that the variables x, $y, z,...$ range over the set of real numbers is in no way essential. We may suppose that x ranges over an *arbitrary* set X, y over a set Y, distinct or not from X, z over a set Z, etc. The Euclidean plane XY should then be replaced by the combinatorial or Cartesian product of X and Y, i.e. by the set of pairs (x, y), where $x \in X$ and $y \in Y$ (in an analogous way, the combinatorial product of X, Y, Z is the set of points (x, y, z), where $x \in X, y \in Y, z \in Z$). The extension of the concept of projection to the general case is obvious.

(2) According to (12) a *continuous* geometrical operation corresponds to the logical operation $\sum$. Numerous topological applications of the logical calculus are connected with this fact (see Kuratowski, C. (39)).

(3) Regarding the operation $\prod$, it is easily seen that the set $\underset{x}{E} \prod_y \phi(x, y)$ is composed of all the x_0 such that the straight line $x = x_0$ is completely contained in the set $\underset{xy}{E} \phi(x, y)$. This geometrical operation, not being in general continuous, is often more advantageously defined by means of the operation $\sum$ with the help of the generalized De Morgan's law:

(13) $$\prod_x \phi(x) \equiv \left[\sum_x (\phi(x))' \right]',$$

which is a generalization of the well-known formula

$$\alpha . \beta \equiv (\alpha' + \beta')'.$$

We thus see that the five logical operations considered can be reduced to three: negation, summation, and the operation $\sum$.

(4) The relation of implication, symbolically represented by $\alpha \rightarrow \beta$ also reduces to the same operations for we have

$$(\alpha \rightarrow \beta) \equiv (\alpha' + \beta).$$

§ 3. PROJECTIVE SETS AND SENTENTIAL FUNCTIONS

Projective sets possess the following fundamental properties:[1]

(a) the complement of a projective set is projective;

(b) the sum (and also the product) of two projective sets is a projective set;

(c) if $\underset{x}{E}\phi(x)$ is projective, so also is $\underset{xy}{E}\phi(x)$;[2]

(d) the projection of a projective set is a projective set.

From this we deduce, by means of the formulas (7)–(9), (12), the principal theorem of this article:

The five logical operations, when applied to projective sentential functions, always lead to projective sentential functions.

First, by remark (3) we can reduce the five operations to three: negation, summation, and the operation $\sum$.

Now, (1) If the function $\phi(x)$ is projective, the set $\underset{x}{E}\phi(x)$ is likewise projective; for by (a) $\left(\underset{x}{E}\phi(x)\right)'$ is projective, from which, by (7), $\underset{x}{E}(\phi(x))'$ is likewise projective, that is to say the function $(\phi(x))'$ is projective.

(2) Let $\phi(x_1,...,x_n) \equiv \psi(x_{k_1},...,x_{k_j}) + \chi(x_{l_1},...,x_{l_m})$, where $k_1,...,k_j$, $l_1,...,l_m \leqslant n$;[3] if the functions ψ and χ are assumed to be

---

[1] See, for example, N. Lusin, *Leçons sur les ensembles analytiques* (Paris, 1930), pp. 276–7.

[2] The set $\underset{xy}{E}\phi(x)$ is obtained by passing a vertical straight line through each point of the set $\underset{x}{E}\phi(x)$.

[3] e.g. $\phi(x, y, z) \equiv \psi(x, y) + \chi(y, z).$

projective, we conclude from (c) that the sets $\underset{x_1 \dots x_n}{E}\, \psi(x_{k_1}, \dots, x_{k_j})$ and $\underset{x_1 \dots x_n}{E}\, \chi(x_{l_1}, \dots, x_{l_m})$ are projective and since, by (8),

$$\underset{x_1 \dots x_n}{E}\, \phi(x_1, \dots, x_n) = \underset{x_1 \dots x_n}{E}\, \psi(x_{k_1}, \dots, x_{k_j}) + \underset{x_1 \dots x_n}{E}\, \chi(x_{l_1}, \dots, x_{l_m}),$$

we conclude from (b) that $\phi(x_1, \dots, x_n)$ is a projective function.

(3) Assuming that $\phi(x_1, \dots, x_n)$ is a projective function, we have to prove that $\sum_{x_k} \phi(x_1, \dots, x_n)$ is also projective.[1] This is evident for the case where $k > n$, for in this case $\sum_{x_k} \phi(x_1, \dots, x_n) \equiv \phi(x_1, \dots, x_n)$. Suppose then that $k \leqslant n$. Now according to (12) the set $\underset{x_1 \dots x_{k-1} x_{k+1} \dots x_n}{E}\, \sum_{x_k} \phi(x_1, \dots, x_n)$ is an orthogonal projection of the set $\underset{x_1 \dots x_n}{E}\, \phi(x_1, \dots, x_n)$; and, by (d), this is a projective set. q.e.d.

EXAMPLES AND NOTES

Let $\phi(x, y)$ be a projective sentential function of two variables. The set $M = \underset{xy}{E}\, \phi(x, y)$ is then a plane projective set. Let Q be the set formed by the union of all the straight lines contained in M. In symbols

$$(x, y) \in Q \equiv \sum_{abc} \Big\{ (ax+by = c) . (a^2+b^2 \neq 0) .$$

$$. \prod_{uv} [(au+bv = c) \to \phi(u, v)] \Big\}.[2]$$

The set $\underset{abcxy}{E}\, (ax+by = c)$ is a closed set (in the space of 5 dimensions), and the set $\underset{ab}{E}\, (a^2+b^2 = 0)$ is composed of a single point $(0, 0)$, thus it results directly from our theorem that Q is a projective set.[3]

We thus see that the operation of deriving Q from M does not take us out of the domain of projective sets. This is the

---

[1] If $k = n = 1$, this is a sentence. Now, each sentence is a projective sentential function, since the whole space as well as the empty space are projective sets; cf. p. 146, footnote.

[2] Q may also be defined as follows:

$$(x, y) \in Q \equiv \sum_t \prod_{uv} \{[(u-x)\sin t = (v-y)\cos t] \to \phi(u, v)\}.$$

[3] Which may, moreover, be non-measurable B when M is open. See Nikodym et Sierpiński, *Fundamenta Mathematicae*, vol. 7, p. 259.

case for the greater part of the operations which are considered in mathematics. In order to convince oneself of this it suffices to write their definitions in logical symbols.

Regarding the formulation of the theorem of § 3, it is to be noted that if, instead of supposing that the given sentential functions are projective, we make more restricted hypotheses, the function which is obtained may also be characterized in a more specific manner. To give an example let us call every sentential function *linear* if it is obtained by addition and multiplication from sentential functions of the form $f(x_1,...,x_n) = 0$ or > 0, where f is a polynomial of degree 1 with integral coefficients. Then the property of being a linear sentential function proves to be invariant under the five logical operations considered. In particular, the sets of real numbers defined by means of linear functions prove to coincide with finite sums of intervals (closed or open) with rational end points.[1]

On the other hand important examples of classes of sentential functions are known which are invariant, not under all, but under some of the logical operations discussed. Such is, for instance, the class of all functions corresponding to closed sets or to Borelian sets, etc.; see Kuratowski, C. (41), pp. 168–9 and 245.

§ 4. IMPLICIT DEFINITIONS

Hitherto we have supposed that the 'unknown' function α is obtained from a system of given functions $\phi_1,..., \phi_n$, by performing upon them logical operations; the type of this definition is

$$\alpha \equiv \Omega(\phi_1,..., \phi_n).$$

These are definitions properly so called, or *explicit* definitions.

If we carry out the five logical operations as well as a change of variables[2] on the system $\phi_1,..., \phi_n$ augmented by the unknown

---

[1] We obtain another example of this kind if we replace, in the example just given, linear sentential functions (corresponding to polynomials of the first degree) by algebraic sentential functions (corresponding to polynomials of an arbitrary degree) and also rational numbers by algebraic numbers. Both examples have been discussed (although in different terms) in VI, see in particular the first footnote on p. 133.

[2] For example, if we pass from $\phi(x, y)$ to $\phi(x, z)$.

function α, and if the equation

$$\Omega(\phi_1,...,\phi_n,\alpha) \equiv 0$$

admits one and only one solution (for α), we can say that this equation defines α *implicitly*.

A very common type of implicit definition is represented by definitions *by induction* (finite or transfinite).

By means of implicit definitions we are able to pass out of the domain of projective sets. They lead to a new class of sets, *implicitly projective sets* which it would be interesting to study.

VIII

THE CONCEPT OF TRUTH IN FORMALIZED LANGUAGES†

INTRODUCTION

THE present article is almost wholly devoted to a single problem—*the definition of truth*. Its task is to construct—with reference to a given language—*a materially adequate and formally correct definition of the term 'true sentence'*. This problem, which belongs to the classical questions of philosophy, raises considerable difficulties. For although the meaning of the term 'true sentence' in colloquial language seems to be quite clear and intelligible, all attempts to define this meaning more precisely have hitherto been fruitless, and many investigations in which this term has been used and which started with apparently evident premises have often led to paradoxes and antinomies (for which, however, a more or less satisfactory solution has been found). The concept of truth shares in this respect the fate of other analogous concepts in the domain of the semantics of language.

The question how a certain concept is to be defined is correctly formulated only if a list is given of the terms by means of which the required definition is to be constructed. If the definition is to fulfil its proper task, the sense of the terms in this list must admit of no doubt. The question thus naturally arises: What terms are we to use in constructing the definition of truth? In the course of these investigations I shall not neglect to clarify this question. In this construction I shall not make use

† BIBLIOGRAPHICAL NOTE. This article was presented (by J. Łukasiewicz) to the Warsaw Scientific Society on 21 March 1931. For reasons beyond the author's control, publication was delayed by two years. The article appeared in Polish in Tarski, A. (73). A German translation was published under the title 'Der Wahrheitsbegriff in den formalisierten Sprachen', in *Studia Philosophica*, vol. 1 (1936) (reprint dated 1935), pp. 261–405; it is provided with a Postscript in which some views which had been stated in the Polish original underwent a rather essential revision and modification. The present English version is based upon the German translation. For earlier publications and historical information concerning the results of this work see p. 154, footnote, p. 247, footnote, and the concluding remarks of the Postscript.

of any semantical concept if I am not able previously to reduce it to other concepts.

A thorough analysis of the meaning current in everyday life of the term 'true' is not intended here. Every reader possesses in greater or less degree an intuitive knowledge of the concept of truth and he can find detailed discussions on it in works on the theory of knowledge. I would only mention that throughout this work I shall be concerned exclusively with grasping the intentions which are contained in the so-called *classical* conception of truth ('true—corresponding with reality') in contrast, for example, with the *utilitarian* conception ('true—in a certain respect useful').[1]

The extension of the concept to be defined depends in an essential way on the particular language under consideration. The same expression can, in one language, be a true statement, in another a false one or a meaningless expression. There will be no question at all here of giving a single general definition of the term. The problem which interests us will be split into a series of separate problems each relating to a single language.

In § 1 colloquial language is the object of our investigations. The results are entirely negative. With respect to this language not only does the definition of truth seem to be impossible, but even the consistent use of this concept in conformity with the laws of logic.

In the further course of this discussion I shall consider exclusively the scientifically constructed languages known at the present day, i.e. the formalized languages of the deductive sciences. Their characteristics will be described at the beginning of § 2. It will be found that, from the standpoint of the present problem, these languages fall into two groups, the division being based on the greater or less stock of grammatical forms in a particular language. In connexion with the 'poorer' languages the problem of the definition of truth has a positive solution: there is a uniform method for the construction of the required

---

[1] Cf. Kotarbiński, T. (37), p. 126 (in writing the present article I have repeatedly consulted this book and in many points adhered to the terminology there suggested).

definition in the case of each of these languages. In §§ 2 and 3 I shall carry out this construction for a concrete language in full and in this way facilitate the general description of the above method which is sketched in § 4. In connexion with the 'richer' languages, however, the solution of our problem will be negative, as will follow from the considerations of § 5. For the languages of this group we shall never be able to construct a correct definition of the notion of truth.† Nevertheless, everything points to the possibility even in these cases—in contrast to the language of everyday life—of introducing a consistent and correct use of this concept by considering it as a primitive notion of a special science, namely of the theory of truth, and its fundamental properties are made precise through axiomatization.

The investigation of formalized languages naturally demands a knowledge of the principles of modern formal logic. For the construction of the definition of truth certain purely mathematical concepts and methods are necessary, although in a modest degree. I should be happy if this work were to convince the reader that these methods already are necessary tools even for the investigation of purely philosophical problems.[1]

§ 1. THE CONCEPT OF TRUE SENTENCE IN EVERYDAY OR COLLOQUIAL LANGUAGE

For the purpose of introducing the reader to our subject, a consideration—if only a fleeting one—of the problem of defining truth in colloquial language seems desirable. I wish especially

[1] This was communicated to the Society of Sciences in Warsaw by J. Łuka-siewicz on 21 March 1931. The results it contains date for the most part from 1929. I have reported on this, among other things, in two lectures which I gave under the title 'On the Concept of Truth in relation to formalized deductive systems' at the logical section of the Philosophical Society in Warsaw (8 October 1930) and at the Polish Philosophical Society in Lwów (15 December 1930), a résumé of which appeared in Tarski, A. (73). For reasons unconnected with me the printing of this work was much delayed. This enabled me to supplement the text with some rather important results (cf. p. 247, footnote). In the meantime a résumé of the chief results has appeared in Tarski, A. (76).

---

† Regarding this statement compare the Postscript.

to emphasize the various difficulties which the attempts to solve this problem have encountered.[1]

Amongst the manifold efforts which the construction of a correct definition of truth for the sentences of colloquial language has called forth, perhaps the most natural is the search for a *semantical definition*. By this I mean a definition which we can express in the following words:

(1) *a true sentence is one which says that the state of affairs is so and so, and the state of affairs indeed is so and so.*[2]

From the point of view of formal correctness, clarity, and freedom from ambiguity of the expressions occurring in it, the above formulation obviously leaves much to be desired. Nevertheless its intuitive meaning and general intention seem to be quite clear and intelligible. To make this intention more definite, and to give it a correct form, is precisely the task of a semantical definition.

As a starting-point certain sentences of a special kind present themselves which could serve as partial definitions of the truth of a sentence or more correctly as explanations of various concrete turns of speech of the type '*x* is a true sentence'. The general scheme of this kind of sentence can be depicted in the following way:

(2) *x is a true sentence if and only if p.*

In order to obtain concrete definitions we substitute in the

---

[1] The considerations which I shall put forward in this connexion are, for the most part, not the result of my own studies. Views are expressed in them which have been developed by S. Leśniewski in his lectures at the University of Warsaw (from the year 1919/20 onwards), in scientific discussions and in private conversations; this applies, in particular, to almost everything which I shall say about expressions in quotation marks and the semantical antinomies. It remains perhaps to add that this fact does not in the least involve Leśniewski in the responsibility for the sketchy and perhaps not quite precise form in which the following remarks are presented.

[2] Very similar formulations are found in Kotarbiński, T. (37), pp. 127 and 136, where they are treated as commentaries which explain approximately the classical view of truth.

Of course these formulations are not essentially new; compare, for example, the well-known words of Aristotle: 'To say of what is that it is not, or of what is not that it is, is false, while to say of what is that it is, or of what is not that it is not, is true.' (Aristotle, *Metaphysica*, Γ, 7, 27; *Works*, vol. 8, English translation by W. D. Ross, Oxford, 1908.)

place of the symbol 'p' in this scheme any sentence, and in the place of 'x' any individual name of this sentence.

If we are given a name for a sentence, we can construct an explanation of type (2) for it, provided only that we are able to write down the sentence denoted by this name. The most important and common names for which the above condition is satisfied are the so-called *quotation-mark names*. We denote by this term every name of a sentence (or of any other, even meaningless, expression) which consists of quotation marks, left- and right-hand, and the expression which lies between them, and which (expression) is the object denoted by the name in question. As an example of such a name of a sentence the name "it is snowing" will serve. In this case the corresponding explanation of type (2) is as follows:

(3) '*it is snowing*' *is a true sentence if and only if it is snowing.*[1]

Another category of names of sentences for which we can construct analogous explanations is provided by the so-called *structural-descriptive names*. We shall apply this term to names which describe the words which compose the expression denoted

---

[1] Statements (sentences) are always treated here as a particular kind of expression, and thus as linguistic entities. Nevertheless, when the terms 'expression', 'statement', etc., are interpreted as names of concrete series of printed signs, various formulations which occur in this work do not appear to be quite correct, and give the appearance of a widespread error which consists in identifying expressions of like shape. This applies especially to the sentence (3), since with the above interpretation quotation-mark names must be regarded as general (and not individual) names, which denote not only the series of signs in the quotation marks but also every series of signs of like shape. In order to avoid both objections of this kind and also the introduction of superfluous complications into the discussion, which would be connected among other things with the necessity of using the concept of likeness of shape, it is convenient to stipulate that terms like 'word', 'expression', 'sentence', etc., do not denote concrete series of signs but whole classes of such series which are of like shape with the series given; only in this sense shall we regard quotation-mark names as individual names of expressions. Cf. Whitehead, A. N., and Russell, B. A. W. (90), vol. 1, pp. 661–6 and—for other interpretations of the term 'sentence'—Kotarbiński, T. (37), pp. 123–5.

I take this opportunity of mentioning that I use the words 'name' and 'denote' (like the words 'object', 'class', 'relation') not in *one*, but in many distinct senses, because I apply them both to objects in the narrower sense (i.e. to individuals) and also to all kinds of classes and relations, etc. From the standpoint of the theory of types expounded in Whitehead, A. N., and Russell, B. A. W. (90) (vol. 1, pp. 139–68) these expressions are to be regarded as systematically ambiguous.

by the name, as well as the signs of which each single word is composed and the order in which these signs and words follow one another. Such names can be formulated without the help of quotation marks. For this purpose we must have, in the language we are using (in this case colloquial language), individual names of some sort, but not quotation-mark names, for all letters and all other signs of which the words and expressions of the language are composed. For example we could use '*A*', '*E*', '*Ef*', '*Jay*', '*Pe*' as names of the letters '*a*', '*e*', '*f*', '*j*', '*p*'. It is clear that we can correlate a structural-descriptive name with every quotation-mark name, one which is free from quotation marks and possesses the same extension (i.e. denotes the same expression) and vice versa. For example, corresponding to the name " 'snow' " we have the name 'a word which consists of the four letters: *Es*, *En*, *O*, Double-*U* following one another'. It is thus clear that we can construct partial definitions of the type (2) for structual-descriptive names of sentences. This is illustrated by the following example:

(4) *an expression consisting of three words of which the first is composed of the two letters I and Te (in that order) the second of the two letters I and Es (in that order) and the third of the seven letters Es, En, O, Double-U, I, En, and Ge (in that order), is a true sentence if and only if it is snowing.*

Sentences which are analogous to (3) and (4) seem to be clear and completely in accordance with the meaning of the word 'true' which was expressed in the formulation (1). In regard to the clarity of their content and the correctness of their form they arouse, in general, no doubt (assuming of course that no such doubts are involved in the sentences which we substitute for the symbol '*p*' in (2)).

But a certain reservation is nonetheless necessary here. Situations are known in which assertions of just this type, in combination with certain other not less intuitively clear premises, lead to obvious contradictions, for example the *antinomy of the liar*. We shall give an extremely simple formulation of this antinomy which is due to J. Łukasiewicz.

For the sake of greater perspicuity we shall use the symbol 'c' as a typographical abbreviation of the expression 'the sentence printed on this page, line 5 from the top'. Consider now the following sentence:

c is not a true sentence.

Having regard to the meaning of the symbol 'c', we can establish empirically:

(α) *'c is not a true sentence' is identical with c.*

For the quotation-mark name of the sentence c (or for any other of its names) we set up an explanation of type (2):

(β) *'c is not a true sentence' is a true sentence if and only if c is not a true sentence.*

The premisses (α) and (β) together at once give a contradiction:

c is a true sentence if and only if c is not a true sentence.

The source of this contradiction is easily revealed: in order to construct the assertion (β) we have substituted for the symbol '*p*' in the scheme (2) an expression which itself contains the term 'true sentence' (whence the assertion so obtained—in contrast to (3) or (4)—can no longer serve as a partial definition of truth). Nevertheless no rational ground can be given why such substitutions should be forbidden in principle.

I shall restrict myself here to the formulation of the above antinomy and will postpone drawing the necessary consequences of this fact till later. Leaving this difficulty on one side I shall next try to construct a definition of true sentence by generalizing explanations of type (3). At first sight this task may seem quite easy—especially for anyone who has to some extent mastered the technique of modern mathematical logic. It might be thought that all we need do is to substitute in (3) any sentential variable (i.e. a symbol for which any sentence can be substituted) in place of the expression 'it is snowing' which occurs there twice, and then to establish that the resulting formula holds for every value of the variable, and thus without further difficulty reach a sentence which includes all assertions of type (3) as special cases:

(5) *for all p, 'p' is a true sentence if and only if p.*

But the above sentence could not serve as a general definition of the expression '*x* is a true sentence' because the totality of possible substitutions for the symbol '*x*' is here restricted to quotation-mark names. In order to remove this restriction we must have recourse to the well-known fact that to every true sentence (and generally speaking to every sentence) there corresponds a quotation-mark name which denotes just that sentence.[1] With this fact in mind we could try to generalize the formulation (5), for example, in the following way:

(6) *for all x, x is a true sentence if and only if, for a certain p, x is identical with 'p' and p.*

At first sight we should perhaps be inclined to regard (6) as a correct semantical definition of 'true sentence', which realizes in a precise way the intention of the formulation (1) and therefore to accept it as a satisfactory solution of our problem. Nevertheless the matter is not quite so simple. As soon as we begin to analyse the significance of the quotation-mark names which occur in (5) and (6) we encounter a series of difficulties and dangers.

Quotation-mark names may be treated like single words of a language, and thus like syntactically simple expressions. The single constituents of these names—the quotation marks and the expressions standing between them—fulfil the same function as the letters and complexes of successive letters in single words. Hence they can possess no independent meaning. Every quotation-mark name is then a constant individual name of a definite expression (the expression enclosed by the quotation marks) and in fact a name of the same nature as the proper name of a man. For example, the name "*p*" denotes one

---

[1] For example, this fact could be formulated in the following way:

(5') *for all x, if x is a true sentence, then—for a certain p—x is identical with 'p';*

from the premises (5) and (5') the sentence (6) given below can be derived as a conclusion.

of the letters of the alphabet. With this interpretation, which seems to be the most natural one and completely in accordance with the customary way of using quotation marks, partial definitions of the type (3) cannot be used for any significant generalizations. In no case can the sentences (5) or (6) be accepted as such a generalization. In applying the rule called the rule of substitution to (5) we are not justified in substituting anything at all for the letter '*p*' which occurs as a component part of a quotation-mark name (just as we are not permitted to substitute anything for the letter '*t*' in the word '*true*'). Consequently we obtain as conclusion not (5) but the following sentence: '*p*' *is a true sentence if and only if it is snowing*. We see at once from this that the sentences (5) and (6) are not formulations of the thought we wish to express and that they are in fact obviously senseless. Moreover, the sentence (5) leads at once to a contradiction, for we can obtain from it just as easily in addition to the above given consequence, the contradictory consequence: '*p*' *is a true sentence if and only if it is not snowing*. Sentence (6) alone leads to no contradiction, but the obviously senseless conclusion follows from it that the letter '*p*' is the only true sentence.

To give greater clarity to the above considerations it may be pointed out that with our conception of quotation-mark names they can be eliminated and replaced everywhere by, for example, the corresponding structural-descriptive names. If, nevertheless, we consider explanations of type (2) constructed by the use of such names (as was done, for example, in (4) above), then we see no way of generalizing these explanations. And if in (5) or (6) we replace the quotation-mark name by the structural-descriptive name '*pe*' (or '*the word which consists of the single letter Pe*') we see at once the absurdity of the resulting formulation.

In order to rescue the sense of sentences (5) and (6) we must seek quite a different interpretation of the quotation-mark names. We must treat these names as syntactically composite expressions, of which both the quotation marks and the expressions within them are parts. Not all quotation-mark expressions will be constant names in that case. The expression " '*p*' " occurring in (5) and (6), for example, must be regarded as a function, the argument of which is a sentential variable and the values of which are constant quotation-mark names of sentences. We shall call such functions *quotation-functions*. The quotation marks then become independent words belonging to the domain of semantics, approximating in their meaning to the word 'name', and from the syntactical point of view, they play the part of functors.[1] But then new complications arise. The sense of the quotation-function and of the quotation marks themselves is not sufficiently clear. In any case such functors are not extensional; there is no doubt that the sentence *"for all p and q, if p if and only if q, then 'p' is identical with 'q' "* is in palpable contradiction to the customary way of using quotation marks. For this reason alone definition (6) would be unacceptable to anyone who wishes consistently to avoid intensional functors and is even of the opinion that a deeper analysis shows it to be impossible to give any precise meaning to such functors.[2] Moreover, the use of the quotation functor exposes us to the danger of becoming involved in various semantical antinomies, such as the antinomy of the liar. This will be so even if—taking every care—we make use only of those properties of quotation-functions which seem almost evident. In contrast to that conception of the antinomy of the liar which has been given above, we can formulate it without using the expression 'true sentence' at all, by introducing the

---

[1] We call such words as 'reads' in the expression '*x* reads' functors (this is a sentence-forming functor with *one* individual name as argument); also 'sees' in the expression '*x* sees *y*' (a sentence-forming functor with *two* name arguments), and 'father' in the expression 'the father of *x*' (a name-forming functor with *one* name argument), as well as 'or' in the expression '*p* or *q*' (a sentence-forming functor with two sentence arguments); quotation marks provide an example of a name-forming functor with *one* sentence argument. The term 'functor' we owe to T. Kotarbiński, the terms 'sentence-forming functor' and 'name-forming functor' to K. Ajdukiewicz; cf. Ajdukiewicz, K. (3).

[2] I shall not discuss the difficult problem of extensionality in more detail here; cf. Carnap, R. (8) where the literature of the problem is given, and especially Whitehead, A. N., and Russell, B. A. W. (90), vol. 1, pp. 659–66. It should be noted that usually the terms 'extensional' and 'intensional' are applied to sentence-forming functors, whilst in the text they are applied to quotation marks and thus to name-forming functors.

quotation-functions with variable arguments. We shall give a sketch of this formulation.

Let the symbol 'c' be a typographical abbreviation of the expression 'the sentence printed on this page, line 6 from the top'. We consider the following statement:

for all p, if c is identical with the sentence 'p', then not p

(if we accept (6) as a definition of truth, then the above statement asserts that c is not a true sentence).

We establish empirically:

(α) *the sentence 'for all p, if c is identical with the sentence 'p', then not p' is identical with c.*

In addition we make only a single supplementary assumption which concerns the quotation-function and seems to raise no doubts:

(β) *for all p and q, if the sentence 'p' is identical with the sentence 'q', then p if and only if q.*

By means of elementary logical laws we easily derive a contradiction from the premisses (α) and (β).

I should like to draw attention, in passing, to other dangers to which the consistent use of the above interpretation of quotation marks exposes us, namely to the ambiguity of certain expressions (for example, the quotation-expression which occurs in (5) and (6) must be regarded in certain situations as a function with variable argument, whereas in others it is a constant name which denotes a letter of the alphabet). Further, I would point out the necessity of admitting certain linguistic constructions whose agreement with the fundamental laws of syntax is at least doubtful, e.g. meaningful expressions which contain meaningless expressions as syntactical parts (every quotation-name of a meaningless expression will serve as an example). For all these reasons the correctness of definition (6), even with the new interpretation of quotation marks, seems to be extremely doubtful.

Our discussions so far entitle us in any case to say that *the attempt to construct a correct semantical definition of the expression 'true sentence meets with very real difficulties.* We know of no general method which would permit us to define the meaning of an arbitrary concrete expression of the type '*x* is a true sentence', where in the place of '*x*' we have a name of some sentence. The method illustrated by the examples (3) and (4) fails us in those situations in which we cannot indicate for a given name of a sentence, the sentence denoted by this name (as an example of such a name 'the first sentence which will be printed in the year 2000' will serve). But if in such a case we seek refuge in the construction used in the formulation of definition (6), then we should lay ourselves open to all the complications which have been described above.

In the face of these facts we are driven to seek other methods of solving our problem. I will draw attention here to only *one* such attempt, namely the attempt to construct a *structural definition*. The general scheme of this definition would be somewhat as follows: *a true sentence is a sentence which possesses such and such structural properties* (i.e. properties concerning the form and arrangement in sequence of the single parts of the expression) *or which can be obtained from such and such structurally described expressions by means of such and such structural transformations.* As a starting-point we can press into service many laws from formal logic which enable us to infer the truth or falsehood of sentences from certain of their structural properties; or from the truth or falsehood of certain sentences to infer analogous properties of other sentences which can be obtained from the former by means of various structural transformations. Here are some trivial examples of such laws: *every expression consisting of four parts of which the first is the word 'if', the third is the word 'then', and the second and fourth are the same sentence, is a true sentence*; *if a true sentence consists of four parts, of which the first is the word 'if', the second a true sentence, the third the word 'then', then the fourth part is a true sentence.* Such laws (especially those of the second type) are very important. With their help every fragmentary definition of truth, the extension of which embraces an arbitrary class of sentences, can be extended to all composite sentences which can be built up from sentences of the given class by combining them by means of such expressions as 'if . . . then', 'if and only if', 'or', 'and', 'not', in short, by means of expressions belonging to the sentential calculus (or theory of deduction). This leads to the idea of setting up sufficiently numerous, powerful, and general laws for every sentence to fall under one of them. In this way we should reach a general structural definition of a true sentence. Yet this way also seems to be almost hopeless, at least as far as natural language is concerned. For this language is not something finished, closed, or bounded by clear limits. It is not laid down what words can be added to this language and thus in a certain sense already belong to it potentially. We are not able to specify structurally those expressions of the language which we call sentences, still less can we distinguish among them the true ones. *The attempt to set up a structural definition of the term 'true sentence'—applicable to colloquial language is confronted with insuperable difficulties.*

The breakdown of all previous attempts leads us to suppose that there is no satisfactory way of solving our problem. Important arguments of a general nature can in fact be invoked in support of this supposition as I shall now briefly indicate.

A characteristic feature of colloquial language (in contrast to various scientific languages) is its universality. It would not be in harmony with the spirit of this language if in some other language a word occurred which could not be translated into it; it could be claimed that 'if we can speak meaningfully about anything at all, we can also speak about it in colloquial language'. If we are to maintain this universality of everyday language in connexion with semantical investigations, we must, to be consistent, admit into the language, in addition to its sentences and other expressions, also the names of these sentences and expressions, and sentences containing these names, as well as such semantic expressions as 'true sentence', 'name', denote', etc. But it is presumably just this universality of everyday language which is the primary source of all semantical antinomies, like the antinomies of the liar or of heterological words. These antinomies seem to provide a proof that every language which is universal in the above sense, and for which the normal laws of logic hold, must be inconsistent. This applies especially to the formulation of the antinomy of the liar which I have given on pages 157 and 158, and which contains no quotation-function with variable argument. If we analyse this antinomy in the above formulation we reach the conviction that no consistent language can exist for which the usual laws of logic hold and which at the same time satisfies the following conditions: (I) for any sentence which occurs in the language a definite name of this sentence also belongs to the language; (II) every expression formed from (2) by replacing the symbol 'p' by any sentence of the language and the symbol 'x' by a name of this sentence is to be regarded as a true sentence of this language; (III) in the language in question an empirically established premiss having the same meaning as (α) can be formulated and accepted as a true sentence.[1]

If these observations are correct, then *the very possibility of a consistent use of the expression 'true sentence' which is in harmony with the laws of logic and the spirit of everyday language seems to be very questionable, and consequently the same doubt attaches to the possibility of constructing a correct definition of this expression.*

§ 2. FORMALIZED LANGUAGES, ESPECIALLY THE LANGUAGE OF THE CALCULUS OF CLASSES

For the reasons given in the preceding section I now abandon the attempt to solve our problem for the language of everyday life and restrict myself henceforth entirely to *formalized languages*.[2] These can be roughly characterized as artificially con-

---

[1] The antinomy of heterological words (which I shall not describe here—cf. Grelling, K., and Nelson, L. (24), p. 307) is simpler than the antinomy of the liar in so far as no empirical premiss analogous to (α) appears in its formulation; thus it leads to the correspondingly stronger consequence: there can be no consistent language which contains the ordinary laws of logic and satisfies two conditions which are analogous to (I) and (II), but differ from them in that they treat not of sentences but of names, and not of the truth of sentences but of the relation of denoting. In this connexion compare the discussion in § 5 of the present article—the beginning of the proof of Th. 1, and in particular p. 248, footnote 2.

[2] The results obtained for formalized language also have a certain validity for colloquial language, and this is owing to its universality: if we translate into colloquial language any definition of a true sentence which has been constructed for some formalized language, we obtain a fragmentary definition of truth which embraces a wider or narrower category of sentences.

structed languages in which the sense of every expression is unambiguously determined by its form. Without attempting a completely exhaustive and precise description, which is a matter of considerable difficulty, I shall draw attention here to some essential properties which all the formalized languages possess: (α) for each of these languages a list or description is given in structural terms of all the *signs with which the expressions of the language are formed*; (β) among all possible expressions which can be formed with these signs those called *sentences* are distinguished by means of purely structural properties. Now formalized languages have hitherto been constructed exclusively for the purpose of studying *deductive sciences* formalized on the basis of such languages. The language and the science grow together to a single whole, so that we speak of the language of a particular formalized deductive science, instead of this or that formalized language. For this reason further characteristic properties of formalized languages appear in connexion with the way in which deductive sciences are built up; (γ) a list, or structural description, is given of the sentences called *axioms* or *primitive statements*; (δ) in special rules, called *rules of inference*, certain operations of a structural kind are embodied which permit the transformation of sentences into other sentences; the sentences which can be obtained from given sentences by one or more applications of these operations are called *consequences* of the given sentences. In particular the consequences of the axioms are called *provable* or *asserted sentences*.[1]

It remains perhaps to add that we are not interested here in 'formal' languages and sciences in one special sense of the word 'formal', namely sciences to the signs and expressions of which no material sense is attached. For such sciences the problem here discussed has no relevance, it is not even meaningful.

---

[1] The formalization of a science usually admits of the possibility of introducing new signs into that science which were not explicitly given at the outset. These signs—called *defined signs* (in contrast to the *primitive* signs)—appear in the science in the first instance in expressions of a special structure called *definitions*, which are constructed in accordance with special rules—the *rules of definition*. Definitions are sometimes regarded as asserted sentences of the science. This feature of the formalization of languages will not be considered in the sequel.

We shall always ascribe quite concrete and, for us, intelligible meanings to the signs which occur in the languages we shall consider.[1] The expressions which we call sentences still remain sentences after the signs which occur in them have been translated into colloquial language. The sentences which are distinguished as axioms seem to us to be materially true, and in choosing rules of inference we are always guided by the principle that when such rules are applied to true sentences the sentences obtained by their use should also be true.[2]

In contrast to natural languages, the formalized languages do not have the universality which was discussed at the end of the preceding section. In particular, most of these languages possess no terms belonging to the theory of language, i.e. no expressions which denote signs and expressions of the same or another language or which describe the structural connexions between them (such expressions I call—for lack of a better term—*structural-descriptive*). For this reason, when we investigate the language of a formalized deductive science, we must always distinguish clearly between the language *about* which we speak and the language *in* which we speak, as well as between the science which is the object of our investigation and the science in which the investigation is carried out. The names of the expressions of the first language, and of the relations between them, belong to the second language, called the *metalanguage* (which may contain the first as a part). The description of these expressions, the definition of the complicated concepts, especially of those connected with the construction of a deductive theory (like the concept of consequence, of provable sentence, possibly of true sentence), the determination of the properties of these concepts, is the task of the second theory which we shall call the *metatheory*.

For an extensive group of formalized languages it is possible

---

[1] Strictly speaking this applies only to the signs called constants. Variables and technical signs (such as brackets, dots, etc.) possess no independent meaning; but they exert an essential influence on the meaning of the expressions of which they form parts.

[2] Finally, the definitions are so constructed that they elucidate or determine the meaning of the signs which are introduced into the language by means of primitive signs or signs previously defined (cf. p. 166, note 1).

to give a method by which a correct definition of truth can be constructed for each of them. The general abstract description of this method and of the languages to which it is applicable would be troublesome and not at all perspicuous. I prefer therefore to introduce the reader to this method in another way. I shall construct a definition of this kind in connexion with a particular concrete language and show some of its most important consequences. The indications which I shall then give in § 4 of this article will, I hope, be sufficient to show how the method illustrated by this example can be applied to other languages of similar logical construction.

I choose, as the object of my considerations, the language of a deductive science of the utmost simplicity which will surely be well known to the reader—that of the *calculus of classes*. The calculus of classes is a fragment of mathematical logic and can be regarded as one of the interpretations of a formal science which is commonly called the *algebra of logic*.[1]

Among the signs comprising the expressions of this language I distinguish two kinds, *constants* and *variables*.[2] I introduce only four constants: the *negation* sign 'N', the sign of *logical sum* (*disjunction*) 'A', the *universal quantifier* 'Π', and finally the *inclusion* sign 'I'.[3] I regard these signs as being equivalent in

---

[1] Cf. Schröder, E. (62), vol. 1 (especially pp. 160–3) and Whitehead, A. N., and Russell, B. A. W. (90), vol. 1, pp. 205–12.

[2] By making use of an idea of Łukasiewicz I avoid introducing any technical signs (like brackets, dots, etc.) into the language, and this is due chiefly to the fact that I always write the functor before the arguments in every meaningful expression; cf. Łukasiewicz, J. (51), especially pp. v and 40.

[3] Usually many other constants occur in the calculus of classes, e.g. the existence sign, the sign of implication, of logical product (conjunction), of equivalence, of identity, as well as of the complement, the sum, and the product of classes (see p. 168, note 1); for that reason only a fragment of the calculus of classes can—formally speaking—be constructed in the language under consideration. It is, however, to be noted that all constants of the calculus of classes could be introduced into this language as defined terms, if we complete its formalization by making the introduction of new signs possible by means of definitions (see p. 166, note 1). Owing to this fact our fragmentary language already suffices for the expression of every idea which can be formulated in the complete language of this science. I would also point out that even the sign of inclusion 'I' can be eliminated from our language by interpreting expressions of the type 'xy' (where any variables occur in the place of 'x' and 'y') in the same way in which in the sequel we shall interpret the expression 'Ixy'.

meaning respectively with the expressions 'not', 'or', 'for all' (in the sense in which this expression was used in statement (6) of § 1, for example) and 'is included in'. In principle any arbitrary symbols could be used as variables, provided only that their number is not limited and that they are distinct in form from the constants. But for the further course of our work it is technically important to specify the form of these signs exactly, and in such a way that they can easily be ordered in a sequence. I shall therefore use as variables only such symbols as '$x_\prime$', '$x_{\prime\prime}$', '$x_{\prime\prime\prime}$', and analogous signs which consist of the symbol 'x' and a number of small strokes added below. The sign which has k such small strokes (k being any natural number distinct from 0) will be called *the k-th variable*. In the intuitive interpretation of the language, which I always have in mind here, the variables represent names of classes of individuals. As *expressions* of the language we have either single constants and variables or complexes of such signs following one another, for example: '$x_\prime N x_{\prime\prime}$', '$N I x_\prime x_{\prime\prime}$', '$A I x_\prime x_{\prime\prime} I x_{\prime\prime} x_\prime$', '$\Pi x_\prime$', '$\Pi x_\prime I x_{\prime\prime} x_{\prime\prime\prime}$', '$I x_{\prime\prime} x_{\prime\prime\prime}$' and so on. Expressions of the type 'Np', 'Apq', 'Πxp', and 'Ixy', where in the place of 'p' and 'q' any sentences or sentential functions (this term will be explained below), and in the place of 'x' and 'y' any variables, appear, are read: 'not p' (or 'it is not true that p'),[1] 'p or q', 'for all classes x we have p', and 'the class x is included in the class y', respectively. Regarding composite expressions, i.e. those which are not signs, we can say that they consist of two or more other simple expressions. Thus the expression '$N I x_\prime x_{\prime\prime}$' is composed of the two successive expressions 'N' and '$I x_\prime x_{\prime\prime}$' or of the expressions '$N I$' and '$x_\prime x_{\prime\prime}$' or finally of the expressions '$N I x_\prime$' and '$x_{\prime\prime}$'.

But the proper domain of the following considerations is not the language of the calculus of classes itself but the corresponding metalanguage. Our investigations belong to the *metacalculus of classes* developed in this metalanguage. From this springs the need to give the reader some account—if only a very brief

---

[1] For stylistic reasons we sometimes use the expression 'it is not true that' instead of the word 'not', the whole expression being regarded as a single word, no independent meaning being given to the separate parts, and in particular to the word 'true', which occur in it.

one—of the structure of the metalanguage and of the metatheory. I shall restrict myself to the two most important points: (1) the enumeration of all the signs and expressions which will be used in the metalanguage, without explaining in more detail their importance in the course of the investigation, and (2) the setting up of a system of axioms which suffices for the establishment of the metatheory or at least will form a foundation for the results obtained in this article. These two points are closely connected with our fundamental problem; were we to neglect them, we should not be able to assert either that we had succeeded in correctly defining any concept on the basis of the metalanguage, or that the definition constructed possesses any particular consequences. But I shall not attempt at all to give the metatheory the character of a strictly formalized deductive science. I shall content myself with saying that—apart from the two points mentioned—the process of formalizing the metatheory shows no specific peculiarity. In particular, the rules of inference and of definition do not differ at all from the rules used in constructing other formalized deductive sciences.

Among the expressions of the metalanguage we can distinguish two kinds. To the first belong *expressions of a general logical character*, obtainable from any sufficiently developed system of mathematical logic.[1] They can be divided into primitive expressions and defined expressions, but this would be pointless in the present case. First we have a series of expressions which have the same meaning as the constants of the science we are considering; thus '*not*' or '*it is not true that*',[2] '*or*', '*for all*', and '*is included in*'—in symbols '$\subseteq$'. Thanks to this circumstance we are able to translate every expression of the language into the metalanguage. For example, the statement 'for all a (or for all classes a) $a \subseteq a$' is the translation of the expression '$\prod x, Ix, x,$'. To the same category belongs a series of analogous

---

[1] For example, from the work Whitehead, A. N., and Russell, B. A. W. (90). (But I do not intend to use here any special logical symbolism. Apart from the exceptions which I shall explicitly mention I shall use expressions of colloquial language.) For the meaning of the general logical expressions given below see Carnap, R. (8).

[2] See p. 169, note 1.

expressions from the domain of the sentential calculus, of the first order functional calculus and of the calculus of classes, for example, *'if . . . , then'*, *'and'*, *'if and only if'*, *'for some x'* (or *'there is an x such that . . .'*), *'is not included in'*—in symbols *'⊄'*, *'is identical with'*—in symbols *'='*, *'is distinct from'*—in symbols *'≠'*, *'is an element of'*—in symbols *'∈'*, *'is not an element of'*—in symbols *'∈̄'*, *'individual'*, *'class'*, *'null class'*, *'class of all x such that'*, and so on. We also find here some expressions from the domain of the theory of the equivalence of classes, and of the arithmetic of cardinal numbers, e.g. *'finite class'*, *'infinite class'*, *'power of a class'*, *'cardinal number'*, *'natural number'* (or *'finite cardinal number'*), *'infinite cardinal number'*, '0', '1', '2', '<', '>', '⩽', '⩾', '+', '−', ... Finally I shall need some terms from the logic of relations. The class of all objects x, to which there corresponds at least *one* object y such that xRy (i.e. x stands in the relation R to y) will be called the *domain of the binary or two-termed relation R*. Analogously, the *counter domain of the relation R* is the set of all objects y for which there is at least *one* object x such that xRy. In the case of many-termed relations we do not speak of domain and counter domain, but of the *1st, 2nd, 3rd,..., n-th domain of the relation*. The relation having *only one* element x in its domain and *only one* element y in its counter domain (a relation which thus holds only between x and y and between no other two objects) is called an *ordered pair, where x is the first and y the second member*. Analogously using many-termed relations we define *ordered triples, quadruples*, and in general *ordered n-tuples*. If, for every object y belonging to the counter domain of a two-termed relation R, there is *only one* object x such that xRy, then the relation R is called *one-many*. The concept of *sequence* will play a great part in the sequel. An *infinite sequence* is a one-many relation whose counter domain is the class of all natural numbers excluding zero. In the same way, the term *'finite sequence of n terms'* denotes every one-many relation whose counter domain consists of all natural numbers k such that $1 \leqslant k \leqslant n$ (where n is any natural number distinct from 0). The unique x which satisfies the formula xRk (for a given sequence R and a given natural number k) is called the *k-th term of the sequence R*, or the *term of the sequence R with index k*, and is denoted by 'R_k'. We say that *the sequences R and S differ in at most the k-th place*, if any two corresponding terms of these sequences R_l and S_l are identical with the exception of the *k*th terms R_k and S_k which may be distinct. In the following pages we shall deal with sequences of classes and of natural numbers, i.e. with sequences all of whose terms are either classes of individuals or natural numbers. In particular, a sequence all of whose terms are classes which are included in a given class *a*, will be called a *sequence of subclasses of the class a*.

In contrast to the first kind of expression, those of the second kind are *specific terms of the metalanguage of a structural-descriptive character*, and thus names of concrete signs or expressions of the language of the calculus of classes. Among these are, in the first place, the terms '*the negation sign*', '*the sign of logical sum*', '*the sign of the universal quantifier*', '*the inclusion sign*', '*the k-th variable*', '*the expression which consists of the expressions x and y following one another*' and '*expression*'. As abbreviations of the first six terms I shall use the symbols '*ng*', '*sm*', '*un*', '*in*', 'v_k', and '$x^\frown y$' (the sign '*v*' thus denotes a sequence, the terms of which are the successive variables $v_1, v_2, v_3, \ldots$). These terms have already been used in introducing the reader to the language of the calculus of classes. I hope that, thanks to the explanations already given, no doubt will remain concerning the meaning of these terms. With the help of these terms (and possibly general logical terms) all other concepts of the metalanguage of a structural-descriptive kind can be defined. It is easy to see that every simple or composite expression of the language under investigation has an individual name in the metalanguage similar to the structural-descriptive names of colloquial language (cf. pp. 156 and 157). For example, the symbolic expression '$((ng^\frown in)^\frown v_1)^\frown v_2$' can serve as a name of the expression '$NIx, x_{''}$'. The fact that the metalanguage contains both an individual name and a translation of every expression (and in particular of every sentence) of the language studied will play a decisive part in the construction of the definition of truth, as the reader will see in the next section.

As variables in the metalanguage I shall use the symbols
(1) 'a', 'b'; (2) 'f', 'g', 'h'; (3) 'k', 'l', 'm', 'n', 'p'; (4) 't', 'u', 'w',
'x', 'y', 'z'; and (5) 'X', 'Y'. In this order they represent the
names of (1) classes of individuals of an arbitrary character,[1]
(2) sequences of such classes, (3) natural numbers and sequences
of natural numbers, (4) expressions, and (5) classes of expressions.

We turn now to the axiom system of the metalanguage. First,
it is to be noticed that—corresponding to the two kinds of expressions in the metalanguage—this system contains two quite
distinct kinds of sentences: the *general logical axioms* which
suffice for a sufficiently comprehensive system of mathematical
logic, and the *specific axioms of the metalanguage* which describe
certain elementary properties of the above structural-descriptive
concepts consistent with our intuitions. It is unnecessary to
introduce explicitly the well-known axioms of the first kind.[2]
As axioms of the second kind we adopt the following statements:[3]

AXIOM 1. *ng, sm, un, and in are expressions, no two of which
are identical.*

AXIOM 2. *v_k is an expression if and only if k is a natural number
distinct from 0; v_k is distinct from ng, sm, un, in, and also from v_l
if $k \neq l$.*

AXIOM 3. *$x^\frown y$ is an expression if and only if x and y are expressions; $x^\frown y$ is distinct from ng, sm, un, in, and from each of the
expressions v_k.*

AXIOM 4. *If x, y, z, and t are expressions, then we have
$x^\frown y = z^\frown t$ if and only if one of the following conditions is satisfied: (α) $x = z$ and $y = t$; (β) there is an expression u such that
$x = z^\frown u$ and $t = u^\frown y$; (γ) there is an expression u such that
$z = x^\frown u$ and $y = u^\frown t$.*

AXIOM 5. (The principle of induction.) *Let X be a class
which satisfies the following conditions: (α) $ng \in X$, $sm \in X$, $un \in X$*

---

[1] Although in the cases (1) and (4) I use distinct variables I here treat
expressions as special classes of individuals, namely as classes of concrete series
of printed signs (cf. p. 156, note 1).

[2] They may again be taken from Whitehead, A. N., and Russell, B. A. W.
(90), cf. p. 156, note 1.

[3] As far as I know the metatheory has never before been given in the form
of an axiomatized system.

and in $\in X$; (β) *if* k *is a natural number distinct from* 0, *then* $v_k \in X$; (γ) *if* $x \in X$ *and* $y \in X$; *then* $x^\frown y \in X$. *Then every expression belongs to the class* X.

The intuitive sense of Axs. 1–4 requires no further elucidation. Ax. 5 gives a precise formulation of the fact that every expression consists of a finite number of signs.

It is possible to prove that the *above axiom system is categorical*. This fact guarantees to a certain degree that it will provide a sufficient basis for the construction of the metalanguage.[1]

Some of the above axioms have a pronounced existential character and involve further consequences of the same kind. Noteworthy among these consequences is the assertion that the class of all expressions is infinite (to be more exact, denumerable). From the intuitive standpoint this may seem doubtful and hardly evident, and on that account the whole axiom-system may be subject to serious criticism. A closer analysis would restrict this criticism entirely to Axs. 2 and 3 as the essential sources of this infinite character of the metatheory. I shall not pursue this difficult problem any further here.[2] The con-

[1] I use the term 'categorical' in the sense given in Veblen, O. (86). I do not propose to explain in more detail why I see in the categoricity of an axiom system an objective guarantee that the system suffices for the establishment of the corresponding deductive science; a series of remarks on this question will be found in Fraenkel, A. (16).

Regarding the interpretation of the term 'categorical' there are certain, although not especially important, differences of opinion. Without going into details I may mention that in the case of one of the possible interpretations the proof that the system is categorical would require the addition of two further axioms to the system given in the text. In these axioms (which otherwise are not of great importance) the specific conception of expressions as classes would occur (cf. p. 156, note 1). The first axiom would state that two arbitrary expressions are disjoint classes (i.e. have no element in common), in the second the number of elements of every expression would be stipulated in some way.

[2] For example, the following truly subtle points are here raised. Normally expressions are regarded as the products of human activity (or as classes of such products). From this standpoint the supposition that there are infinitely many expressions appears to be obviously nonsensical. But another possible interpretation of the term 'expression' presents itself: we could consider all physical bodies of a particular form and size as expressions. The kernel of the problem is then transferred to the domain of physics. The assertion of the infinity of the number of expressions is then no longer senseless and even forms a special consequence of the hypotheses which are normally adopted in physics or in geometry.

sequences mentioned could of course be avoided if the axioms were freed to a sufficient degree from existential assumptions. But the fact must be taken into consideration that the elimination or weakening of these axioms, which guarantee the existence of all possible expressions, would considerably increase the difficulties of constructing the metatheory, would render impossible a series of the most useful consequences and so introduce much complication into the formulation of definitions and theorems. As we shall see later this will become clear even in the present investigations. For these reasons it seems desirable, at least provisionally, to base our work on the axiom system given above in its initial unweakened form.

Making use of the expressions and symbols of the metalanguage which have now been enumerated, I shall define those concepts which establish the calculus of classes as a formalized deductive science. These are the concepts of *sentence*, *axiom* (*primitive sentence*), *consequence* and *provable sentence*. But first I introduce a series of auxiliary symbols which will denote various simple types of expression and greatly facilitate the later constructions.

DEFINITION 1. *x is an* inclusion *with v_k as first and v_l as second term—in symbols $x = \iota_{k,l}$—if and only if $x = (in^\frown v_k)^\frown v_l$.*

DEFINITION 2. *x is a* negation *of the expression y—in symbols $x = \bar{y}$—if and only if $x = ng^\frown y$.*

DEFINITION 3. *x is a* logical sum (disjunction) *of the expressions y and z—in symbols $x = y+z$—if and only if $x = (sm^\frown y)^\frown z$.*

DEFINITION 4. *x is a* logical sum *of the expressions $t_1, t_2,..., t_n$ (or a logical sum of a finite n-termed sequence t of expressions)— in symbols $x = \sum_{k}^{n} t_k$—if and only if t is a finite n-termed sequence of expressions which satisfies one of the following conditions: (α) $n = 1$ and $x = t_1$, (β) $n > 1$ and $x = \sum_{k}^{n-1} t_k + t_n$.*[1]

---

[1] As will be seen, Def. 4 is a recursive definition which, as such, raises certain methodological misgivings. It is, however, well known that with the help of a general method, the idea of which we owe to G. Frege and R. Dedekind, every recursive definition can be transformed into an equivalent normal definition

DEFINITION 5. x *is a* logical product (conjunction) *of the expressions* y *and* z—*in symbols* $x = y.z$—*if and only if* $x = \overline{\bar{y}+\bar{z}}$.

DEFINITION 6. x *is a* universal quantification *of the expression* y *under the variable* v_k—*in symbols* $x = \bigcap_k y$—*if and only if* $x = (un^\frown v_k)^\frown y$.

DEFINITION 7. x *is a* universal quantification *of the expression* y *under the variables* v_{p_1}, v_{p_2},..., v_{p_n}—*in symbols* $x = \bigcap_{p_k}^{k \leqslant n} y$—*if and only if* p *is a finite n-termed sequence of natural numbers which satisfies one of the following conditions*: (α) $n = 1$ *and* $x = \bigcap_{p_1} y$, (β) $n > 1$ *and* $x = \bigcap_{p_k}^{k \leqslant n-1} \bigcap_{p_n} y$.

DEFINITION 8. x *is a* universal quantification *of the expression* y *if and only if either* $x = y$ *or there is a finite n-termed sequence* p *of natural numbers such that* $x = \bigcap_{p_k}^{k \leqslant n} y$.

DEFINITION 9. x *is an* existential quantification *of the expression* y *under the variable* v_k—*in symbols* $x = \bigcup_k y$—*if and only if* $x = \overline{\bigcap_k \bar{y}}$.

We have thus introduced three fundamental operations by means of which compound expressions are formed from simpler ones: negation, logical addition, and universal quantification. (Logical addition is, of course, the operation which consists in forming logical sums of given expressions. The terms 'negation' and 'universal quantification' are thus used to refer both to certain operations on expressions and to expressions resulting from these operations.) If, beginning with the inclusions $\iota_{k,l}$, we apply to them the above operations any number of times we obtain an extensive class of expressions which are called *sentential functions*. We obtain the concept of *sentence* as a special case of this notion.

(cf. Dedekind, R. (15), pp. 33–40, and Whitehead, A. N., and Russell, B. A. W. (90), vol. 1, pp. 550–7, and vol. 3, p. 244). This, however, is unpractical in so far as the formulations so obtained have a more complicated logical structure, are less clear as regards their content, and are less suitable for further derivations. For these reasons I do not propose to avoid recursive definitions in the sequel.

DEFINITION 10. *x is a* sentential function *if and only if x is an expression which satisfies one of the four following conditions*: (α) *there exist natural numbers k and l such that* $x = \iota_{k,l}$; (β) *there exists a sentential function y such that* $x = \bar{y}$; (γ) *there exist sentential functions y and z such that* $x = y+z$; (δ) *there exists a natural number k and a sentential function y such that* $x = \bigcap_k y$.[1]

The following expressions will serve as examples of sentential functions according to Def. 10: '$Ix_, x_{n}$', '$NIx_, x_{m}$', '$AIx_, x_{m} Ix_{m} x_{i}$', '$\prod x_, NIx_, x_{n}$', and so on. On the other hand the expressions 'I', '$Ix_,$', '$AIx_, x_{m}$', '$\prod Ix_, x_{n}$', etc., are not sentential functions. It is easily seen that for every sentential function in the language we can automatically construct a structural-descriptive name of this function in the metalanguage, by making use exclusively of symbols which were introduced in Defs. 1, 2, 3, and 5. For example, the following symbolic

---

[1] Def. 10 is a recursive definition of a somewhat different type from that of Def. 4 since the usual 'transition from $n-1$ to n' is lacking in it. In order to reduce this to an ordinary inductive definition we must first inductively define the expressions '*x is a sentential function of the nth degree*' (inclusions $\iota_{k,l}$ would then be functions of the 0th degree, the negations and logical sums of these inclusions, as well as their generalizations for any variable, functions of the 1st degree, and so on), and then simply stipulate that '*x is a sentential function*' means the same as '*there is a natural number n such that x is a sentential function of the nth degree*'. Def. 10 could also be transformed into an equivalent normal definition in the following way:

x is a sentential function *if and only if the formula* $x \in X$ *holds for every class X which satisfies the following four conditions*: (α) *if k and l are natural numbers distinct from* 0, *then* $\iota_{k,l} \in X$; (β) *if* $y \in X$, *then* $\bar{y} \in X$; (γ) *if* $y \in X$ *and* $z \in X$ *then* $y+z \in X$; (δ) *if k is a natural number distinct from* 0 *and* $y \in X$, *then* $\bigcap_k y \in X$.

It should be emphasized that recursive definitions of the type of Def. 10 are open to much more serious methodological objections than the usual inductive definitions, since in contrast to the latter, statements of this type do not always admit of a transformation into equivalent normal definitions (see p. 175, note 1). The fact that such a transformation is possible in the present case is owing to the special nature of the concepts occurring in the definition (to the fact, namely, that every expression consists of a finite number of signs and that the operations given in conditions (β)–(δ) always lead from shorter to longer expressions). If, nevertheless, I sometimes give definitions of this kind in the present article in the place of equivalent normal definitions (Defs. 10, 11, 14, 22, and 24), I do so because these definitions have important advantages of quite another kind: they bring out the content of the concept defined more clearly than the normal definition does, and—in contrast to the usual recursive definition—they require no previous introduction of accessory concepts which are not used elsewhere (e.g. the accessory concept of a sentential function of the nth degree).

expressions function as names of the above examples of sentential functions: '$\iota_{1,2}$', '$\overline{\iota_{1,3}}$', '$\iota_{1,3}+\iota_{3,1}$', and '$\bigcap_1 \overline{\iota_{1,2}}$'.

DEFINITION 11. v_k *is a* free variable *of the sentential function x if and only if k is a natural number distinct from 0, and x is a sentential function which satisfies one of the following four conditions*: (α) *there is a natural number l such that* $x = \iota_{k,l}$ *or* $x = \iota_{l,k}$; (β) *there is a sentential function y such that* v_k *is a free variable of y and* $x = \bar{y}$; (γ) *there are sentential functions y and z such that* v_k *is a free variable of y and* $x = y+z$ *or* $x = z+y$; (δ) *there is a number l distinct from k and a sentential function y such that* v_k *is a free variable of y and* $x = \bigcap_l y$.

Variables which occur in a sentential function but are not free variables of this function, are usually called *bound (apparent) variables*.[1]

DEFINITION 12. *x is a* sentence (*or a* meaningful sentence)— *in symbols* $x \in S$—*if and only if x is a sentential function and no variable* v_k *is a free variable of the function x.*

Thus the expressions: $\bigcap_1 \iota_{1,1}$, $\bigcap_1 \bigcap_2 \iota_{1,2}$, $\bigcap_1 \bigcup_2 \iota_{1,2}$, $\bigcap_1 (\iota_{1,1}+\bigcap_1 \bigcup_2 \iota_{2,1})$ are sentences, but the functions: $\iota_{1,1}$, $\bigcap_2 \iota_{1,2}$, $\iota_{1,1}+\bigcap_1 \bigcup_2 \iota_{2,1}$ are not sentences because they contain the free variable v_1. By virtue of the above definition the symbol 'S' denotes the class of all meaningful sentences.

The system of primitive sentences of the calculus of classes will contain two kinds of sentences.[2] The sentences of the first kind are obtained by taking any axiom system which suffices as a basis for the sentential calculus and contains the signs of negation and logical addition as the only constants. For example, the axiom system consisting of the following four axioms:

'*ANAppp*', '*ANpApq*', '*ANApqAqp*',

and '*ANANpqANArpArq*'.[3]

---

[1] Cf. Hilbert, D., and Ackermann, W. (30), pp. 52–54.

[2] Concepts which I shall discuss in the further course of § 2 do not occur in the definition of true sentence itself. I shall, however, make use of them in the preparatory discussions at the beginning of § 3 which establish the definitive form of the definition. I shall also use them in the formulation of certain consequences of this definition (Ths. 3–6 of § 3) which express characteristic and materially important properties of true sentences.

[3] This axiom system is the result of a modification and simplification of the axiom system which is found in Whitehead, A. N., and Russell, B. A. W. (90), vol. 1, pp. 96–97; cf. Hilbert, D., and Ackermann, W. (30), p. 22.

In these axioms we replace the sentential variables 'p', 'q', and 'r' by any sentential functions, and then to the expressions thus obtained, if they are not already sentences, we apply the operation of universal quantification a sufficient number of times until all the free variables have disappeared. The following will serve as examples:

$$\textit{'ANA} \prod x, Ix, x, \prod x, Ix, x, \prod x, Ix, x,\text{'},$$
$$\textit{'} \prod x, \prod x, ANIx, x, AIx, x, Ix, x,\text{'}, \text{ etc.}$$

In order to obtain the sentences of the second kind we shall take as our starting-point some axiom system of the as yet unformalized calculus of classes which contains the inclusion sign as the only undefined sign,[1] and we then translate the axioms of this system into the language of the present article. Naturally we must first eliminate all constants which are defined by means of the inclusion sign, as well as all terms belonging to the sentential calculus and the functional calculus which are distinct in meaning from the universal quantifier, the negation sign and the sign of logical addition. As examples of sentences of this second kind we have

$$\textit{'} \prod x, Ix, x,\text{'} \text{ and } \textit{'} \prod x, \prod x, \prod x_{m} ANIx, x, ANIx, x_{m} Ix, x_{m}\text{'}.$$

DEFINITION 13. *x is an axiom (primitive sentence) if and only if x satisfies one of the two following conditions:* (α) $x \in S$ *and there exist sentential functions y, z, and u such that x is a universal quantification of one of the four functions* $\overline{y+y}+y$, $\bar{y}+(y+z)$, $\overline{y+z}+(z+y)$, *and* $\bar{\bar{y}}+z+(\overline{u+y}+(u+z))$); ($\beta$) *x is identical with one of the five sentences*

$$\bigcap_1 \overline{\iota_{1,1}}, \qquad \bigcap_1 \bigcap_2 \bigcap_3 (\overline{\iota_{1,2}}+\overline{\iota_{2,3}}+\iota_{1,3}),$$
$$\bigcap_1 \bigcap_2 \bigcup_3 (\iota_{1,3} \cdot \iota_{2,3} \cdot \bigcap_4 (\overline{\iota_{1,4}}+\overline{\iota_{2,4}}+\iota_{3,4})),$$
$$\bigcap_1 \bigcap_2 \bigcup_3 (\iota_{3,1} \cdot \iota_{3,2} \cdot \bigcap_4 (\overline{\iota_{4,1}}+\overline{\iota_{4,2}}+\iota_{4,3})),$$

and

$$\bigcap_1 \bigcup_2 (\bigcap_3 \bigcap_4 ((\overline{\iota_{3,1}}+\overline{\iota_{3,2}}+\iota_{3,4}) \cdot (\overline{\iota_{1,3}}+\overline{\iota_{2,3}}+\iota_{4,3})).$$
$$\bigcap_5 (\overline{\iota_{5,2}}+\bigcup_6 (\iota_{6,1} \cdot \iota_{6,2} \cdot \iota_{6,5}))).$$

---

[1] I have chosen here the system of postulates which is given in Huntington, E. V. (32), p. 297 (this system has, however, been simplified by the elimination of, among others, certain assumptions of an existential nature).

In the formulation of the definition of the concept of consequence I shall use, among others, the following expression: '*u is an expression obtained from the sentential function w by substituting the variable v_k for the variable v_l*'. The intuitive meaning of this expression is clear and simple, but in spite of this the definition has a somewhat complicated form:

DEFINITION 14. *x is an expression obtained from the sentential function y by* substituting the *(free) variable v_k for the (free) variable v_l if and only if k and l are natural numbers distinct from 0, and x and y are sentential functions which satisfy one of the following six conditions:* (α) $x = \iota_{k,k}$ *and* $y = \iota_{l,l}$; (β) *there exists a natural number m distinct from l, such that* $x = \iota_{k,m}$ *and* $y = \iota_{l,m}$ *or* $x = \iota_{m,k}$ *and* $y = \iota_{m,l}$; (γ) v_l *is not a free variable of the function y and* $x = y$; (δ) *there exist sentential functions z and t such that* $x = \bar{z}, y = \bar{t}$, *and z is an expression obtained from t by substituting the variable v_k for the variable v_l;* (ϵ) *there exist sentential functions z, t, u, and w, such that* $x = z + u$, $y = t + w$, *where z and u are obtained from t and w respectively by substituting the variable v_k for the variable v_l;* (ζ) *there exist sentential functions z, t and a natural number m distinct from k and l such that* $x = \bigcap_m z, y = \bigcap_m t$, *and z is obtained from t by substituting the variable v_k for the variable v_l.*[1]

For example, it follows from this definition that the expressions $\iota_{1,1}$, $\bigcap_3 (\iota_{3,1} + \iota_{1,3})$ and $\iota_{1,3} + \bigcap_2 \iota_{2,3}$ are obtained from the functions: $\iota_{2,2}$, $\bigcap_3 (\iota_{3,2} + \iota_{2,3})$ and $\iota_{2,3} + \bigcap_2 \iota_{2,3}$ respectively by substituting v_1 for v_2. But the expression $\bigcap_1 \iota_{1,3}$ cannot be obtained in this way from the function $\bigcap_2 \iota_{2,3}$ nor the expression $\bigcap_1 \iota_{1,1}$ from the function $\bigcap_2 \iota_{2,1}$.

---

[1] The following is a normal definition which is equivalent to the above recursive one (cf. p. 177, note 1):

x is an expression obtained from the sentential function y by substituting the variable v_k for the variable v_l if and only if k and l are natural numbers distinct from 0 and if the formula xRy holds for every relation R which satisfies the following six conditions: (α) $\iota_{k,k} R \iota_{l,l}$; ($\beta$) *if m is a natural number distinct from 0 and l, then* $\iota_{k,m} R \iota_{l,m}$ *and* $\iota_{m,k} R \iota_{m,l}$; ($\gamma$) *if z is a sentential function and v_l is not a free variable of z, then zRz;* (δ) *if zRt, then $\bar{z} R \bar{t}$;* (ϵ) *if zRt and uRw, then* $z + uRt + w$; (ζ) *if m is a natural number distinct from 0, k, and l and zRt, then* $\bigcap_m z R \bigcap_m t$.

The definitions of substitution in Leśniewski, S. (46), p. 73 (T.E. XLVII), and (47), p. 20 (T.E. XLVII°) depend on a totally different idea.

Among the consequences of a given class of sentences we include first all the sentences belonging to this class, and all the sentences which can be obtained from these by applying, an arbitrary number of times, the four operations of *substitution, detachment,* and *introduction* and *removal of the universal quantifier*.[1] If we had wished to apply these operations not only to sentences, but to arbitrary sentential functions, obtaining thereby sentential functions as results, then the meaning of the operation of substitution would be completely determined by Def. 14, the operation of detachment would correlate the function z with the functions y and $\bar{y}+z$, the operation of introduction of the universal quantifier would consist in forming the function $y+\bigcap_k z$ from the function $y+z$ (provided that v_k is not a free variable of the function y), the operation of removal of the universal quantifier would proceed in the opposite direction—from the function $y+\bigcap_k z$ to the function $y+z$.[1]

In order to simplify the construction I first define the auxiliary concept of *consequence of the n-th degree.*

DEFINITION 15. *x is a* consequence of the *n*th degree *of the class X of sentences if and only if* $x \in S$, $X \subseteq S$, *n is a natural number and either* (α) $n = 0$ *and* $x \in X$, *or* $n > 0$ *and one of the following five conditions is satisfied:* (β) *x is a consequence of the* $n-1$*th degree of the class X;* (γ) *there exist sentential functions u and w, a sentence y and natural numbers k and l such that x is the universal quantification of the function u, y is the universal quantification of the function w, u is obtainable from the function w by substituting the variable* v_k *for the variable* v_l, *and y is a consequence of the class X of the* $n-1$*th degree;* (δ) *there exist sentential functions u and w as well as sentences y and z such that x, y, and z are universal quantifications of the functions u,* $\bar{w}+u$, *and w respectively, and y and z are consequences of the class X of the* $n-1$*th degree;* (ε) *there exist sentential functions u and w, a sentence y and a natural number k such that x is a universal quantification of the function* $u+\bigcap_k w$, *y is a universal quantification of the function* $u+w$, v_k *is not a free variable of u, and y is a consequence of the class X of the* $n-1$*th*

[1] Cf. Łukasiewicz, J. (51), pp. 159–63; IV, p. 56.

degree; (ζ) *there exist sentential functions u and w, a sentence y and a natural number k, such that x is a universal quantification of the function* $u+w$, *y is a universal quantification of the function* $u + \bigcap_k w$ *and y is a consequence of the class X of the* $n-1$*th degree.*

DEFINITION 16. *x is a* consequence *of the class X of sentences— symbolically* $x \in Cn(X)$—*if and only if there is a natural number n such that x is a consequence of the nth degree of the class X.*[1]

DEFINITION 17. *x is a* provable (accepted) sentence *or a* theorem—*in symbols* $x \in Pr$—*if and only if x is a consequence of the set of all axioms.*

From this definition, it is easy to see that we shall have, among the provable sentences, not only all the sentences which can be obtained from the theorems of the sentential calculus in the same way in which the axioms of the first kind (i.e. those satisfying the condition (α) of Def. 13) were obtained from the axioms of the sentential calculus, but also all known theorems of the unformalized calculus of classes, provided they are first translated into the language under investigation. In order to become convinced of this we imitate in the metatheory, in every particular case, the corresponding proof from the domain of the sentential calculus or of the calculus of classes. For example, it is possible in this way to obtain the sentence $\bigcap_1 \overline{(\iota_{1,1}+\iota_{1,1})}$ from the well-known theorem '$ANpp$' of the

---

[1] The concept of consequence could also be introduced directly (i.e. without the help of consequence of the nth degree) in the following way:

$x \in Cn(X)$ *if and only if* $X \subseteq S$ *and if the formula* $x \in Y$ *holds for every class Y which satisfies the following conditions:* (α) $X \subseteq Y$; (β) *if* $y \in S$ *and is a universal quantification of the function u, z is a universal quantification of the function w, u is obtainable from the function w by substituting the variable* v_k *for the variable* v_l *and* $z \in Y$, *then* $y \in Y$; (γ) *if* $y \in S$, y, z, *and t are universal quantifications of the functions* $u, \overline{w}+u$, *and w respectively and* $z \in Y$ *and* $t \in Y$, *then* $y \in Y$; (δ) *if* $y \in S$, *u and w are sentential functions, y is a universal quantification of the function* $u + \bigcap_k$, *z a universal quantification of the function* $u+w$, v_k *is not a free variable of the function u and* $z \in Y$, *then* $y \in Y$; (ϵ) *if* $y \in S$, *u and w are sentential functions, y is a universal quantification of the function* $u+w$, *z a universal quantification of the function* $u+ \bigcap_k w$ *and* $z \in Y$, *then* $y \in Y$.

It is, however, to be noted that by transformation of the definition just given into a recursive sentence of the type of Def. 10 we obtain a sentence which is equivalent neither with the above definition nor with any other normal definition (cf. p. 177, note 1).

sentential calculus. Translating the proof of this theorem,[1] we show successively from Def. 13 that

$$\bigcap_1(\overline{\iota_{1,1}+\iota_{1,1}+\iota_{1,1}}), \qquad \bigcap_1(\overline{\iota_{1,1}}+(\iota_{1,1}+\iota_{1,1})),$$

and
$$\bigcap_1(\overline{\overline{\iota_{1,1}+\iota_{1,1}+\iota_{1,1}}+(\overline{\iota_{1,1}}+(\iota_{1,1}+\iota_{1,1}))+(\overline{\iota_{1,1}}+\iota_{1,1})})$$

are axioms; consequently by Def. 15

$$\bigcap_1(\overline{\overline{\iota_{1,1}}+(\iota_{1,1}+\iota_{1,1})}+(\overline{\iota_{1,1}}+\iota_{1,1}))$$

is a consequence of the 1st degree and $\bigcap_1(\overline{\iota_{1,1}}+\iota_{1,1})$ is a consequence of the second degree of the class of all axioms. Hence by Defs. 16 and 17 $\bigcap_1(\overline{\iota_{1,1}}+\iota_{1,1})$ is a provable sentence.

From examples of such inferences the difficulties can be imagined which would at once arise if we wished to eliminate from the axioms of the metatheory the assumptions which are of an existential nature. The fact that the axioms would no longer guarantee the existence of some particular sentence, which we wish to demonstrate, is not of much consequence. Serious importance attaches only to the fact that, even assuming the existence of some concrete sentence, we could not establish its provability; since in the proof it would be necessary to refer to the existence of other, as a rule more complicated, sentences (as is seen in the proof of the theorem '$\bigcap_1(\overline{\iota_{1,1}}+\iota_{1,1}) \in Pr$' which was sketched above). So long as we are dealing with special theorems of the type '$x \in Pr$', we can take measures to provide these sentences with premisses which guarantee the existence of the sentences necessary for the proof. The difficulties would increase significantly if we passed to sentences of a general character which assert that all sentences of a certain kind are provable—or, still more generally, are consequences of the given class of sentences. It would then often be necessary to include among the premisses general existential assumptions which would not be weaker than those which, for intuitive reasons, we had eliminated from the axioms.[2]

---

[1] Cf. Whitehead, A. N., and Russell, B. A. W (90), vol. 1, p. 101, *2.1.

[2] This is easily seen from the examples of Ths. 11, 12, 24, and 28 in § 3.

For these reasons the standpoint might be taken that Def. 17, in case the existential assumptions are rejected, would no longer embrace all the properties which we ascribe to the concept of *theorem*. The problem of a suitable 'correction' of the above definition would then arise. More precisely expressed, it would be a question of constructing a definition of *theorem* which would be equivalent to Def. 17 under the existential assumptions and yet—quite independently of these assumptions—would have as consequences all theorems of the type '*if the sentence x exists, then $x \in Pr$*', provided the corresponding theorem '$x \in Pr$' could be proved with the help of the existential assumptions. I shall give here a brief sketch of an attempt to solve this problem.

It can easily be shown that the axiom system adopted in the metatheory possesses an interpretation in the arithmetic of the natural numbers. A one-one correspondence can be set up between expressions and natural numbers where operations on numbers having the same formal properties are correlated with the operations on expressions. If we consider this correspondence, we can pick out, from the class of all numbers, those which are correlated with sentences; among these will be the 'primitive' numbers. We can introduce the concept of a 'consequence' of a given class of numbers, and finally define the 'accepted' numbers as 'consequences' of the class of all 'primitive' numbers. If we now eliminate the existential assumptions from the axioms, the one-one correlation disappears: to every expression a natural number still corresponds, but not to every number, an expression. But we can still preserve the concept of 'accepted' number previously established and define the theorems as those which are correlated with 'accepted' numbers. If we try, on the basis of this new definition, to prove that a concrete sentence is a theorem, we shall no longer be compelled—as is easily seen—to refer to the existence of any other sentences. Nevertheless the proof will still require—and this must be emphasized—an existential assumption, the assumption, namely, that there exist sufficiently many natural numbers or—what amounts to the same thing—sufficiently many distinct individuals. Thus in order to derive all desired conclusions from the new definition, it would be necessary to include in the metatheory the *axiom of infinity*, i.e. the assumption that the class of all individuals is infinite.[1] I know of no method, be it even less natural and more complicated than the one just discussed, which would lead to a satisfactory solution of our problem which is independent of the above axiom.

In connexion with the concepts of consequence and of theorem I have mentioned rules of inference. When we have in mind the construction of a deductive science itself, and not the investigation of such a science carried out on the basis of the metatheory, we give, instead of Def. 17, a rule by which we may add to the science as a theorem every consequence of the axioms. In our case this rule can be divided into four rules—corresponding to the four operations which we use in the construction of consequences.

By means of the concepts of sentence and of consequence all the most important methodological concepts can be introduced into the metatheory, in particular the concepts of *deductive system*, of *consistency* and of *completeness*.[2]

DEFINITION 18. X *is a* deductive system *if and only if*

$$Cn(X) \subseteq X \subseteq S.$$

DEFINITION 19. X *is a* consistent *class of sentences if and only if* $X \subseteq S$ *and if, for every sentence* x, *either* $x \equiv Cn(X)$ *or* $\bar{x} \equiv Cn(X)$.

DEFINITION 20. X *is a* complete *class of sentences if and only if* $X \subseteq S$ *and if, for every sentence* x, *either* $x \in Cn(X)$ *or* $\bar{x} \in Cn(X)$.

In the sequel yet another concept will prove useful:

DEFINITION 21.   *The sentences* x *and* y *are* equivalent *with respect to the class* X *of sentences if and only if* $x \in S, y \in S, X \subseteq S$ *and both* $\bar{x}+y \in Cn(X)$ *and* $\bar{y}+x \in Cn(X)$.

A more detailed analysis of the concepts introduced in this section would exceed the limits of the present work.

---

[1] Cf. Whitehead, A. N., and Russell, B. A. W. (90), vol. 2, p. 203.

[2] Cf. pp. 70, 90, and 93 of the present volume.

§ 3. The Concept of True Sentence in the Language of the Calculus of Classes

I pass on now to the chief problem of this article—the construction of the definition of *true sentence*, the language of the calculus of classes still being the object of investigation.

It might appear at first sight that at the present stage of our discussion this problem can be solved without further difficulty, that 'true sentence' with respect to the language of a formalized deductive science means nothing other than 'provable theorem', and that consequently Def. 17 is already a definition of truth and moreover a purely structural one. Closer reflection shows, however, that this view must be rejected for the following reason: no definition of true sentence which is in agreement with the ordinary usage of language should have any consequences which contradict the principle of the excluded middle. This principle, however, is not valid in the domain of provable sentences. A simple example of two mutually contradictory sentences (i.e. such that one is the negation of the other) neither of which is provable is provided by Lemma E below. The extension of the two concepts is thus not identical. From the intuitive standpoint all provable sentences are without doubt true sentences (the Defs. 13–17 of § 2 were formulated with that in mind). Thus the definition of true sentence which we are seeking must also cover sentences which are not provable.[1]

---

[1] The fact must also be taken into consideration that—in contrast to the concept of true sentence—the concept of provable sentence has a purely accidental character when applied to some deductive sciences, which is chiefly connected with the historical development of the science. It is sometimes difficult to give objective grounds for narrowing or widening the extension of this concept in a particular direction. For example, when we are dealing with the calculus of classes the sentence $\overline{\bigcap_1 \bigcap_2 \iota_{1,2}}$, which stipulates the existence of at least two distinct classes, is not accepted on the basis of the definitions of § 2—which will be expressed in Lemma E. Moreover, this sentence cannot be derived from the formal hypotheses upon which the work of Schröder is based, although in this case the matter is not quite clear (cf. Schröder, E. (62), vol. 1, pp. 245 and 246; vol. 2, Part 1, p. 278; vol. 3, Part 1, pp. 17 and 18); but in many works this sentence occurs as an axiom of the algebra of logic or forms an obvious consequence of these axioms (cf. Huntington, E. V. (32), p. 297, Post. 10). For quite different reasons, which will be discussed below in connexion with Th. 24 (cf. especially p. 207, footnote), it would be desirable to include the sentence $\bigcap_1(\bigcap_2 \iota_{1,2} + \bigcup_2(\iota_{2,1} \cdot \bigcap_3(\bigcap_4 \iota_{3,4} + \iota_{3,2} + \iota_{2,3})))$ among the

Let us try to approach the problem from quite a different angle, by returning to the idea of a semantical definition as in § 1. As we know from § 2, to every sentence of the language of the calculus of classes there corresponds in the metalanguage not only a name of this sentence of the structural-descriptive kind, but also a sentence having the same meaning. For example, corresponding to the sentence '$\prod x, \prod x_{\prime\prime} A I x, x_{\prime\prime} I x_{\prime\prime} x,$' is the name '$\bigcap_1 \bigcap_2 (\iota_{1,2} + \iota_{2,1})$' and the sentence 'for any classes a and b we have $a \subseteq b$ or $b \subseteq a$'. In order to make clear the content of the concept of truth in connexion with some one concrete sentence of the language with which we are dealing we can apply the same method as was used in § 1 in formulating the sentences (3) and (4) (cf. p. 156). We take the scheme (2) and replace the symbol 'x' in it by the name of the given sentence, and 'p' by its translation into the metalanguage. All sentences obtained in this way, e.g. '$\bigcap_1 \bigcap_2 (\iota_{1,2} + \iota_{2,1})$ *is a true sentence if and only if for any classes a and b we have $a \subseteq b$ or $b \subseteq a$*', naturally belong to the metalanguage and explain in a precise way, in accordance with linguistic usage, the meaning of phrases of the form 'x is a true sentence' which occur in them. Not much more in principle is to be demanded of a general definition of true sentence than that it should satisfy the usual conditions of methodological correctness and include all partial definitions of this type as special cases; that it should be, so to speak, their logical product. At most we can also require that only sentences are to belong to the extension of the defined concept, so that, on the basis of the definition constructed, all sentences of the type '*x is not a true sentence*', in which in the place of 'x' we have the name of an arbitrary expression (or of any other object) which is not a sentence, can be proved.

Using the symbol 'Tr' to denote the class of all true sentences, the above postulate can be expressed in the following convention:

CONVENTION **T**. *A formally correct definition of the symbol*

---

theorems, although this is not usually done. In the course of this work I shall have several occasions to return to the problem of the mutual relations of these two concepts: of theorem and of true sentence.

'Tr', *formulated in the metalanguage, will be called an* adequate definition of truth *if it has the following consequences*:

(α) *all sentences which are obtained from the expression '$x \in Tr$ if and only if p' by substituting for the symbol 'x' a structural-descriptive name of any sentence of the language in question and for the symbol 'p' the expression which forms the translation of this sentence into the metalanguage*;

(β) *the sentence 'for any x, if $x \in Tr$ then $x \in S$' (in other words* '$Tr \subseteq S$').[1]

It should be noted that the second part of the above convention is not essential; so long as the metalanguage already has the symbol 'Tr' which satisfies the condition (α), it is easy to define a new symbol 'Tr'' which also satisfies the condition (β). It suffices for this purpose to agree that Tr' is the common part of the classes Tr and S.

If the language investigated only contained a finite number of sentences fixed from the beginning, and if we could enumerate all these sentences, then the problem of the construction of a correct definition of truth would present no difficulties. For this purpose it would suffice to complete the following scheme: $x \in Tr$ *if and only if either $x = x_1$ and p_1, or $x = x_2$ and p_2,... or $x = x_n$ and p_n,* the symbols 'x_1', 'x_2',..., 'x_n' being replaced by structural-descriptive names of all the sentences of the language investigated and 'p_1', 'p_2',..., 'p_n' by the corresponding translation of these sentences into the metalanguage. But the situation is not like this. Whenever a language contains infinitely many sentences, the definition constructed automatically according to the above scheme would have to consist of infinitely many words, and such sentences cannot be formulated either in the metalanguage

---

[1] If we wished to subject the metalanguage and the metatheory expressed in it to the process of formalization, then the exact specification of the meaning of various expressions which occur in the convention **T** would present no great difficulties, e.g. the expressions '*formally correct definition of the given symbol*', '*structural-descriptive name of a given expression of the language studied*', '*the translation of a given sentence (of the language studied) into the metalanguage*'. After unimportant modifications of its formulation the convention itself would then become a normal definition belonging to the metatheory.

or in any other language. Our task is thus greatly complicated.

The idea of using the recursive method suggests itself. Among the sentences of a language we find expressions of rather varied kinds from the point of view of logical structure, some quite elementary, others more or less complicated. It would thus be a question of first giving all the operations by which simple sentences are combined into composite ones and then determining the way in which the truth or falsity of composite sentences depends on the truth or falsity of the simpler ones contained in them. Moreover, certain elementary sentences could be selected, from which, with the help of the operations mentioned, all the sentences of the language could be constructed; these selected sentences could be explicitly divided into true and false, by means, for example, of partial definitions of the type described above. In attempting to realize this idea we are however confronted with a serious obstacle. Even a superficial analysis of Defs. 10–12 of § 2 shows that in general composite sentences are in no way compounds of simple *sentences*. Sentential functions do in fact arise in this way from elementary functions, i.e. from inclusions; sentences on the contrary are certain special cases of sentential functions. In view of this fact, no method can be given which would enable us to define the required concept directly by recursive means. The possibility suggests itself, however, of introducing a more general concept which is applicable to any sentential function, can be recursively defined, and, when applied to sentences, leads us directly to the concept of truth. These requirements are met by the notion of the *satisfaction of a given sentential function by given objects*, and in the present case by given classes of individuals.

Let us try first to make clear by means of some examples the usual meaning of this notion in its customary linguistic usage. The way in which we shall do this represents a natural generalization of the method which we have previously used for the concept of truth.

The simplest and clearest case is that in which the given sentential function contains only *one* free variable. We can then significantly say of every single object that it does or does not satisfy the given function.[1] In order to explain the sense of this phrase we consider the following scheme:

for all a, a satisfies the sentential function x if and only if p

and substitute in this scheme for '*p*' the given sentential function (after first replacing the free variable occurring in it by '*a*') and for '*x*' some individual name of this function. Within colloquial language we can in this way obtain, for example, the following formulation:

for every a, we have a satisfies the sentential function 'x is white' if and only if a is white

(and from this conclude, in particular, that snow satisfies the function '*x* is white'). A similar construction will be familiar to the reader from school algebra, where sentential functions of a special type, called *equations*, are considered together with the numbers which satisfy these functions, the so-called *roots* of the equations (e.g. 1 is the only root of the equation '$x+2 = 3$').

When, in particular, the function belongs to the language of the calculus of classes, and the corresponding explanation of the expression '*a* satisfies the given sentential function' is to be formulated wholly in the terms of the metalanguage, then in the above scheme we insert for '*p*' not the sentential function itself, but the expression of the metalanguage having the same meaning, and for '*x*' we substitute an individual name of this function which likewise belongs to the metalanguage. For example, this method gives the following formulation in connexion with the function '$\prod x_{n} \, I x_{\prime} x_{n}$':

for all a, a satisfies the sentential function $\bigcap_{2} \iota_{1,2}$ if and only if for all classes b we have $a \subseteq b$

(whence it follows at once that the only class which satisfies the function '$\prod x_{n} \, I x_{\prime} x_{n}$' is the null class).

In cases where the sentential function has two distinct free variables we proceed in an exactly analogous manner. The only

---

[1] Provisionally I ignore problems connected with semantical categories (or logical types); these problems will be discussed in § 4.

difference is that the concept of satisfaction now refers not to single objects but to pairs (more accurately to ordered pairs) of objects. In this way we reach the following formulations:

for all a and b, a and b satisfy the sentential function 'x sees y' if and only if a sees b; for all a and b, a and b satisfy the sentential function $\iota_{2,3}$ (i.e. 'Ix$_{\prime\prime}$ x$_{\prime\prime\prime}$') if and only if a $\subseteq$ b.

Finally we pass to the general case, where the given sentential function contains an arbitrary number of free variables. For the sake of a uniform mode of expression we shall from now on not say that given objects but that *a given infinite sequence of objects satisfies a given sentential function*. If we restrict ourselves to functions from the calculus of classes, then the establishment of an unambiguous explanation of this expression is facilitated by the fact that all the variables which occur in the language of this science are ordered (enumerated) in a sequence. In considering the question of which sequences satisfy a given sentential function, we shall always have in mind a one-many correspondence of certain terms of a sequence f with the free variables of the sentential function, where with every variable corresponds the term of the sequence with the same index (i.e. the term f_k will be correlated with the variable v_k). No account will be taken of the terms which are not correlated with any variable.[1] We can explain the procedure best by means of concrete examples. Consider the function $\bigcap_2 \iota_{1,2}$ already mentioned. This function contains only *one* free variable v_1, so that we consider only the first terms of sequences. We say that the *infinite sequence f of classes satisfies the sentential function* $\bigcap_2 \iota_{1,2}$ *if and only if the*

---

[1] This is a simplification of a purely technical nature. Even if we could not order all the variables of a given language in a sequence (e.g. because we used symbols of arbitrary shapes as variables), we could still number all the signs, and thus all the variables, of every given expression, e.g. on the basis of the natural order in which they follow one another in the expression: the sign standing on the extreme left could be called the first, the next the second, and so on. In this way we could again set up a certain correlation between the free variables of a given function and the terms of the sequence. This correlation (in contrast to the one described in the text) would obviously vary with the form of the function in question; this would carry with it rather serious complications in the formulation of Def. 22 given below and especially of conditions (γ) and (δ).

class f_1 satisfies this function in the former sense, i.e. if for all classes b, we have $f_1 \subseteq b$. In an analogous way the *infinite sequence f of classes satisfies the sentential function $\iota_{2,3}$ if and only if the classes f_2 and f_3 satisfy the function in the previous sense, i.e. if $f_2 \subseteq f_3$.* This process may be described in general terms as follows:

We consider the following scheme:

The sequence f satisfies the sentential function x if and only if f is an infinite sequence of classes and p. If we have a sentential function from the calculus of classes, then in the above we replace the symbol 'x' by an individual (structural-descriptive) name of this function formulated in the metalanguage, but 'p' by a translation of the function into the metalanguage, where all free variables v_k, v_l, etc. are replaced by corresponding symbols 'f_k', 'f_l', etc.

We shall use a recursive method in order to formulate a general definition of satisfaction of a sentential function by a sequence of classes, which will include all partial definitions of this notion as special cases which are obtained from the given scheme in the way described above. For this purpose it will suffice, bearing in mind the definition of sentential function, to indicate which sequences satisfy the inclusions $\iota_{k,l}$ and then to specify how the notion we are defining behaves when the three fundamental operations of negation, disjunction, and universal quantification are performed on sentential functions.

The operation of universal quantification calls for special consideration. Let x be any sentential function, and assume that we already know which sequences satisfy the function x. Considering the meaning of the operation of universal quantification, we shall say that the sequence f satisfies the function $\bigcap_k x$ (where k is a particular natural number) only if this sequence itself satisfies the function x and does not cease to satisfy it even when the kth term of this sequence varies in any way; in other words, if every sequence which differs from the given sequence in at most the kth place also satisfies the function. For example, the function $\bigcap_2 \iota_{1,2}$ is satisfied by those, and only those, sequences f for which the formula $f_1 \subseteq f_2$ holds without regard to the way in which the second term of this sequence is allowed to vary (as is easily seen, this is only possible when the first term is the null class).

After these explanations the understanding of the following definition should not be difficult.

DEFINITION 22. *The sequence f satisfies the sentential function x if and only if f is an infinite sequence of classes and x is a sentential function and these satisfy one of the following four conditions:* (α) *there exist natural numbers k and l such that* $x = \iota_{k,l}$ *and* $f_k \subseteq f_l$; (β) *there is a sentential function y such that* $x = \bar{y}$ *and f does not satisfy the function y;* (γ) *there are sentential functions y and z such that* $x = y + z$ *and f either satisfies y or satisfies z;* (δ) *there is a natural number k and a sentential function y such that* $x = \bigcap_k y$ *and every infinite sequence of classes which differs from f in at most the k-th place satisfies the function y.*[1]

The following are examples of the application of the above definition to concrete sentential functions: the infinite sequence f satisfies the inclusion $\iota_{1,2}$ if and only if $f_1 \subseteq f_2$, and the function $\overline{\iota_{2,3}} + \overline{\iota_{3,2}}$ if and only if $f_2 \neq f_3$; the functions $\bigcap_2 \iota_{1,2}$ and $\bigcap_2 \iota_{2,3}$ are satisfied by those, and only those, sequences f in which f_1 is the null class and f_3 the universal class (i.e. the class of all individuals) respectively; finally, every infinite sequence of classes satisfies the function $\iota_{1,1}$ and no such sequence satisfies the function $\iota_{1,2} . \overline{\iota_{1,2}}$.

The concept just defined is of the greatest importance for investigations into the semantics of language. With its help the meaning of a whole series of concepts in this field can easily be

---

[1] The normal definition, which is equivalent to the above recursive one, is as follows (cf. pp. 70, 90, and 93):

The sequence f satisfies the sentential function x if and only if we have fRx for every relation R which satisfies the following condition:

For any g and y, in order that gRy it is necessary and sufficient that g is an infinite sequence of classes, y is a sentential function and either (α) *there are natural numbers k and l such that* $y = \iota_{k,l}$ *and* $g_k \subseteq g_l$ *or* (β) *there is a sentential function z such that* $y = \bar{z}$ *and the formula gRz does not hold; or* (γ) *there are sentential functions z and t such that* $y = z + t$ *and gRz or gRt; or finally* (δ) *there is a natural number k and a sentential function z such that* $y = \bigcap_k z$ *and hRz for every infinite sequence h of classes which is distinct from g at the k-th place at most.*

O

defined, e.g. the concepts of denotation, definability,[1] and truth, with the last of which we are especially concerned here.

The concept of truth is reached in the following way. On the basis of Def. 22 and the intuitive considerations which preceded it, it is easy to realize that whether or not a given sequence satisfies a given sentential function depends only on those terms of the sequence which correspond (in their indices) with the free variables of the function. Thus in the extreme case, when the function is a sentence, and so contains no free variable (which is in no way excluded by Def. 22), the satisfaction of a function by a sequence does not depend on the properties of the terms of the sequence at all. Only two possibilities then remain: either every infinite sequence of classes satisfies a given sentence, or no sequence satisfies it (cf. the Lemmas A and B given below). The sentences of the first kind, e.g. $\bigcup_1 \iota_{1,1}$, are the *true sentences*; those of the second kind, e.g. $\bigcap_1 \overline{\iota_{1,1}}$, can correspondingly be called the *false sentences*.†

[1] To say that the name x denotes a given object a is the same as to stipulate that the object a (or every sequence of which a is the corresponding term) satisfies a sentential function of a particular type. In colloquial language it would be a function which consists of three parts in the following order: a variable, the word 'is' and the given name x. As regards the concept of definability, I shall try to explain its content only in a particular case. If we consider which properties of classes we regard as definable (in reference to the system of the calculus of classes discussed here), we reach the following formulations:

We say that the sentential function x defines the property P of classes if and only if for a natural number k (α) x contains v_k as its only free variable, and (β) in order that an infinite sequence f of classes should satisfy x, it is necessary and sufficient that f_k should have the property P; we say that the property P of classes is definable if and only if there is a sentential function x which defines P.

On the basis of these stipulations it can be shown, for example, that such properties of classes as emptiness, of containing only one, two, three, etc., elements are definable. On the other hand the property of containing infinitely many elements is not definable (cf. the remarks given below in connexion with Ths. 14–16). It will also be seen that with this interpretation the concept of definability does not depend at all on whether the formalization of the science investigated admits of the possibility of constructing definitions. More exact discussions of definability will be found in articles VI and X of the present volume.

† A method of defining truth which is essentially equivalent to the method developed in this work, but is based upon a different idea, has recently been suggested by J. C. C. McKinsey in his paper 'A new definition of truth', *Synthèse*, vol. 7 (1948–9), pp. 428–33.

DEFINITION 23. *x is a* true sentence—*in symbols* $x \in Tr$—*if and only if* $x \in S$ *and every infinite sequence of classes satisfies* x.[1]

The question now arises whether this definition, about the formal correctness of which there is no doubt, is also materially correct—at least in the sense previously laid down in the convention **T**. It can be shown that the answer to this question is affirmative: *Def.* 23 *is an adequate definition of truth in the sense of convention* **T**, since its consequences include all those required by this convention. Nevertheless it can be seen without difficulty (from the fact that the number of these consequences is infinite) that the exact and general establishment of this fact has no place within the limits of the considerations so far brought forward. The proof would require the setting up of an entirely new apparatus: in fact it involves the transition to a level one step higher—to the meta-metatheory, which would have to be preceded by the formalization of the metatheory which forms the foundation of our investigations.[2] If we do not wish to depart from the level of our previous discussions, only one

---

[1] In the whole of the above construction we could operate with finite sequences with a variable number of terms instead of with infinite sequences. It would then be convenient to generalize the concept of finite sequence. In the usual interpretation of this term a sequence which has an nth term must also have all terms with indices less than n—we must now relinquish this postulate and regard any many-one relation as a finite sequence if its counter domain consists of a finite number of natural numbers distinct from 0. The modification of the construction would consist in eliminating from the sequences which satisfy the given sentential function all 'superfluous' terms, which have no influence on the satisfaction of the function. Thus if v_k, v_l, etc., occur as free variables in the function (of course in finite number), only those terms with the indices k, l, etc., would remain in the sequence which satisfies this function. For example, those, and only those, sequences f of classes would satisfy the function $\iota_{2,4}$ which consist of only two terms f_2 and f_4 verifying the formula $f_2 \subseteq f_4$. The value of such a modification from the standpoint of naturalness and conformity with the usual procedure is clear, but when we come to carry it out exactly certain defects of a logical nature show themselves: Def. 22 then takes on a more complicated form. Regarding the concept of truth, it is to be noted that—according to the above treatment—only *one* sequence, namely the 'empty' sequence which has no member at all, can satisfy a sentence, i.e. a function without free variables; we should then have to call those sentences true which are actually satisfied by the 'empty' sequence. A certain artificiality attaching to this definition will doubtless displease all those who are not sufficiently familiar with the specific procedures which are commonly used in mathematical constructions.

[2] See p. 188, footnote.

method, the empirical method, remains—the verification of the properties of Def. 23 in a series of concrete examples.

Consider, for example, the sentence $\bigcap_1 \bigcup_2 \iota_{1,2}$, i.e. '$\prod x_\prime N \prod x_{\prime\prime} N I x_\prime x_{\prime\prime}$'. According to Def. 22 the sentential function $\iota_{1,2}$ is satisfied by those, and only those, sequences f of classes for which $f_1 \subseteq f_2$ holds, but its negation, i.e. the function $\overline{\iota_{1,2}}$, only by those sequences for which $f_1 \nsubseteq f_2$ holds. Consequently a sequence f satisfies the function $\bigcap_2 \overline{\iota_{1,2}}$, if every sequence g which differs from f in at most the 2nd place satisfies the function $\overline{\iota_{1,2}}$ and thus verifies the formula $g_1 \nsubseteq g_2$. Since $g_1 = f_1$ and the class g_2 may be quite arbitrary, only those sequences f satisfy the function $\bigcap_2 \overline{\iota_{1,2}}$, which are such that $f_1 \nsubseteq b$ for any class b. If we proceed in an analogous way, we reach the result that the sequence f satisfies the function $\bigcup_2 \iota_{1,2}$, i.e. the negation of the function $\bigcap_2 \overline{\iota_{1,2}}$, only if there is a class b for which $f_1 \subseteq b$ holds. Moreover, the sentence $\bigcap_1 \bigcup_2 \iota_{1,2}$ is only satisfied (by an arbitrary sequence f) if there is for an arbitrary class a, a class b for which $a \subseteq b$. Finally by applying Def. 23 we at once obtain one of the theorems which were described in the condition (α) of the convention $\mathbf{T}$:

$\bigcap_1 \bigcup_2 \iota_{1,2} \in Tr$ *if and only if for all classes a there is a class b such that* $a \subseteq b$.

From this we infer without difficulty, by using the known theorems of the calculus of classes, that $\bigcap_1 \bigcup_2 \iota_{1,2}$ is a true sentence.

We can proceed in an exactly analogous way with every other sentence of the language we are considering. If for such a sentence we construct a corresponding assertion described in the condition (α) and then apply the mode of inference used above, we can prove without the least difficulty that this assertion is a consequence of the definition of truth which we have adopted. In many cases, with the help of only the simplest laws of logic (from the domain of the sentential calculus and the calculus of classes), we can draw definitive conclusions from theorems obtained in this way about the truth or falsity of the sentences in question. Thus, for example, $\bigcap_1 \bigcup_2 (\iota_{1,2} + \overline{\iota_{1,2}})$ is shown to be a true and $\bigcap_1 \bigcap_2 \iota_{1,2}$ a false sentence. With respect to other sentences, e.g. the sentence $\bigcap_1 \bigcap_2 \bigcap_3 (\iota_{1,2} + \iota_{2,3} + \iota_{3,1})$ or its negation, the analogous question cannot be decided (at least so long as we do not have recourse to the special existential assumptions of the metatheory, cf. p. 174): Def. 23 alone gives no general criterion for the truth of a sentence.[1] Nevertheless, through the theorems obtained, the meaning of the corresponding expressions of the type '$x \in Tr$' becomes intelligible and unambiguous. It should also be noted that the theorem expressed in the condition (β) of the convention **T** is also an obvious consequence of our definition.

With these discussions the reader will doubtless have reached the subjective conviction that Def. 23 actually possesses the property which it is intended to have: it satisfies all the conditions of convention **T**. In order to fix the conviction of the material correctness of the definition which has been reached in this way, it is worth while studying some characteristic general theorems that can be derived from it. With a view to avoiding encumbering this work with purely deductive matter, I shall give these theorems without exact proofs.[2]

THEOREM 1 (The principle of contradiction).   *For all sentences x, either* $x \,\overline{\in}\, Tr$ *or* $\bar{x} \,\overline{\in}\, Tr$.

This is an almost immediate consequence of Defs. 22 and 23.

THEOREM 2 (The principle of excluded middle).   *For all sentences x, either* $x \in Tr$ *or* $\bar{x} \in Tr$.

---

[1] At least when it is regarded from the methodological viewpoint this is not a defect of the definition in question; in this respect it does not differ at all from the greater part of the definitions which occur in the deductive sciences.

[2] The proofs are based on the general laws of logic, the specific axioms of the metascience and the definitions of the concepts occurring in the theorems. In some cases the application of the general properties of the concepts of consequence, of deductive system, etc., which are given in article V of the present volume is indicated. We are able to use the results obtained there because it can easily be shown that the concepts of sentence and consequence introduced here satisfy all the axioms upon which the above-mentioned work was based.

In the proof the following lemma, which follows from Defs. 11 and 22, plays an essential part:

LEMMA A. *If the sequence f satisfies the sentential function x, and the infinite sequence g of classes satisfies the following condition: for every $k, f_k = g_k$ if v_k is a free variable of the function x; then the sequence g also satisfies the function x.*

As an immediate consequence of this lemma and Def. 12 we obtain Lemma B which, in combination with Defs. 22 and 23 easily leads to Th. 1:

LEMMA B. *If $x \in S$ and at least one infinite sequence of classes satisfies the sentence x, then every infinite sequence of classes satisfies x.*

THEOREM 3. *If $X \subseteq Tr$ then $Cn(X) \subseteq Tr$; thus in particular $Cn(Tr) \subseteq Tr$.*

This theorem is proved by strong induction based chiefly on Defs. 15, 16, 22, and 23; the following simple lemma is also useful in this connexion:

LEMMA C. *If y is a universal quantification of the sentential function x, then in order that every infinite sequence of classes should satisfy x, it is necessary and sufficient that every infinite sequence of classes satisfies y.*

The results contained in Ths. 1–3 may be summarized in the following (obtained with the help of Defs. 18–20):

THEOREM 4. *The class Tr is a consistent and complete deductive system.*

THEOREM 5. *Every provable sentence is a true sentence, in other words, $Pr \subseteq Tr$.*

This theorem follows immediately from Def. 17, from Th. 3, and from Lemma D, the proof of which (on the basis of Def. 13 and Lemma C among others) presents no difficulty.

LEMMA D. *Every axiom is a true sentence.*

Th. 5 cannot be inverted:

THEOREM 6. *There exist true sentences which are not provable, in other words, $Tr \nsubseteq Pr$.*

This is an immediate consequence of Th. 2 and the following lemma, the exact proof of which is not quite easy:

LEMMA E.   *Both* $\bigcap_1 \bigcap_2 \iota_{1,2} \bar{\in} Pr$ *and* $\overline{\bigcap_1 \bigcap_2 \iota_{1,2}} \bar{\in} Pr.$[1]

As a corollary from Ths. 1, 5, and 6, I give finally the following theorem:

THEOREM 7.   *The class Pr is a consistent, but not a complete deductive system.*

In the investigations which are in progress at the present day in the methodology of the deductive sciences (in particular in the work of the Göttingen school grouped around Hilbert) another concept of a relative character plays a much greater part than the absolute concept of truth and includes it as a special case. This is the concept of *correct or true sentence in an individual domain a.*[2,3] By this is meant (quite generally and roughly speaking) every sentence which is true in the usual sense if we restrict the extension of the individuals considered to a given class a, or—somewhat more precisely—when we agree to interpret the terms 'individual', 'class of individuals', etc., as 'element of the class a', 'subclass of the class a', etc., respectively. Where we are dealing with the concrete case of sentences from the calculus of classes we must interpret expressions of the type '$\prod xp$' as '*for every subclass x of the*

---

[1] If we wish to include the sentence $\bigcap_1 \bigcap_2 \iota_{1,2}$ among the acceptable sentences (as is often the case, cf. p. 186, footnote) we could use here, instead of Lemma E, the following Lemma E':

Both $\bigcap_1 \bigcap_2 (\iota_{1,2} + \iota_{2,1}) \bar{\in} Pr$ *and* $\overline{\bigcap_1 \bigcap_2 (\iota_{1,2} + \iota_{2,1})} \bar{\in} Pr.$

The idea of the proof of both of these lemmas is the same as that of the proofs of the consistency and incompleteness of the lower functional calculus which is found in Hilbert, D., and Ackermann, W. (30), pp. 65–68.

[2] The discussion of these relativized notions is not essential for the understanding of the main theme of this work and can be omitted by those readers who are not interested in special studies in the domain of the methodology of the deductive sciences (only the discussions on pp. 208–9 are in closer connexion with our main thesis).

[3] In this connexion see Hilbert, D., and Ackermann, W. (30), especially pp. 72–81, and Bernays, P., and Schönfinkel, M. (5 a). But it should be emphasized that the authors mentioned relate this concept not to sentences but to sentential functions with free variables (because in the language of the lower functional calculus which they use there are no sentences in the strict sense of the word) and, connected with this, they use the term 'generally valid' instead of the term 'correct' or 'true'; cf. the second of the works cited above, pp. 347–8.

class a we have p', and expressions of the type *'Ixy'* as *'the sub-class x of the class a is contained in the subclass y of the class a'*. We obtain a precise definition of this concept by means of a modification of Defs. 22 and 23. As derived concepts we introduce the notion of a *correct sentence in an individual domain with k elements* and the notion of a *correct sentence in every individual domain*. It is worthy of note that—in spite of the great importance of these terms for metamathematical investigations—they have hitherto been used in a purely intuitive sense without any attempt to define their meaning more closely.[1]

DEFINITION 24. *The sequence f satisfies the sentential function x* in the individual domain *a if and only if a is a class of individuals, f an infinite sequence of subclasses of the class a and x a sentential function satisfying one of the following four conditions*: (α) *there exist natural numbers k and l such that* $x = \iota_{k,l}$ *and* $f_k \subseteq f_l$; (β) *there is a sentential function y such that* $x = \bar{y}$ *and the sequence f does not satisfy y in the individual domain a*; (γ) *there are sentential functions y and z such that* $x = y+z$ *and f satisfies either y or z in the individual domain a*; (δ) *there is a natural number k and a sentential function y such that* $x = \bigcap_k y$ *and every infinite sequence g of subclasses of the class a which differs from f in at most the k-th place satisfies y in the individual domain a.*

DEFINITION 25. *x is a* correct (true) sentence in the individual domain *a if and only if* $x \in S$ *and every infinite sequence of subclasses of the class a satisfies the sentence x in the individual domain a.*

DEFINITION 26. *x is a* correct (true) sentence in an individual domain with k elements—*in symbols* $x \in Ct_k$—*if and only if there exists a class a such that k is the cardinal number of the class a and x is a correct sentence in the individual domain a.*

---

[1] An exception is furnished by Herbrand, J. (26) in which the author defines the concept of true sentence in a finite domain (pp. 108–12). A comparison of Herbrand's definition with Defs. 25 and 26 given in the text will lead the reader at once to the conclusion that we have to do here with like-sounding terms rather than with a relationship of content. Nevertheless, it is possible that with respect to certain concrete deductive sciences, and under special assumptions for the corresponding metatheory, Herbrand's concept has the same extension (and also the same importance for metamathematical investigations) as a certain special case of the concept introduced in Def. 25.

DEFINITION 27. x *is a* correct (true) sentence in every individual domain—*in symbols* $x \in Ct$—*if and only if for every class* a, x *is a correct sentence in the individual domain* a.

If we drop the formula '$x \in S$' from Def. 25, and thereby modify the content of Defs. 26 and 27, we reach concepts of a more general nature which apply not only to sentences but also to arbitrary sentential functions.

Examples of the application to concrete sentences of the concepts defined will be given below. In the interest of more convenient formulation of various properties of these concepts, I introduce some further symbolical abbreviations.

DEFINITION 28. $x = \epsilon_k$ *if and only if*

$$x = \overline{\bigcap_{k+1} \iota_{k,k+1} \cdot \bigcap_{k+1}\left(\bigcap_{k+2} \iota_{k+1,k+2} + \overline{\iota_{k+1,k}} + \iota_{k,k+1}\right)}.$$

DEFINITION 29. $x = \alpha$ *if and only if*

$$x = \bigcap_1\left(\bigcap_2 \iota_{1,2} + \bigcup_2(\iota_{2,1} \cdot \epsilon_2)\right).$$

As is easily seen, the sentential function ϵ_k states that the class denoted by the variable v_k consists of only *one* element; the sentence α, which plays a great part in subsequent investigations, states that every non-null class includes a one-element class as a part.

DEFINITION 30. $x = \beta_n$ *if and only if either* $n = 0$ *and* $x = \bigcap_1 \overline{\epsilon_1}$, *or* $n \neq 0$ *and* $x = \bigcap_k^{k \leqslant n+1}\left(\sum_k^{n+1} \overline{\epsilon_k} + \sum_l^n \sum_k^l (\iota_{k,l+1} \cdot \iota_{l+1,k})\right)$.

DEFINITION 31. $x = \gamma_n$ *if and only if either* $n = 0$ *and* $x = \beta_0$, *or* $n \neq 0$ *and* $x = \overline{\beta_{n-1}} \cdot \beta_n$.

It follows from these definitions that the sentences β_n and γ_n (where n is any natural number) establish that there are at most n, and exactly n, distinct one-element classes respectively, or, what amounts to the same thing, that there are n distinct individuals.

DEFINITION 32. x *is a* quantitative sentence (*or a* sentence about the number of individuals) *if and only if there exists a finite sequence* p *of* n *natural numbers such that either* $x = \sum_k^n \gamma_{p_k}$ *or* $x = \overline{\sum_k^n \gamma_{p_k}}$.

I shall now give a series of characteristic properties of the defined concepts and the more important connexions which relate them with notions already introduced. This is the place for some results of a more special nature which are connected with the particular properties of the calculus of classes and cannot be extended to other disciplines of related logical structure (e.g. Ths. 11–13, 24, and 28).

THEOREM 8. *If a is a class of individuals and k the cardinal number of this class, then in order that x should be a correct sentence in the individual domain a it is necessary and sufficient that* $x \in Ct_k$.

The proof is based on the following lemma (among other things) which follows from Def. 24:

LEMMA F. *Let a and b be two classes of individuals and R a relation which satisfies the following conditions:* (α) *for any f' and g', if f'Rg' then f' is an infinite sequence of subclasses of a, and g' of subclasses of b;* (β) *if f' is any infinite sequence of subclasses of a, then there is a sequence g' such that f'Rg';* (γ) *if g' is any infinite sequence of subclasses of b, then there is a sequence f' such that f'Rg';* (δ) *for all f', g', f'', g'', k and l, if f'Rg', f''Rg'', and k and l are natural numbers distinct from 0, then* $f'_k \subseteq f''_l$ *if and only if* $g'_k \subseteq g''_l$. *If fRg and the sequence f satisfies the sentential function x in the individual domain a, then the sequence g also satisfies this function in the individual domain b.*

From this lemma, with the help of Def. 25, we easily obtain Lemma G which, together with Def. 26, at once gives Th. 8:

LEMMA G. *If the classes a and b of individuals have the same cardinal number, and x is a correct sentence in the individual domain a, then x is also a correct sentence in the individual domain b.*

According to Th. 8 (or Lemma G) the extension of the concept 'a sentence which is correct in the individual domain a' depends entirely on one property of the class a, namely on its cardinal number. This enables us to neglect in the sequel all results concerning this concept, because they can be derived immediately from the corresponding theorems relating to the classes Ct_k.

With the help of Defs. 24 and 25 the Ths. 1–6 and Lemmas A–D can be generalized by replacing the expressions '*infinite sequence of classes*', '*the sequence . . . satisfies the sentential function . . .*', '*true sentence*', and so on, by '*infinite sequence of subclasses of the class a*', '*the sequence . . . satisfies the sentential function . . . in the individual domain a*', '*correct sentence in the individual domain a*', and so on, respectively. As a consequence of Th. 8 the results so obtained can be extended to sentences which belong to the classes Ct_k. In this way we reach, among other things, the following generalizations of Ths. 4–6:

THEOREM 9. *For every cardinal number k the class Ct_k is a consistent and complete deductive system.*

THEOREM 10. *For every cardinal number k we have $Pr \subseteq Ct_k$, but $Ct_k \nsubseteq Pr$.*

In reference to Th. 10 the following problem presents itself: how is the list of axioms in Def. 13 to be completed, so that the class of all consequences of this extended class of axioms may coincide with the given class Ct_k? Ths. 11 and 12 which follow immediately below contain the solution of this problem and also prove that—with respect to the language of the calculus of classes —the definition of a correct sentence in a domain with k elements (Def. 26) can be replaced by another equivalent one which is analogous to the definition of provable sentence (Def. 17) and therefore has a structural character.

THEOREM 11. *If k is a natural number, and X the class consisting of all the axioms together with the sentences α and γ_k, then $Ct_k = Cn(X)$.*

THEOREM 12. *If k is an infinite cardinal number, and X the class consisting of all the axioms together with the sentence α and all sentences $\bar{\gamma}_l$ (where l is any natural number), then $Ct_k = Cn(X)$.*

The proof of these theorems is based chiefly on Ths. 9 and 10 and the three following lemmas:

LEMMA H. *For every cardinal number k we have $\alpha \in Ct_k$.*

LEMMA I. *If k is a natural number and l a cardinal number distinct from k, then $\gamma_k \overline{\in} Ct_k$ and $\gamma_k \in Ct_l$, but $\overline{\gamma_k} \overline{\in} Ct_k$ and $\overline{\gamma_k} \in Ct_l$.*

LEMMA K. *If $x \in S$ and X is the class consisting of all the axioms together with the sentence α, then there is a sentence y which is equivalent to the sentence x with respect to the class X and such that either y is a quantitative sentence, or $y \in Pr$ or $\bar{y} \in Pr$.*

Lemmas H and I are almost immediately evident, but the proof of the very important and interesting Lemma K is rather difficult.[1]

By means of Th. 9 and Lemma I it is possible from Th. 12 to derive the following consequence which combined with Th. 11 brings out the essential differences existing in the logical structure of the classes Ct_k according to whether the cardinal number k is finite or infinite:

THEOREM 13. *If k is an infinite cardinal number, then there is no class X which contains only a finite number of sentences which are not axioms, and also satisfies the formula*

$$Ct_k = Cn(X).^2$$

From Lemma I and Ths. 11 and 12 we easily obtain the following consequences:

THEOREM 14. *If k is a natural number and l a cardinal number distinct from k, then $Ct_k \nsubseteq Ct_l$ and $Ct_l \nsubseteq Ct_k$.*

THEOREM 15. *If k and l are infinite cardinal numbers, then $Ct_k = Ct_l$.*

THEOREM 16. *If k is an infinite cardinal number and $x \in Ct_k$, then there is a natural number l such that $x \in Ct_l$ (in other words, the class Ct_k is included in the sum of all the classes Ct_l).*

According to Ths. 14–16 (or Lemma I) there exists for every natural number k a sentence which is correct in every domain

---

[1] In its essentials this lemma is contained in the results to be found in Skolem, Th. (64), pp. 29–37.

[2] The idea of the proof of this theorem is the same as that of the proofs of Ths. 24 and 25 in article V of the present volume, pp. 78–9. If we take over from the latter Def. 3, p .76, and at the same time extend our present concept of consequence by adding the words '*or x is an axiom*' to the condition (α) of Def. 15, then we could derive the following consequence from Ths. 11 and 13:

In order that the class Ct_k should be an axiomatizable deductive system, it is necessary and sufficient that k be a natural number.

with k elements and in no domain with any other cardinal number. On the other hand, every sentence which is correct in one infinite domain is also correct in every other infinite domain (without reference to its cardinal number) as well as in certain finite domains. From this we infer that the language in question allows us to express such a property of classes of individuals as their being composed of exactly k elements, where k is any natural number; but we find in this language no means by which we can distinguish a special kind of infinity (e.g. denumerability), and we are unable, either with the help of a single or of a finite number of sentences, to distinguish two such properties of classes as finiteness and infinity.[1]

By means of Ths. 9, 11, and 12 we can prove

THEOREM 17. *If X is a consistent class of sentences which contains all the axioms together with the sentence α, then there is a cardinal number k such that $X \subseteq Ct_k$; if X is a complete deductive system, then $X = Ct_k$.*

If we combine this theorem with Ths. 11 and 12, we obtain a structural description of all complete deductive systems which contain all the axioms and the sentence α. It should be noted that the presence of the sentence α is essential here; the multiplicity of the systems which do not contain this sentence is significantly greater and their exhaustive description would not at all be a simple matter.[2]

The remaining considerations concern sentences which are correct in every individual domain, i.e. belong to the class Ct.

---

[1] These results, as well as Th. 19 given below, we owe to Löwenheim; cf. Löwenheim, L. (49) (especially Th. 4, p. 459) and Skolem, Th. (64).

[2] I have occupied myself in the years 1926–8 with problems of this type, i.e. with the structural description of all complete systems of a given science, in application to various elementary deductive sciences (algebra of logic, arithmetic of real numbers, geometry of straight lines, theory of order, theory of groups); on the results of these investigations, reports were made in the seminar exercises in the methodology of the deductive sciences which I conducted in Warsaw University in the years 1927/8 and 1928/9. Cf. Presburger, M. (61) (especially note 4 on p. 95), and XII, § 5. For a detailed discussion of certain closely related problems (as well as for further bibliographical references) see also the recent publications of the author, Tarski, A. (84) and (84 a).

THEOREM 18. *In order that $x \in Ct$ it is necessary and sufficient that, for every cardinal number k, $x \in Ct_k$ (in other words, the class Ct is the product of all the classes Ct_k).*

This theorem, which is an immediate consequence of Def. 27 and Th. 8, can be essentially sharpened by means of Ths. 9 and 16:

THEOREM 19. *In order that $x \in Ct$ it is necessary and sufficient that, for every natural number k, $x \in Ct_k$.*

The correctness of a sentence in all finite domains thus entails its correctness in every individual domain.

The following two corollaries are derivable from Ths. 9, 14, and 18:

THEOREM 20. *For every cardinal number k we have $Ct \subseteq Ct_k$, but $Ct_k \nsubseteq Ct$.*

THEOREM 21. *The class Ct is a consistent but not a complete deductive system.*

THEOREM 22. *$Pr \subseteq Ct$, but $Ct \nsubseteq Pr$.*

This theorem follows from Ths. 10 and 18 and Lemma L:

LEMMA L. *$\alpha \in Ct$ but $\alpha \bar\in Pr$.*

That $\alpha \in Ct$ follows at once from Lemma H and Th. 18. The exact proof of the second part of the lemma is considerably more difficult.

THEOREM 23. *If x is a quantitative sentence then $x \bar\in Ct$.*

The proof, which is based on Lemma I, Th. 18, and Def. 32, offers no difficulties.

THEOREM 24. *If X is the class consisting of all the axioms together with the sentence α, then $Ct = Cn(X)$.*

This theorem is most easily proved with the help of Ths. 11, 12, and 18. By using Lemma K we obtain from it at once:

THEOREM 25. *If $x \in S$, $x \bar\in Ct$ and $\bar x \bar\in Ct$, then there is a quantitative sentence y, which is equivalent to the sentence x with respect to the class Ct.*

By reference to Lemma L and Th. 24 we notice that we have the following situation: the concept of a sentence which is correct in every individual domain has a larger extension than the concept of provable sentence, since the sentence α belongs to the extension of the first concept but not to that of the second. But if we increase the system of axioms by just this single sentence α, the two concepts become identical in extension. Because it seems to me desirable that, with respect to the calculus of classes, the concepts of theorem and of correct sentence in every individual domain should not be distinct in extension,[1] I would advocate the inclusion of the sentence α among the axioms of this science.

The problem still remains of clarifying the relation of the absolute concept of truth defined in Def. 23 to the concepts we have just investigated.

If we compare Defs. 22 and 23 with Defs. 24 and 25 and apply Th. 8, we easily obtain the following result:

THEOREM 26. *If a is the class of all individuals then $x \in Tr$ if and only if x is a correct sentence in the domain a; thus if k is the cardinal number of the class a, then $Tr = Ct_k$.*

As an immediate consequence of Ths. 20 and 26 we have:

THEOREM 27. $Ct \subseteq Tr$, *but* $Tr \nsubseteq Ct$.

If we bring together Ths. 26 and 14 or Ths. 11 and 12, we reach the conclusion that those assumptions of the metatheory which determine the cardinal number of the class of all individuals (and which do not intervene in the proof of Th. 26 itself) exert an essential influence on the extension of the term 'true sentence'. The extension of this term is different according to whether that class is finite or infinite. In the first case the extension even depends on how big the cardinal number of this class is.

---

[1] This tendency will be discussed in the next paragraph. It should be mentioned that Schröder, although beginning with other ideas, has made the suggestion of completing the system of hypotheses of the calculus of classes with the sentence α (and even with still other sentences which, however, as can easily be shown, follow in a simple way from the sentence α); cf. Schröder, E. (62), vol. 2, Part 1, pp. 318–49. In this connexion I may remark that it seems to me that the inclusion of the sentence α in the 'formal' system of the algebra of logic (of which the calculus of classes is an interpretation) would not be useful, for many interpretations of this system are known in which the sentence in question is not satisfied.

Because we can show, on the basis of the system of assumptions here adopted, that the class of all individuals is infinite, Th. 26 in combination with Th. 12 makes a structural characterization of true sentences possible:

THEOREM 28. *In order that $x \in Tr$, it is necessary and sufficient that x is a consequence of the class which consists of all the axioms together with the sentence α and all sentences $\bar{\gamma}_l$, where l is any natural number.*

This sentence could, in its form, obviously be regarded as a definition of true sentence. It would then be a purely structural definition, completely analogous to Def. 17 of provable theorem. But it must be strongly emphasized that the possibility of constructing a definition of such a kind is purely accidental. We owe it to the specific peculiarities of the science in question (to those peculiarities which, among others, have been expressed in Lemma K, which is the most essential premiss in the proof of Ths. 12 and 28) as well as—in some degree—to the strong existential assumptions adopted in the metatheory. On the other hand—in contrast to the original definition—we have here no general method of construction which could be applied to other deductive sciences.

It is worth noticing that by analysing the proof of Th. 28 and of the lemmas from which this theorem follows, we can obtain a general structural criterion of truth for all sentences of the language investigated. From Th. 28 such a criterion for quantitative sentences is easily derivable, and the proof of Lemma K allows us effectively to correlate with every sentence of the language a sentence which is equivalent to it and which, if it is not quantitative, is manifestly true or manifestly false. An analogous remark holds for the concept of correctness in a given, or in every, individual domain.

Summarizing the most important results obtained in this section we can say:

We have succeeded in doing for the language of the calculus of classes what we tried in vain to do for colloquial language: namely

*to construct a formally correct and materially adequate semantical
definition of the expression 'true sentence'.*

Moreover, by making use of the special peculiarities of the
calculus of classes, we have been able to transform this definition
into an equivalent structural definition which even yields a
general criterion of truth for the sentences of the language
of this calculus.

§ 4. THE CONCEPT OF TRUE SENTENCE IN LANGUAGES OF FINITE ORDER

The methods of construction which I have used in the previous
section for the investigation of the language of the calculus of
classes can be applied, without very important changes, to
many other formalized languages, even to those with a considerably more complicated logical structure. In the following pages
the generality of these methods will be emphasized, the limits
of their applicability will be determined, and the modifications
which they undergo in their various concrete applications will be
briefly described.

It is by no means my intention, in these investigations, to consider all languages that can conceivably be imagined, or which
any one at any time could or might wish to construct; such an
attempt would be condemned to failure from the start. In what
I shall say here I shall consider exclusively languages of the same
structure as those which are known to us at the present day (in
the perhaps unfounded conviction that they will form in the
future, as they have done hitherto, a sufficient basis for the
foundation of the whole of deductive knowledge). And even
these languages show such great differences in their construction
that their investigation in a perfectly general, but at the same
time precise, way must encounter serious difficulties. These
differences are, of course, rather of a 'calligraphical' nature. In
some languages, for example, only constants and variables
occur, in others it is not possible to avoid the use of so-called
technical signs (brackets, points, and so on). In some languages
symbols of an exactly specified form are used as variables, so
that the form of the variables depends on the part they play

and their significance. In others quite arbitrary symbols may be used as variables, so long as they are distinguished by their form from the constants. In some languages every expression is a system of linearly ordered signs, i.e. signs following one another in a line, but in others the signs may lie at different levels, not only alongside but also below one another. This calligraphy of the language nevertheless exerts a fairly strong influence on the form of the constructions in the domain of the metalanguage, as will doubtless be seen from a brief survey of the preceding paragraphs.[1] For those reasons alone the following exposition will have the nature of a sketch; wherever it takes a more precise form, it is dealing with concretely described languages which are constructed in the same way as the language of the calculus of classes (i.e. languages without technical signs, with variables of an exactly specified form, with linear arrangement of the signs in every expression and so on).[2]

Before we approach our principal task—the construction of the definition of true sentence—we must undertake, in every concrete case, the construction of a corresponding metalanguage and the establishment of the metatheory which forms the proper field of investigation. A metalanguage which meets our requirements must contain three groups of primitive expressions: (1) expressions of a general logical kind; (2) expressions having the same meaning as all the constants of the language to be discussed or which suffice for the definition of such expressions (taking as a basis the rules of definition adopted in the meta-

---

[1] Cf., for example, p. 191, footnote.

[2] In order to give the following exposition a completely precise, concrete, and also sufficiently general form, it would suffice if we chose, as the object of investigation, the language of some one complete system of mathematical logic. Such a language can be regarded as a universal language in the sense that all other formalized languages—apart from 'calligraphical' differences—are either fragments of it, or can be obtained from it or from its fragments by adding certain constants, provided that the semantical categories of these constants (cf. below, pp. 215 ff.) are already represented by certain expressions of the given language. The presence or absence of such constants exerts, as we shall show, only a minimal influence on the solution of the problem in which we are interested. As such a language we could choose the language of the general theory of sets which will be discussed in § 5, and which might be enriched by means of variables representing the names of two- and of many-termed relations (of arbitrary semantical categories).

names. The first sign of such a complex is always the name of a class or a relation or a corresponding variable, and is called a (*sentence forming*) *functor of the given primitive sentential function*;[1] the remaining signs are called *arguments*, namely 1st, 2nd,..., kth argument—according to the place they occupy. For every constant and variable of the language studied—with the exception of the constants of the sentential calculus and the universal and existential quantifiers—a primitive function can be constructed which contains this sign (the sentential variables, even when they appear in the language, do not occur in the primitive functions as functors or arguments, but each is regarded as an independent primitive function). Next we introduce the *fundamental operations on expressions* by means of which composite expressions are formed from simpler ones. In addition to the operations of negation, logical addition and universal quantification, which we have met with in § 2 (Defs. 2, 3, and 6), we consider here other analogously defined operations, such as logical multiplication, formation of implications and equivalences, as well as existential quantification. Each of these operations consists in putting in front of the expression considered, or in front of two successive expressions (according to the kind of operation), either one of the constants of the sentential calculus which belongs to the language, or one of the two quantifiers together with the variables immediately following it. The expressions which we obtain from the primitive functions by applying to them any number of times and in any order any of

[1] Thus sentence-forming functors which have names as arguments are here identified with the names of classes or relations (in fact the one-argument functors with names of classes and the rest with names of two- or many-termed relations). This interpretation seems artificial with the interpretation of the term 'functor' which was given by some examples on p. 161, note 1; in any case it certainly does not agree with the spirit and formal structure of the language of everyday life. Without going into details, it seems to me for various reasons to be neither necessary nor useful to distinguish between these two categories of expressions (i.e. sentence-forming functors and names of classes or relations). Moreover, the whole question is rather of a terminological nature and is without influence on subsequent developments. We may either regard the definition of functor given in the text as purely formal and disregard the current interpretation of the term, or so extend the interpretation of terms like 'name of a class', 'name of a relation' that we include expressions which are not names in the usual sense.

the fundamental operations, we call sentential functions. Among the variables which occur in a given sentential function we can distinguish—by means of recursive definitions—*free* and *bound* variables. Sentential functions without free variables are called sentences (cf. Defs. 10–12 in § 2).

Next we define yet other concepts which are closely connected with the deductive character of the science under investigation, namely the concepts of *axiom*, *consequence*, and *theorem*. Among the axioms we include as a rule certain logical sentences which are constructed in a manner similar to that used for the first kind of axioms of the calculus of classes (cf. § 2, Def. 13). Moreover the definition of axiom depends wholly on the individual peculiarities of the science investigated, sometimes even on accidental factors which are connected with its historical development. In the definition of the concept of consequence we follow—*mutatis mutandis*—the pattern of § 2. The operations by means of which we form the consequences of a given class of sentences differ in no essential points from the operations which were given in Def. 15. The consequences of the axioms are called *provable sentences* or *theorems*.

After this preliminary work we turn now to our principal task —the construction of a correct definition of *true sentence*. As we saw in § 3, the method of construction available to us presupposes first a definition of another concept of a more general kind which is of fundamental importance for investigations in the semantics of language. I mean the notion of the *satisfaction of a sentential function by a sequence of objects*. In the same section I have attempted to clarify the customary meaning of this expression in its ordinary usage. I have pointed out that in drawing up a correct definition of the concept of satisfaction use can be made of recursive definition. For this purpose it suffices—recalling the recursive definition of sentential function and bearing in mind the intuitive sense of the primitive sentential functions and the fundamental operations on expressions—to establish two facts: (1) which sequences satisfy the fundamental functions, and (2) how the concept of satisfaction behaves under the application of any of the fundamental operations (or to put it more exactly: which sequences satisfy the sentential functions which are obtained from given sentential functions by means of one of the fundamental operations, assuming that it has already been established which sequences satisfy the sentential functions to which the operation is applied). As soon as we have succeeded in making precise the sense of this concept of satisfaction, the definition of truth presents no further difficulty: the true sentences may be defined as those sentences which are satisfied by an arbitrary sequence of objects.

In carrying out the plan just sketched in connexion with various concrete languages we nevertheless meet with obstacles of a fundamental kind; in fact, just at the point where we try finally to formulate the correct definition of the concept of satisfaction. In order to make clear the nature of these difficulties a concept must first be discussed which we have not hitherto had an opportunity of introducing, namely the concept of *semantical category*.

This concept, which we owe to E. Husserl, was introduced into investigations on the foundations of the deductive sciences by Leśniewski. From the formal point of view this concept plays a part in the construction of a science which is analogous to that played by the notion of type in the system *Principia Mathematica* of Whitehead and Russell. But, so far as its origin and content are concerned, it corresponds (approximately) rather to the well-known concept of part of speech from the grammar of colloquial language. Whilst the theory of types was thought of chiefly as a kind of prophylactic to guard the deductive sciences against possible antinomies, the theory of semantical categories penetrates so deeply into our fundamental intuitions regarding the meaningfulness of expressions, that it is scarcely possible to imagine a scientific language in which the sentences have a clear intuitive meaning but the structure of which cannot be brought into harmony with the above theory.[1]

---

[1] Cf. Leśniewski, S. (46), especially pp. 14 and 68; Ajdukiewicz, K. (3), pp. 9 and 148. From the formal point of view the theory of semantical categories is rather remote from the original theory of types of Whitehead, A. N., and Russell, B. A. W. (90), vol. 1, pp. 37 ff.; it differs less from the simplified theory of types (cf. Chwistek, L. (12), pp. 12–14; Carnap, R. (8), pp. 19–22) and is an extension of the latter. Regarding the views expressed in the last paragraph of the text, compare the Postscript to this article (p. 268).

For reasons mentioned at the beginning of this section we cannot offer here a precise structural definition of semantical category and will content ourselves with the following approximate formulation: two expressions *belong to the same semantical category* if (1) there is a sentential function which contains one of these expressions, and if (2) no sentential function which contains one of these expressions ceases to be a sentential function if this expression is replaced in it by the other. It follows from this that the relation of belonging to the same category is reflexive, symmetrical, and transitive. By applying the principle of abstraction,[1] all the expressions of the language which are parts of sentential functions can be divided into mutually exclusive classes, for two expressions are put into one and the same class if and only if they belong to the same semantical category, and each of these classes is called a semantical category. Among the simplest examples of semantical categories it suffices to mention the category of the sentential functions, further the categories which include respectively the names of individuals, of classes of individuals, of two-termed relations between individuals, and so on. Variables (or expressions with variables) which represent names of the given categories likewise belong to the same category.

In connexion with the definition of semantical category the following question arises: in order to establish the fact that two given expressions belong to one and the same semantical category, is it necessary to consider all possible sentential functions which contain one of the given expressions and to investigate their behaviour when one of these expressions is replaced by the other, or does it suffice to make this observation in some or even in only *one* case? From the standpoint of the ordinary usage of language the second possibility seems much more natural; in order that two expressions shall belong to the same semantical category, it suffices if there exists *one* function which contains one of these expressions and which remains a function when this expression is replaced by the other. This principle, which can be called the *first principle of the theory of semantical categories*, is

[1] Cf. Carnap, R. (8), pp. 48–50.

taken strictly as a basis for the construction of the formalized languages here investigated.[1] It is especially taken into account in the definition of the concept of sentential function. It also exerts an essential influence on the definition of the operation of substitution, i.e. one of those operations with the help of which we form the consequences of a class of sentences. For if we wish that this operation, when carried out on any sentence, should always give a new sentence as a result, we must restrict ourselves to substituting for the variables only those expressions which belong to the same semantical category as the corresponding variables.[2] Closely connected with this principle is a general law concerning the semantical categories of sentence-forming functors: the functors of two primitive sentential functions belong to the same category if and only if the number of arguments in the two functions is the same, and if any two arguments which occupy corresponding places in the two functions also belong to the same category. From this it follows that, in particular, no sign can be simultaneously a functor of two functions which possess a different number of arguments, or of two such functions (even if they possess the same number of arguments)

---

[1] When applied to concrete languages the formulations given in the text— both the definition of semantical category and the above-mentioned principle— require various corrections and supplementations. They are in any case too general, for they also include expressions to which we do not usually ascribe independent meaning, and which we often include in the same semantical categories to which meaningful expressions belong (for example, in the language of the calculus of classes, the expressions 'N', '$\Pi x_{\prime}$', and '$AIx_{\prime}x_{\prime\prime}$' would belong to the same semantical category); in the case of these meaningless expressions, it can easily be shown that even the first principle of semantical categories loses its validity. This fact is of no essential importance for our investigations, for we shall apply the concept of semantical category, not to composite expressions, but exclusively to variables. On the other hand, the examples which we shall encounter in the sequel show that the above formulations admit of very far-reaching simplifications in concrete cases. Thanks to a suitable choice of the signs used in the construction of the expressions of the language, the mere shape of the sign (and even of the composite expression) decides to which category it belongs. Consequently it is possible that in methodological and semantical investigations concerning a concrete language, the concept of semantical category does not explicitly occur at all.

[2] In the language of the calculus of classes, and in the languages which I shall describe in more detail in the sequel, such expressions can only be other variables; this explains the formulation of Def. 14 in § 2.

in which two arguments which occupy corresponding places belong to different categories.

We require a classification of the semantical categories; to every category a particular natural number is assigned called the *order of the category*. This order is also assigned to all expressions which belong to this category.[1] The meaning of this term can be determined recursively. For this purpose we adopt the following convention (in which we have in mind only those languages which we shall deal with here and we take account only of the semantical categories of the variables): (1) the 1st order is assigned only to the names of individuals and to the variables representing them; (2) among expressions of the $n+1$th order, where n is any natural number, we include the functors of all those primitive functions all of whose arguments are of at most the nth order, where at least one of them must be of exactly the nth order. Thanks to the above convention all expressions which belong to a given semantical category have the same order assigned to them, which is therefore called the order of that category.[2] On the other hand the category

[1] Cf. Carnap, R. (8), pp. 31–32.

[2] This classification by no means includes all semantical categories which are to be found in formalized languages. For example, it does not include sentential variables and functors with sentences as arguments—i.e. signs which occur in the sentential calculus—neither does it include functors which, together with the corresponding arguments, form expressions which belong to one of the categories distinct from sentential functions, such as the name-forming functors mentioned on p. 213, footnote.

In view of this, the definition of order given in the text could be widened in the following way: (1) to the 1st order belong sentences, names of individuals and expressions representing them; (2) among expressions of the $n+1$th order we include those functors with an arbitrary number of arguments of order $\leqslant n$, which together with these arguments form expressions of order $\leqslant n$, but are not themselves expressions of the nth order. Even this definition does not yet cover all meaningful expressions which occur in the deductive sciences. No signs which 'bind' variables fall under this definition (thus such signs as the universal and existential quantifiers, the signs 'Σ' and 'Π' of the theory of sets and analysis or the sign of integration), signs which—in contrast to the functors—can be called *operators*. (von Neumann speaks of *abstractions* in this connexion, see Neumann, J. v. (54).) On the other hand the latter classification is completely adapted to the system invented by Leśniewski and sketched by him in Leśniewski (46) and (47). This system contains no operators except the universal quantifier which belongs to no semantical category. I may add that, in my view, the lack of operators in Leśniewski's system constitutes a deficiency which restricts its 'universal' character (in the sense of p. 210, note 2) to a certain degree.

is by no means specified by the order: every natural number which is greater than 1 can be the order of many different categories. Thus, for example, both the names of classes of individuals and the names of two-, three-, and many-termed relations between individuals are expressions of the 2nd order.

It is desirable to classify the sentential functions of the language according to the semantical categories of the free variables occurring in them. We shall say of two functions that they *possess the same semantical type* if the number of free variables of every semantical category in the two functions is the same (or, in other words, if the free variables of the one function can be put into one-one correspondence with the free variables of the other in such a way that to every variable a variable of the same category corresponds). The class of all sentential functions which possess the same type as a given function we can call a *semantical type*.

We sometimes use the term 'semantical category' in a derivative sense, by applying it, not to the expressions of the language, but to the objects which they denote. Such 'hypostatizations' are not quite correct from a logical standpoint, but they simplify the formulation of many ideas. We say, for example, that all individuals belong to the same semantical category, but that no classes or relations belong to this category. From the general law stated above concerning sentence-forming functors we conclude that two classes belong to the same category if and only if all their elements belong to one and the same category. Two two-termed relations belong to the same category if and only if their domains belong to the same category and their counter domains belong to the same category. In particular, two sequences belong to the same category if and only if all their terms belong to the same category. A class and a relation, or two relations having different numbers of terms never belong to the same category. It also follows that there can be no class whose elements belong to two or more semantical categories; in an analogous way there can be no sequence whose terms belong to distinct semantical categories. Individuals are sometimes called objects of the 1st order, classes of individuals and relations between individuals objects of the 2nd order, and so on.

The language of a complete system of logic should contain—actually or potentially—all possible semantical categories which occur in the languages of the deductive sciences. Just this fact gives to the language mentioned a certain 'universal' character, and it is one of the factors to which logic owes its fundamental importance for the whole of deductive knowledge. In various fragmentary systems of logic, as well as in other deductive sciences, the multiplicity of the semantical categories may undergo a significant restriction in both their number and their order. As we shall see, the degree of difficulty which we have to overcome in the construction of a correct definition of truth for a given concrete language, depends in the first place on this multiplicity of the semantical categories appearing in the language, or, more exactly, on whether the expressions and especially the variables of the language belong to a finite or an infinite number of categories, and in the latter case on whether the orders of all these categories are bounded above or not. From this point of view we can distinguish four kinds of languages: (1) languages in which all the variables belong to one and the same semantical category; (2) languages in which the number of categories in which the variables are included is greater than 1 but finite; (3) languages in which the variables belong to infinitely many different categories but the order of these variables does not exceed a previously given natural number n; and finally (4) languages which contain variables of arbitrarily high order. We shall call the languages of the first three kinds *languages of finite order*, in contrast to languages of the fourth kind, the *languages of infinite order*. The languages of finite order could be further divided into languages of the 1st, 2nd order, and so on, according to the highest order of the variables occurring in the language. By way of supplementation of the sketch given at the beginning of this section of the construction of a metatheory, it must be noted here that the metalanguage, on the basis of which the investigation is conducted, is to be furnished with at least all the semantical categories which are represented in the language studied. This is necessary if it is to be possible to translate any expression of the language into the metalanguage.[1]

From the point of view of their logical structure the languages of the 1st kind are obviously the simplest. The language of the calculus of classes is a typical example. We have seen in § 3 that for this language the definition of the satisfaction of a sentential function by a sequence of objects, and hence the definition of true sentence, presents no great difficulties. The method of construction sketched there can be applied as a whole to other languages of the 1st order. It is clear that in doing this certain small deviations in detail may occur. Among other things it may be necessary to operate not with sequences of classes but with sequences of other kinds, e.g. with sequences of individuals or relations, according to the intended interpretation and the semantical categories of the variables occurring in the language.[2]

A particularly simple example of a language of the 1st kind which is worthy of attention is the language of the ordinary sentential calculus enlarged by the introduction of the universal and existential quantifiers. The simplicity of this language lies, among other things, in the fact that the concept of variable coincides with that of primitive sentential function. In the metatheory of the sentential calculus two different definitions can be given of provable theorem, the equivalence of which is in no way evident: the one is based on the concept of consequence and is analogous to Defs. 15–17 of § 2, the second is connected with the concept of the two-valued matrix. By virtue of this second definition we can easily determine whether any sentence is provable provided its structure is known.[3] If we now construct for this language a definition of true sentence strictly according

---

[1] Here—*mutatis mutandis*—the remarks of p. 211, footnote, also apply.

[2] Certain complications, which I shall not discuss here, arise if in addition to variables, composite expressions of the same semantical category also occur in the language investigated; the complete language of the calculus of classes which was mentioned on p. 168, note 3, will serve as an example, or the language of a system of arithmetic investigated in Presburger, M. (61) (cf. also p. 212, footnote).

[3] Cf. Hilbert, D., and Ackermann, W. (30), pp. 84–85; Łukasiewicz, J. (51), pp. 154 ff.; IV, § 4.

to the pattern given in § 3, we can easily convince ourselves that it represents a simple transformation of the second of these definitions of provable sentence, and thus the two terms 'provable theorem' and 'true sentence' in this case have the same extension. This fact provides us, among other things, with a general structural criterion for the truth of the sentences of this language. The method of construction laid down in the present work could thus be regarded, in a certain sense, as a generalization of the matrix method familiar in investigations on the sentential calculus.

Serious difficulties only arise when we consider languages of more complicated structure, e.g. languages of the 2nd, 3rd, and 4th kinds. We must now analyse these difficulties and describe the methods which enable us at least partially to overcome them. In order to make the exposition as clear and precise as possible I shall discuss in somewhat greater detail some concrete formalized languages, one of each kind. I shall try to choose examples which are as simple as possible, are free from all less essential, subordinate complications, and are at the same time sufficiently typical to exhibit the difficulties mentioned to the fullest extent and in the most striking form.

The language of the *logic of two-termed relations* will serve as an example of a language the 2nd order.[1] The only constants of this language are: the sign of negation 'N', the sign of logical sum 'A' and the universal quantifier '$\prod$'. As variables we can use the signs '$x_{,}$', '$x_{,,}$', '$x_{,,,}$', ... and '$X_{,}$', '$X_{,,}$', '$X_{,,,}$', ... The sign composed of the symbol 'x' and of k small additional strokes is called the *k-th variable of the 1st order*, and is denoted by the symbol 'v_k'. The sign analogously constructed with the symbol 'X' is called the *k-th variable of the 2nd order*, symbolically 'V_k'. The variables of the 1st order represent names of individuals, those of the 2nd order names of two-termed relations between individuals. From the material and also—in agreement with the further description of the language—from the formal

---

[1] This is a fragment of the language of the algebra of relations, the foundations of which are given in Schröder, E. (62), vol. 3—a fragment which nevertheless suffices to express every idea which can be formulated in this language.

point of view, the signs 'v_k' and 'V_k' belong to two distinct semantical categories. Expressions of the form 'Xyz' are regarded as primitive sentential functions, where in the place of 'X' any variable of the 2nd order, and in the place of 'y' and 'z' any variables of the 1st order may appear. These expressions are read: 'the individual y stands in the relation X to the individual z' and they are denoted—according to the form of the variables—by the symbols '$\rho_{k,l,m}$'. By the use of the sign '$^\frown$' from § 2 we specify that $\rho_{k,l,m} = (V_k {}^\frown v_l) {}^\frown v_m$. The definitions of the fundamental operations on expressions, as well as those of sentential function, sentence, consequence, provable sentence, and so on, are all quite analogous to the definitions of § 2. But it must always be borne in mind that in this language two distinct categories of variables appear and that the expressions $\rho_{k,l,m}$ play the part of the inclusions $\iota_{k,l}$. In connexion with the first of these facts we have to consider not *one* operation of quantification (Defs. 6 and 9) but *two* analogous operations: with respect to a variable of the 1st order as well as with respect to a variable of the 2nd order, the results of which are denoted by the symbols '$\bigcap_k' x$', and '$\bigcap_k'' x$' or '$\bigcup_k' x$' and '$\bigcup_k'' x$' respectively. Correspondingly there will be two operations of substitution. Among the axioms of the logic of relations we include the sentences which satisfy the condition (α) of Def. 13, i.e. substitutions of the axioms of the sentential calculus, and universal quantifications of these substitutions, and also all sentences which are universal quantifications of expressions of the type

$$\bigcup_k'' \bigcap_l' \bigcap_m' (\rho_{k,l,m} \cdot y + \overline{\rho_{k,l,m}} \cdot \bar{y}),$$

where k, l, and m are any natural numbers ($l \neq m$) and y any sentential function in which the free variable V_k does not occur. Considering their intuitive meaning the axioms of the last category may be called *pseudodefinitions*.[1]

---

[1] This term we owe to Leśniewski, who has drawn attention to the necessity of including pseudodefinitions among the axioms of the deductive sciences in those cases in which the formalization of the science does not admit the possibility of constructing suitable definitions (cf. p. 166, footnote). Pseudodefinitions can be regarded as a substitute for the *axiom of reducibility* of Whitehead, A. N., and Russell, B. A. W. (90), vol. 1, pp. 55 ff. It would not be difficult to show the connexion between these sentences and a group of axioms adopted in Neumann, J. v. (54), p. 18.

To obtain a correct definition of satisfaction in connexion with the language we are considering we must first extend our knowledge of this concept. In the first stage of operating with it we spoke of the satisfaction of a sentential function by one, two, three objects, and so on, according to the number of free variables occurring in the given function (cf. pp. 189 ff.). From the semantical standpoint the concept of satisfaction had there a strongly ambiguous character; it included relations in which the number of terms was diverse, relations whose last domain was a class of sentential functions, whilst the other domains—in the case of the language of the calculus of classes—consisted of objects of one and the same category, namely classes of individuals. Strictly speaking we were dealing not with *one* concept, but with an infinite number of analogous concepts, belonging to different semantical categories. If we had formalized the meta-language it would have been necessary to use infinitely many distinct terms instead of the *one* term 'satisfies'. The semantical ambiguity of this concept increases still more when we pass to languages of more complicated logical structure. If we continue the intuitive considerations of § 3, analyse the examples given there and construct new ones after the same pattern, it soon becomes clear that a strict semantical correlation exists between the free variables of the sentential function and the objects which satisfy these functions: every free variable belongs to the same semantical category as the name of the object corresponding to it. If, therefore, at least two different categories occur among the variables of the language—as in the case we are investigating —it does not suffice to restrict consideration to only a single category of objects in dealing with the concept of satisfaction. The domains of the single relations which are covered by the term 'satisfaction', thus cease to be semantically unambiguous (only the last domain consists as before exclusively of sentential functions). But since the semantical category of a relation not only depends on the number of domains, i.e. the number of terms standing in the relation to one another, but also on the categories of these domains, the category of the concept of satisfaction, or rather the category of each single one of these concepts, also depends on two circumstances. It depends on the number and also on the categories of the free variables which appear in the sentential functions to which the concept of satisfaction relates. In brief, it depends on what we have called the semantical type of the sentential function. To functions which belong to two distinct types two semantically distinct concepts of satisfaction always correspond.[1] Some examples will make this clear. We shall say that the objects R, a, and b satisfy the function $\rho_{1,2,3}$ if and only if R is a relation and a and b are individuals and we have aRb (i.e. a stands in the relation R to b). The function $\rho_{1,2,2} \cdot \rho_{3,2,2}$ is satisfied by the objects R, a, and S if and only if R and S are relations, a is an individual and we have both aRa and aSa. The function $\bigcap_2' \bigcap_3' (\overline{\rho_{1,2,3} + \rho_{1,3,2}})$ is satisfied by symmetrical relations and only by them, i.e. by relations such that, for all individuals a and b, if we have aRb we also always have bRa. The function $\bigcap_1'' (\overline{\rho_{1,2,3} + \rho_{1,3,2}})$ is satisfied by those and only those individuals a and b which satisfy the following condition: for every relation R, if aRb, then bRa, i.e. individuals which are identical. In the above examples we have sentential functions belonging to four different semantical types, and we are, therefore, dealing with four different relations of satisfaction, in spite of the fact that the number of free variables and also the number of terms in the relations is the same in the first two examples.

The semantical ambiguity attaching to the concept of satisfaction in its original conception renders an exact characterization of this concept in a single sentence, or even in a finite number of sentences, impossible, and so denies us the use of the only method so far known to us of constructing a definition of a true sentence. In order to avoid this ambiguity, in dealing with the calculus of classes we had recourse to an artifice which is used by logicians and mathematicians in similar situations. Instead of using infinitely many concepts of satisfaction of a sentential

---

[1] Moreover, functions of *one* semantical type can correspond to several semantically distinct concepts of satisfaction, provided the free variables of these functions belong to at least two distinct semantical categories; in addition to the number and the categories of the variables their arrangement must also be taken into consideration.

function by single objects, we tried to operate with the semantically uniform, if somewhat artificial, concept of the satisfaction
of a function by a sequence of objects. It happened that this
concept is sufficiently more general than the previous one to
include it—intuitively speaking—as a special case (to define the
logical nature of this inclusion would, however, be a little difficult). It will easily be seen that this method cannot be applied to
the present problem without further difficulty. Satisfaction in
its new form is a two-termed relation, whose domain consists of
sequences and counter domain of sentential functions. As
before, there exists between the free variables of a sentential
function and the corresponding terms of the sequences which
satisfy it, a strict semantical correlation. Thus if the language
of the logic of relations contains variables of two different
semantical categories, we must likewise use two categories of
sequences in our investigations. For example, the function
$\bigcap'_2 \bigcap'_3 (\overline{\rho_{1,2,3} + \rho_{1,3,2}})$ is satisfied exclusively by sequences of two-
termed relations between individuals (namely by those and only
those sequences F whose first term F_1 is a symmetrical relation).
But the function $\bigcap''_1 (\overline{\rho_{1,2,3} + \rho_{1,3,2}})$ is satisfied exclusively by
sequences of individuals (i.e. by sequences f for which $f_2 = f_3$
holds). The domain of the relation of satisfaction and *eo ipso*
the relation itself thus again becomes semantically ambiguous.
Again we are dealing not with *one*, but with at least two different
concepts of satisfaction. But still worse, a closer analysis shows
that the new interpretation of the concept of satisfaction can no
longer as a whole be maintained. For one and the same sentential
function often contains free variables of two different categories.
To deal with such functions we must operate with sequences
whose terms likewise belong to two categories. The first term, for
example, of the sequence which satisfies the function $\rho_{1,2,3}$ must
be a relation, but the two following ones must be individuals.
But it is known that the theory of semantical categories does
not permit the existence of such heterogeneous sequences.
Consequently the whole conception collapses. Thus changing
the original interpretation of the concept of satisfaction has
removed only *one* subsidiary cause of its ambiguity, namely the

diversity in the number of terms in the relations which are the object of the concept; another far more important factor, the semantical diversity of the terms of the relations, has lost none of its force.

Nevertheless the methods used in § 3 can be applied to the language now being investigated, although with certain modifications. In this case also it is possible to find an interpretation of the concept of satisfaction in which this notion loses its semantical ambiguity and at the same time becomes so general that it includes all special cases of the original concept. In fact, two different methods are available; I shall call them the *method of many-rowed sequences* and the *method of semantical unification of the variables*.

The first method requires that we should treat satisfaction not as a two-termed, but as a three-termed relation which holds between sequences of individuals, between sequences of two-termed relations and between sentential functions. We use the following mode of expression: 'the sequence f of individuals and the sequence F of relations together satisfy the sentential function x'. The content of this phrase can easily be visualized by means of concrete examples. For example, the sequence f of individuals and the sequence F of relations together satisfy the function $\rho_{1,2,3}$ if and only if the individual f_2 stands in the relation F_1 to the individual f_3. In order to formulate a general definition we proceed exactly in the manner of Def. 22 in § 3, care being taken to remember that, in the language we are considering, the expressions $\rho_{k,l,m}$ play the part of primitive sentential functions and that instead of *one* operation of universal quantification *two* related operations occur. The definition of true sentence is completely analogous to Def. 23.

This method can now be modified to some extent by treating satisfaction as a two-termed relation between so-called two-rowed sequences and sentential functions. Every ordered pair which consists of two sequences f and F is called a *two-rowed sequence* (or *two-rowed matrix*), where the kth term of the sequence f or of the sequence F is called the kth term of the first or second row respectively of the two-rowed sequence. In the present case we have to deal with ordered pairs which consist of a sequence of individuals and a sequence of relations. It is easily seen that this modification is a purely formal one and has no essential effect on the construction as a whole. It is to this modification of the method that the term 'method of many-rowed sequences' is adapted.

To understand the method of semantical unification of the variables we begin with certain considerations which are not immediately connected with the language we are at present investigating. It is known that with every individual a a definite two-termed relation $a*$ can be correlated in such a way that to distinct individuals distinct relations correspond. For this purpose it suffices to take as $a*$ an ordered pair whose terms are identical with a, i.e. the relation R which holds between any two individuals b and c if and only if $b = a$ and $c = a$. On the basis of this correlation we can now correlate in a one-one fashion with every class of individuals a class of relations, with every many-termed relation between individuals a corresponding relation between relations, and so on. For example, to any class A of individuals there corresponds a class $A*$ of all those relations $a*$ which are correlated with the elements a of the class A. In this way every sentence about individuals can be transformed into an equivalent sentence about relations.

Bearing these facts in mind we return to the language of the logic of relations and change the intuitive interpretation of the expressions of this language without in any way touching their formal structure. All constants will retain their previous meaning, whilst all variables both of the 1st and 2nd order are from now on to represent names of two-termed relations. To the primitive sentential functions of the type 'Xyz', where instead of 'X' some variable V_k and instead of 'y' and 'z' any two variables v_l and v_m occur, we assign the following meaning: 'there exist individuals a and b such that a stands in the relation X to b, $y = a*$, and $z = b*$.' In this way the meaning of the composite sentential functions will likewise be modified. It is almost immediately evident that every true or false sentence in the earlier interpretation will remain true or false respectively in

the new one. By virtue of this new interpretation all the variables of the language now belong to one and the same semantical
category, not indeed from the formal but from the intuitive
point of view; they represent words of the same 'part of speech'.
Consequently the language we are considering can be investigated by exactly the same methods as all languages of the 1st
kind; in particular, satisfaction can be treated as a two-termed
relation between sequences of relations and sentential functions.
At the same time a complication of a technical nature—although
an unimportant one—presents itself. Since two free variables of
different orders but the same indices, e.g. v_l and V_l, may occur in
the same sentential function, it is not clear without supplementary stipulations which terms of the sequence are to correspond
to the variables of the 1st, and which to those of the 2nd order.
To overcome this difficulty we shall stipulate that to every
variable v_k a term of the sequence with an uneven index $2.k-1$
corresponds, and to every variable V_k a term with even index $2.k$
corresponds. For example, the sequence F of relations satisfies
the function $\rho_{k,l,m}$ if and only if there are individuals a and b
such that a stands in the relation $F_{2.k}$ to b, $F_{2.l-1} = a^*$, and
$F_{2.m-1} = b^*$. Apart from this detail the definitions of satisfaction
and of true sentence differ in no essential point from the definitions given in § 3.

The two methods described can be applied to all languages of
the 2nd kind.[1] If the variables of the language studied belong to
n different semantical categories, we regard satisfaction—under
the method of many-rowed sequences—as an $n+1$-termed
relation holding between n sequences of the corresponding
categories and the semantical functions, or as a two-termed
relation whose domain consists of n-rowed sequences (i.e. ordered

---

[1] This holds even for languages in which variables occur which are not
included in the classification on p. 218 (cf. p. 218, note 2). I shall not deal with
certain (not particularly important) difficulties which may occur here. But
I take this opportunity of mentioning that sentential variables, even if they
occur in the language, do not complicate the construction at all, and that,
in particular, it would not be worth while to include them in the process of
semantical unification. Sentences which contain such variables can be excluded by correlating with each of them, in one-many fashion, an equivalent
sentence which does not contain sentential variables (cf. Hilbert, D., and
Ackermann, W. (30), pp. 84–85).

n-tuples of ordinary sequences) and whose counter domain consists of sentential functions. Constructions based on this method form the most natural generalization of the constructions in § 3 and their material correctness appears to leave no doubts.

In applying the method of semantical unification of the variables, the choice of the *unifying category* plays an essential part, i.e. that semantical category in which all the variables of the language studied can be interpreted. Only one thing is required of the unifying category: that with all objects of every semantical category which is represented by the variables of the given language, effective objects of the chosen category can be correlated in a one-one fashion (i.e. so that to distinct objects, distinct objects correspond). Nevertheless, the choice of the unifying category is not always so simple as in the example discussed above in connexion with the language of the logic of relations; this choice cannot always be made from the categories which occur in the language. If, for example, the variables of the language represent names of two-termed relations between individuals and names of classes which consist of classes of individuals, then the simplest unifying category seems to be the category of two-termed relations between classes of individuals. I do not propose to enter into a further analysis of this problem (it would presuppose a knowledge of certain facts belonging to set theory). I add only the following remarks: (1) the unifying category cannot be of lower order than any one category among those occurring in the language; (2) for every language of the 2nd kind a unifying category can be found, even infinitely many such categories and in fact among categories of the nth order, where n is the highest order of the variables occurring in the language. As soon as the unifying category is specified, and the primitive sentential functions correspondingly interpreted, the further course of the work does not differ at all from the methods of construction used for languages of the 1st kind.

In contrast to the method of many-rowed sequences, there is no doubt that the second method is somewhat artificial. Nevertheless the definitions constructed by this method prove, on closer analysis, to be intuitively evident to a scarcely less degree than the constructions based on the first method. At the same time they have the advantage of greater logical simplicity. In particular, when we are dealing with the definition of true sentence the proof of the equivalence of its two formulations presents no difficulty in any concrete case. The essential advantages of the method of unification of the variables only become clear, however, in the investigation of languages of the 3rd kind, since the method of many-rowed sequences here proves to be quite useless.

As a typical example of a language of the 3rd kind we choose the language of the *logic of many-termed relations*.[1] In this science we deal with the same constants 'N', 'A', and '$\prod$' and with the same variables of the 1st order v_k, as in the logic of two-termed relations. But we also find here variables of the 2nd order in greater multiplicity than before. As variables of this kind we shall use such signs as '$X'_{\prime}$', '$X'_{\prime\prime}$', '$X'_{\prime\prime\prime}$',..., '$X''_{\prime}$', '$X''_{\prime\prime}$', '$X''_{\prime\prime\prime}$',..., '$X'''_{\prime}$', '$X'''_{\prime\prime}$', '$X'''_{\prime\prime\prime}$',... and so on. The composite symbol constructed from the sign 'X' with k small strokes below and l such strokes above will be called the *kth variable functor with l arguments*, and denoted by 'V^l_k'. Intuitively interpreted, the variables v_k represent, as before, names of individuals, whilst the variables V^l_k represent names of l-termed relations between individuals, in particular for $l = 1$ names of one-termed relations, i.e. names of classes. Both from the intuitive and the formal points of view the signs v_k, V^1_k, V^2_k,... belong to infinitely many distinct semantical categories of the 1st and 2nd orders respectively. The fundamental sentential functions are expressions of the type '$Xxy...z$', where in place of 'X' any variable functor with l arguments and in place of 'x', 'y',..., 'z' variables of the 1st order, l in number, occur. These expressions are read as follows: 'the l-termed relation X holds between the l individuals $x, y,..., z$.' According to the number and form of the variables we denote the primitive functions by the symbols '$\rho_{k,m}$', '$\rho_{k,m,n}$',...,

[1] This is a language which resembles the language of the lower predicate calculus of Hilbert, D., and Ackermann, W. (30), pp. 43 ff., but is richer than the latter because variable functors can occur in it both as free and as bound variables.

putting $\rho_{k,m} = V_k^1 \frown v_m$, $\rho_{k,m,n} = (V_k^2 \frown v_m) \frown v_n$, and so on. In order to obtain a unified symbolism, which is independent of the number of variables, we shall use symbols of the type '$\rho_{k,p}^l$' (where 'p' represents the name of a finite sequence of natural numbers), the meaning of which is determined by the formula $\rho_{k,p}^l = (((V_k^l \frown v_{p_1}) \frown v_{p_2}) \frown ...) \frown v_{p_l}$.[1] The further definitions of the metatheory do not differ at all from the analogous definitions relating to the logic of two-termed relations and even to the calculus of classes. As operations of quantification we introduce quantification with respect to the variables v_k and the variables V_k^l and denote the result of the operations by the symbols '$\bigcap_k x$' and '$\bigcap_k^l x$' respectively. The list of axioms includes those which satisfy the condition (α) of Def. 13 of § 2, and pseudodefinitions which form a natural generalization of the pseudodefinitions from the logic of two-termed relations. Their more detailed description seems to be unnecessary.

We turn now to the problem of how the concept of satisfaction is to be conceived and the definition of truth to be constructed for the language we are now considering. Any attempt to apply the method of many-rowed sequences in this case fails completely. In this method the term 'satisfaction'—in whatever form—expresses the relation of dependence between n sequences of various categories and the sentential functions, where n is exactly equal to the number of semantical categories represented by the variables of the given language. In the case we are investigating the number n is indefinitely large and the metalanguage we are using—like all other actually existing formalized languages—provides no means for dealing with the relation of mutual dependence between objects which belong to infinitely many distinct semantical categories.[2]

---

[1] Strictly speaking the meaning of the symbol '$\rho_{k,p}^l$' should be defined recursively.

[2] In those cases in which, in logical and mathematical constructions, we deal with the mutual dependence between an arbitrary, not previously determined number of objects of one and the same semantical category, we mostly use ordinary sequences. For objects which belong to a finite number of distinct categories many-rowed sequences fulfil the analogous function. But on the basis of the known languages we find nothing like 'sequences with infinitely many rows' (of distinct semantical categories).

The method of semantical unification of the variables can, however, be applied to this language with complete success. To see
this it suffices to note that we can correlate in a one-one fashion,
with every n-termed relation R between individuals, a class R^*
which consists of n-termed sequences of individuals, namely the
class of all sequences f which satisfy the following condition: the
relation R holds between the individuals $f_1, f_2, ..., f_n$. For example, the class of all sequences f with two terms f_1 and f_2 such
that $f_1 R f_2$ corresponds to the two-termed relation R. Consequently every sentence concerning many-termed relations
can be transformed into an equivalent sentence which asserts
something about classes of sequences. It will be remembered
that by sequences of individuals we mean two-termed relations
between individuals and natural numbers. Accordingly all
sequences of individuals, whatever the number of their terms,
belong to one and the same semantical category and therefore
the classes of these sequences, in contrast to many-termed
relations, likewise belong to one and the same category.

On the basis of these considerations we now partially unify
the semantical categories of the variables in the following way.
To the variables v_k we give—at least provisionally—the same
significance as before. But the variables V_k^l now represent the
names of any classes which consist of finite sequences of individuals or of other objects of the same category (i.e. the names
of objects of at least the 3rd order, according to the order which
we assign to the natural numbers).[1] The primitive functions of
the form '$Xxy...z$', which begin with a functor with l arguments
and hence contain l variables of the 1st order, are interpreted
by phrases of the type: 'the sequence of individuals the first term
of which is x, the second y,... and the lth (the last) is z, belongs to

---

[1] In systems of mathematical logic, e.g. in Whitehead, A. N., and Russell,
B. A. W. (90), vol. 2, pp. 4 ff., the cardinal numbers and in particular the
natural numbers are usually treated as classes consisting of classes of individuals
(or other objects), namely as the classes of all those classes which are similar
(in the *Principia Mathematica* sense) to a given class. For example, the number
1 is defined as the class of all those classes which have exactly *one* element.
With this conception the natural numbers are thus objects of (at least) the 3rd,
sequences of individuals of the 4th, and classes of these sequences of the 5th
order.

the class X which consists of l-termed sequences'. From the intuitive, although not from the formal, standpoint, the variables from now on still belong to only two distinct semantical categories; in view of this circumstance we can use, in the further course of our work, the same methods as we employed in investigating languages of the 2nd kind.

By means of the phrase: 'the sequence f of individuals and the sequence F, whose terms form classes of finite sequences of individuals, together satisfy the given sentential function', we can bring into service the method of many-rowed sequences. To use this concept consistently we must first set up a one-one correlation between the variables V_k^l and the terms of the sequence F in such a way that terms with different indices correspond to different variables. This is most easily done by putting every variable V_k^l in correspondence with a term having the index $(2.k-1).2^{l-1}$. For example, the terms F_1, F_3, F_5, F_2, F_6, F_4,... correspond to the variables V_1^1, V_2^1, V_3^1, V_1^2, V_2^2, V_1^3,....[1] With this convention the establishment of the meaning of the above phrase in its application to any concrete sentential function, and even the construction of a general definition of the concept in question, presents no further difficulties. Thus concerning the primitive functions, those and only those sequences f and F (of the categories given above) will together satisfy the function $\rho_{k,m}$ which satisfy the following condition: the sequence g of individuals, whose single term g_1 is identical with f_m, belongs to the class $F_{2.k-1}$. In an analogous way, those functions f and F will together satisfy the function $\rho_{k,m,n}$ which satisfy the following condition: the sequence g of individuals with two terms, where $g_1 = f_m$ and $g_2 = f_n$, belongs to the class $F_{(2.k-1).2}$. In general, in order that the sequences f and F should together satisfy the function $\rho_{k,p}^l$, it is necessary and sufficient that the sequence g of individuals with l terms, where $g_1 = f_{p_1}$, $g_2 = f_{p_2}$,..., $g_l = f_{p_l}$, should belong to the class $F_{(2.k-1).2^{l-1}}$ (which consists of sequences with the same number of terms).

[1] Instead of the function $f(k, l) = (2.k-1).2^{l-1}$ we could use any other function $f(k, l)$ which correlates the natural numbers in one-one fashion with ordered pairs of natural numbers. Set theory offers many examples of such correlations; cf. Fraenkel, A. (16), pp. 30 ff. and 96 ff.

If we wish to apply the method of unification of the variables we again make use of the fact that a one-one correlation can be set up between any individuals and certain classes of finite sequences, and in such a way that to every individual a there corresponds the class a^* containing as its only element a sequence whose only member is just the given individual. Beginning in this way we next modify the interpretation of the variables of 1st order in exactly the same direction in which we formerly modified the interpretation of the variables of the 2nd order. The primitive functions of the form '$Xxy...z$', containing $l+1$ signs, we now regard as having the same meaning as expressions of the type 'the l-termed sequence g of individuals which satisfies the conditions: $g_1^* = x$, $g_2^* = y$,..., $g_l^* = z$, belongs to the class X, which consists of sequences with l terms'. With this intuitive interpretation all variables now belong to the same semantical category. The further construction contains no essentially new features and the reader will encounter no serious difficulties in carrying it out.

The method of semantical unification of the variables can be applied with equal success to the investigation of any language of the 3rd kind. Determining the unifying category may sometimes be more difficult. As in the case of languages of the 2nd kind it is here impossible to restrict consideration to categories occurring in the language studied. In contrast to those languages it is never possible to make the choice from among the categories of one of the orders represented in the language. This difficulty is not, however, essential and exclusively concerns languages of the lowest order. For it is possible to prove that for those languages in which the order of the variables does not exceed a given number n, where $n > 3$, any category of the nth order can serve as the unifying category.

In this way the various methods at our disposal enable us to define the concept of satisfaction and with it to construct a correct definition of truth for any language of finite order. We shall see in the next section that these methods do not extend further; the totality of languages of finite order exhausts the domain of applicability of our methods. This is therefore the place in which to summarize the most important consequences which follow from the definitions we have constructed.†

First, *the definition of true sentence is a correct definition of truth in the sense of convention* **T** of § 3. It embraces, as special cases, all partial definitions which were described in condition (α) of this convention and which elucidate in a more precise and materially correct way the sense of expressions of the type '*x* is a true sentence'. Although this definition alone provides no general criterion of truth, the partial definitions mentioned do permit us definitely to decide in many cases the question of the truth or falsity of the sentences investigated.

In particular, it can be proved—on the basis of the axioms of the second group adopted in the metatheory (cf. p. 211)—that *all axioms of the science under investigation are true sentences*. In a similar manner we can prove, making essential use of the fact that the rules of inference employed in the metatheory are not logically weaker than the corresponding rules of the science itself, *that all consequences of true sentences are true*. These two facts together enable us to assert that *the class of true sentences contains all provable sentences of the science investigated* (cf. Lemma D and Ths. 3 and 5 of § 3).

Among the most important consequences of a general nature which follow from the definition of truth must be reckoned the *principle of contradiction* and the *principle of the excluded middle*.

These two theorems, together with the theorem on the consequences of true sentences already mentioned, show that *the class of all true sentences forms a consistent and complete deductive system* (Ths. 1, 2, and 4).

As an immediate, although a somewhat subsidiary, consequence of these facts we obtain the theorem that *the class of all provable sentences likewise forms a consistent* (although not necessarily complete) *deductive system*. In this way we are able to produce a proof of the consistency of every science for which we can construct the definition of truth. The proof carried out by

† Some further consequences of this type are discussed in the article of the author 'On undecidable statements in enlarged systems of logic and the concept of truth', *Journal of Symbolic Logic*, vol. 4 (1939), pp. 105–12; cf. in particular sect. 9, p. 111.

means of this method does not, of course, add much to our knowledge, since it is based upon premises which are at least as strong
as the assumptions of the science under investigation.[1] Nevertheless it seems to be worthy of note that such a general method of
proof exists, which is applicable to an extensive range of deductive sciences. It will be seen that from the deductive standpoint
this method is not entirely trivial, and in many cases no simpler,
and in fact no other, method is known.†

   In those cases in which the class of provable sentences is not
only consistent but also complete, it is easy to show that it coincides with the class of true sentences. If, therefore, we identify
the two concepts—that of true sentence and that of provable
sentence—we reach a new definition of truth of a purely structural nature and essentially different from the original semantical
definition of this notion.[2] Even when the provable sentences

   [1] As Ajdukiewicz has rightly pointed out in a somewhat different connexion
(cf. Ajdukiewicz, K. (2), pp. 39–40) it does not at all follow from this that this
proof is not correct from the methodological standpoint—that it contains in
some form a *petitio principii*. The assertion which we prove, i.e. the consistency
of the science, does not occur in any way among the hypotheses of the proof.
   [2] In the course of this work I have several times contrasted semantical
definitions of true sentence with structural definitions. But this does not mean
that I intend to specify the distinction between the two kinds of definitions
in an exact way. From the intuitive standpoint these differences seem to be
tolerably clear. Def. 23 in § 3—as well as other definitions constructed in the
same way—I regard as a semantical definition because in a certain sense

   † In connexion with the problem discussed in the last three paragraphs see
the recent publications: Mostowski, A. (53 e) as well as Wang, H. (87 c). From
the results of these authors it is seen that in some cases, having succeeded
in constructing an adequate definition of truth for a theory T in its metatheory, we may still be unable to show that all the provable sentences of T
are true in the sense of this definition, and hence we may also be unable to
carry out the consistency proof for T in M. This phenomenon can roughly
be explained as follows: in the proof that all provable sentences of T are
true a certain form of mathematical induction is essentially involved, and the
formalism of M may be insufficiently powerful to secure the validity of this
inductive argument. Hence a certain clarification of the assumptions (on pp.
174 ff.) concerning foundations of the metatheory may be desirable. In
particular the phrase 'from any sufficiently developed system of mathematical
logic' (p. 170) should be understood in a way which does not deprive the
metatheory of any normally applied modes of inference. If the theory T is
of finite order our purpose will be fully achieved if we decide to provide the
metatheory M with a logical basis as strong as the general theory of classes
discussed in the following section.

do not form a complete system the question of the construction of a structural definition is not *a priori* hopeless. Sometimes it is possible, by adding certain structurally described sentences, to extend the axiom system of the science in a suitable way so that it becomes a system in which the class of all its consequences coincides with the class of all true sentences. But there can be no question of a general method of construction. I suspect that the attempt to construct a structural definition, even in relatively simple cases—e.g. in connexion with the logic of two-termed relations studied in the preceding section—would encounter serious difficulties. These difficulties would certainly become much greater when it came to the question of giving a general structural criterion of truth, although we have already dealt with two languages, that of the calculus of classes and that of

(which would be difficult to define) it represents a 'natural generalization', so to speak an 'infinite logical product', of those partial definitions which were described in convention **T** and which establish a direct correlation between the sentences of the language and the names of these sentences. Among the structural definitions, on the other hand, I include those which are constructed according to the following scheme: a class of sentences or other expressions is described in such a way that from the form of every expression it is possible to know whether it belongs to the given class or not. Further operations on expressions are given of such a kind that if certain expressions in finite number are given and if the form of an arbitrary other expression is given, then we can decide whether it can be obtained from the given expressions by means of the given operations. Finally the true sentences are defined as those which are obtained by applying the given operation to the expressions of the given class any number of times (it is to be noted that such a structural definition still in no way provides a general criterion of truth). Certain differences of a formal nature can be recognized between these two kinds of definitions. The semantical definition requires the use of terms of higher order than all variables of the language investigated, e.g. the use of the term 'satisfies'; but for the formulation of a structural definition the terms of perhaps two or three of the lowest orders suffice. In the construction of a semantical definition we use—explicitly or implicitly—those expressions of the metalanguage which are of like meaning with the expressions of the language investigated, whilst they play no part in the construction of a structural definition; it is easy to see that this distinction vanishes when the language studied is a fragment of logic. Moreover, the distinction as a whole is not very clear and sharp, as is shown by the fact that with respect to the sentential calculus the semantical definition can be regarded as a formal transformation of the structural definition based on the matrix method. At the same time it must be remembered that the construction of semantical definitions, based on the methods at present known to us, is essentially dependent upon the structural definitions of sentence and sentential function.

the sentential calculus, for which this problem could be relatively easily solved.[1]

In all cases in which we are able to define satisfaction and the notion of true statement, we can—by means of a modification of these definitions—also define two still more general concepts of a relative kind, namely the concepts of *satisfaction* and *correct sentence*—both *with respect to a given individual domain a*.[2] This modification depends on a suitable restriction of the domain of objects considered. Instead of operating with arbitrary individuals, classes of individuals, relations between individuals, and so on, we deal exclusively with the elements of a given class a of individuals, subclasses of this class, relations between elements of this class, and so on. It is obvious that in the special case when a is the class of all individuals, the new concepts coincide with the former ones (cf. Defs. 24 and 25, and Th. 26). As I have already emphasized in § 3 the general concept of correct sentence in a given domain plays a great part in present day methodological researches. But it must be added that this only concerns researches whose object is mathematical logic and its parts. In connexion with the special sciences we are interested in correct sentences in a quite specific individual domain for which the general concept loses its importance. Likewise it is only in connexion with sciences which are parts of logic that some general properties of these concepts, which were proved in § 3 for the language of the calculus of classes, preserve their validity. For example, it happens that in these sciences the extension of the term 'correct sentence in the individual domain a' depends exclusively on the cardinal number of the class a. Thus in these investigations we can replace this term by the more convenient term 'correct sentence in a domain with k elements' (Def. 26, Th. 8). The theorems previously discussed concerning the concept of truth, such as the principles of contradiction and the excluded middle can be extended to the concept of correct sentence in a given domain. The concept of correct sentence in every

---

[1] Cf. the remarks on pp. 207 f. and 221; I shall return to this problem in § 5 (cf. p. 254, footnote).

[2] See p. 199, note 2.

individual domain (Def. 27) deserves special consideration. In its extension it stands midway between the concept of provable sentence and that of true sentence; the class of correct sentences in every domain contains all theorems and consists exclusively of true sentences (Ths. 22 and 27). This class is therefore in general narrower than the class of all true sentences; it contains, for example, no sentences whose validity depends on the magnitude of the number of all individuals (Th. 23). If it is desired to transform the system of the provable sentences of every science into a complete one, it is necessary at the outset to add sentences to the system which decide the question how many individuals exist. But for various reasons another point of view seems to be better established, namely the view that the decision regarding such problems should be left to the specific deductive sciences, whilst in logic and its parts we should try to ensure only that the extension of the concept of provable sentence coincides with that of correct sentence in every individual domain. For a supporter of this standpoint the question whether the extension of these two concepts is actually identical is of great importance. In the case of a negative answer the problem arises of completing the axiom system of the science studied in such a way that the class of provable sentences thus extended now coincides with the class of sentences which are correct in every domain. This problem, which properly is equivalent to the question of structurally characterizing the latter concept, can be positively decided only in a few cases (cf. Th. 24).[1] Generally speaking the difficulties presented by this question are no less essential than those connected with the analogous problem of a structural definition of true sentence. We meet with similar difficulties when we attempt to define structurally the concept of correct sentence in a domain with k elements. Only in the case where k is a finite number is it easy to give a general method, modelled on the method of matrices from investigations on the extended sentential calculus, which makes a structural definition

---

[1] In the case of the lower functional calculus this problem, which is raised in Hilbert, D., and Ackermann, W. (30), p. 68, has recently been decided by Gödel, see Gödel, K. (20).

of this concept possible. In this way we even obtain a general criterion which enables us to decide from the form of any sentence whether it is correct in a domain with a previously given finite number of elements.[1]

I do not wish to enter here into a more detailed discussion of special investigations on the concepts just considered. Some results which are relevant here, relating to the calculus of classes, have already been given as examples in § 3. I will only mention that in recent years numerous results have been obtained which enable us to infer from the correctness of certain sentences in special individual domains or from their structural properties their correctness in every domain and thus their truth.[2] It is evident that all these results only receive a clear content and can only then be exactly proved, if a concrete and precisely formulated definition of correct sentence is accepted as a basis for the investigation.

§ 5. The Concept of True Sentence in Languages of Infinite Order

We come now to languages of the 4th kind, hence to those of infinite order and so lying beyond the scope of the methods of construction sketched in the preceding section. The language of the *general theory of classes* will serve as an example. This language is noteworthy because, in spite of its elementary structure and its poverty in grammatical forms, it suffices for

---

[1] Cf. Bernays, P., and Schönfinkel, M. (5 a), p. 352.

[2] According to the well-known theorems of Löwenheim and Skolem, certain categories of sentences are correct in every domain provided they are correct in all finite and denumerable domains. These sentences include, for example, all sentences of the logic of two- or many-termed relations, described in this section, which are generalizations of sentential functions in which variables of the 2nd order occur exclusively as free variables. In the case of the sentences of the calculus of classes this result—as is shown in Ths. 15 and 19 of § 3—can be essentially sharpened. Certain results of Bernays, Schönfinkel, and Ackermann have a narrower domain of application. They allow us to correlate a particular natural number k with sentences of a special structure in such a way that from the correctness of a given sentence in the domain with k elements (thus—as we already know—from purely structural properties of the sentence) its correctness in every domain follows. Cf. Ackermann, W. (1), Bernays, P., and Schönfinkel, M. (5 a), Herbrand, J. (26), Löwenheim, L. (49), Skolem, Th. (64), (65), and (66). For a systematic presentation of the results in this direction including more recent ones, see Church, A. (11 a).

the formulation of every idea which can be expressed in the whole language of mathematical logic. It is difficult to imagine a simpler language which can do this.[1]

In the general theory of classes the same constants occur as in the previously investigated sciences, i.e. the signs of negation and of logical sum, as well as the universal quantifier. As variables we use such symbols as 'X'_i', 'X''_i', 'X'''_i', and so on, i.e. signs composed of the symbol 'X' and a number of small strokes above and below. The sign having n strokes above and k below is called the *k-th variable of the n-th order* and is denoted by the symbol 'V^n_k'. The variables $V^1_k, V^2_k, V^3_k, \ldots$ represent respectively names of individuals, objects of the 1st order; classes of individuals, objects of the 2nd order; classes of such classes, objects of the 3rd order, and so on. These variables obviously belong to infinitely many semantical categories. As primitive sentential functions we have expressions of the type 'XY' where in the place of 'X' any variable of the $n+1$th order, and instead of 'Y' a variable of the nth order occurs. This expression is

---

[1] The language of the general theory of classes is much inferior to the language of Whitehead, A. N., and Russell, B. A. W. (90) in its stock of semantical categories, and still more inferior in this respect to the language used by Leśniewski in his system (cf. p. 210, note 2; p. 218, note 2). In particular, in this language no sentential variables and neither names of two- or many-termed relations, nor variables representing these names, occur. The dispensability of sentential variables depends on the fact mentioned on p. 229, footnote: to every sentence which contains sentential variables there is a logically equivalent sentence which does not contain such variables. The results of § 2, especially Defs. 13–17, suffice to show how such variables are to be avoided in setting up lists of axioms and in the derivation of theorems; cf. also Neumann, J. v. (54) (especially note 9, p. 38). The possibility of eliminating two-termed relations results from the following consideration. With every relation R we can correlate, in one-one fashion, a class of ordered pairs, namely, the class of all ordered pairs whose terms x and y satisfy the formula, xRy. If the relation is homogeneous, i.e. if the domain and counter domain of this relation belong to the same semantical category, then the ordered pair can be interpreted otherwise than we have done on p. 171, namely as classes having two classes as elements: the class whose only element is x and the class consisting of the two elements x and y. In order to apply an analogous method to inhomogeneous relations we must first correlate homogeneous relations with them in one-one fashion, and this presents no great difficulty. We proceed in an analogous way with many-termed relations. In this way every statement about two- or many-termed relations of arbitrary category can be transformed into an equivalent statement about individuals, classes of individuals, classes of such classes, and so on. Cf. Kuratowski, C. (38), p. 171, and Chwistek, L. (13), especially p. 722.

read: 'the class X (of $n+1$th order) has as an element the object Y (of nth order)', or 'the object Y has the property X'. For the designation of the primitive functions we employ the symbol '$\epsilon^n_{k,l}$', setting $\epsilon^n_{k,l} = V^{n+1}_k \frown V^n_l$. The further development of the science differs in no essential way from that of the logic of two- or many-termed relations. The quantifications of the sentential functions x with respect to the variable V^n_k are denoted by the symbols '$\bigcap^n_k x$' and '$\bigcup^n_k x$'. The axioms consist of (1) sentences which satisfy the condition (α) of Def. 13 of § 2, which are thus derived from the axioms of the sentential calculus by substitution, sometimes also followed by generalization; (2) pseudodefinitions, i.e. statements which are quantifications of sentential functions of the type

$$\bigcup^{n+1}_k \bigcap^n_l (\epsilon^n_{k,l} \cdot y + \overline{\epsilon^n_{k,l}} \cdot \bar{y}),$$

where y is any sentential function which does not contain the free variable V^{n+1}_k; (3) the *laws of extensionality*, i.e. sentences of the form

$$\bigcap^{p+2}_k \bigcap^{p+1}_l \bigcap^{p+1}_m (\bigcup^p_n (\epsilon^p_{l,n} \cdot \overline{\epsilon^p_{m,n}} + \overline{\epsilon^p_{l,n}} \cdot \epsilon^p_{m,n}) + \overline{\epsilon^{p+1}_{k,l}} + \epsilon^{p+1}_{k,m}),$$

which state that two classes which do not differ in their elements do not differ in any of their properties and are thus identical. In order to obtain in this science a sufficient basis for the establishment of various parts of mathematics and in particular of the whole of theoretical arithmetic, we must add to the above still one more axiom: (4) the *axiom of infinity*, i.e. the sentence

$$\bigcup^3_1 (\bigcup^2_1 \epsilon^1_{1,1} \cdot \bigcap^2_1 (\overline{\epsilon^2_{1,1}} + \bigcup^2_2 (\epsilon^1_{1,2} \cdot \bigcap^1_1 (\epsilon_{1,1} + \overline{\epsilon^1_{2,1}}) \cdot \bigcup^1_1 (\epsilon^1_{1,1} \cdot \overline{\epsilon^1_{2,1}})))),$$

which guarantees the existence of infinitely many individuals.[1] In the derivation of consequences from the axioms we apply the operations of substitution, detachment, and the introduction and removal of the universal quantifier, analogous to the operations described in conditions (γ)–(ζ) of Def. 15 in § 3.

When we try to define the concept of satisfaction in connexion with the present language we encounter difficulties which we cannot overcome. In the face of the infinite diversity of seman-

---

[1] In adopting the axiom of infinity we admittedly give up the postulate according to which only the sentences which are correct in every individual domain are to be provable sentences of logic (cf. p. 240).

tical categories which are represented in the language, the use of the method of many-rowed sequences is excluded from the beginning, just as it was in the case of the logic of many-termed relations. But the situation here is still worse, because the method of semantical unification of the variables also fails us. As we learnt in § 4, the unifying category cannot be of lower order than any one of the variables of the language studied. Sequences whose terms belong to this category, and still more the relation of satisfaction, which holds between such sequences and the corresponding sentential functions, must thus be of higher order than all those variables. In the language with which we are now dealing variables of arbitrarily high (finite) order occur: consequently in applying the method of unification it would be necessary to operate with expressions of 'infinite order'. Yet neither the metalanguage which forms the basis of the present investigations, nor any other of the existing languages, contains such expressions. It is in fact not at all clear what intuitive meaning could be given to such expressions.

These considerations seem to show that it is impossible to construct a general, semantically unambiguous concept of satisfaction for the language we are studying which will be applicable to all sentential functions without regard to their semantical type. On the other hand there appear to be no difficulties which would render impossible in principle a consistent application of the concept of satisfaction in its original formulation, or rather—in view of the semantical ambiguity of that formulation—of an infinite number of such concepts. Each of these concepts is, from the semantical standpoint, already specified and would relate exclusively to functions of a specific semantical type (e.g. to functions which contain a variable of the 1st order as the only free variable). Actually—independently of the logical structure of the language—the intuitive sense of none of these expressions raises any doubt. For every particular sentential function we can in fact define this meaning exactly by constructing for every phrase of the type 'the objects a, b, c,... satisfy the given sentential function' an intuitively equivalent phrase which is expressed wholly in terms of the metalanguage. Nevertheless the problem of the construction of a correct definition for each of these concepts again presents us with difficulties of an essential nature. On the basis of the languages which we have previously studied it was easy to obtain each special concept of satisfaction by a certain specialization of the general concept; in the present case this way is clearly not open to us. A brief reflection shows that the idea of using the recursive method analogously to the definition of sentential function proves, in spite of its naturalness, to be unsuitable. It is easily seen that the composite functions of a particular semantical type cannot always be formed from simpler functions of the same type. On the contrary, if we are to be able to construct arbitrary functions of a given type, we must use for that purpose all possible semantical types.[1] It would, therefore, be necessary, in the recursive definition of any one of the special concepts of satisfaction, to cover, in one and the same recursive process, infinitely many analogous concepts, and this is beyond the possibilities of the language.

The central problem of our work, the construction of the definition of truth, is closely connected with these considerations. If we were successful in defining, if not the general, at least any one of the special concepts of satisfaction, then this problem would not offer the least difficulty.[2] On the other

---

[1] An external expression of this state of affairs is that in the definition of satisfaction not only is it essential to take free variables into account but also all the bound variables of the function in question, although these variables have no influence on the semantical type of the function; and whether the relation of satisfaction holds or not does not depend in any way on the terms of the sequence which correspond to these variables (cf. Def. 22 of § 3, condition (δ)). It is to be remembered that analogous difficulties to those mentioned in the text appeared earlier in the attempt to construct a recursive definition of truth by a direct route (cf. p. 189).

[2] For example, let us imagine that we have succeeded in some way in defining the concept of satisfaction in the case of sentential functions which contain a variable of 1st order as the only free variable. We could then operate freely with phrases of the type 'the individual a satisfies the sentential function y'. If we now consider some one concrete sentential function, e.g. $\bigcup_1^2 \epsilon_{1,1}^1$, which is satisfied by every arbitrary individual, we obtain at once the following definition of true sentence: x *is a true sentence if and only if every individual a satisfies the function $x \cdot \bigcup_1^2 \epsilon_{1,1}^1$* (*i.e. the conjunction of the sentence x and the function $\bigcup_1^2 \epsilon_{1,1}^1$*). In an exactly analogous way we can pass from every other specific concept of satisfaction to the concept of truth.

hand we know of no method of construction which would not—directly or indirectly—presuppose a previous definition of the concept of satisfaction. Therefore we can say—considering the failure of previous attempts—that at present we can construct no correct and materially adequate definition of truth for the language under investigation.†

In the face of this state of affairs the question arises whether our failure is accidental and in some way connected with defects in the methods actually used, or whether obstacles of a fundamental kind play a part which are connected with the nature of the concepts we wish to define, or of those with the help of which we have tried to construct the required definitions. If the second supposition is the correct one all efforts intended to improve the methods of construction would clearly be fruitless. If we are to answer this question we must first give it a rather less indefinite form. It will be remembered that in the convention **T** of § 3 the conditions which decide the material correctness of any definition of true sentence are exactly stipulated. The construction of a definition which satisfies these conditions forms in fact the principal object of our investigation. From this standpoint the problem we are now considering takes on a precise form: it is a question of *whether on the basis of the metatheory of the language we are considering the construction of a correct definition of truth in the sense of convention* **T** *is in principle possible*. As we shall see, the problem in this form can be definitely solved, but in a *negative sense*.

It is not difficult to see that this problem exceeds the bounds of our previous discussion. It belongs to the field of the meta-metatheory. Its definitive solution, even its correct formulation, would require new equipment for investigation and especially the formalization of the metalanguage and the metatheory which uses it. But without going so far, and still avoiding

† The problem of the possibility of defining satisfaction and truth for the language under investigation will be considerably clarified by the discussion in the Postscript. It should be mentioned that the method of defining truth recently suggested in McKinsey, J. C. C. (53 *b*) is not based on a preliminary definition of satisfaction. Instead, McKinsey has to consider formalized languages with non-denumerably many constants and has to use a metalanguage which is provided with a very strong set-theoretical apparatus.

various technical complications, I believe I am able to give a fairly clear account of everything of a positive nature that can at present be established in connexion with the above problem.

In operating with the metalanguage we shall adhere to the symbolism given in §§ 2 and 3. To simplify the further developments and avoid possible misunderstandings we shall suppose the metalanguage to be so constructed that the language we are studying forms a fragment of it; every expression of the language is at the same time an expression of the metalanguage, but not vice versa. This enables us in certain cases (e.g. in the formulation of condition (α) of convention T) to speak simply of the expressions of the language itself, instead of expressions of the metalanguage which have the same meaning.

After these reservations and conventions we turn to the formulation and proof of the fundamental result.

THEOREM I. (α) *In whatever way the symbol 'Tr', denoting a class of expressions, is defined in the metatheory, it will be possible to derive from it the negation of one of the sentences which were described in the condition (α) of the convention* T;

(β) *assuming that the class of all provable sentences of the metatheory is consistent, it is impossible to construct an adequate definition of truth in the sense of convention* T *on the basis of the metatheory.*

The idea of the proof of this theorem can be expressed in the following words:[1] (1) a particular interpretation of the meta-

[1] We owe the method used here to Gödel, who has employed it for other purposes in his recently published work, Gödel, K. (22), cf. especially pp. 174–5 or 187–90 (proof of Th. VI). This exceedingly important and interesting article is not directly connected with the theme of our work—it deals with strictly methodological problems: the consistency and completeness of deductive systems; nevertheless we shall be able to use the methods and in part also the results of Gödel's investigations for our purpose.

I take this opportunity of mentioning that Th. I and the sketch of its proof was only added to the present work after it had already gone to press. At the time the work was presented at the Warsaw Society of Sciences (21 March 1931), Gödel's article—so far as I know—had not yet appeared. In this place therefore I had originally expressed, instead of positive results, only certain suppositions in the same direction, which were based partly on my own investigations and partly on the short report, Gödel, K. (21), which had been published some months previously.

After I had become acquainted with the above mentioned article I convinced myself, among other things, that the deductive theory which Gödel

language is established in the language itself and in this way with every sentence of the metalanguage there is correlated, in one-many fashion, a sentence of the language which is equivalent to it (with reference to the axiom system adopted in the meta-theory); in this way the metalanguage contains as well as every particular sentence, an individual name, if not for that sentence at least for the sentence which is correlated with it and equivalent to it. (2) Should we succeed in constructing in the metalanguage a correct definition of truth, then the metalanguage—with reference to the above interpretation—would acquire that universal character which was the primary source of the semantical antinomies in colloquial language (cf. p. 164). It would then be possible to reconstruct the antinomy of the liar in the metalanguage, by forming in the language itself a sentence x such that the sentence of the metalanguage which is correlated with x asserts that x is not a true sentence. In doing this it would be possible, by applying the diagonal procedure[1] from the theory of sets, to avoid all terms which do not belong to the meta-language, as well as all premisses of an empirical nature which have played a part in the previous formulations of the antinomy of the liar.[2]

had chosen as the object of his studies, which he called the 'system P', was strikingly similar to the general theory of classes considered in the present section. Apart from certain differences of a 'calligraphical' nature, the only distinction lies in the fact that in the system P, in addition to three logical constants, certain constants belonging to the arithmetic of the natural numbers also occur (a far-reaching analogy also exists between the system P and the system of arithmetic sketched in VI (see pp. 113–16)). Consequently the results obtained for the system P can easily be carried over to the present discussion. Moreover, the abstract character of the methods used by Gödel renders the validity of his results independent to a high degree of the specific peculiarities of the science investigated.

[1] Cf. Fraenkel, A. (16), pp. 48 ff.

[2] If we analyse the sketch of the proof given below we easily note that an analogous reconstruction could be carried out even on the basis of colloquial language, and that in consequence of this reconstruction the antinomy of the liar actually approximates to the antinomy of the expression 'hetero-logical'. For a rather simple reconstruction of the antinomy of the liar in this direction see Tarski, A. (82), note 11, p. 371. It seems interesting that in this reconstruction all the technical devices are avoided which are used in the proof of Th. 1 (such as interpretation of the metalanguage in arithmetic or the diagonal procedure). In connexion with the last paragraph of the text cf. the concluding remarks of § 1, pp. 164 f., and in particular p. 165, note 1.

We shall sketch the proof a little more exactly.[1]

Let us agree for the moment to use the symbol 'n' instead of 'X_l''''. The existential quantification of the sentential function y with respect to the variable 'n' will be denoted by the symbol '$\bigcup_1^3 y$' as before. The variable 'n' thus represents names of classes the elements of which are classes of individuals. Among these classes we find, among other things, the natural numbers and generally speaking the cardinal numbers.[2]

I have already mentioned that all facts belonging to the arithmetic of the natural numbers can be expressed in the language of the general theory of classes. In particular, if a natural number k is given, a sentential function ι_k is easily constructed in this language containing the symbol 'n' as the only free variable and which asserts that the class whose name is represented by this symbol is identical with the number k (and thus consists of just those classes of individuals which have exactly k elements).[2] For example:

$$\iota_1 = \bigcap_1^2(\epsilon_{1,1}^2 \cdot \bigcup_1^1 \bigcap_2^1 \bigcap_2^2(\epsilon_{1,1}^1 \cdot (\overline{\epsilon_{1,2}^1 + \epsilon_{2,1}^1} + \epsilon_{2,2}^1)) +$$
$$+ \overline{\epsilon_{1,1}^2} \cdot \bigcap_1^1 \bigcup_2^1 \bigcup_2^2(\overline{\epsilon_{1,1}^1} + \epsilon_{1,2}^1 \cdot \epsilon_{2,1}^1 \cdot \overline{\epsilon_{2,2}^1})).$$

A general recursive definition of the sequence of functions ι_k within the metalanguage presents no great difficulty.

As I have already pointed out in § 2 (p. 184) a one-one correspondence can be set up without difficulty between the expressions of the language and the natural numbers; we can define in the metalanguage an infinite sequence ϕ of expressions in which every expression of the language occurs once and only once. With the help of this correlation we can correlate with every operation on expressions an operation on natural numbers (which possesses the same formal properties), with every class of expressions a class of natural numbers, and so on. In this way the metalanguage receives an interpretation in the arithmetic of the

---

[1] For the sake of simplicity we shall in many places express ourselves as though the demonstration which follows belonged to the metatheory and not to the meta-metatheory; in particular, instead of saying that a given sentence is provable in the metatheory, we shall simply assert the sentence itself. In any case it must not be forgotten that only a sketch of the proof is given here and one which is far from complete.

[2] See p. 233, note 1.

natural numbers and indirectly in the language of the general theory of classes.

Let us suppose that we have defined the class Tr of sentences in the metalanguage. There would then correspond to this class a class of natural numbers which is defined exclusively in the terms of arithmetic. Consider the expression '$\bigcup_1^3(\iota_n \cdot \phi_n) \bar{\in} Tr$'. This is a sentential function of the metalanguage which contains 'n' as the only free variable. From the previous remarks it follows that with this function we can correlate another function which is equivalent to it for any value of 'n', but which is expressed completely in terms of arithmetic. We shall write this new function in the schematic form '$\psi(n)$'. Thus we have:

(1)   *for any n,* $\bigcup_1^3(\iota_n \cdot \phi_n) \bar{\in} Tr$ *if and only if* $\psi(n)$.

Since the language of the general theory of classes suffices for the foundation of the arithmetic of the natural numbers, we can assume that '$\psi(n)$' is one of the functions of this language. The function '$\psi(n)$' will thus be a term of the sequence ϕ, e.g. the term with the index k, '$\psi(n)$' $= \phi_k$. If we substitute 'k' for 'n' in the sentence (1) we obtain:

(2) $\bigcup_1^3(\iota_k \cdot \phi_k) \bar{\in} Tr$ *if and only if* $\psi(k)$.

The symbol '$\bigcup_1^3(\iota_k \cdot \phi_k)$' denotes, of course, a sentence of the language under investigation. By applying to this sentence condition (α) of the convention **T** we obtain a sentence of the form '$x \in Tr$ *if and only if* p', where 'x' is to be replaced by a structural-descriptive or any other individual name of the statement $\bigcup_1^3(\iota_k \cdot \phi_k)$, but '$p$' by this statement itself or by any statement which is equivalent to it. In particular we can substitute '$\bigcup_1^3(\iota_k \cdot \phi_k)$' for '$x$' and for '$p$'—in view of the meaning of the symbol 'ι_k'—the statement 'there is an n such that $n = k$ and $\psi(n)$' or, simply '$\psi(k)$'. In this way we obtain the following formulation:

(3) $\bigcup_1^3(\iota_k \cdot \phi_k) \in Tr$ *if and only if* $\psi(k)$.

The sentences (2) and (3) stand in palpable contradiction to one another; the sentence (2) is in fact directly equivalent to the negation of (3). In this way we have proved the first part of the theorem. We have proved that among the consequences of the definition of the symbol 'Tr' the negation of one of the sentences mentioned in the condition (α) of the convention T must appear. From this the second part of the theorem immediately follows.

The assumption of consistency appearing in the part (β) of this theorem is essential. If the class of all provable sentences of the metatheory contained a contradiction, then every definition in the metatheory would have among its consequences all possible sentences (since they all would be provable in the metatheory), in particular those described in the convention T. On the other hand, as we now know,[1] there is no prospect of proving the consistency of the metatheory which we are working with, on the basis of the meta-metatheory. It is to be noted that, in view of the existence of an interpretation of the metatheory in the science itself (a fact which has played such an essential part in the proof sketched above), the assumption of the second part of Th. I is equivalent to the assumption of the consistency of the science investigated itself and from the intuitive standpoint is just as evident.

The result reached in Th. I seems perhaps at first sight uncommonly paradoxical. This impression will doubtless be weakened as soon as we recall the fundamental distinction between the content of the concept to be defined and the nature of those concepts which are at our disposal for the construction of the definition.

The metalanguage in which we carry out the investigation contains exclusively structural-descriptive terms, such as names of expressions of the language, structural properties of these expressions, structural relations between expressions, and so on, as well as expressions of a logical kind among which (in the present case) we find all the expressions of the language studied. What we call metatheory is, fundamentally, the *morphology of language*—a science of the form of expressions—a correlate of such parts of traditional grammar as morphology, etymology, and syntax.

[1] Cf. Gödel, K. (22), p. 196 (Th. XI).

The fact that the language studied and the deductive science carried out in this language are formalized has brought about an interesting phenomenon; it has been possible to reduce to structural-descriptive concepts certain other notions of a totally different kind, which are distinguished from the former both in their origin and in their usual meaning, namely the concept of consequence together with a series of related notions.[1] It has been possible to establish what may be called the *logic of the given science* as a part of morphology.

Encouraged by this success we have attempted to go further and to construct in the metalanguage definitions of certain concepts belonging to another domain, namely that called the *semantics of language*—i.e. such concepts as satisfaction, denoting, truth, definability, and so on. A characteristic feature of the semantical concepts is that they give expression to certain relations between the expressions of language and the objects about which these expressions speak, or that by means of such relations they characterize certain classes of expressions or other objects. We could also say (making use of the *suppositio materialis*) that these concepts serve to set up the correlation between the names of expressions and the expressions themselves.

For a long time the semantical concepts have had an evil reputation among specialists in the study of language. They have resisted all attempts to define their meaning exactly, and the properties of these concepts, apparently so clear in their content, have led to paradoxes and antinomies. For that reason the tendency to reduce these concepts to structural-descriptive ones must seem quite natural and well-founded. The following fact seemed to favour the possibility of realizing this tendency: it has always been possible to replace every phrase which contains these semantical terms, and which concerns particular

---

[1] The reduction of the concept of consequence to concepts belonging to the morphology of language is a result of the deductive method in its latest stages of development. When, in everyday life, we say that a sentence follows from other sentences we no doubt mean something quite different from the existence of certain structural relations between these sentences. In the light of the latest results of Gödel it seems doubtful whether this reduction has been effected without remainder.

structurally described expressions of the language, by a phrase which is equivalent in content and is free from such terms. In other words it is possible to formulate infinitely many partial definitions for every semantical concept, which in their totality exhaust all cases of the application of the concept to concrete expressions and of which the sentences adduced in condition (α) of convention **T** are examples. It was with just this end in view that, as a rule, we included in the metalanguage, with regard to the content of the semantical concepts, not only the names of expressions but all expressions of the language itself or expressions having the same meaning (even when these expressions were not of a logical kind, cf. pp. 210 f.), although such an enrichment of the metalanguage has no advantages for the pursuit of the 'pure' morphology of language.

In the abstract the fact mentioned has no decisive importance; it offers no path by which an automatic transition from the partial definitions to a general definition is possible, which embraces them all as special cases and would form their infinite logical product.[1] Only thanks to the special methods of construction which we developed in §§ 3 and 4 have we succeeded in carrying out the required reduction of the semantical concepts, and then only for a specified group of languages which are poor in grammatical forms and have a restricted equipment of semantical categories—namely the languages of finite order. Let it be remembered that the methods there applied required the use in the metalanguage of categories of higher order than all categories of the language studied and are for that reason fundamentally different from all grammatical forms of this language. The analysis of the proof of Th. I sketched above shows that this

---

[1] In the course of our investigation we have repeatedly encountered similar phenomena: the impossibility of grasping the simultaneous dependence between objects which belong to infinitely many semantical categories; the lack of terms of 'infinite order'; the impossibility of including, in *one* process of definition, infinitely many concepts, and so on (pp. 188 f., 232 f., 243, 245). I do not believe that these phenomena can be viewed as a symptom of the formal incompleteness of the actually existing languages—their cause is to be sought rather in the nature of language itself: language, which is a product of human activity, necessarily possesses a 'finitistic' character, and cannot serve as an adequate tool for the investigation of facts, or for the construction of concepts, of an eminently 'infinitistic' character.

circumstance is not an accidental one. Under certain general assumptions, it proves to be impossible to construct a correct definition of truth if only such categories are used which appear in the language under consideration.[1] For that reason the situation had fundamentally changed when we passed to the 'rich' languages of infinite order. The methods used earlier proved to be inapplicable; all concepts and all grammatical forms of the metalanguage found an interpretation in the language and hence we were able to show conclusively that the semantics of the language could not be established as a part of its morphology. The significance of the results reached reduces just to this.

But, apart from this, Th. I has important consequences of a methodological nature. It shows that it is impossible to define in the metatheory a class of sentences of the language studied which consists exclusively of materially true sentences and is at the same time complete (in the sense of Def. 20 in § 3). In particular, if we enlarge the class of provable sentences of the science investigated in any way—whether by supplementing the list of axioms or by sharpening the rules of inference—then we either add false sentences to this class or we obtain an incomplete system. This is all the more interesting inasmuch as the enlarge-

---

[1] From this, or immediately from certain results contained in Gödel, K. (22) (pp. 187–91), it can easily be inferred that a structural definition of truth—in the sense discussed on pp. 236 ff., especially on p. 237, note 2—cannot be constructed even for languages of finite order which are in some degree richer. From other investigations of this author (op. cit., p. 193; Th. IX) it follows that in certain elementary cases in which we can construct such a definition, it is nevertheless impossible to give a general structural criterion of the truth of a sentence. The first of these results applies, for instance, to the logic of two-termed and many-termed relations discussed in § 4. The second result applies, for example, to the lower predicate calculus ('engere Funktionenkalkül') of Hilbert-Ackermann (30), pp. 43 ff.; in this case, however, the result is applied, not to the notion of a true sentence, but to the related notion of a universally valid ('allgemeingültig') sentential function.

At this point we should like to call attention to the close connexion between the notions of 'structural definition of truth', and of 'general structural criterion of truth' discussed in this work, and the notions of recursive enumerability and general recursiveness known from the recent literature (see, for example, Mostowski, A. (53ƒ), chap. 5). In fact, by saying that there is a 'structural definition of truth' for a given formalized theory we essentially mean that the set of all true sentences of this theory is recursively enumerable; when we say that there is a 'general structural criterion of truth' we mean that the set of all true sentences is general recursive.

ment of the class of provable sentences to form a complete and consistent system in itself presents no difficulties.[1]

An interpretation of Th. I which went beyond the limits given would not be justified. In particular it would be incorrect to infer the impossibility of operating consistently and in agreement with intuition with semantical concepts and especially with the concept of truth. But since one of the possible ways of constructing the scientific foundations of semantics is closed we must look for other methods. The idea naturally suggests itself of setting up semantics as a special deductive science with a system of morphology as its logical substructure. For this purpose it would be necessary to introduce into morphology a given semantical notion as an undefined concept and to establish its fundamental properties by means of axioms. The experience gained from the study of semantical concepts in connexion with colloquial language, warns us of the great dangers that may accompany the use of this method. For that reason the question of how we can be certain that the axiomatic method will not in this case lead to complications and antinomies becomes especially important.

In discussing this question I shall restrict myself to the theory of truth, and in the first place I shall establish the following theorem, which is a consequence of the discussion in the preceding section:

THEOREM II. *For an arbitrary, previously given natural number k, it is possible to construct a definition of the symbol 'Tr' on the basis of the metatheory, which has among its consequences all those sentences from the condition (α) of the convention* T *in which in the place of the symbol 'p' sentences with variables of at most the k-th order occur (and moreover, the sentence adduced in the condition (β) of this convention).*

By way of proof it suffices to remark that this theorem no longer concerns the language studied in its whole extent but only a fragment of it which embraces all those expressions which contain no variable of higher order than the kth. This fragment

---

[1] Cf. V, Th. 56, a result of Lindenbaum's (see p. 98 of the present volume).

is clearly a language of finite order and in fact a language of the 2nd kind. We can therefore easily construct the required definition by applying one of the two methods described in § 4. It is to be noted that the definition obtained in this way (together with the consequences given in Th. II) yields a series of theorems of a general nature, like the Ths. 1–5 in § 3, for example, if the formulations of these theorems are suitably weakened by restricting the domain of their applicability to sentences with variables of at most the kth order.

Hence it will be seen that, in contrast to the theory of truth in its totality, the single fragments of this theory (the objects of investigation of which are sentences which contain only variables whose order is bounded above) can be established as fragments of the metatheory. If, therefore, the metatheory is consistent we shall not find a contradiction in these fragments. This last result can be extended in a certain sense to the whole theory of truth, as the following theorem shows:

THEOREM III. *If the class of all provable sentences of the meta-theory is consistent and if we add to the metatheory the symbol 'Tr' as a new primitive sign, and all theorems which are described in conditions (α) and (β) of the convention* T *as new axioms, then the class of provable sentences in the metatheory enlarged in this way will also be consistent.*

To prove this theorem we note that the condition (α) contains infinitely many sentences which are taken as axioms of the theory of truth. A finite number of these axioms—even in union with the single axiom from condition (β)—cannot lead to a contradiction (so long as there is no contradiction already in the metatheory). Actually in the finite number of axioms obtained from (α) only a finite number of sentences of the language studied appears and in these sentences we find a finite number of variables. There must, therefore, be a natural number k such that the order of none of these variables exceeds k. From this it follows, by Th. II, that a definition of the symbol 'Tr' can be constructed in the metatheory such that the axioms in question become consequences of this definition. In other words: these axioms, with

a suitable interpretation of the symbol 'Tr', become provable
sentences of the metatheory (this fact can also be established
directly, i.e. independently of Th. II). If any class of sentences
contains a contradiction, it is easy to show that the contradiction must appear in a finite part of this class.[1] Since, however,
no finite part of the axiom system described in Th. III contains
a contradiction, the whole system is consistent, which was to be
proved.

The value of the result obtained is considerably diminished
by the fact that the axioms mentioned in Th. III have a very
restricted deductive power. A theory of truth founded on them
would be a highly incomplete system, which would lack the most
important and most fruitful general theorems. Let us show this
in more detail by a concrete example. Consider the sentential
function '$x \in Tr$ or $\bar{x} \in Tr$'. If in this function we substitute for
the variable 'x' structural-descriptive names of sentences, we
obtain an infinite number of theorems, the proof of which on
the basis of the axioms obtained from the convention **T** presents
not the slightest difficulty. But the situation changes fundamentally as soon as we pass to the generalization of this sentential function, i.e. to the general principle of contradiction.
From the intuitive standpoint the truth of all those theorems is
itself already a proof of the general principle; this principle
represents, so to speak, an 'infinite logical product' of those
special theorems. But this does not at all mean that we can
actually derive the principle of contradiction from the axioms or
theorems mentioned by means of the normal modes of inference
usually employed. On the contrary, by a slight modification in
the proof of Th. III it can be shown that the principle of contradiction is not a consequence (at least in the existing sense of the
word) of the axiom system described.

We could, of course, now enlarge the above axiom system by
adding to it a series of general sentences which are independent
of this system. We could take as new axioms the principles of
contradiction and excluded middle, as well as those sentences
which assert that the consequences of true sentences are true,

---

[1] Cf. V, Th. 48, p. 91 of the present volume.

and also that all primitive sentences of the science investigated belong to the class of true sentences. Th. III could be extended to the axiom system enlarged in this way.[1] But we attach little importance to this procedure. For it seems that every such enlargement of the axiom system has an accidental character, depending on rather inessential factors such, for example, as the actual state of knowledge in this field. In any case, various objective criteria which we should wish to apply in the choice of further axioms prove to be quite inapplicable. Thus it seems natural to require that the axioms of the theory of truth, together with the original axioms of the metatheory, should constitute a categorical system.[2] It can be shown that this postulate co-incides in the present case with another postulate, according to which the axiom system of the theory of truth should un-ambiguously determine the extension of the symbol 'Tr' which occurs in it, and in the following sense: if we introduce into the metatheory, alongside this symbol, another primitive sign, e.g. the symbol 'Tr''' and set up analogous axioms for it, then the statement '$Tr = Tr''$' must be provable. But this postulate cannot be satisfied. For it is not difficult to prove that in the contrary case the concept of truth could be defined exclusively by means of terms belonging to the morphology of language, which would be in palpable contradiction to Th. I. For other reasons of a more general nature there can be no question of an axiom system that would be complete and would consequently suffice for the solution of every problem from the domain of the theory under consideration. This is an immediate methodo-logical consequence of Th. I applied not to the language of the general theory of classes but to the richer language of the meta-theory and the theory of truth (cf. the remarks on p. 254).

There is, however, quite a different way in which the founda-tions of the theory of truth may be essentially strengthened.

[1] For this purpose we must nevertheless to some extent sharpen the pre-misses of the theorem by assuming that the class of all provable sentences of the metatheory is not only consistent, but also ω-consistent in the sense of Gödel, K. (22), p. 187, or in other words, that this class remains consistent after a single application of the rule of infinite induction, which will be dis-cussed below.

[2] See p. 174, note 1.

The fact that we cannot infer from the correctness of all substitutions of such a sentential function as '$x \in Tr$ or $\bar{x} \in Tr$' the correctness of the sentence which is the generalization of this function, can be regarded as a symptom of a certain imperfection in the rules of inference hitherto used in the deductive sciences. In order to make good this defect we could adopt a new rule, the so-called *rule of infinite induction*, which in its application to the metatheory may be formulated somewhat as follows: if a given sentential function contains the symbol 'x', which belongs to the same semantical category as the names of expressions, as its only free variable, and if every sentence, which arises from the given function by substituting the structural-descriptive name of any expression of the language investigated for the variable 'x', is a provable theorem of the metatheory, then the sentence which we obtain from the phrase '*for every x, if x is an expression then p*' by substituting the given function for the symbol 'p', may also be added to the theorems of the metatheory. This rule can also be given another formulation which differs from the foregoing only by the fact that in it, instead of speaking about expressions, we speak of natural numbers; and instead of structural-descriptive names of expressions, the so-called specific symbols of natural numbers are dealt with, i.e. such symbols as '0', '1', '1+1', '1+1+1', and so on. In this form the rule of infinite induction recalls the principle of complete induction, which it exceeds considerably in logical power. Since it is possible to set up effectively a one-one correspondence between expressions and the natural numbers (cf. the proof of Th. I) it is easy to see that the two formulations are equivalent on the basis of the metatheory. But in the second formulation no specific concepts of the metalanguage occur at all, and for this reason it is applicable to many other deductive sciences. In the case where we are dealing with a science in the language of which there are no specific symbols for the natural numbers this formulation requires certain external modifications. For example, in order to formulate the rule for the general theory of classes, instead of substitutions of a given sentential function we must operate with expressions of the type '$\bigcup_{1}^{3}(\iota_k . p)$', where, in the place of

'p' the function in question occurs and the symbol 'ι_k' has the same meaning as in the proof of Th. I.[1]

On account of its non-finitist nature the rule of infinite induction differs fundamentally from the normal rules of inference. On each occasion of its use infinitely many sentences must be taken into consideration, although at no moment in the development of a science is such a number of previously proved theorems effectively given. It may well be doubted whether there is any place for the use of such a rule within the limits of the existing conception of the deductive method. The question whether this rule does not lead to contradictions presents no less serious difficulties than the analogous problem regarding the existing rules, even if we assume the consistency of the existing rules and permit the use of the new rule not only in the theory but also in the corresponding metatheory and in particular in any attempted proof of consistency. Nevertheless from the intuitive standpoint the rule of infinite induction seems to be as reliable as the rules normally applied: it always leads from true sentences to true sentences. In connexion with languages of finite order this fact can be strictly proved by means of the definition of truth constructed for these languages. The fact that this rule enables many problems to be solved which are not solvable on the basis of the old rules is in favour of the acceptance of the new rule, not only in the theory but also in the metatheory. By the introduction of this rule the class of provable sentences is enlarged by a much greater extent than by any supplementation of the list of axioms.[2] In the case of certain elementary deductive sciences, this enlargement is so great that the class of theorems becomes a complete system and coincides

---

[1] I have previously pointed out the importance of the rule of infinite induction in the year 1926. In a report to the Second Polish Philosophical Congress, in 1927, I have given, among other things, a simple example of a consistent deductive system which after a single application of this rule ceases to be consistent, and is therefore not ω-consistent (cf. p. 258, note 1; see also IX, p. 282, note 2). Some remarks on this rule are to be found in Hilbert, D. (29), pp. 491–2.

[2] Thus, for example, if we adopt this rule in the metalanguage without including it in the language, we can prove that the class of provable sentences of the science is consistent, which previously was not possible. In connexion with this problem cf. Gödel, K. (22), pp. 187–91 and 196.

with the class of true sentences. Elementary number theory provides an example, namely, the science in which all variables represent names of natural or whole numbers and the constants are the signs from the sentential and predicate calculi, the signs of zero, one, equality, sum, product and possibly other signs defined with their help.†

If it is decided to adopt the rule of infinite induction in the metatheory, then the system of axioms to which Th. III refers already forms a sufficient foundation for the development of the theory of truth. The proof of any of the known theorems in this field will then present no difficulty, in particular the Ths. 1–6 in § 3 and the theorem according to which the rule of infinite induction when applied to true sentences always yields true sentences. More important still, these axioms, together with the general axioms of the metatheory, form a categorical (although not a complete) system, and determine unambiguously the extension of the symbol 'Tr' which occurs in them.

Under these circumstances the question whether the theory erected on these foundations contains no inner contradiction acquires a special importance. Unfortunately this question cannot be finally decided at present. Th. I retains its full validity: in spite of the strengthening of the foundations of the metatheory the theory of truth cannot be constructed as a part of the morphology of language. On the other hand for the present we cannot prove Th. III for the enlarged metalanguage. The premiss which has played the most essential part in the original proof, i.e. the reduction of the consistency of the infinite axiom system to the consistency of every finite part of this system, now completely loses its validity—as is easily seen—on account of the content of the newly adopted rule. The possibility that the question cannot be decided in any direction is not excluded (at least on the basis of a 'normal' system of the meta-metatheory, which is constructed

† This last remark enables us to construct a rather simple definition of truth for elementary number theory without using our general method. The definition thus constructed can be further simplified. In fact we can first structurally describe all true sentences which contain no variables (or quantifiers), and then define an arbitrary sentence to be true if and only if it can be obtained from those elementary true sentences by applying the rule of infinite induction arbitrarily many times.

according to the principles given at the beginning of § 4 and does not contain the semantics of the metalanguage). On the other hand the possibility of showing Th. III to be false in its new interpretation seems to be unlikely from the intuitive viewpoint. One thing seems clear: the antinomy of the liar cannot be directly reconstructed either in the formulation met with in § 1 or in the form in which it appeared in the proof of Th. I. For here the axioms adopted in the theory of truth clearly possess, in contrast to colloquial language, the character of partial definitions. Through the introduction of the symbol 'Tr' the metalanguage does not in any way become semantically universal, it does not coincide with the language itself and cannot be interpreted in that language (cf. pp. 158 and 248).[1]

No serious obstacles stand in the way of the application of the results obtained to other languages of infinite order. This is especially true of the most important of these results—Th. I. The languages of infinite order, thanks to the variety of meaningful expressions contained in them, provide sufficient means for the formulation of every sentence belonging to the arithmetic

---

[1] This last problem is equivalent to a seemingly more general problem of a methodological nature which can be formulated as follows. We presuppose the consistency of the metatheory supplemented by the rule of infinite induction. We consider an infinite sequence t of sentences of the metatheory; further we take into the metatheory a new primitive sign 'N', and add as axioms those and only those sentences which are obtained from the scheme '$n \in N$ if and only if p' by substituting for the sign 'n' the kth specific symbol of the natural numbers (i.e. the expression composed of k signs '1' separated from one another by the signs '$+$') and for the sign 'p' the kth term of the sequence t (k being here an arbitrary natural number). The question now arises whether the class of provable sentences of the metatheory, when enlarged in this way, remains consistent. This problem may be called the *problem of infinite recursive definitions*. The axiom system described in it can—from the intuitive standpoint—be regarded as a definition *sui generis* of the symbol 'N', which is distinguished from normal definitions only by the fact that it is formulated in infinitely many sentences. In view of this character of the axioms the possibility of a negative solution of the problem does not seem very probable. From Th. II and the interpretation of the metatheory in the theory itself, it is not difficult to infer that this problem can be solved in a positive sense in those cases in which the order of all variables which occur in the sentences of the sequence t is bounded above. It is then even possible to construct a definition of the symbol 'N' in the metatheory such that all the axioms mentioned follow from it. This problem obviously does not depend on the specific properties of the metatheory as such; it can also be presented in the same or in a somewhat modified form for other deductive sciences, e.g. for the general theory of classes.

of natural numbers and consequently enable the metalanguage to be interpreted in the language itself. It is thanks to just this circumstance that Th. I retains its validity for all languages of this kind.[1]

Some remarks may be added about those cases in which not single languages but whole classes of languages are investigated. As I have already emphasized in the Introduction, the concept of truth essentially depends, as regards both extension and content, upon the language to which it is applied. We can only meaningfully say of an expression that it is true or not if we treat this expression as a part of a concrete language. As soon as the discussion is about a large number of languages the expression 'true sentence' ceases to be unambiguous. If we are to avoid this ambiguity we must replace it by the relative term 'a true sentence with respect to the given language'. In order to make the sense of this term precise we apply to it essentially the same procedure as before: we construct a common metalanguage for all the languages of the given class; within the metalanguage we try to define the expression in question with the help of the methods developed in §§ 3 and 4. If we are not successful we add this term to the fundamental expressions of the metalanguage and by the axiomatic method determine its meaning according to the instructions of Th. III of this section. On account of the relativization of this term we should nevertheless expect *a priori* that in carrying out the plan sketched above the earlier difficulties would be significantly increased and quite new complications might arise (connected for example with the necessity of

---

[1] A reservation is necessary here: if we choose as our starting-point the classification of semantical categories sketched on p. 218, note 2, then we again encounter languages of infinite order for which Th. I loses its validity. A typical example is furnished by the language of Leśniewski's *Prototethic* (cf. Leśniewski, S. (46)). In consequence of the 'finitistic' character of all the semantical categories of this language, it is easy to construct, in the metalanguage, a correct definition of truth, by choosing as model the matrix method from the extended sentential calculus. Moreover, such a definition can be obtained in other ways: as Leśniewski has shown, the class of provable sentences of the prototethic is complete, and therefore the concept of provable sentence coincides in its extension with that of true sentence. Th. I on the other hand applies without restriction to all languages in which the order of the semantical categories from the domain of Leśniewski's *Ontology* (cf. Leśniewski, S. (47)) is not bounded above.

defining the word 'language'). I do not propose to discuss the problem touched upon in more detail in this place. The prospects for such investigations at the present time seem to be rather limited. In particular it would be incorrect to suppose that the relativization of the concept of truth—in the direction mentioned above—would open the way to some general theory of this concept which would embrace all possible or at least all formalized languages. The class of languages which is chosen as the object of simultaneous study must not be too wide. If, for example, we include in this class the metalanguage, which forms the field of the investigations and already contains the concept of truth, we automatically create the conditions which enable the antinomy of the liar to be reconstructed. The language of the general theory of truth would then contain a contradiction for exactly the same reason as does colloquial language.

In conclusion it may be mentioned that the results obtained can be extended to other semantical concepts, e.g. to the concept of satisfaction. For each of these concepts a system of postulates can be set up which (1) contains partial definitions analogous to the statements described in condition (α) of the convention **T** which determine the meaning of the given concept with respect to all concrete, structurally described expressions of a given class (e.g. with respect to sentences or sentential functions of a specific semantical type), and (2) contains a further postulate which corresponds to the sentence from the condition (β) of the same convention and stipulates that the concept in question may be applied only to expressions of the given class. We should be prepared to regard such a definition of the concept studied as a materially adequate one if its consequences included all the postulates of the above system. Methods which are similar to those described in §§ 3 and 4 enable the required definition to be constructed in all cases where we are dealing with languages of finite order, or, more generally, in which the semantical concept studied concerns exclusively linguistic expressions in which the order of the variables is bounded above (cf. Th. II). In the remaining cases it can be shown—after the pattern of the proof of Th. I—that no definition with the properties mentioned can be formulated in the metalanguage.[1] In order to construct the theory of the concept studied in these cases also, it must be included in the system of primitive concepts, and the postulate described above must be included in the axiom system of the metatheory. A procedure analogous to the proof of Th. III proves that the system of the metalanguage supplemented in this way remains internally consistent. But the deductive power of the added postulates is very restricted. They do not suffice for the proof of the most important general theorems concerning the concept in question. They do not determine its extension unambiguously and the system obtained is not complete, nor is it even categorical. To remove this defect we must strengthen the foundations of the metatheory itself by adding the rule of infinite induction to its rules of inference. But then the proof of consistency would present great difficulties which we are not able at present to overcome.

§ 6. SUMMARY

The principal results of this article may be summarized in the following theses:

A. *For every formalized language of finite order a formally correct and materially adequate definition of true sentence can be constructed in the metalanguage, making use only of expressions of a general logical kind, expressions of the language itself as well as terms belonging to the morphology of language, i.e. names of linguistic expressions and of the structural relations existing between them.*

B. *For formalized languages of infinite order the construction of such a definition is impossible.*

---

[1] This especially concerns the concept of definability (although in this case both the formulation of the problem itself, as well as the method of solution, require certain modifications in comparison with the scheme put forward in the text). In VI, I have expressed the conjecture that it is impossible to define this concept in its full extent on the basis of the metalanguage. I can now prove this conjecture exactly. This fact is all the more noteworthy in that it is possible—as I have shown in the article mentioned—to construct the definitions of the particular cases of the concept of definability which apply, not to the whole language, but to any of its fragments of finite order, not only in the metalanguage but also in the language itself.

C. *On the other hand, even with respect to formalized languages of infinite order, the consistent and correct use of the concept of truth is rendered possible by including this concept in the system of primitive concepts of the metalanguage and determining its fundamental properties by means of the axiomatic method* (the question whether the theory of truth established in this way contains no contradiction remains for the present undecided).

Since the results obtained can easily be extended to other semantical concepts the above theses can be given a more general form:

A′. *The semantics of any formalized language of finite order can be built up as a part of the morphology of language, based on correspondingly constructed definitions.*

B′. *It is impossible to establish the semantics of the formalized languages of infinite order in this way.*

C′. *But the semantics of any formalized language of infinite order can be established as an independent science based upon its own primitive concepts and its own axioms, possessing as its logical foundation a system of the morphology of language* (although a full guarantee that the semantics constructed by this method contains no inner contradiction is at present lacking).

From the formal point of view the foregoing investigations have been carried out within the boundaries of the methodology of the deductive sciences. Some so to speak incidental results will perhaps be of interest to specialists in this field. I would draw attention to the fact that with the definition of true sentence for deductive sciences of finite order a general method has been obtained for proving their consistency (a method which, however, does not add greatly to our knowledge). I would point out also that it has been possible to define, for languages of finite order, the concepts of correct sentence in a given and in an arbitrary individual domain—concepts which play a great part in recent methodological studies.

But in its essential parts the present work deviates from the main stream of methodological investigations. Its central problem—the construction of the definition of true sentence and establishing the scientific foundations of the theory of truth —belongs to the theory of knowledge and forms one of the chief problems of this branch of philosophy. I therefore hope that this work will interest the student of the theory of knowledge above all and that he will be able to analyse the results contained in it critically and to judge their value for further researches in this field, without allowing himself to be discouraged by the apparatus of concepts and methods used here, which in places have been difficult and have not hitherto been used in the field in which he works.

One word in conclusion. Philosophers who are not accustomed to use deductive methods in their daily work are inclined to regard all formalized languages with a certain disparagement, because they contrast these 'artificial' constructions with the one natural language—the colloquial language. For that reason the fact that the results obtained concern the formalized languages almost exclusively will greatly diminish the value of the foregoing investigations in the opinion of many readers. It would be difficult for me to share this view. In my opinion the considerations of § 1 prove emphatically that the concept of truth (as well as other semantical concepts) when applied to colloquial language in conjunction with the normal laws of logic leads inevitably to confusions and contradictions. Whoever wishes, in spite of all difficulties, to pursue the semantics of colloquial language with the help of exact methods will be driven first to undertake the thankless task of a reform of this language. He will find it necessary to define its structure, to overcome the ambiguity of the terms which occur in it, and finally to split the language into a series of languages of greater and greater extent, each of which stands in the same relation to the next in which a formalized language stands to its metalanguage. It may, however, be doubted whether the language of everyday life, after being 'rationalized' in this way, would still preserve its naturalness and whether it would not rather take on the characteristic features of the formalized languages.

§ 7. POSTSCRIPT

In writing the present article I had in mind only formalized languages possessing a structure which is in harmony with the theory of semantical categories and especially with its basic principles. This fact has exercised an essential influence on the construction of the whole work and on the formulation of its final results. It seemed to me then that 'the theory of the semantical categories penetrates so deeply into our fundamental intuitions regarding the meaningfulness of expressions, that it is hardly possible to imagine a scientific language whose sentences possess a clear intuitive meaning but whose structure cannot be brought into harmony with the theory in question in one of its formulations' (cf. p. 215). Today I can no longer defend decisively the view I then took of this question. In connexion with this it now seems to me interesting and important to inquire what the consequences would be for the basic problems of the present work if we included in the field under consideration formalized languages for which the fundamental principles of the theory of semantical categories no longer hold. In what follows I will briefly consider this question.

Although in this way the field to be covered is essentially enlarged, I do not intend—any more than previously—to consider all possible languages which someone might at some time construct. On the contrary I shall restrict myself exclusively to languages which—apart from differences connected with the theory of semantical categories—exhibit in their structure the greatest possible analogy with the languages previously studied. In particular, for the sake of simplicity, I shall consider only those languages in which occur, in addition to the universal and existential quantifiers and the constants of the sentential calculus, only individual names and the variables representing them, as well as constant and variable sentence-forming functors with arbitrary numbers of arguments. After the manner of the procedure in §§ 2 and 4 we try to specify for each of these languages the concepts of primitive sentential function, fundamental operations on expressions, sentential function in general, axiom, consequence, and provable theorem. Thus, for example, we include as a rule among the axioms—just as in the language of the general theory of classes in § 5—the substitutions of the axioms of the sentential calculus, the pseudo-definitions and the law of extensionality (perhaps also other sentences, according to the specific peculiarities of the language). In determining the concept of consequence we take as our model Def. 15 in § 2.

The concept introduced in § 4 of the order of an expression plays a part which is no less essential than before in the construction of the language we are now considering. It is advisable to assign to names of individuals and to the variables representing them the order 0 (and not as before the order 1). The order of a sentence-forming functor of an arbitrary (primitive) sentential function is no longer unambiguously determined by the orders of all arguments of this function. Since the principles of the theory of the semantical categories no longer hold, it may happen that one and the same sign plays the part of a functor in two or more sentential functions in which arguments occupying respectively the same places nevertheless belong to different orders. Thus in order to fix the order of any sign we must take into account the orders of all arguments in all sentential functions in which this sign is a sentence-forming functor. If the order of all these arguments is smaller than a particular natural number n, and if there occurs in at least *one* sentential function an argument which is exactly of order $n-1$, then we assign to the symbol in question the order n. All such sentence-forming functors—as well as the names of individuals and the variables representing them—are included among the signs of finite order. But account must also be taken of the possibility that yet other sentence-forming functors may occur in the language to which an infinite order must be assigned. If, for example, a sign is a sentence-forming functor of only those sentential functions which have all their arguments of finite order, where, however, these orders are not bounded above by any natural number, then this sign will be of infinite order.

In order to classify the signs of infinite order we make use of the notion of *ordinal number*, taken from the theory of sets, which is a generalization of the usual concept of natural number.[1] As is well known, the natural numbers are the smallest ordinal numbers. Since, for every infinite sequence of ordinal numbers, there are numbers greater than every term of the sequence, there are, in particular, numbers which are greater than all natural numbers. We call them *transfinite ordinal numbers*. It is known that in every non-empty class of ordinal numbers there is a smallest number. In particular there is a smallest transfinite number which is denoted by the symbol 'ω'. The next largest number is $\omega+1$, then follow the numbers $\omega+2$, $\omega+3$,..., $\omega.2$, $\omega.2+1$, $\omega.2+2$,..., $\omega.3$,..., and so on. To those signs of infinite order which are functors of sentential functions containing exclusively arguments of finite order we assign the number ω as their order. A sign which is a functor in only those sentential functions in which the arguments are either of finite order or of order ω (and in which at least *one* argument of a function is actually of order ω), is of the order $\omega+1$. The general recursive definition of order is as follows: the order of a particular sign is the smallest ordinal number which is greater than the orders of all arguments in all sentential functions in which the given sign occurs as a sentence-forming functor.[2]

Just as in § 4, we can distinguish languages of finite and infinite order. We can in fact assign to every language a quite specific ordinal number as its order, namely the smallest ordinal number which exceeds the orders of all variables occurring in this language (the former languages of the nth order—as can easily be shown—retain their former order under this convention because the order of the names of individuals has been diminished. The language of the general theory of classes has the order ω).

It does not at all follow from these stipulations that every variable in the languages in question is of a definite order. On the contrary it seems to me (by reason of trials and other considerations) almost certain that we cannot restrict ourselves to the use of variables of definite order if we are to obtain languages

[1] Cf. Fraenkel, A. (16), pp. 185 ff.

[2] Cf. the introduction of the system of levels in Carnap, R. (10), pp. 139 ff. (p. 186 in English translation).

which are actually superior to the previous languages in the abundance of the concepts which are expressible by their means, and the study of which could throw new light on the problems in which we are here interested. We must introduce into the languages variables of indefinite order which, so to speak, 'run through' all possible orders, which can occur as functors or arguments in sentential functions without regard to the order of the remaining signs occurring in these functions, and which at the same time may be both functors and arguments in the same sentential functions. With such variables we must proceed with the greatest caution if we are not to become entangled in antinomies like the famous antinomy of the class of all classes which are not members of themselves. Special care must be taken in formulating the rule of substitution for languages which contain such variables and in describing the axioms which we have called pseudodefinitions. But we cannot go into details here.[1]

There is obviously no obstacle to the introduction of variables of transfinite order not only into the language which is the object investigated, but also into the metalanguage in which the investigation is carried out. In particular it is always possible to construct the metalanguage in such a way that it contains variables

---

[1] From the languages just considered it is but a step to languages of another kind which constitute a much more convenient and actually much more frequently applied apparatus for the development of logic and mathematics. In these new languages all the variables are of indefinite order. From the formal point of view these are languages of a very simple structure; according to the terminology laid down in § 4 they must be counted among the languages of the first kind, since all their variables belong to one and the same semantical category. Nevertheless, as is shown by the investigations of E. Zermelo and his successors (cf. Skolem, Th. (66), pp. 1–12), with a suitable choice of axioms it is possible to construct the theory of sets and the whole of classical mathematics on the basis provided by this language. In it we can express so to speak every idea which can be formulated in the previously studied languages of finite and infinite order. For the languages here discussed the concept of order by no means loses its importance; it no longer applies, however, to the expressions of the language, but either to the objects denoted by them or to the language as a whole. Individuals, i.e. objects which are not sets, we call objects of order 0; the order of an arbitrary set is the smallest ordinal number which is greater than the orders of all elements of this set; the order of the language is the smallest ordinal number which exceeds the order of all sets whose existence follows from the axioms adopted in the language. Our further exposition also applies without restriction to the languages which have just been discussed.

of higher order than all the variables of the language studied. The metalanguage then becomes a language of higher order and thus one which is essentially richer in grammatical forms than the language we are investigating. This is a fact of the greatest importance from the point of view of the problems in which we are interested. For with this the distinction between languages of finite and infinite orders disappears—a distinction which was so prominent in §§ 4 and 5 and was strongly expressed in the theses A and B formulated in the Summary. In fact, the setting up of a correct definition of truth for languages of infinite order would in principle be possible provided we had at our disposal in the metalanguage expressions of higher order than all variables of the language investigated. The absence of such expressions in the metalanguage has rendered the extension of these methods of construction to languages of infinite order impossible. But now we are in a position to define the concept of truth for any language of finite or infinite order, provided we take as the basis for our investigations a metalanguage of an order which is at least greater by 1 than that of the language studied (an essential part is played here by the presence of variables of indefinite order in the metalanguage). It is perhaps interesting to emphasize that the construction of the definition is then to a certain degree simplified. We can adhere strictly to the method outlined in § 3 without applying the artifice which we were compelled to use in § 4 in the study of languages of the 2nd and 3rd kinds. We need neither apply many-rowed sequences nor carry out the semantical unification of the variables, for having abandoned the principles of the theory of semantical categories we can freely operate with sequences whose terms are of different orders. On the other hand the considerations brought forward in § 5 in connexion with Th. I lose none of their importance and can be extended to languages of any order. It is impossible to give an adequate definition of truth for a language in which the arithmetic of the natural numbers can be constructed, if the order of the metalanguage in which the investigations are carried out does not exceed the order of the language investigated (cf. the relevant remarks on p. 253).

Finally, the foregoing considerations show the necessity of revising, to a rather important extent, the Theses A and B given in the conclusions of this work and containing a summary of its chief results:

A. *For every formalized language a formally correct and materially adequate definition of true sentence can be constructed in the metalanguage with the help only of general logical expressions, of expressions of the language itself, and of terms from the morphology of language—but under the condition that the metalanguage possesses a higher order than the language which is the object of investigation.*

B. *If the order of the metalanguage is at most equal to that of the language itself, such a definition cannot be constructed.*

From a comparison of the new formulation of the two theses with the earlier one it will be seen that the range of the results obtained has been essentially enlarged, and at the same time the conditions for their application have been made more precise.

In view of the new formulation of Thesis A the former Thesis C loses its importance. It possesses a certain value only when the investigations are carried out in a metalanguage which has the same order as the language studied and when, having abandoned the construction of a definition of truth, the attempt is made to build up the theory of truth by the axiomatic method. It is easy to see that a theory of truth built up in this way cannot contain an inner contradiction, provided there is freedom from contradiction in the metalanguage of higher order on the basis of which an adequate definition of truth can be set up and in which those theorems which are adopted in the theory of truth as axioms can be derived.[1]

Just as in the conclusion of this work, the Theses A and B can be given a more general formulation by extending them to other semantical concepts:

A′. *The semantics of any formalized language can be established as a part of the morphology of language based on suitably constructed*

---

[1] In particular, the question broached on p. 261 has a positive answer. The same also holds for the problem of infinite inductive definitions mentioned on p. 262, footnote.

definitions, provided, however, that the language in which the mor-phology is carried out has a higher order than the language whose morphology it is.

B'. *It is impossible to establish the semantics of a language in this way if the order of the language of its morphology is at most equal to that of the language itself.*

The Thesis A in its new generalized form is of no little impor-tance for the methodology of the deductive sciences. Its con-sequences run parallel with the important results which Gödel has reported in this field in recent years. The definition of truth allows the consistency of a deductive science to be proved on the basis of a metatheory which is of higher order than the theory itself (cf. pp. 199 and 236). On the other hand, it follows from Gödel's investigations that it is in general impossible to prove the consistency of a theory if the proof is sought on the basis of a metatheory of equal or lower order.[1] Moreover Gödel has given a method for constructing sentences which—assuming the theory concerned to be consistent—cannot be decided in any direction in this theory. All sentences constructed according to Gödel's method possess the property that it can be established whether they are true or false on the basis of the metatheory of higher order having a correct definition of truth. Consequently it is possible to reach a decision regarding these sentences, i.e. they can be either proved or disproved. Moreover a decision can be reached within the science itself, without making use of the concepts and assumptions of the metatheory—provided, of course, that we have previously enriched the language and the logical foundations of the theory in question by the introduction of variables of higher order.[2]

Let us try to explain this somewhat more exactly. Consider an arbitrary deductive science in which the arithmetic of natural numbers can be constructed, and provisionally begin the investigation on the basis of a metatheory of the same order as the theory itself. Gödel's method of constructing undecidable sentences has been outlined implicitly in the proof of Th. I in

---

[1] Cf. Gödel, K. (22), p. 196 (Th. XI).

[2] Cf. Gödel, K. (22), pp. 187 ff., and in particular, p. 191, note 48 *a*.

§ 5 (p. 249 ff.). Everywhere, both in the formulation of the theorem and in its proof, we replace the symbol 'Tr' by the symbol 'Pr' which denotes the class of all provable sentences of the theory under consideration and can be defined in the metatheory (cf. e.g. Def. 17 in § 2). In accordance with the first part of Th. I we can obtain the negation of one of the sentences in condition (α) of convention T of § 3 as a consequence of the definition of the symbol 'Pr' (provided we replace 'Tr' in this convention by 'Pr'). In other words we can construct a sentence x of the science in question which satisfies the following condition:

$$\textit{it is not true that } x \in Pr \textit{ if and only if } p$$

or in equivalent formulation:

(1) $$x \mathbin{\overline{\in}} Pr \textit{ if and only if } p$$

where the symbol 'p' represents the whole sentence x (in fact we may choose the sentence $\bigcup_1^3 (\iota_k . \phi_k)$ constructed in the proof of Th. I as x).

We shall show that the sentence x is actually undecidable and at the same time true. For this purpose we shall pass to a metatheory of higher order; Th. I then obviously remains valid. According to Thesis A we can construct, on the basis of the enriched metatheory, a correct definition of truth concerning all the sentences of the theory studied. If we denote the class of all true sentences by the symbol 'Tr' then—in accordance with convention T—the sentence x which we have constructed will satisfy the following condition:

(2) $$x \in Tr \textit{ if and only if } p;$$

from (1) and (2) we obtain immediately

(3) $$x \mathbin{\overline{\in}} Pr \textit{ if and only if } x \in Tr.$$

Moreover, if we denote the negation of the sentence x by the symbol '$\bar{x}$' we can derive the following theorems from the definition of truth (cf. Ths. 1 and 5 in § 3):

(4) $$\textit{either } x \mathbin{\overline{\in}} Tr \textit{ or } \bar{x} \mathbin{\overline{\in}} Tr;$$

(5) $$\textit{if } x \in Pr, \textit{ then } x \in Tr;$$

(6) $$\textit{if } \bar{x} \in Pr, \textit{ then } \bar{x} \in Tr;$$

From (3) and (5) we infer without difficulty that

(7) $$x \in Tr$$

and that

(8) $$x \,\bar{\in}'\, Pr.$$

In view of (4) and (7) we have $\bar{x} \in Tr$, which together with (6) gives the formula

(9) $$\bar{x} \,\bar{\in}\, Pr.$$

The formulas (8) and (9) together express the fact that x is an undecidable sentence; moreover from (7) it follows that x is a true sentence.

By establishing the truth of the sentence x we have *eo ipso* —by reason of (2)—also proved x itself in the metatheory. Since, moreover, the metatheory can be interpreted in the theory enriched by variables of higher order (cf. p. 184) and since in this interpretation the sentence x, which contains no specific term of the metatheory, is its own correlate, the proof of the sentence x given in the metatheory can automatically be carried over into the theory itself: the sentence x which is undecidable in the original theory becomes a decidable sentence in the enriched theory.

I should like to draw attention here to an analogous result. For every deductive science in which arithmetic is contained it is possible to specify arithmetical notions which, so to speak, belong intuitively to this science, but which cannot be defined on the basis of this science. With the help of methods which are completely analogous to those used in the construction of the definition of truth, it is nevertheless possible to show that these concepts can be so defined provided the science is enriched by the introduction of variables of higher order.[1]

In conclusion it can be affirmed that the definition of truth and, more generally, the establishment of semantics enables us to match some important negative results which have been obtained

---

[1] Cf. my review, 'Über definierbare Mengen reeller Zahlen,' *Annales de la Société Polonaise de Mathématique*, t. ix, année 1930, Kraków, 1931, pp. 206–7 (report on a lecture given on 16 December 1930 at the Lemberg Section of the Polish Mathematical Society); the ideas there sketched were in part developed later in VI.

in the methodology of the deductive sciences with parallel positive results, and thus to fill up in some measure the gaps thereby revealed in the deductive method and in the edifice of deductive knowledge itself.

HISTORICAL NOTES. In the course of the years 1929 to 1935, in which I reached the final definition of the concept of truth and most of the remain. ing results described here, and in the last year of which the whole work appeared for the first time in a universal language, the questions here discussed have been treated several times. In the German language, in addition to my summary, Tarski, A. (76), works by Carnap have appeared in which quite similar ideas were developed (cf. R. Carnap, 'Die Antinomien und die Unvollständigkeit der Mathematik', *Monatshefte f. Math. u. Phys.* vol. 41 (Leipzig, 1934), pp. 263–84 and 'Ein Gültigkeitskriterium für die Sätze der klassischen Mathematik, ibid., vol. 42, pp. 163–90; both articles being supplementations of R. Carnap (10)). The two articles have been incorporated in the English edition of Carnap's book, entitled *Logical Syntax of Language* (London, 1937).

It was to be expected that, in consequence of this lapse of six years, and of the nature of the problem and perhaps also of the language of the original text of my work, errors regarding the historical connexions might occur. And in fact Carnap writes in the second of the above-mentioned articles regarding my investigations that they have been carried out '. . . in connexion with those of Gödel . . .'. It will therefore not be superfluous if I make some remarks in this place about the dependence or independence of my studies.

I may say quite generally that all my methods and results, with the exception of those at places where I have expressly emphasized this— cf. footnotes, pp. 154 and 247—were obtained by me quite independently. The dates given in footnote, p. 154, provide, I believe, sufficient basis for testing this assertion. I may point out further that my article which appeared in French (VI), about which I had already reported in December 1930 (cf. the report in German in A. Tarski (74)) contains precisely those methods of construction which were used there for other purposes but in the present work for the construction of the definition of truth.

I should like to emphasize the independence of my investigations regarding the following points of detail: (1) the general formulation of the problem of defining truth, cf. especially pp. 187–8; (2) the positive solution of the problem, i.e. the definition of the concept of truth for the case where the means available in the metalanguage are sufficiently rich (for logical languages this definition becomes that of the term 'analytical' used by Carnap). Cf. pp. 194 and 236; (3) the method of proving consistency on the basis of the definition of truth, cf. pp. 199 and 236; (4) the axiomatic construction of the metasystem, cf. pp. 173 ff., and in connexion with this (5) the discussions on pp. 184 f. on the interpretation of the metasystem in arithmetic, which already contain the so-called 'method of arithmetizing the metalanguage' which was developed far more completely and quite independently by Gödel. Moreover, I should like to draw attention to results not relating to the concept of truth but to another semantical concept, that of definability reported on p. 276.

In the one place in which my work is connected with the ideas of Gödel—in the negative solution of the problem of the definition of truth for the case where the metalanguage is not richer than the language investigated—I have naturally expressly emphasized this fact (cf. p. 247, footnote); it may be mentioned that the result so reached, which very much completed my work, was the only one subsequently added to the otherwise already finished investigation.

IX

SOME OBSERVATIONS ON THE CONCEPTS OF ω-CONSISTENCY AND ω-COMPLETENESS†

IN an extremely interesting article Gödel[1] introduces the concept of ω-consistency, and constructs an example of a deductive system which is consistent in the usual sense, but is not ω-consistent. In the present article I propose to give another simple example of such a system, together with some general remarks on the concept mentioned as well as on the corresponding concept of ω-completeness.[2]

The symbolical language in which I shall construct this system is closely related to the language of the system P used by Gödel. It is also the result of an exact formalization and simplification, as far as possible, of the language in which the system of *Principia Mathematica* of Whitehead and Russell[3] is constructed. In spite of its great simplicity this language suffices for the expression of every idea which can be formulated in *Principia Mathematica*.[4]

---

[1] See Gödel, K. (22).

[2] Already, in the year 1927, at the Second Conference of the Polish Philosophical Society in Warsaw in the lecture 'Remarks on some notions of the methodology of the deductive sciences' (listed by title in *Ruch Filozoficzny*, vol. 10 (1926–7), p. 96), I had pointed out the importance of these two concepts, and the rule of transfinite induction which is closely related to them and about which more is said in the text, but I had not suggested special names for these concepts. I also communicated the example of a consistent and yet not ω-consistent system which I give in the present article in a slightly altered form. Naturally it is not hereby claimed that I already knew then the results later obtained by Gödel or had even foreseen them. On the contrary, I had personally felt that the publication of the work of Gödel cited above was a most exciting scientific event.

[3] Whitehead, A. N., and Russell, B. A. W. (90).

[4] Cf. articles VI and VIII of the present work, where I have used the same or a very similar language.

---

† BIBLIOGRAPHICAL NOTE. This article first appeared under the title 'Einige Betrachtungen über die Begriffe ω-Widerspruchsfreiheit und der ω-Vollständigkeit', *Monatshefte für Mathematik und Physik*, vol. 40 (1933), pp. 97–112. For the historical information about the results of this article see footnote 2 above.

In our language two kinds of signs occur: constants and variables. Three constants suffice: the negation sign 'N',[1] the sign of implication 'C' and the universal quantifier '$\prod$'. On the other hand we shall operate with infinitely many variables. As variables we shall use the symbols 'x_i'', 'x_{ii}'', 'x_i''', etc., i.e. symbols composed of the sign 'x' and an arbitrary series of small strokes both above and below. The constants and variables are the simplest expressions of the language. By writing down these signs one after another in any number and arrangement we obtain more complicated expressions. Every composite expression, i.e. one containing more than one sign, can profitably be regarded as the result of putting together two (or more) successive simple expressions. For example: '$\prod x_i' x_i' x_i''$' consists of '$\prod$' and '$x_i' x_i' x_i''$' or of '$\prod x_i'$' and '$x_i' x_i''$' or also of '$\prod x_i' x_i'$' and 'x_i'''. An especially important category of expressions will be distinguished below, namely that of (meaningful) sentences.

The constants have their usual interpretation: expressions of the form 'Np', 'Cpq' and '$\prod xp$', in which, in place of 'p' and 'q' any sentences and instead of 'x' any variables, occur, are read: *not p*, *if p then q* and *for every x, p* respectively. Variables with one upper stroke are regarded as names of individuals (objects of 1st order), those with two strokes as naming classes of individuals (objects of 2nd order) and so on. The use of lower strokes with the variable serves the same purpose as the customary use of letters of different shapes in the function of the variables. An expression of the form 'xy', where instead of 'x' any variable occurs and instead of 'y' a variable with one more upper stroke, is read: *x is an element of the class y* or *x has the property y*.

The language here briefly described may be made the object of a special investigation. This task presupposes that we consider it from the standpoint of another language which in contrast to the original one can be called the *matalanguage*, and on the basis of this metalanguage we can establish a special discipline, the so-called *metadiscipline*. In order to avoid unnecessary

---

[1] The negation sign can be dispensed with and serves only to simplify the discussion. Regarding symbolism see IV.

complication we shall not undertake the exact formalization of the metalanguage. It suffices to note that in the metalanguage the following two categories of concepts occur: (1) concepts of a general logical nature, belonging to an arbitrary but sufficiently developed system of mathematical logic (e.g. to *Principia Mathematica*), and in particular those from the domain of the calculus of classes, and the arithmetic of natural numbers;[1] (2) specific concepts of a structural-descriptive character which designate the expressions of the original language, their structural properties, and their mutual structural relations. The following terms are regarded as primitive concepts of the metalanguage: 'expression' or rather 'the class of all expressions', in symbols 'A'; 'negation sign'—'ν'; 'implication sign'— 'ι'; 'universal quantifier'—'π'; 'variable with k strokes below and l above' (or 'the kth variable of the lth order')—'ϕ_k^l'; finally 'expression which consists of the two successive expressions ζ and η' symbolically '$\zeta^\frown\eta$'.[2] It can easily be established that the above concepts suffice for the exact description of every expression of the original language. With their help we can correlate with every such expression a particular individual name in the metalanguage. For example, the expression '$\prod x' x' x''$' is denoted by the individual name '$\pi^\frown\phi_1^1{}^\frown\phi_1^1{}^\frown\phi_1^2$'.[3]

Corresponding to the two categories of concepts of the metalanguage, the system of assumptions which suffice for the deductive establishment of the metadiscipline consists of two kinds of statements: (1) the *axioms of mathematical logic*; and (2) such theses *as suffice to establish the basic properties of the specific primitive concepts of the metalanguage*. It will not perhaps

---

[1] I shall make free use of various symbols which are in common use in works on the theory of sets. For example, formulas of the form '$x, y,... \in Z$' and '$x, y,... \bar{\in} Z$' express the fact that $x, y,...$ do and do not belong to the set Z respectively. '0' denotes the null set and '$\{x, y,..., z\}$' the set which contains only $x, y,..., z$ as elements. By the symbol 'Nt' I denote the set of all natural numbers. $Nt-\{0\}$ is thus the set of all natural numbers distinct from 0.

[2] This last primitive concept can be replaced by the concept 'the nth sign from the beginning of the expression ζ'. The symbol 'A' can be defined by means of the remaining undefined concepts.

[3] By reference to the associative law which the operation $\zeta^\frown\eta$ obeys (see below, Ax. 4) the omission of the brackets in expressions of the form '$(\zeta^\frown\eta)^\frown\delta$', '$((\zeta^\frown\eta)^\frown\delta)^\frown\xi$' cannot give rise to any misunderstandings.

be out of place here to introduce explicitly the full system of axioms of the second kind (although it will not be possible within the limits of the present sketchy account to bring these axioms into practical use).

AXIOM 1. $\nu, \iota, \pi \in A$, where $\nu \neq \iota, \nu \neq \pi$, and $\iota \neq \pi$.

AXIOM 2. If $k, l \in Nt-\{0\}$, then $\phi_k^l \in A$, where $\phi_k^l \neq \nu$, $\phi_k^l \neq \iota$, and $\phi_k^l \neq \pi$; if, moreover, $k_1, l_1 \in Nt-\{0\}$ and $k \neq k_1$ or $l \neq l_1$, then $\phi_k^l \neq \phi_{k_1}^{l_1}$.

AXIOM 3. If $\zeta, \eta \in A$, then $\zeta^\frown\eta \in A$, where $\zeta^\frown\eta \neq \nu$, $\zeta^\frown\eta \neq \iota$, $\zeta^\frown\eta \neq \pi$, and $\zeta^\frown\eta \neq \phi_k^l$ for any $k, l \in Nt-\{0\}$.

AXIOM 4. If $\zeta, \eta, \delta, \xi \in A$, then the formula: $\zeta^\frown\eta = \delta^\frown\xi$ holds if and only if either $\zeta = \delta$ and $\eta = \xi$, or if there is a $\tau \in A$ such that $\zeta = \delta^\frown\tau$ and $\xi = \tau^\frown\eta$, or finally if there is a $\tau \in A$ such that $\delta = \zeta^\frown\tau$ and $\eta = \tau^\frown\xi$.

AXIOM 5. (the principle of induction). Let X be any set which satisfies the following conditions: (1) $\nu, \iota, \pi \in X$; (2) if $k, l \in Nt-\{0\}$, then $\phi_k^l \in X$; (3) if $\zeta, \eta \in X$, then $\zeta^\frown\eta \in X$. Then $A \subseteq X$.

It is not difficult to prove that *the above axiom system is categorical,*[1] *has an interpretation in the arithmetic of the natural numbers and consists of mutually independent statements.*[2]

In order to give the further developments a simpler form a series of symbolic abbreviations is now introduced into the metalanguage.

---

[1] In the sense of Veblen, O. (86).

[2] The intuitive obviousness of the axioms is not uncontested. Certain objections may be raised in connexion with the existential assumptions which are implicit in Axs. 2 and 3 (on the basis of these assumptions it can be proved, among other things, that the set A has the cardinal number $\aleph_0$, and is thus infinite). Without going into the analysis of these difficult questions I should like to say only that the objections mentioned are weakened significantly if it is assumed: (i) that we regard as expressions of the language not concrete inscriptions but whole classes of inscriptions of like form, and correspondingly modify the intuitive sense of the remaining primitive concepts of the metalanguage; (ii) that we interpret the concept of inscription as widely as possible, to include not only manuscript inscriptions, but also all material bodies (or geometrical figures) of definite form. I discuss this point in more detail in my article VIII on the concept of truth, where I have given the above axiom system for the first time. It may be mentioned that other metasciences can, *mutatis mutandis,* be axiomatized after the pattern of this axiom system.

DEFINITION 1. (a) $\zeta \in V$ (ζ is a variable), if there are

$$k, l \in Nt - \{0\}$$

such that $\zeta = \phi_k^l$;

(b) $\zeta = \eta_0 ... \eta_n$ (ζ is an expression consisting of the successive expressions $\eta_0, \eta_1, ..., \eta_n$), if $n \in Nt$, $\eta_0, \eta_1, ..., \eta_n \in A$ and either $n = 0$, $\zeta = \eta_0$ or $n > 0$, $\zeta = (\eta_0 \frown ... \frown \eta_{n-1}) \frown \eta_n$.

DEFINITION 2. (a) $\zeta = \bar{\eta}$ (ζ is the negation of η), if $\eta \in A$, $\zeta = \nu \frown \eta$;

(b) $\zeta = \eta \rightarrow \delta$ (ζ is the implication with η as antecedent and δ as consequent), if η, $\delta \in A$, $\zeta = \iota \frown \eta \frown \delta$;

(c) $\zeta = \eta \vee \delta$ (ζ is the logical sum or disjunction of η and δ), if η, $\delta \in A$, $\zeta = \bar{\eta} \rightarrow \delta$;

(d) $\zeta = \eta_0 \vee ... \vee \eta_n$ (ζ is the logical sum of $\eta_0, \eta_1, ..., \eta_n$), if $n \in Nt$, $\eta_0, \eta_1, ..., \eta_n \in A$ and either $n = 0$, $\zeta = \eta_0$ or $n > 0$, and $\zeta = ((\iota \frown \bar{\eta}_0) \frown ... \frown (\iota \frown \bar{\eta}_{n-1})) \frown \eta_n$;

(e) $\zeta = \eta \wedge \delta$ (ζ is the logical product or conjunction of η and δ), if η, $\delta \in A$, $\zeta = \overline{\bar{\eta} \vee \bar{\delta}}$;

(f) $\zeta = \eta_0 \wedge ... \wedge \eta_n$ (ζ is the logical product of $\eta_0, \eta_1, ..., \eta_n$), if $n \in Nt$, $\eta_0, \eta_1, ..., \eta_n \in A$ and $\zeta = \overline{\bar{\eta}_0 \vee ... \vee \bar{\eta}_n}$;

(g) $\zeta = \eta \Leftrightarrow \delta$ (ζ is the logical equivalence of η and δ), if η, $\delta \in A$, $\zeta = (\eta \rightarrow \delta) \wedge (\delta \rightarrow \eta)$.

DEFINITION 3. (a) $\zeta = \bigcap_k^l \eta$ (ζ is the universal quantification of η with respect to the variable ϕ_k^l), if k, $l \in Nt - \{0\}$, $\eta \in A$ and $\zeta = \pi \frown \phi_k^l \frown \eta$;

(b) $\zeta = \bigcap_{k_0}^{l_0} ... \bigcap_{k_n}^{l_n} \eta$ (ζ is the universal quantification of η with respect to the variables $\phi_{k_0}^{l_0}, \phi_{k_1}^{l_1}, ..., \phi_{k_n}^{l_n}$), if

$$\eta \in A \quad and \quad n, k_0, l_0, k_1, l_1, ..., k_n, l_n \in Nt - \{0\}$$

and $\zeta = ((\pi \frown \phi_{k_0}^{l_0}) \frown ... \frown (\pi \frown \phi_{k_n}^{l_n})) \frown \eta$;

(c) $\zeta = \bigcup_k^l \eta$ (ζ is the existential quantification of η with respect to the variable ϕ_k^l), if k, $l \in Nt - \{0\}$, $\eta \in A$ and $\zeta = \overline{\bigcap_k^l \bar{\eta}}$;

(d) $\zeta = \bigcup_{k_0}^{l_0} ... \bigcup_{k_n}^{l_n} \eta$ (ζ is the existential quantification of η with respect to the variables $\phi_{k_0}^{l_0}, \phi_{k_1}^{l_1}, ..., \phi_{k_n}^{l_n}$), if

$$\eta \in A \quad and \quad k_0, l_0, k_1, l_1, ..., k_n, l_n \in Nt - \{0\}$$

and $\zeta = \overline{\bigcap_{k_0}^{l_0} ... \bigcap_{k_n}^{l_n} \bar{\eta}}$.

DEFINITION 4. (a) $\zeta = \epsilon_{k,l}^{m}$ (ζ is the atomic sentence with the terms ϕ_k^m and ϕ_l^{m+1}), if $k, l, m \in Nt - \{0\}$, $\zeta = \phi_k^m {}^\frown \phi_l^{m+1}$;

(b) $\zeta = \tau_{k,l}^m$ (ζ is the identity between ϕ_k^m and ϕ_l^m), if

$$k, l, m \in Nt - \{0\} \quad \text{and} \quad \zeta = \bigcap_1^{m+1}(\epsilon_{k,1}^m \to \epsilon_{l,1}^m).$$

Among the most important concepts of the metadiscipline are those of *meaningful sentence* (or simply *sentence*) and of the *consequences* of a set of sentences. The definition of the first presents no difficulties:

DEFINITION 5. $\zeta \in S$ (ζ is a meaningful sentence),[1] if ζ belongs to every set X which satisfies the following conditions: (1) $\epsilon_{k,l}^m \in X$ for every $k, l, m \in Nt - \{0\}$; (2) if $\eta \in X$, then $\bar{\eta} \in X$; (3) if $\eta, \delta \in X$, then $\eta \to \delta \in X$; (4) if $k, l \in Nt - \{0\}$ and $\eta \in X$ then $\bigcap_k^l \eta \in X$.

Meaningful sentences are thus the expressions we obtain from elementary sentences $\epsilon_{k,l}^m$ by performing upon them, any number of times and in any order, the three operations of negation, implication, and quantification.

For the definition of the concept of consequence certain preliminary definitions will be useful:

DEFINITION 6. (a) $\zeta \underset{m}{F} \eta$ (ζ occurs at the m-th place in η as a free variable), if $m \in Nt$, $\zeta \in V$ and if η can be represented in the form: $\eta = \delta_0 {}^\frown ... {}^\frown \delta_n$, where $n \in Nt, \delta_0, \delta_1, ..., \delta_n \in \{v, \iota, \pi\} + V, m \leqslant n$, $\delta_m = \zeta$ and there are no numbers $k, l \in Nt$ such that

$$k \leqslant m \leqslant k+l \leqslant n, \qquad \delta_k = \pi, \delta_{k+1} = \zeta \quad \text{and} \quad \delta_k {}^\frown ... {}^\frown \delta_{k+l} \in S.$$

(b) $\zeta \in Fr(\eta)$ (ζ is a free variable of η), if there is an $m \in Nt$ such that $\zeta \underset{m}{F} \eta$.

DEFINITION 7. $\zeta \in L$ (ζ is a logical axiom), if one of the following conditions is satisfied: (1) there exist statements $\eta, \delta, \xi \in S$ such that either $\zeta = (\eta \to \delta) \to ((\delta \to \xi) \to (\eta \to \xi))$ or $\zeta = (\bar{\eta} \to \eta) \to \eta$ or $\zeta = \eta \to (\bar{\eta} \to \xi)$; (2) there is a number $k \in Nt - \{0\}$ and a sentence $\eta \in S$ such that $\zeta = \bigcup_1^{k+1} \bigcap_1^k (\epsilon_{1,1}^k \Leftrightarrow \eta)$, where $\phi_1^{k+1} \bar{\in} Fr(\eta)$; (3) there is a number $k \in Nt - \{0\}$ such that

$$\zeta = \left(\bigcap_1^k (\epsilon_{1,1}^k \Leftrightarrow \epsilon_{1,2}^k)\right) \to \tau_{1,2}^{k+1}.$$

---

[1] The expression 'sentential function' would be more suitable here; the expression 'sentence' should be reserved for those sentential functions which contain no free variables (see below, Def. 6).

The sentences which satisfy condition (1) in the above definition are the *axioms of the sentential calculus* of Łukasiewicz[1] (or rather substitutions of these axioms); the sentences under (2), clearly related to the axiom of reducibility of *Principia Mathematica*, can be called *pseudodefinitions* in accordance with the proposal of S. Leśniewski; finally we recognize the sentences under (3) as the so-called laws of extensionality.

DEFINITION 8. $\zeta Sb_{k|l}^m \eta$ (ζ *can be obtained from* η *by the substitution of the free variable* ϕ_k^m *for the free variable* ϕ_l^m), *if*

$$k, l, m \in Nt - \{0\},$$

and if ζ *and* η *can be represented in the form*

$$\zeta = \delta_0 \frown ... \frown \delta_n, \qquad \eta = \xi_0 \frown ... \frown \xi_n,$$

where (1) $n \in Nt$, *and* $\delta_0, \xi_0, \delta_1, \xi_1, ..., \delta_n, \xi_n \in \{\nu, \iota, \pi\} + V$; (2) *if* $\phi_l^m \underset{i}{F} \eta$, *then* $\phi_k^m \underset{i}{F} \zeta$; (3) *if* $i \in Nt$, $i \leqslant n$ *and the formula* $\phi_l^m \underset{i}{F} \eta$ *does not hold, then* $\delta_i = \xi_i$.

DEFINITION 9. $\zeta \in Cn(X)$ (ζ *is a* consequence *of the set* X *of sentences*), *if* ζ *belongs to every set* Y *which satisfies the following conditions*: (1) $X + L \subseteq Y$; (2) *if* $\delta Sb_{k|l}^m \eta$ *and* $\eta \in Y$, *then* $\delta \in Y$; (3) *if* η, $\eta \to \delta \in Y$, *then* $\delta \in Y$; (4) *if* k, $l \in Nt - \{0\}$, η, $\delta \in S$, $\phi_k^l \in Fr(\eta)$ *and* $\eta \to \delta \in Y$, *then* $\eta \to \bigcap_k^l \delta \in Y$; (5) *if* $k, l \in Nt - \{0\}$, η, $\delta \in S$ *and* $\eta \to \bigcap_k^l \delta \in Y$, *then* $\eta \to \delta \in Y$.

All sentences are thus regarded as consequences of the set X which can be obtained from the sentences of this set and the logical axioms by means of the four operations of substitution, detachment, and the introduction and elimination of the universal quantifier (these rules are commonly called the rules of inference).[2]

From Defs. 5 and 9 various elementary properties of the concept of consequence are easily derived, e.g.:[2]

---

[1] Cf. IV.

[2] Cf. article V of the present work where the terms 'S' and 'Cn' occur as the only undefined concepts and the parts (*a*), (*b*), (*d*) of Th. 1 belong to the axiom system. In connexion with Ths. 2 and 3, cf. III.

THEOREM 1. (a) *If $X \subseteq S$, then $X \subseteq Cn(X) \subseteq S$;*
(b) *if $X \subseteq S$, then $Cn(Cn(X)) = Cn(X)$;*
(c) *if $X \subseteq Y \subseteq S$ or $X \subseteq Cn(Y)$ and $Y \subseteq S$, then*

$$Cn(X) \subseteq Cn(Y);$$

(d) *let $X \subseteq S$; then in order that $\zeta \in Cn(X)$ it is necessary and sufficient that there exists a finite set $Y \subseteq X$ such that $\zeta \in Cn(Y)$;*
(e) *if $X_n \subseteq X_{n+1}$ for every $n \in Nt$, then*

$$Cn\left(\sum_{n=0}^{\infty} X_n\right) = \sum_{n=0}^{\infty} Cn(X_n).$$

THEOREM 2. (a) *If $X \subseteq S$, $\zeta, \eta \in S$, $Fr(\zeta) = 0$ and $\eta \in Cn(X + \{\zeta\})$, then $\zeta \to \eta \in Cn(X)$;*

(b) *if $X \subseteq S$, $\zeta \in S$ and $Fr(\zeta) = 0$, then*

$$Cn(X + \{\zeta\}) . Cn(X + \{\bar{\zeta}\}) = Cn(X);$$

(c) *if $X \subseteq S$ and $\zeta \in S$, then $Cn(X + \{\zeta, \bar{\zeta}\}) = S$;*
(d) *if $\zeta \in S$ and $\eta = \bigcap_{k_0}^{l_0} \dots \bigcap_{k_n}^{l_n} \zeta$, where*

$$n \in Nt \quad and \quad k_0, l_0, k_1, l_1, \dots, k_n, l_n \in Nt - \{0\},$$

then $Cn(\{\zeta\}) = Cn(\{\eta\})$; to every $\zeta \in S$ there thus corresponds a sentence $\eta \in S$ such that $Fr(\eta) = 0$, $Cn(\{\zeta\}) = Cn(\{\eta\})$ and even more generally $Cn(X + \{\zeta\}) = Cn(X + \{\eta\})$ for every set $X \subseteq S$.

A certain difficulty may arise solely in connexion with the proof of Th. 2a. We consider here the set Y of all sentences δ, such that $\zeta \to \delta \in Cn(X)$ and show that this set satisfies the conditions (1)–(5) of the definition, if '$X + \{\zeta\}$' is inserted instead of 'X'. From this it follows immediately that the set Y contains all sentences $\eta \in Cn(X + \{\zeta\})$, which was to be proved. From Th. 2a we easily obtain 2b.

As is well known, a series of fundamental concepts belonging to the methodology of the deductive sciences can be defined by means of the concepts of meaningful sentence and of consequence. In particular this is the case for the concepts of a *deductive system* and of *consistency* and *completeness*:

DEFINITION 10. (a) $X \in \mathfrak{S}$ (X is a deductive or closed system), if $Cn(X) = X \subseteq S$;

(b) $X \in \mathfrak{W}$ (X is a consistent set of sentences), if $X \subseteq S$ and $Cn(X) \neq S$;

(c) $X \in \mathfrak{V}$ (X is a complete set of sentences), if $X \subseteq S$ and at the same time we have, for any $\zeta \in S$, either $\zeta \in Cn(X)$ or $X + \{\zeta\} \bar{\in} \mathfrak{W}$.

Some transformations of Defs. 10 b and c are worth emphasizing, since they are equivalent to the original formulations but in many cases are easier to manipulate:

THEOREM 3. (a) In order that $X \in \mathfrak{W}$, it is necessary and sufficient that $X \subseteq S$ and that for every $\zeta \in S$ we have either $\zeta \bar{\in} Cn(X)$ or $\bar{\zeta} \bar{\in} Cn(X)$.

(b) Let $Fr(\zeta) = 0$; in order that $X + \{\zeta\} \in \mathfrak{W}$, it is necessary and sufficient that $X \subseteq S$, $\zeta \in S$, and $\bar{\zeta} \bar{\in} Cn(X)$;

(c) in order that $X \in \mathfrak{V}$, it is necessary and sufficient that $X \subseteq S$ and that for every $\zeta \in S$ such that $Fr(\zeta) = 0$, we have either $\zeta \in Cn(X)$ or $\bar{\zeta} \in Cn(X)$.

(a) follows from Def. 10 b and Ths. 1 a and c and 2 c; (b) is obtained from (a) with the help of Th. 2 b; finally, to obtain (c) we make use of (b), Def. 10 b and c and Th. 2 d.

In contrast to the ordinary consistency and completeness, the concepts of ω-consistency and ω-completeness cannot be expressed exclusively in terms of 'S' and 'Cn'. In defining these two concepts the following supplementary symbols will be used:

DEFINITION 11. (a) $\zeta = \beta_k^n$, if $k \in Nt - \{0\}$, $n \in Nt$ and either $n = 0$, $\zeta = \bigcap_1^1 \overline{\epsilon_{1,k}^1}$ or

$n > 0$, $\zeta = \bigcup_1^1 \cdots \bigcup_n^1 \bigcap_{n+1}^1 (\epsilon_{n+1,k}^1 \to (\tau_{1,n+1}^1 \vee \cdots \vee \tau_{n,n+1}^1))$;

(b) $\zeta = \gamma_{k,l}$, if $k, l \in Nt - \{0\}$ and $\zeta = \bigcup_1^1 \bigcap_2^1 (\epsilon_{2,l}^1 \leftrightarrow (\epsilon_{2,k}^1 \vee \tau_{1,2}^1))$;

(c) $\zeta = \delta_k$, if $k \in Nt - \{0\}$ and

$$\zeta = \bigcap_1^3 (\bigcap_1^2 \bigcap_2^2 ((\beta_1^0 \vee (\epsilon_{2,1}^2 \wedge \gamma_{2,1})) \to \epsilon_{1,1}^2) \to \epsilon_{k,1}^2).$$

The meaning of the above symbolical expressions is clear: the sentence $\gamma_{k,l}$ states that the set of individuals denoted by the symbol ϕ_l^2 is obtained from the set denoted by ϕ_k^2 by the addition of a single element; the sentence β_k^n means that the set denoted by the symbol ϕ_k^2 contains at most n elements; likewise δ_k means that this set is finite (inductive in the sense of *Principia Mathematica*).

DEFINITION 12. (*a*) $X \in \mathfrak{W}_\omega$ (*X is an ω-consistent set of sentences*), *if* $X \subseteq S$ *and if for every* $\zeta \in S$ *such that* $Fr(\zeta) = \{\phi_1^2\}$[1] *the condition* $\bigcap_1^2(\beta_1^n \to \zeta) \in Cn(X)$ *for every* $n \in Nt$ *always implies the formula* $\overline{\bigcap_1^2(\delta_1 \to \zeta)} \bar{\in} Cn(X)$.

(*b*) $X \in \mathfrak{B}_\omega$ (*X is an ω-complete set of sentences*), *if* $X \subseteq S$ *and if for every* $\zeta \in S$ *such that* $Fr(\zeta) = \{\phi_1^2\}$, *the condition*

$$\bigcap_1^2(\beta_1^n \to \zeta) \in Cn(X)$$

for every $n \in Nt$ *always implies the formula* $\bigcap_1^2(\delta_1 \to \zeta) \in Cn(X)$.

In order to make the content of these concepts more comprehensible, let us consider any property P which sets of individuals can have and which can be formulated in the present language. Let $\xi_n(P)$ (where $n \in Nt$), or $\xi_\omega(P)$ be sentences of the language which assert respectively that every set which contains at most n individuals, or every finite set of individuals possesses the property P. A set X of sentences will be called ω-consistent if it cannot be the case that for some property P all sentences of the infinite sequence $\xi_n(P)$ are consequences of the set X but nevertheless the negation of the sentence $\xi_\omega(P)$ is also a consequence of this set. On the other hand, a set X will be called ω-complete if, whenever all sentences $\xi_n(P)$ belong to the consequence of the set X, the sentence $\xi_\omega(P)$ is also a consequence of the same set.[2]

The following relations hold between the classes $\mathfrak{W}$, $\mathfrak{W}_\omega$, $\mathfrak{B}$, and $\mathfrak{B}_\omega$:

THEOREM 4. (*a*) *if* $X \in \mathfrak{W}_\omega$, *then* $X \in \mathfrak{W}$ ($\mathfrak{W}_\omega \subseteq \mathfrak{W}$);

(*b*) *if* $X \in \mathfrak{W}$ *and* $X \in \mathfrak{B}_\omega$, *then* $X \in \mathfrak{W}_\omega$ ($\mathfrak{W} . \mathfrak{B}_\omega \subseteq \mathfrak{W}_\omega$);

(*c*) *if* $X \in \mathfrak{B}$ *and* $X \in \mathfrak{W}_\omega$, *then* $X \in \mathfrak{B}_\omega$ ($\mathfrak{B} . \mathfrak{W}_\omega \subseteq \mathfrak{B}_\omega$);

(*d*) *if* $X \subseteq S$ *and* $X \bar{\in} \mathfrak{B}_\omega$, *then* $X \in \mathfrak{W}$ *and there is a sentence* $\zeta \in S$ *such that* $X + \{\zeta\} \in \mathfrak{W} - \mathfrak{W}_\omega$.

[1] The formula $Fr(\zeta) = \{\phi_1^2\}$ can here be omitted.

[2] Between the definition of ω-consistency given above and that of Gödel there exist certain purely formal differences. They are accounted for by the fact that, in the formal language here in question, specific symbols to denote the natural numbers do not occur. It is easy to show that with regard to Gödel's system P (and all more comprehensive systems) the two formulations are equivalent.

This theorem is easily obtained from Defs. 10 b and 12 a and b, and Th. 3.

As was mentioned at the beginning, the converse of Th. 4 a does not hold: there exist consistent deductive systems which are not ω-consistent. Th. 4 d shows that such systems can be obtained by extending systems which are not ω-complete. We shall adopt this method and construct a simple example of a system which is not ω-complete—it will be the system T_ω; by extending this system we shall obtain a consistent system T'_ω which, however, is not ω-consistent.

DEFINITION 13. (a) $\zeta = \alpha^n$, if $n \in Nt$, $\zeta = \bigcap_1^2 (\beta_1^n \to \bigcup_1^1 \overline{\epsilon_{1,1}^1})$;
(b) $\zeta = \alpha^\infty$, if $\zeta = \bigcap_1^2 (\delta_1 \to \bigcup_1^1 \overline{\epsilon_{1,1}^1})$.

By the assertion of α^n the existence of at least $n+1$ distinct individuals is obviously established, and in the sentence α^∞ we recognize the *axiom of infinity*, according to which the set of all individuals is infinite.

DEFINITION 14. (a) $\zeta \in T_\omega$, if $\zeta \in Cn\left(\sum_{n=0}^{\infty} \{\alpha^n\} \right)$;

(b) $\zeta \in T'_\omega$, if $\zeta \in Cn\left(\sum_{n=0}^{\infty} \{\alpha^n\} + \overline{\{\alpha^\infty\}} \right)$.

In order to establish the desired properties of the systems T_ω and T'_ω, I next introduce the following lemma:

THEOREM 5. α^{n+1}, $\alpha^\infty \in Cn\left(\sum_{i=0}^{n} \{\alpha^i\} \right)$ *for all* $n \in Nt$.

Sketch of a proof. We make use of the well-known matrix method from investigations on the sentential calculus.[1]

Let S_1 be the set of all sentences without free variables, and S_2 the set of all sentences without universal quantifiers, i.e. those containing only free variables. We put $\psi_1(\zeta) = \zeta$ for $\zeta \in S_1$ and

$$\psi_1(\zeta) = \bigcap_{k_0}^{l_0} \cdots \bigcap_{k_n}^{l_n} \zeta$$

for $\zeta \in S - S_1$, where $\phi_{k_0}^{l_0}$, $\phi_{k_1}^{l_1}$,..., $\phi_{k_n}^{l_n}$ are all free variables of the sentence ζ (the order in which we carry out the operations of quantification is indifferent, provided that it is established once and for all for any sentences $\zeta \in S - S_1$). The function ψ_1 obtained

---

[1] See IV, p. 48.

establishes a one-many correspondence between elements of the set S and the set S_1.

We shall call *designated variables* the first $n+1$ variables of the 1st order (i.e. the variables $\phi_1^1, \phi_2^1,..., \phi_{n+1}^1$), the first 2^{n+1} variables of the 2nd order, the first $2^{2^{n+1}}$ variables of the 3rd order, and so on. We denote by 'S'' the set of all sentences which contain no designated variables, and by 'S'''' the set of all sentences in which, in addition to constants, only designated variables occur. Further, with every sentence $\zeta \in S$ will be correlated a sentence $\psi_2(\zeta) \in S'$ obtained by replacing every variable $\phi_k^1, \phi_k^2,...$ by $\phi_{k+(n+1)}^1, \phi_{k+2^{n+1}}^2....$. In this way, to every sentence $\zeta \in S_1$ there corresponds a sentence $\psi_2(\zeta) \in S_1 . S'$.

Further, we define recursively a certain function ψ_3 for all sentences $\zeta \in S'$. We put, namely, $\psi_3(\epsilon_{k,l}^m) = \epsilon_{k,l}^m$, $\psi_3(\bar\zeta) = \overline{\psi_3(\zeta)}$, $\psi_3(\zeta \to \eta) = \psi_3(\zeta) \to \psi_3(\eta)$, $\psi_3(\bigcap_k^1 \zeta) = \zeta_1 \wedge ... \wedge \zeta_{n+1}$, where

$$\zeta_1 \, Sb_{1|k}^1 \, \zeta, \quad \zeta_2 \, Sb_{2|k}^1 \, \zeta, \quad ..., \quad \zeta_{n+1} \, Sb_{n+1|k}^1 \, \zeta,$$

and in general $\psi_3(\bigcap_k^l \zeta) = \zeta_1 \wedge ... \wedge \zeta_p$, where $\zeta_1, \zeta_2,..., \zeta_p$ are sentences which are obtained from ζ by substituting for the free variables ϕ_k^l all successive designated variables of the lth order, $\phi_1^l, \phi_2^l,..., \phi_p^l$. As will easily be seen, the function ψ_3 maps the set S' onto a part of the set S_2; in particular the sentences $\psi_3(\zeta) \in S_2 . S''$ correspond to the sentences $\zeta \in S_1 . S'$. If, therefore, we put $\psi(\zeta) = \psi_3(\psi_2(\psi_1(\zeta)))$ we obtain a function which correlates a sentence $\psi(\zeta) \in S_2 . S''$ with every sentence $\zeta \in S$.

Finally we consider a function f which assumes only one of the two values 0 and 1. We define this function for all sentences $\zeta \in S_2$ by recursion:

$$f(\epsilon_{k,l}^m) = \left[\frac{l-1}{2^{k-1}}\right] - 2 \cdot \left[\frac{l-1}{2^k}\right],$$

where $[x]$ is by definition the greatest integer not exceeding x, (in other words $f(\epsilon_{k,l}^m)$ is equal to the kth figure in the dyadic expansion of the number $l-1$),

$$f(\bar\zeta) = 1 - f(\zeta),$$

$$f(\zeta \to \eta) = 1 - f(\zeta) . (1 - f(\eta))$$

(in other words $f(\zeta \to \eta) = 0$, if $f(\zeta) = 1$ and $f(\eta) = 0$, and in the remaining cases $f(\zeta \to \eta) = 1$). In particular, all sentences of the form $\psi(\zeta)$, where $\zeta \in S$, can occur as arguments of the function f. Let S^* be the set of sentences ζ for which $f(\psi(\zeta)) = 1$ holds.[1] If 'X' in Def. 9 is replaced by '$\sum\limits_{i=0}^{n} \{\alpha^i\}$', it follows without great difficulty that the set $Y = S^*$ satisfies the conditions (1)–(5) of the above definition; whence we conclude that

$$Cn\Big(\sum_{i=0}^{n} \{\alpha^i\} \Big) \subseteq S^*.$$

On the other hand, we establish immediately that

$$f(\psi(\alpha^{n+1})) = f(\psi(\alpha^\infty)) = 0,$$

whence α^{n+1}, $\alpha^\infty \bar{\in} S^*$. Consequently we have

$$\alpha^{n+1}, \ \alpha^\infty \bar{\in} Cn\Big(\sum_{i=0}^{n} \{\alpha^i\} \Big),$$

which was to be proved.

THEOREM 6. (a) $T_\omega \in \mathfrak{S} - \mathfrak{B}_\omega$; (b) $T'_\omega \in \mathfrak{S} . \mathfrak{W} . - \mathfrak{W}_\omega$.

Proof. From Th. 1a, b, and Def. 14a and b we obtain

(1) $\qquad \sum\limits_{n=0}^{\infty} \{\alpha^n\} \subseteq Cn\Big(\sum\limits_{n=0}^{\infty} \{\alpha^n\} \Big) = T_\omega = Cn(T_\omega) \subseteq S$

and

(2) $\quad \sum\limits_{n=0}^{\infty} \{\alpha^n\} + \{\overline{\alpha^\infty}\} \subseteq Cn\Big(\sum\limits_{n=0}^{\infty} \{\alpha^n\} + \{\overline{\alpha^\infty}\} \Big) = T'_\omega = Cn(T'_\omega) \subseteq S.$

In accordance with Def. 10a the above formulas yield

(3) $\qquad\qquad\qquad T_\omega, \ T'_\omega \in \mathfrak{S}.$

By putting $X_n = \sum\limits_{i=0}^{n} \{\alpha^i\}$ in Th. 1e, we obtain

$$Cn\Big(\sum_{n=0}^{\infty} \{\alpha^n\} \Big) = \sum_{n=0}^{\infty} Cn\Big(\sum_{i=0}^{n} \{\alpha^i\} \Big);$$

since, in view of Th. 5, $\alpha^\infty \bar{\in} Cn\Big(\sum\limits_{i=0}^{n} \{\alpha^i\} \Big)$ for every natural number n, we infer that

(4) $\qquad\qquad\qquad \alpha^\infty \bar{\in} Cn\Big(\sum\limits_{n=0}^{\infty} \{\alpha^n\} \Big).$

---

[1] S^* may be called the set of all sentences which are generally valid in every individual domain of $n+1$ elements.

With the help of (1) and (4), Def. 12b with $X = T_\omega$ and $\zeta = \bigcup_1^1 \overline{\epsilon_{1,1}^1}$ together with 13a, b leads to

$$(5) \qquad\qquad T_\omega \;\bar{\in}\; \mathfrak{B}_\omega.$$

From (4) it follows that $\overline{\overline{\alpha^\infty}} \;\bar{\in}\; Cn\Big(\sum_{n=0}^\infty \{\alpha^n\}\Big)$, whence on the basis of Th. 3b $\Big($with $X = \sum_{n=0}^\infty \{\alpha^n\}$ and $\zeta = \overline{\alpha^\infty}\Big)$ we obtain the formula $\sum_{n=0}^\infty \{\alpha^n\} + \{\overline{\alpha^\infty}\} \in \mathfrak{W}$. Hence from (2) and Def. 10b we conclude that

$$(6) \qquad\qquad T'_\omega \in \mathfrak{W}.$$

Finally, (2), 13a, b, and Def. 12a (for $X = T'_\omega$ and $\zeta = \bigcup_1^1 \overline{\epsilon_{1,1}^1}$) give

$$(7) \qquad\qquad T'_\omega \;\bar{\in}\; \mathfrak{W}_\omega.$$

By virtue of (3) and (5)–(7) the two formulas desired, $T_\omega \in \mathfrak{S} - \mathfrak{B}_\omega$ and $T'_\omega \in \mathfrak{S} \cdot \mathfrak{W} - \mathfrak{W}_\omega$, are proved.

In conclusion, I should like to draw attention to the system T_∞, which is characterized in the following way:

DEFINITION 15. $\zeta \in T_\infty$, if $\zeta \in Cn(\{\alpha^\infty\})$.

In its extension this system coincides almost completely with the system P of Gödel mentioned earlier. Thus, using Gödel's methods of proof, we can extend all the results obtained for the system P to the system T_∞. In particular we can prove

THEOREM 7. If $T_\infty \in \mathfrak{W}$, then $T_\infty \in \mathfrak{S} - \mathfrak{B}_\omega$.

For this theorem even an effective proof is known; we can construct, namely, a sentence ζ_0 (of relatively simple logical structure), which contains ϕ_1^2 as its only free variable, and for which $\bigcap_1^2(\beta_1^n \to \zeta_0) \in Cn(T_\infty)$ holds for every $n \in Nt$, but at the same time $\bigcap_1^2(\delta_1 \to \zeta_0) \;\bar{\in}\; Cn(T_\infty)$. If now we put

$$T'_\infty = Cn(\{\alpha^\infty, \overline{\bigcap_1^2(\delta_1 \to \zeta_0)}\}),$$

we obtain a system with the properties $T'_\infty \in \mathfrak{S} \cdot \mathfrak{W} - \mathfrak{W}_\omega$, analogous to those of T'_ω, it being assumed that $T_\infty \in \mathfrak{W}$.

The system T_∞ stands in close structural relations to T_ω and T'_ω; we have in fact

$$T_\omega = Cn\Big(\sum_{n=0}^{\infty}\{\alpha^n\}\Big), \qquad T'_\omega = Cn\Big(\sum_{n=0}^{\infty}\{\alpha^n\}+\{\overline{\alpha^\infty}\}\Big)$$

and $$T_\infty = Cn(\{\alpha^\infty\}) = Cn\Big(\sum_{n=0}^{\infty}\{\alpha^n\}+\{\alpha^\infty\}\Big),$$

whence by means of Th. 2 b the formula $T_\omega = T'_\omega . T_\infty$ follows. Nevertheless, in spite of the formal relationship, there exists a fundamental essential distinction between these systems. In a sense which is not defined more closely here, not one of the three systems possesses a 'finitistic interpretation', they are not wholly valid in any finite individual domain; thus we have to do with 'infinitistic' systems. Nevertheless, every sentence of the system T_ω or T'_ω, taken by itself, holds in some finite domain of individuals. These systems can be represented as a sum of an infinitely increasing sequence of 'finitistic' systems, e.g. systems $T_n = Cn(\{\alpha^n\})$. On the other hand, the system T_∞ contains sentences, e.g. the axiom of infinity α^∞, which in themselves admit of no 'finitistic interpretation'. In consequence this system is much more comprehensive than T_ω and—in contrast to the latter—is sufficient for the establishment of the arithmetic of the natural numbers and also of analysis in its whole extent. The system T_ω merits the title 'potentially infinitistic', whilst T_∞ on the contrary might be called 'actually infinitistic'.

In connexion with what has been said, the ω-incompleteness of the system T_∞ appears as a significantly more profound phenomenon than the analogous property of the system T_ω. The latter is in a certain degree *ex definitione* ω-incomplete, for while we include in it all statements α^n, we have not included the statement α^∞. For this reason Th. 6 a is also intuitively clear, if not trivial, and its proof is quite simple of its kind. By contrast, the ω-incompleteness of the system T_∞ discovered by Gödel is an entirely unexpected fact for modern mathematics and logic. The establishment of this fact required subtle and ingenious methods of inference. But that is not all. The proof of consistency for the 'potentially infinitistic' system T_ω is already implicit in the proof of Th. 6a (cf. Th. 4a) and so presents no special difficulties. But from Gödel's investigations it follows that the consistency of the system T_∞—and likewise that of the other 'actually infinitistic' systems—cannot be established at all on the basis of the metadiscipline. From this we can understand the presence of the condition '$T_\infty \in \mathfrak{W}$' in the hypothesis of Th. 7.[1]

The facts here brought forward are noteworthy for many reasons. Formerly it could be assumed that the formalized concept of consequence coincides in extension with that concept in everyday language, or at least that all purely structural operations, which unconditionally lead from true statements to true statements, could be reduced without exception to the rules of inference employed in the deductive disciplines. It might also be thought that the consistency of a deductive system is in itself a sufficient guarantee against the appearance of statements in the system which—on account of their mutual structural relations—cannot both be true. Since, however, there are systems which are on the one hand ω-incomplete and on the other consistent, but not ω-consistent, the basis of both of these assumptions is removed. In fact, from the intuitive standpoint the truth of the sentence $\bigcap_1^2(\delta_1 \to \zeta)$ is without doubt a consequence of the truth of all sentences of the form $\bigcap_1^2(\beta_1^n \to \zeta)$, and hence follows the falsehood of the negation of the former, namely $\overline{\bigcap_1^2(\delta_1 \to \zeta)}$. But it is clear that the sentence $\bigcap_1^2(\delta_1 \to \zeta)$ need not at all be a consequence of all sentences $\bigcap_1^2(\beta_1^n \to \zeta)$, and that the sentences $\bigcap_1^2(\beta_1^n \to \zeta)$ and $\overline{\bigcap_1^2(\delta_1 \to \zeta)}$ can occur in one and the same consistent system.

These disadvantages can be avoided if we add to the rules of inference the so-called *rule of infinite induction*, i.e., if we enlarge Def. 9 by this condition: (6) *if* $\eta \in S$, $Fr(\eta) = \{\phi_1^2\}$ *and* $\bigcap_1^2(\beta_1^n \to \eta) \in Y$ *for any* $n \in Nt$, *then* $\bigcap_1^2(\delta_1 \to \eta) \in Y$. Every deductive system will then be ω-complete by definition,

---

[1] Moreover, we find the same difficulties in the case of the problem of the ω-consistency of the system T_ω: from the positive solution of the problem the consistency of the system T_∞ would follow, whilst a negative solution is quite improbable.

and the usual consistency will coincide with ω-consistency. But this would have no fundamental advantage. It may be remarked in passing that such a rule, on account of its 'infinitistic' character, departs significantly from all rules of inference hitherto used, that it cannot easily be brought into harmony with the current view of the deductive method, and finally that the possibility of its practical application in the construction of deductive systems seems to be problematical in the highest degree.[1] But the following is worthy of emphasis: the profound analysis of Gödel's investigations shows that whenever we have undertaken a sharpening of the rules of inference, the facts, for the sake of which this sharpening was felt to be necessary, still persist, although in a more complicated form, and in connexion with sentences of a more complicated logical structure. The formalized concept of consequence will, in extension, never coincide with the ordinary one, the consistency of the system will not prevent the possibility of 'structural falsehood'. However liberally we interpret the concept of the deductive method, its essential feature has always been (at least hitherto) that in the construction of the system and in particular in the formulation of its rules of inference, use is made exclusively of general logical and structurally descriptive concepts. If now we wish to regard as the ideal of deductive science the construction of a system in which all the true statements (of the given language) and only such are contained, then this ideal unfortunately cannot be combined with the above view of the deductive method.[2]

[1] In contrast to all other rules of inference the rule of infinite induction is only applicable if we have succeeded in showing that all sentences of a particular infinite sequence belong to the system constructed. But since in every phase of development of the system only a finite number of sentences is 'effectively' given to us, this fact can only be established by means of metamathematical considerations. The rule of inference mentioned has recently been discussed by D. Hilbert, see Hilbert, D. (29).

[2] The validity of the remarks stated in the last paragraph essentially depends on the decision not to use in metamathematical discussion any devices which cannot be formalized within the framework of the theory of types of *Principia Mathematica*. But as soon as we abandon this decision and allow ourselves to use stronger devices in metamathematical discussion, most of the remarks as originally formulated lose their validity and the whole paragraph requires a thorough revision. Compare, in this connexion, article VIII, Postscript, pp. 274 ff.

X

SOME METHODOLOGICAL INVESTIGATIONS ON THE DEFINABILITY OF CONCEPTS†

Introductory Remarks

In the methodology of the deductive sciences two groups of concepts occur which, although rather remote from one another in content, nevertheless show considerable analogies, if we consider their role in the construction of deductive theories, as well as the inner relations between concepts within each of the two groups themselves. To the first group belong such concepts as *'axiom'*, *'derivable sentence'* (or *'theorem'*), *'rule of inference'*, *'proof'*, to the second—*'primitive (undefined) concept'* (or *'primitive term'*), *'definable concept'*, *'rule of definition'*, *'definition'*. A far-reaching parallelism can be established between the concepts of the two groups: the primitive concepts correspond to the axioms, the defined concepts to the derivable sentences, the process and rules of definition to the process and rules of proof.

Hitherto, investigations in the domain of the methodology of the deductive sciences have chiefly dealt with the concepts of the first group. Nevertheless, in considering the second group of concepts many interesting and important problems force themselves upon us, some of which are quite analogous to those which arise in connexion with the first group. In the present article I propose to discuss two problems in this domain, namely the *problem of the definability and the mutual independence of*

† Bibliographical Note. This article is based upon the text of an address given by the author in 1934 to the Conference for the Unity of Science in Prague. It first appeared in print in Polish under the title 'Z badań metodologicznych nad definiowalnością terminów' in *Przegląd Filozoficzny*, vol. 37 (1934), pp. 438–60. Later it was published (in a somewhat condensed form) with the title 'Einige methodologische Untersuchungen über die Definierbarkeit der Begriffe' in *Erkenntnis*, vol. 5 (1935), pp. 80–100. The present version is a translation of the German one supplemented by some passages translated from the Polish original. Further historical references are contained in footnote 1, p. 297.

concepts and the *problem of the completeness of concepts* of an arbitrary deductive theory.[1]

We shall be concerned here only with those deductive theories which are based upon a sufficiently developed system of mathematical logic. Problems concerning concepts of logic itself will not be considered. In order to make the discussion more specific we shall have in mind, as the logical basis of the deductive theories discussed, the system of *Principia Mathematica* of A. N. Whitehead and B. Russell,[2] modified in certain respects. These modifications are the following: (1) it is assumed that the ramified theory of types is replaced by the simple theory, and that the axiom of reducibility is set aside; (2) in the system of logical axioms we include the axiom of extensionality (for all logical types), and consequently we identify class-signs with one-place predicates, relation-signs with two-place predicates, and so on;[3] (3) we use no defined operators such, for example, as '$\hat{x}(\phi x)$' or '$(\imath x)(\phi x)$'; (4) as sentences, and in particular as provable sentences, of the system only those sentential functions are considered which contain no real (free) variables.

Deductive theories based upon a system of logic can be roughly characterized as follows: In addition to the logical constants and variables, such a theory contains new terms, the so-called *extra-logical constants* or the *specific terms* of the theory in question; to each of these terms a definite logical type is assigned. Among the *sentential functions* of the theory we include the sentential functions of logic, as well as all expressions which are obtainable from them by replacing real variables (not necessarily all) by extra-logical constants of the corresponding

---

[1] The results concerning the first problem were communicated by me to the session of the Warsaw Section of the Polish Mathematical Society on 17 December 1926; cf. A. Tarski et A. Lindenbaum (85). The discussion and results relating to the second problem were first communicated by me to the session of the Logical Section of the Warsaw Philosophical Society on 15 June 1932.

[2] Cf. (90). We have taken this system as a basis for our discussion chiefly because it is more developed than other systems of logic and because it is widely known. It should be mentioned, however, that *Principia Mathematica* is not very suitable for methodological investigations because the formalization of this system does not satisfy the present-day requirements of methodology.

[3] In connexion with these modifications cf. Carnap, R. (8), pp. 21–22, and (10), pp. 98–101.

type. Sentential functions in which no real variables occur are called *sentences*. Among sentences *logically provable sentences* are distinguished; these will be the sentences which are provable in logic, as well as those which can be obtained from them by legitimate substitution of extra-logical constants in the place of variables (it would be superfluous to specify here in what a legitimate substitution consists).

Let X be any set of sentences and y any sentence of the given theory. We shall say that y is a *consequence of the set X* of sentences, or that y *is derivable from the set X*, if either y is logically provable or if there is a logically provable implication having a sentence belonging to X, or a conjunction of such sentences, as its antecedent, and having y as its consequent.

As is well known, various other concepts belonging to the methodology of the deductive sciences can be defined in terms of the concept of derivability, such, for example, as the concepts of a *deductive* (or *closed*) *system*, the (*logical*) *equivalence* of two sets of sentences, the *axiom system* of a set of sentences, the *axiomatizability*, the *independence*, the *consistency* and the *completeness* of a set of sentences. We shall assume that all these concepts are understood.[1] In accordance with customary usage we shall identify the set which consists of a single sentence with the sentence itself; thus we shall speak of the consequences of a sentence or of the consistency of a sentence.

§ 1. The Problem of the Definability and of the Mutual Independence of Concepts

The problems to be discussed in this article concern the specific terms of any deductive theory.[2]

---

[1] Cf. for example articles III and V.

[2] It has been my aim to keep this article short and to make it generally intelligible. I have not attempted, therefore, to give my exposition a strictly deductive character or to present it in a precise form. I have, for example, used expressions in quotation marks, I have often contented myself with a schematic representation, rather than an exact description, of expressions, etc. No essential difficulties would be encountered in modifying this presentation so as to meet at least those standards of precision at which I have aimed in some of my other methodological writings (e.g. IX).

Let 'a' be some extra-logical constant and B any set of such constants. Every sentence of the form:

(I) $(x) : x = a . \equiv . \phi(x; b', b'', ...),$

where '$\phi(x; b', b'', ...)$' stands for any sentential function which contains 'x' as the only real variable, and in which no extra-logical constants other than 'b''', 'b''',... of the set B occur, will be called a *possible definition* or simply a *definition of the term* '*a*' *by means of the terms of the set B*. We shall say that the term '*a*' *is definable by means of the terms of the set B on the basis of the set X of sentences*, if '*a*' and all terms of B occur in the sentences of the set X and if at the same time at least one possible definition of the term '*a*' by means of the terms of B is derivable from the sentences of X.[1]

With the help of the concept of definability we can explain the meaning of various other methodological concepts which are exactly analogous to those defined in terms of the concept of derivability. For example, the concepts of the *equivalence of two sets of terms*, of the *system of undefined terms for a given set of terms*, etc. It is clear that all these concepts must be relativized to a set X of sentences.[1] In particular, B will be called an *independent set of terms* or a *set of mutually independent terms, with respect to a set X of sentences*, if no term of the set B is definable by means of the remaining terms of this set on the basis of X.

Attention will be confined here to the case where the set X of sentences, and consequently the set B of terms, is finite; the extension of the results reached to any axiomatizable set of sentences presents no difficulties.

Many years ago A. Padoa sketched a method which enables

----

[1] It is not difficult to see why the concept of definability, as well as all derived concepts, must be related to a set of sentences: there is no sense in discussing whether a term can be defined by means of other terms before the meaning of those terms has been established, and on the basis of a deductive theory we can establish the meaning of a term which has not previously been defined only by describing the sentences in which the term occurs and which we accept as true.

us to establish, in particular cases, the undefinability of a term by means of other terms.[1] In order, by this method, to show that a term 'a' cannot be defined by means of the terms of a set B on the basis of a set X of sentences, it suffices to give two interpretations of all extra-logical constants which occur in the sentences of X, such that (1) in both interpretations all sentences of the set X are satisfied and (2) in both interpretations all terms of the set B are given the same sense, but (3) the sense of the term 'a' undergoes a change. We shall here present some results which provide a theoretical justification for the method of Padoa and, apart from this, appear to throw an interesting light on the problem of definability.[2]

We consider any finite set X of sentences, an extra-logical constant 'a' which occurs in the sentences of X, and a set B of such constants which does not, however, include 'a' among its members. The conjunction of all sentences of the set X (in an arbitrary order) we represent in the schematic form '$\psi(a; b', b'',...; c', c'',...)$'; here '$b'$', '$b''$',... are terms of the set B, and 'c'', 'c''',... are extra-logical constants which occur in the sentences of X but are distinct from 'a', 'b'', 'b''',... (if there are no such constants the formulation and proof of the theorems given below are somewhat simplified). It sometimes happens that in all sentences of the set X variables 'x', 'y'', 'y''', 'z'', 'z''',... are substituted for (some or all) extra-logical constants, it being assumed that none of these variables become apparent (bound) variables as a result of the substitution.[3] In order to represent schematically the conjunction of all sentential functions formed in this way, the same substitution is made in the expression '$\psi(a; b', b'',...; c', c'',...)$'. On the basis of these assumptions and conventions the following theorem can now be proved:

[1] Cf. Padoa, A. (57), pp. 249–56 and (58), pp. 85–91.

[2] An essential advance in this domain has recently been made by E. W. Beth, who has extended Ths. 2 and 3 of this article to a much wider class of deductive theories, in fact, to all theories based upon the lower functional calculus (which is a much weaker logical system than *Principia Mathematica*), and has thus shown that Padoa's method can be applied to all theories of this class. Cf. Beth (6a).

[3] A similar assumption is tacitly made in all the analogous situations which occur in this article.

THEOREM 1. *In order that the term 'a' should be definable by means of the terms of the set B on the basis of the set X of sentences, it is necessary and sufficient that the formula*

(II) $(x) : x = a . \equiv . (\exists z', z'',...) . \psi(x; b', b'',...; z', z'',...)$

should be derivable from the sentences of X.

Proof. It follows directly from the definition of the concept of definability that the condition of the theorem is sufficient; for (II) is a formula of the type (I) and thus a possible definition of the term 'a' by means of the terms of B.

It remains to be shown that this condition is at the same time necessary. Let us assume that the term 'a' can be defined by means of the terms of B on the basis of the set X, and thus that at least one formula of type (I) can be derived from the sentences of X. This formula is at the same time a consequence of the conjunction of all sentences of X, i.e. of the sentence

$$\text{'}\psi(a; b', b'',...; c', c'',...)\text{'}.$$

In accordance with the definition of the concept of consequence we conclude from this that the formula

(1) $\psi(a; b', b'',..., c', c'',...) . \supset : (x) : x = a . \equiv . \phi(x; b', b'',...)$

is logically provable. Since the formula (1) is a logically provable sentence with extra-logical constants, it must be obtainable by substitution from a logically provable sentence without such constants, and in fact from the formula

(2) $(x', y', y'',..., z', z'',...) : . \psi(x'; y', y'',...; z', z'',...) . \supset :$
$$: (x) : x = x' . \equiv . \phi(x'; y', y'',...).$$

By means of easy transformations we obtain from (2) further logically provable formulas:

(3) $(x, x', y', y'',..., z', z'',...) : . \psi(x'; y', y'',...; z', z'',...) . \supset :$
$$: x = x' . \equiv . \phi(x; y', y'',...);$$

(4) $(x, y', y'',..., z', z'',...) : . \psi(x; y', y'',...; z', z'',...) . \supset :$
$$: x = x . \equiv . \phi(x; y', y'',...);$$

(5) $(x, y', y'',..., z', z'',...) : \psi(x; y', y'',...; z', z'',...) . \supset .$
$$. \phi(x; y', y'',...);$$

(6) $(x, y', y'',...) : (\exists z', z'',...) . \psi(x; y', y'',...; z', z'',...) . \supset .$
$$. \phi(x; y', y'',...).$$

From (3) and (6) we derive

(7) $(x, x', y', y'', ..., z', z'', ...) : . \psi(x'; y', y'', ...; z', z'', ...) :$

$: (\exists z', z'', ...) . \psi(x; y', y'', ...; z', z'', ...) : \supset . x = x',$

whence

(8) $(x, x', y', y'', ..., z', z'', ...) : . \psi(x'; y', y'', ...; z', z'', ...) . \supset :$

$: (\exists z', z'', ...) . \psi(x; y', y'', ...; z', z'', ...) . \supset . x = x'.$

On the other hand the logical definition of the identity sign yields

(9) $(x, x', y', y'', ..., z', z'', ...) : . \psi(x'; y', y'', ...; z', z'', ...) . \supset :$

$: x = x' . \supset . \psi(x; y', y'', ...; z', z'', ...)$

whence

(10) $(x, x', y', y'', ..., z', z'', ...) : . \psi(x'; y', y'', ...; z', z'', ...) . \supset :$

$: x = x' . \supset . (\exists z', z'', ...) . \psi(x; y', y'', ...; z', z'', ...).$

From (8) and (10) we obtain step by step the formulas:

(11) $(x, x', y', y'', ..., z', z'', ...) : . \psi(x'; y', y'', ...; z', z'', ...) . \supset :$

$: x = x' . \equiv . (\exists z', z'', ...) . \psi(x; y', y'', ...; z', z'', ...);$

(12) $(x', y', y'', ..., z', z'', ...) : . \psi(x'; y', y'', ...; z', z'', ...) . \supset :$

$: (x) : x = x' . \equiv . (\exists z', z'', ...) . \psi(x; y', y'', ...; z', z'', ...);$

(13) $\psi(a; b', b'', ...; c', c'', ...) . \supset : (x) : x = a .$

$. \equiv . (\exists z', z'', ...) . \psi(x; b', b'', ...; z', z'', ...).$

Since (13) is logically provable, the consequent of this implication, i.e. formula (II), is derivable from the antecedent '$\psi(a; b', b'', ...; c', c'', ...)$', and hence from the sentences of the set X. Consequently the condition of the theorem is not only sufficient, but also necessary, which was to be proved.

The importance of the above theorem lies in the following: if a term can be defined at all by means of other terms, then our theorem enables us to construct 'effectively' a possible definition of this term. Moreover, the theorem shows that the problem of the definability of terms is reducible without remainder to the problem of the derivability of sentences.

The next theorem is essentially a transformation of Th. 1.

THEOREM 2. *In order that the term 'a' should be definable by means of the terms of the set B on the basis of the set X of sentences, it is necessary and sufficient that the formula*

(III) $(x', x'', y', y'',..., z', z'',..., t', t'',...) : \psi(x'; y', y'',...; z', z'',...).$
$$.\psi(x''; y', y'',...; t', t'',...). \supset .x' = x''$$

should be logically provable.

Proof. If the term 'a' is definable with the help of the terms of the set B, then, by Th. 1, sentence (II) is a consequence of the set X of sentences and hence is also a consequence of the conjunction of all sentences of this set, i.e. of the sentence

$$\psi(a; b', b'',...; c', c'',...).$$

Therefore we conclude that the sentence

(1) $\psi(a; b', b'',...; c', c'',...). \supset : (x) : x = a. \equiv .(\exists z', z'',...).$
$$\psi(x; b', b'',...; z', z'',...)$$

is logically provable.

Thus the sentence

(2) $(x'', y', y'',..., t', t'',...) :. \psi(x''; y', y'',...; t', t'',...).$
$$\supset : (x) : x = x''. \equiv .(\exists z', z'',...).\psi(x; y', y'',...; z', z'',...),$$

of which (1) is a substitution, must also be logically provable.

From (2), by means of simple transformations, we successively derive some further logically provable sentences:

(3) $(x, x'', y', y'',..., t', t'',...) :. \psi(x''; y', y'',...; t', t'',...).$
$$\supset : x = x''. \equiv .(\exists z', z'',...).\psi(x; y', y'',...; z', z'',...);$$

(4) $(x, x'', y', y'',..., t', t'',...) :. \psi(x''; y', y'',...; t', t'',...).$
$$\supset : (\exists z', z'',...).\psi(x; y', y'',...; z', z'',...). \supset .x = x'';$$

(5) $(x, x'', y', y'',..., t', t'',...) :. (\exists z', z'',...).$
$$.\psi(x; y', y'',...; z', z'',...) : \psi(x''; y', y'',...; t', t'',...) :$$
$$: \supset .x = x'';$$

(6) $(x, x'', y', y'',..., z', z'',..., t', t'',...) : \psi(x; y', y'',...; z', z'',...).$
$$\psi(x''; y', y'',...; t', t'',...) : \supset .x = x'';$$

(7) $(x', x'', y', y'',..., z', z'',..., t', t'',...) : \psi(x'; y', y'',...; z', z'',...).$
$$\psi(x''; y', y'',...; t', t'',...). \supset .x' = x''.$$

Sentence (7) coincides with (III). Hence we have shown that the condition of the theorem is necessary for the term 'a' to be definable by means of the terms of the set B. We have still to prove that the condition is sufficient.

For this purpose we reverse the reasoning sketched above. We assume that the sentence (III), i.e. the sentence (7), is logically provable. By successive transformations we show that sentences (6), (5), and (4) are logically provable. In addition, it follows from the definition of the symbol '$=$' that the sentence

(8) $(x, x'', y', y'',..., t', t'',...) : . \psi(x''; y', y'',...; t', t'',...).$

$. \supset : x = x'' . \supset . \psi(x; y', y'',...; t', t'',...)$

is also logically provable, and this implies

(9) $(x, x'', y', y'',..., t', t'',...) : . \psi(x''; y', y'',...; t', t'',...).$

$. \supset : x = x'' . \supset . (\exists z', z'',...) . \psi(x; y', y'',...; z', z'',...).$

From (4) and (9) we immediately obtain (3), and then (2) and (1).

Since (1) is logically provable, the consequent (II) of the implication (1) must be a consequence of '$\psi(a; b', b'',...; c', c'',...)$', the sentence which is the antecedent of (1), and hence (II) must be a consequence of the set X of sentences. Hence, by Th. 1, the term 'a' is definable by means of the terms of the set B on the basis of the set X. Thus the theorem has been proved in both directions.

From the above theorem we obtain as an immediate corollary

THEOREM 3. *In order that the term 'a' should not be definable by means of the terms of the set B on the basis of the set X of sentences, it is necessary and sufficient that the formula*

(IV) $(\exists x', x'', y', y'',..., z', z'',...., t', t'',...).$

$. \psi(x'; y', y'',...; z', z'',...) . \psi(x''; y', y'',..., t', t'',...) . x' \neq x''$

should be consistent.

In order to show this, it suffices to note that formula (IV) is equivalent to the negation of formula (III) of Th. 2, and then to apply the general methodological principle according to which a formula is not logically provable if and only if its negation is consistent.[1]

---

[1] Cf. for example Th. 13* in article III; we substitute in this theorem '0' in the place of 'X' keeping in mind the remark which follows Th. 4 in that article.

Th. 3 constitutes the proper theoretical foundation for the method of Padoa. In fact, in order to establish, on the basis of this theorem, that the term 'a' is not definable by means of the terms of the set B, we can apply the following procedure. We consider a deductive system Y, the consistency of which has previously been established or is assumed. We then seek certain constants (not necessarily extra-logical) '$\bar{a}$', '$\bar{\bar{a}}$', '$\bar{b}''$', '$\bar{b}'''$',..., '$\bar{c}''$', '$\bar{c}'''$',..., '$\bar{\bar{c}}''$', '$\bar{\bar{c}}'''$',..., which occur in the sentences of Y and satisfy the following conditions: (1) if in all sentences of the set X we replace the symbols 'a', 'b''', 'b'''',..., 'c''', 'c'''',... by '$\bar{a}$', '$\bar{b}''$', '$\bar{b}'''$',..., '$\bar{c}''$', '$\bar{c}'''$',... respectively, then the resulting sentences belong to the system Y; (2) the same holds also if we substitute the symbols '$\bar{b}''$', '$\bar{b}'''$',... exactly as before for the terms 'b''', 'b'''',... of B, but replace the remaining terms 'a', 'c''', 'c'''',... by '$\bar{\bar{a}}$', '$\bar{\bar{c}}''$', '$\bar{\bar{c}}'''$',..., respectively; (3) the system Y contains the formula '$\bar{a} \neq \bar{\bar{a}}$'. From the foregoing conditions it results that the following conjunction belongs to the system Y:

$$\psi(\bar{a}; \bar{b}', \bar{b}'',...; \bar{c}', \bar{c}'',...).\psi(\bar{\bar{a}}; \bar{b}', \bar{b}'',...; \bar{\bar{c}}', \bar{\bar{c}}'',...).\bar{a} \neq \bar{\bar{a}}.$$

Hence it easily follows that formula (IV) of Th. 3 which is an immediate consequence of this conjunction also belongs to the system Y. According to a general methodological principle, every part of a consistent set of sentences is consistent; in particular, the formula (IV) as an element of the consistent system Y must be consistent. Thus the condition of Th. 3 is satisfied and the term 'a' is not definable exclusively by means of the terms of B. In the above procedure we easily recognize the method of Padoa, in an extended form, inasmuch as provision is made for the possible occurrence of terms which are identical neither with 'a' nor with any of the terms of B.

The method of Padoa, or Th. 3, is often successfully applied in establishing the mutual independence of the terms of an arbitrary set B. The method must obviously be applied as many times as there are elements in the given set of terms. The problem of independence has a certain practical importance in those cases when the set B involved is the system of primitive terms of a deductive theory. For, should the primitive terms prove to be not mutually independent, then the unnecessary (i.e. definable) terms can be eliminated and this sometimes makes possible a simplification of the axiom system. It is perhaps worth noting that under the terminology accepted here the situation can be described in precise terms as follows: If the set B is divided into two disjoint sets C and D, and every term of the set D is definable by means of the terms of the set C on the basis of the axiom system X, then the system X can be replaced by an equivalent axiom system which again consists of two parts Y and Z such that the sentences of the set Y, so-called *proper axioms*, contain extra-logical constants of the set C exclusively, whilst every sentence of the set Z is a possible definition of a certain term of the set D by means of the terms of the set C.

Let us now give an example. Consider the system of n-dimensional Euclidean geometry (where n is any natural number). Let us assume that the symbols 'a' and 'b' denote certain relations between points, 'a' denoting the four-termed relation of congruence of two pairs of points, and 'b' the three-termed relation of lying between. The expressions '$a(x,y,z,t)$' and '$b(x,y,z)$' are thus read: '*the point x is the same distance from the point y as the point z is from the point t*' and '*the point y lies between the points x and z*', respectively. The term 'a' may be called the *metrical primitive term* and 'b' the *descriptive primitive term* of geometry. As is well known, the set consisting simply of the two terms 'a' and 'b' can be taken as the system of primitive terms of geometry. Accordingly an axiom system X of geometry can be constructed in which these two terms constitute the only extra-logical constants. The question now arises whether the terms 'a' and 'b' are mutually independent with respect to the set X of sentences. The question is to be answered in the negative. This is so despite the fact that, by Padoa's method, the term 'a' can be shown not to be definable by means of the term 'b'; in fact, assuming the consistency of geometry it can be proved that formula (IV) of Th. 3 is in this case consistent. (It is interesting that in essence this proof appears in certain investigations on the foundations of geometry which apparently serve a directly opposite goal, in fact the establishment of metrical geometry as a part of descriptive.)[1] Nonetheless it is found that, if the system of geometry is at least two-dimensional, then the term 'b' is definable by means of the term 'a' on the basis of the axiom system X. The definition can be constructed according to the scheme (II) of Th. 1 (although other much simpler definitions of the term 'b' are also known). In accordance with this, the system X can be replaced by another equivalent axiom system, in the axioms of which the term 'a' occurs as the only extra-logical constant, apart from a single 'improper' axiom, namely a possible definition of the term 'b'. As A. Lindenbaum has proved[2] (with the help of the axiom of choice), it is only in the case of one-dimensional geometry, i.e. the geometry of the straight line, that the two terms 'a' and 'b' are independent of one another.

The following is also worth emphasizing. As a starting-point for this discussion we have used a special form of definition, namely the scheme (I). Nevertheless the results obtained remain valid for other known forms of definition. In order to illustrate this by an example, let us suppose that 'a' is a two-place predicate. As a possible definition of the term 'a' by means of the terms of the set B we can consider, for example, every sentence of the form

(Ia) $(u, v) : a(u, v). \equiv . \phi(u, v; b', b'', ...)$

where '$\phi(u, v; b', b'', ...)$' stands for any sentential function which contains no real variable apart from 'u' and 'v', and no specific terms apart from the terms 'b'', 'b''',... of the set B.[3] It can

---

[1] Cf. for example Veblen, O. (86).

[2] Cf. A. Tarski and A. Lindenbaum (85).

[3] The form of definition (Ia) is much more special than (I), but it has the advantage that the consistency of the definition is already ensured by its structure. In order to be sure that within a deductive theory definitions of type (I) will lead to no contradiction, a stipulation is usually made not to introduce such a definition without first having proved the sentence:

(A) $(\exists x). \phi(x; b', b'', ...) : (x', x'') : \phi(x'; b', b'', ...) . \phi(x''; b', b'', ...) . \supset . x' = x''$.

If this sentence is proved, then we can even replace (I) by the simpler equivalent formula

(B) $\phi(a; b', b'', ...)$,

for the two conditions for a correct definition—consistency and re-translatability—are guaranteed by sentence (A).

now easily be shown that the definition of the concept of definability, relativized to sentences of the form (I*a*), is equivalent to the original definition (although naturally only with respect to two-termed predicates). From this it follows immediately that Ths. 1–3 also remain valid on the basis of the new definition. There is no difficulty in transforming Th. 1 in such a way that a formula of type (I*a*) occurs in it instead of formula (II). It suffices, in fact, to replace (II) by

(II*a*) $(u, v) :. a(u, v) . \equiv : (x, z', z'', ...) : \psi(x; b', b'', ...; z', z'', ...) .$

$$\supset . x(u, v).$$

§ 2. The Problem of the Completeness of Concepts

The problem of the definability and the mutual independence of concepts is an exact correlate of the problem of the derivability and the mutual independence of sentences. The problem of the *completeness of concepts*, to which we now turn, is also closely related to certain problems concerning systems of sentences, and indeed to the problems of completeness and categoricity, although the analogy does not extend so far as in the previous case.

In order to make the problem of completeness precise, we first introduce an auxiliary concept. Let X and Y be any two sets of sentences. We shall say that the set Y is *essentially richer than the set X with respect to specific terms*, if (1) every sentence of the set X also belongs to the set Y (and therefore every specific term of X also occurs in the sentences of Y) and if (2) in the sentences of Y there occur specific terms which are absent from the sentences of X and cannot be defined, even on the basis of the set Y, exclusively by means of those terms which occur in X.

If now there existed a set X of sentences for which it is impossible to construct an essentially richer set Y of sentences with respect to specific terms, then we should be inclined to say that the set X is complete with respect to its specific terms. It appears, however, that there are in general no such complete sets of sentences, apart from some trivial cases. In fact, for every set X of sentences which is consistent and which does not contain all extra-logical constants, we can construct a set Y which is essentially richer in specific terms. For this purpose it suffices to add to the set X an arbitrary logically provable sentence which contains an extra-logical constant not occurring in the sentences of X.[1] For this reason the proposed definition of completeness requires modification. In this the concept of *categoricity* will play an essential part.

As is well known, a set of sentences is called categorical if any two interpretations (realizations) of this set are isomorphic.[2] In order to formulate this more exactly, let us suppose that symbolic expressions of the form '$x' \underset{R}{\widetilde{}} x'''$' (in words: '*the relation R maps x' onto x'''*') have been introduced into the system of logic. The variable 'R' always denotes a two-termed relation between individuals, but 'x''' and 'x'''' may be of any logical type, so long as they are both of the same type. The precise sense of the expression '$x' \underset{R}{\widetilde{}} x'''$' will depend upon the logical type of the variables 'x''' and 'x''''. We shall explain it for only a few cases. For example, if x' and x'' are individuals, then '$x' \underset{R}{\widetilde{}} x'''$' has the same meaning as '$x'Rx'''$' (i.e. 'x' *stands in the relation R to x'''*'). If 'x''' and 'x'''' denote classes of individuals, then the expression has the same sense as the sentential function

$$(u') : u' \in x' . \supset . (\exists u'') . u'' \in x'' . u' \underset{R}{\widetilde{}} u'' :. (u'') : u'' \in x'' .$$

$$\supset . (\exists u') . u' \in x' . u' \underset{R}{\widetilde{}} u''$$

(this involves no circularity since 'u''' and 'u'''' are individual variables, so that the sense of the expression '$u' \underset{R}{\widetilde{}} u'''$' has already been determined). In an exactly analogous manner the expression '$x' \underset{R}{\widetilde{}} x'''$' can be defined for class terms of higher type. Consider again the case when 'x''' and 'x'''' are two-termed predicates with individual variables as arguments (thus denoting

---

[1] In connexion with what is said on p. 299, note 1, it is obvious that the new extra-logical constant cannot be defined by means of the terms of X. In fact the only sentence of Y in which this constant occurs is logically provable, and thus true independently of the meaning of the specific terms contained in it. It can thus be asserted that the meaning of the constant in question is not at all determined by the set Y of sentences.

[2] Cf. for example Veblen, O. (86).

binary relations between individuals); then the formula '$x' \underset{R}{\simeq} x''$' has the same significance as

$$(u',v') : x'(u',v') . \supset . (\exists u'',v'') . x''(u'',v'') . u' \underset{R}{\simeq} u'' . v' \underset{R}{\simeq} v'' :.$$

$$(u'',v'') : x''(u'',v'') . \supset . (\exists u',v') . x'(u',v') . u' \underset{R}{\simeq} u'' . v' \underset{R}{\simeq} v''.$$

The above examples will probably suffice to explain the sense to be given to the expression discussed with respect to variables of arbitrary logical type. Moreover, recalling that, in the notation of *Principia Mathematica*, V is the class of all individuals and $1 \to 1$ the class of all one-one relations, let us say that the formula

$$R \frac{x', y', z',...}{x'', y'', z'',...}$$

is to have the same meaning as the conjunction

$$R \in 1 \to 1 . V \underset{R}{\simeq} V . x' \underset{R}{\simeq} x'' . y' \underset{R}{\simeq} y'' . z' \underset{R}{\simeq} z'' ...$$

(in words: '*R is a one-one mapping of the class of all individuals onto itself, by which x', y', z',... are mapped onto x'', y'', z'',..., respectively*').[1]

Consider now any finite set Y of sentences; 'a', 'b', 'c',... are all specific terms which occur in the sentences of Y, and '$\psi(a,b,c,...)$' is the conjunction of all these sentences. The set Y is called *categorical* if the formula

(V) $(x',x'',y',y'',z',z'',...) : \psi(x',y',z',...) . \psi(x'',y'',z'',...).$

$$\supset . (\exists R) . R \frac{x',y',z',...}{x'',y'',z'',...}$$

is logically provable.[2]

---

[1] From the standpoint of the theory of types the mode of introduction of the symbolic expressions '$x' \underset{R}{\simeq} x''$' and '$R \frac{x', y', z',...}{x'', y'', z'',...}$' is not quite free from objections. It should therefore be emphasized that these symbols are only to be regarded as schemata of expressions. In a precise presentation of this discussion the use of such symbols becomes unnecessary and all difficulties connected with them disappear.

[2] We use the word 'categorical' in a different, somewhat stronger sense than is customary: usually it is required of the relation R which occurs in (V) only that it maps x', y', z',... onto x'', y'', z'',... respectively, but not that it maps the class of all individuals onto itself. The sets of sentences which are cate-

On various grounds, which will not be entered into further, great importance is ascribed to categoricity. A non-categorical set of sentences (especially if it is used as an axiom system of a deductive theory) does not give the impression of a closed and organic unity and does not seem to determine precisely the meaning of the concepts contained in it. We shall therefore subject the original definition of the concept of completeness to the following modification: a set X of sentences is said to be *complete with respect to its specific terms* if it is impossible to construct a categorical set Y of sentences which is essentially richer than X with respect to specific terms. In order to establish the incompleteness of a set of sentences it is from now onwards requisite to construct a set of sentences which is not only essentially richer but also categorical. Trivial constructions in which the meaning of the newly introduced specific terms is quite indeterminate are thus excluded from the beginning.[1]

With this conception of the notion of completeness the existence of both complete and incomplete sets of sentences can be established. For example, axiom systems of various geometrical theories are incomplete with respect to their specific terms, but various axiom systems of arithmetic are complete.[2]

Let us consider this more closely with reference to specific examples. Consider the system of descriptive one-dimensional geometry, i.e. the totality of all geometrical sentences concerning points and sub-sets of a straight line, in which the only extra-logical concepts are the relation of lying between, and the concepts definable by means of it. For this geometry

gorical in the usual (Veblen's) sense can be called *intrinsically categorical,* those in the new sense *absolutely categorical.* The axiom systems of various deductive theories are for the most part intrinsically but not absolutely categorical. It is, however, easy to make them absolutely categorical. It suffices, for example, to add a single sentence to the axiom system of geometry which asserts that every individual is a point (or more generally one which determines the number of individuals which are not points). It will become clear later (cf. p. 314 footnote) what it is that leads us to such a modification of the concept of categoricity. It may be mentioned that the concept of categoricity which we use belongs to the a-concepts in the sense of R. Carnap; but the corresponding f-concept can also be defined.

[1] Cf. p. 309, note 1 and the corresponding place in the text.

[2] This is only exact if the axiom system of arithmetic contains sentences to the effect that every individual is a number (cf. p. 310, note 2).

it is easy to formulate a categorical axiom system X_1 which contains the term of that relation as the only primitive term. The system X_1 is not complete with respect to its specific terms. As we have already mentioned in the previous section in connexion with the problem of independence, various concepts occur in metrical geometry which cannot be defined with the help of the descriptive ones exclusively, e.g. the relation of the congruence of two pairs of points. If we now take the symbol of this relation as a new primitive term and adequately extend the axiom system X_1, we obtain an axiom system X_2 for the full metrical geometry of the straight line. This system X_2 is again categorical and at the same time essentially richer than X_1 with respect to specific terms. But X_2 is also incomplete with respect to its specific terms. It can, for example, be proved that on the basis of this system not a single symbol can be defined which denotes an individual point. For that reason it is easy to set up a new categorical set X_3 of sentences which is essentially richer than X_2 with respect to specific terms. For this purpose it suffices to introduce any two symbols, say '0' and '1', as new primitive terms, and to supplement X_2 with a single new axiom to the effect that these symbols denote two distinct points. The deductive theory founded on the axiom system X_3 is identical, from a formal standpoint, with the arithmetic of real numbers; formally the arithmetic of real numbers is nothing but the geometry of the straight line in which two points have been 'effectively' distinguished. The question now arises whether the above process can be continued *ad infinitum*, i.e. whether it is possible to construct for X_3 a new set X_4 of sentences which is categorical and essentially richer with respect to specific terms, for this set X_4 a new set X_5, and so on. This proves to be impossible, the set X_3 of sentences is already complete with respect to its specific terms.

No method is apparent *a priori* which would enable us to establish the completeness of any particular set of sentences, e.g. the set X_3 discussed above. For a set of sentences can usually be extended in infinitely many ways, and to prove the completeness of this set it would be necessary to show that none of these extensions, if it is categorical, is essentially richer in specific terms than the set discussed. We shall give here a result which makes our task much easier in all cases actually known. In fact we shall show that the completeness of sets of sentences, with respect to the sets of terms occurring in them, is a consequence of another property of such sets which may be called *monotransformability*. The problem whether a given set of sentences actually has this new property presents, in general, no greater difficulties.

The notion of monotransformability is closely related to that of categoricity. A set of sentences is categorical if for any two realizations of this set at least one relation exists which establishes the isomorphism of the two realizations; it is monotransformable if there is at most one such relation. In precise terms, the set X of sentences is monotransformable if the sentence

(VI) $(x', x'', y', y'', ..., R', R'') : \phi(x', y', ...) \cdot \phi(x'', y'', ...)$.

$$\cdot R' \frac{x', y', ...}{x'', y'', ...} \cdot R'' \frac{x', y', ...}{x'', y'', ...} \cdot \supset \cdot R' = R''$$

is logically provable, where '$\phi(a, b, ...)$' represents the conjunction of all the sentences of X.[1]

A set of sentences which is both categorical and monotransformable is called *strictly* (or *uniquely*) *categorical*. The isomorphism of any two realizations of such a set can thus be established in exactly one way. For example, the various axiom systems of arithmetic are strictly categorical—both the arithmetic of natural numbers (e.g. the axiom system of Peano), and that of real numbers.[2] But the axiom systems of various geometrical theories—of topology, projective, descriptive, metrical geometry, etc.—although usually categorical, are not monotransformable and hence not strictly categorical.

This applies, in particular, to the sets X_1 and X_2 discussed above. By way of example we should like to explain this in detail for the axiom system X_1. As we know the only specific term occurring in this set is the name of the relation of lying

[1] We can again distinguish two senses of 'monotransformable', intrinsic and absolute monotransformability. Here the term is always used in the second sense (cf. p. 310, note 2).

[2] This is exact only if the axiom system of arithmetic contains sentences to the effect that every individual is a number (cf. p. 310, note 2).

between. The system X_1 is categorical; if we find any two interpretations of this system, we can always establish a one-one correspondence between points a, b, c,... in the first interpretation and points a', b', c',... in the second in such a way that, whenever b lies between a and c, b' lies between a' and c', and conversely. It is found that a correspondence of this kind can be established in infinitely many different ways. Therefore the set X_1 is not monotransformable and hence not strictly categorical. On the other hand the axiom system X_3 (of the arithmetic of real numbers) is monotransformable and strictly categorical. In this instance the one-one correspondence between points (or numbers) for any two interpretations can be established in exactly one way.

We shall now prove the following:

THEOREM 4. *Every monotransformable set of sentences is complete with respect to its specific terms.*[1]

Proof. Let X be any monotransformable set of sentences, and '$\phi(a, b,...)$' the conjunction of all sentences of X. Formula (VI) is thus logically provable. As a substitution into this formula we obtain

$$(1) \ (x, y,..., R) : \phi(x, y,...) \cdot \phi(x, y,...) \cdot R\frac{x, y,...}{x, y,...} \cdot I\frac{x, y,...}{x, y,...} \cdot \supset \cdot R = I;$$

the symbol 'I' here denotes, as in *Principia Mathematica*, the relation of identity (and thus has the same meaning as the sign '$=$').

From the definitions of the expression '$R\dfrac{x', y',...}{x'', y'',...}$' and of the identity sign we easily see that the formula

$$(2) \qquad\qquad (x, y,...) \cdot I\frac{x, y,...}{x, y,...}$$

---

[1] If we wish to use the words 'categorical' and 'monotransformable' in the intrinsic sense (cf. p. 310, note 2 and p. 313, note 1), then Th. 4 in the above form would be false. In order that it should remain valid the definition of completeness must be weakened, *absolute completeness* must be replaced by *intrinsic completeness* or by *completeness with respect to intrinsic concepts*. For this purpose it would be necessary to give first an explanation of what is meant by 'intrinsic concept'. All this would complicate our exposition considerably. In this connexion compare the concluding remarks in XIII.

is logically provable, independently of the type of the variables
'x', 'y',.... In view of (2), formula (1) can be simplified as
follows:

(3) $(x, y,..., R) : \phi(x, y,...) . R \dfrac{x, y,...}{x, y,...} . \supset . R = I.$

(It is worthy of note that, not only does the formula (3) follow
from the formula (VI), but also, conversely, (VI) follows from
(3); the two formulas are therefore equivalent. The definition
of monotransformability can thus be simplified by replacing
(VI) by (3). The new definition can be expressed as follows:
a set of sentences is monotransformable if there are non-identical
automorphisms in none of its interpretations.)

Consider now any categorical set Y of sentences which includes the set X as a part. We assume that in the sentences of
Y new extra-logical constants 'g', 'h',...—as well as the old ones
'a', 'b',...—occur, and we represent the conjunction of all these
sentences in the form '$\psi(a, b, ..., g, h,...)$'. It is to be proved that
the set Y is not essentially richer than X with respect to specific
terms.

From the categoricity of Y it follows that the formula

(4) $(x', x'', y', y'',..., u', u'', v', v'',...) : \psi(x', y',..., u', v',...).$

$\psi(x'', y'',..., u'', v'',...) . \supset . (\exists R) . R \dfrac{x', y',..., u', v',...}{x'', y'',..., u'', v'',...}$

is logically provable. From this by substitution we obtain

(5) $(x, y,..., u', u'', v', v'',...) : \psi(x, y,..., u', v',...) . \psi(x, y,..., u'' . v'').$

$\supset . (\exists R) . R \dfrac{x, y,..., u', v',...}{x, y,..., u'', v'',...}.$

Since Y contains all sentences of X, the formula

(6) $(x, y,..., u', v',...) : \psi(x, y,..., u', v',...) . \supset . \phi(x, y,...)$

must also be logically provable. From (5) and (6) we obtain
immediately

(7) $(x, y,..., u', u'', v', v'',...) : \psi(x, y,..., u', v',...).$

$\psi(x, y,..., u'', v'',...) . \supset . (\exists R) . \phi(x, y,...) . R \dfrac{x, y,..., u', v',...}{x, y,..., u'', v'',...};$

from (7), recalling the meaning of the expression

$$R \frac{x,\, y,...,\, u',\, v',...}{x,\, y,...,\, u'',\, v'',...}$$

we derive

(8) $(x, y,..., u', u'', v', v'',...) : \psi(x, y,..., u', v',...).$

$$\psi(x, y,..., u'', v'',...).$$

$$\supset . (\exists R). \phi(x, y,...). R \frac{x,\, y,...}{x,\, y,...} . R \frac{u',\, v',...}{u'',\, v'',...}.$$

The formulas (3) and (8) imply

(9) $(x, y,..., u', u'', v', v'',...) : \psi(x, y,..., u', v',...).$

$$\psi(x, y,..., u'', v'',...). \supset . (\exists R). R = I. R \frac{u',\, v',...}{u'',\, v'',...}$$

whence, in accordance with the definition of the identity sign,

(10) $(x, y,..., u', u'', v', v'',...) : \psi(x, y,..., u', v',...).$

$$\psi(x, y,..., u'', v'',...). \supset . I \frac{u',\, v',...}{u'',\, v'',...}.$$

Finally, it is easy to see that the sentence

(11) $(u', u'', v', v'',...) : I \dfrac{u',\, v',...}{u'',\, v'',...}. \supset . u' = u''. v' = v''....$

is logically provable, and in fact independently of the logical type of the variables 'u'', 'u''', 'v'', 'v''',... . From formulas (10) and (11) we obtain directly

(12) $(x, y,..., u', u'', v', v'',...) : \psi(x, y,..., u', v',...).$

$$\psi(x, y,..., u'', v'',...). \supset . u' = u''. v' = v''.... .$$

We have thus shown that (12) is logically provable. Since '$\psi(a, b,..., g, h,...)$' is the conjunction of all sentences of the set Y, we conclude by applying Th. 2, that each of the terms 'g', 'h',..., which are lacking in the sentences of X, can be defined on the basis of Y by means of the terms 'a', 'b',... occurring in X. Consequently the set Y is not essentially richer than X with respect to specific terms. Hence it is impossible to construct for the set X a categorical set of sentences which is essentially richer than X. The set X of sentences is thus complete with respect to its specific terms, which was to be proved.

The practical importance of the above theorem is clear. In order to establish that a given set of sentences is complete with respect to its specific terms, it suffices to show that this set is monotransformable and hence that the formula (VI) can be proved in logic. But this demands no special method of investigation and, in the cases so far considered, presents no special difficulties.

The question naturally arises whether the converse of Th. 4 holds: in other words, whether the monotransformability of a set of sentences is not only a sufficient but also a necessary condition for its completeness with respect to specific terms. The problem is not yet solved; no one has so far succeeded either in establishing the converse theorem or in constructing a counter-example.

In connexion with the theorem last given I should like to make some observations which exceed the strict limits of this article; they deal with the prospects of deductively constructing theoretical physics either in its entirety or in a particular part. Suppose we have adopted the system of four-dimensional Euclidean geometry—the geometry of the space-time continuum —as a foundation for the construction of a physical theory. Let the system of axioms and primitive terms of geometry be strengthened in such a way that on this basis a particular co-ordinate system can be distinguished. Geometry then formally coincides with the arithmetic of hypercomplex numbers; its axiom system is strictly categorical and hence, according to Th. 4, complete with respect to undefined terms.

Let us now consider how some part of theoretical physics, e.g. mechanics, could be deductively built up on this basis. Two different procedures may be considered here. In the first method we attempt to define the concepts of mechanics by means of geometrical concepts. If this were successful, then mechanics would, from the methodological standpoint, simply become a special chapter of geometry. It is easily seen that all attempts in this direction must be confronted with quite essential difficulties.

The specific concepts of mechanics are of a manifold logical character. They are, for example, properties of individual space-time points, or properties of sets of space-time points (like the concept of the rigid body), or single-valued functions which correlate numbers with space-time points (e.g. the concept of density). We should be in a position to define such a concept, e.g. a function, by means of geometrical concepts, only if—loosely speaking—the behaviour of the function over the whole world were known to us, and if we were able to derive every function value from the mere position of the corresponding point in space and time. We have good reasons for regarding this task as unfeasible from the start. We turn therefore to the second procedure. In this method some specific concepts of mechanics are taken as primitive concepts and some of its laws as axioms. Mechanics would then assume the character of an independent deductive theory for which geometry would be only the foundation. Although the axiom system of mechanics would be incomplete initially, we could extend and complete it in accordance with the progress of empirical science. If, however, we wish to apply to mechanics the same criteria as to every other deductive theory, then we should regard the problem of setting up deductive foundations for mechanics as exhausted only if we obtained a categorical axiom system for this theory. But at this stage—and this is the most important point—Th. 4 would come into action. Since we have adopted a strictly categorical system of geometry as a basis, mechanics cannot be essentially richer than geometry with respect to specific concepts. The concepts of mechanics would have to be definable by means of geometrical concepts, and thus we are brought back to the first method. *The problem of defining the concepts of mechanics (or any other physical theory) exclusively by means of geometrical concepts, and the problem of basing mechanics on a categorical axiom system, are completely equivalent*; whoever regards the first problem as hopeless must take an entirely analogous attitude to the second.

The above remarks just made do not apply to certain new physical theories in which the basic system of geometry is neither strictly categorical nor categorical in the usual sense and in which the whole of physics or some of its parts are conceived just as special chapters of geometry. Naturally the difficulties do not here vanish, only the centre of gravity is shifted: it lies in the problem of providing the basic geometry with a categorical axiom system.

XI

ON THE FOUNDATIONS OF
BOOLEAN ALGEBRA†

BOOLEAN ALGEBRA, also called the *algebra of logic*, is a formal system with a series of important interpretations in various fundamental departments of logic and mathematics. The most important and best known interpretation is the *calculus of classes*.

The present article contains a study of some alternative systems of Boolean algebra. Of the results I have obtained concerning these problems I shall give here only those which can be established with the help of the usual mathematical methods and demand no special metamathematical apparatus.

§ 1. THE ORDINARY AND THE EXTENDED (OR COMPLETE) SYSTEMS OF BOOLEAN ALGEBRA

At present several equivalent systems of primitive expressions and postulates are known which suffice for the setting up of Boolean algebra. In order to make the subsequent developments more specific I first explicitly introduce one of these systems.

In this system eight primitive expressions occur: 'B'—'*the universe of discourse*', '$x < y$'—'*the element x is included in the element y*', '$x = y$'—'*the element x stands in the relation of equality to the element y*' ('*the element x is equal to the element y*'), '$x+y$'—'*the sum of the elements x and y*', '$x.y$'—'*the product of the elements x and y*', '0'—'*the zero element*', '1'—'*the unit element*' and 'x'''—'*the complement of the element x*'. The symbol 'B' denotes a certain set of elements, the symbols '0' and '1' certain elements of this set, the symbols '$<$' and '$=$' certain relations between the elements of this set; finally, the symbols

---

† BIBLIOGRAPHICAL NOTE. This article first appeared in German under the title 'Zur Grundlegung der Booleschen Algebra. I', in *Fundamenta Mathematicae*, vol. 24 (1935), pp. 177–98. (The unpublished Part II is referred to in the second footnote on p. 341.)

'+', '.', and '′' denote certain operations which correlate, with one or two elements of this set, another element of this set.[1]

The following sentences are taken as postulates:

POSTULATE $\mathfrak{A}_1$. (a) If $x \in B$, then $x < x$;

(b) if $x, y, z \in B$, $x < y$, and $y < z$, then $x < z$.

POSTULATE $\mathfrak{A}_2$. If $x, y \in B$, then $x = y$ if and only if both $x < y$ and $y < x$.

POSTULATE $\mathfrak{A}_3$. If $x, y \in B$, then

(a) $x+y \in B$;

(b) $x < x+y$ and $y < x+y$;

(c) if, moreover, $z \in B$, $x < z$, and $y < z$, then $x+y < z$.

POSTULATE $\mathfrak{A}_4$. If $x, y \in B$, then

(a) $x.y \in B$;

(b) $x.y < x$ and $x.y < y$;

(c) if, moreover, $z \in B$, $z < x$, and $z < y$, then $z < x.y$.

POSTULATE $\mathfrak{A}_5$. If $x, y, z \in B$, then

(a) $x.(y+z) = x.y+x.z$;

(b) $x+y.z = (x+y).(x+z)$.

POSTULATE $\mathfrak{A}_6$. (a) $0, 1 \in B$;

(b) if $x \in B$, then $0 < x$ and $x < 1$.

POSTULATE $\mathfrak{A}_7$. If $x \in B$, then

(a) $x' \in B$;

(b) $x.x' = 0$;

(c) $x+x' = 1$.

---

[1] In addition to the specific symbols of Boolean algebra the usual symbols belonging to set theory will also be used. For example, the formula '$x, y,... \in X$' means that the elements $x, y,...$ belong to the set X. The formula '$X \subseteq Y$' will assert that the set X is included in the set Y; the symbol '$E[\phi(x)]$' denotes the set of all these elements x which satisfy the condition (sentential function) $\phi(x)$; the symbol '$\{x, y,...\}$' denotes the set which consists only of the elements $x, y,...$, etc. If it is desired to express that the relation R holds between x and y, then 'xRy' will be written. If the denial of this is to be expressed, we shall write 'x non-Ry'.

The totality of the Posts. $[\mathfrak{A}_1-\mathfrak{A}_7]^1$ and the theorems which follow from them we call *the ordinary system of Boolean algebra.*

By introducing two operations of an infinite character, one of addition and one of multiplication, this system of Boolean algebra can be extended. For this purpose we add the following expressions to the list of primitive concepts : ' $\sum\limits_{y \in X} y$ '—'*the sum of all elements of the set X*' and ' $\prod\limits_{y \in X} y$ '—'*the product of all elements of the set X*'.[2] The system of postulates is correspondingly supplemented by the following sentences:

POSTULATE $\mathfrak{A}_8$. *If $X \subseteq B$, then*

(a) $\sum\limits_{y \in X} y \in B$;

(b) $x < \sum\limits_{y \in X} y$ *for every $x \in X$*;

(c) *if, moreover, $z \in B$ and $x < z$ for every $x \in X$, then $\sum\limits_{y \in X} y < z$.*

POSTULATE $\mathfrak{A}_9$. *If $X \subseteq B$, then*

(a) $\prod\limits_{y \in X} y \in B$;

(b) $\prod\limits_{y \in X} y < x$ *for every $x \in X$*;

(c) *if, moreover, $z \in B$ and $z < x$ for every $x \in X$, then $z < \prod\limits_{y \in X} y$.*

---

[1] With slight modifications this is the system of postulates given by L. Couturat in (14a). Regarding the changes which we have introduced into this system the following may be said: (1) the expressions '$x = y$', '$x+y$', '$x.y$', '0', '1', and 'x'' which are defined expressions for Couturat are treated as primitive, the postulates being correspondingly modified; (2) the symbol 'B', which denotes the universe of discourse, is included in the list of primitive expressions; (3) the postulate of existence: '$0 \neq 1$' (op. cit., p. 28) is excluded from the system. In Couturat's book the reader will learn how all theorems of Boolean algebra which are used in the further considerations are derived from the adopted postulate system.

[2] In addition to the expressions ' $\sum\limits_{y \in X} y$ ' and ' $\prod\limits_{y \in X} y$ ' symbolic expressions of a more general nature will be used, ' $\sum\limits_{\phi(x)} f(x)$ ' and ' $\prod\limits_{\phi(x)} f(x)$ ' they will denote *the sum and the product of all elements y of the form $y = f(x)$, which correspond to elements satisfying the condition $\phi(x)$*. These expressions can easily be defined by means of the primitive expressions: Let X be the set of all elements y of the form: $y = f(x)$, where x is any element which satisfies the condition $\phi(x)$; then the formulas hold: $\sum\limits_{\phi(x)} f(x) = \sum\limits_{y \in X} y$ and $\prod\limits_{\phi(x)} f(x) = \prod\limits_{y \in X} y$. Those more general expressions appear in Post. $\mathfrak{A}_{10}$.

POSTULATE $\mathfrak{A}_{10}$. *If* $x \in B$ *and* $X \subseteq B$, *then*

(a) $x . \sum\limits_{y \in X} y = \sum\limits_{y \in X} (x.y);$

(b) $x + \prod\limits_{y \in X} y = \prod\limits_{y \in X} (x+y).$

The postulate system $[\mathfrak{A}_1 - \mathfrak{A}_{10}]$, together with the totality of the theorems which can be derived from it, forms the *extended* (or *complete*) *system of Boolean algebra*.

It must be emphasized that the postulates of this system are not independent of one another. Thus, for example, any three of the four postulates $\mathfrak{A}_5^a$, $\mathfrak{A}_5^b$, $\mathfrak{A}_{10}^a$, $\mathfrak{A}_{10}^b$ can be derived from the remaining postulates of the system.† Moreover, the primitive expressions of the system are not independent. For example, the inclusion sign '$<$' can be defined in various ways by means of the other expressions. This is seen from the following theorem which is easily proved with the help of Posts. $\mathfrak{A}_1 - \mathfrak{A}_7$.

If $x, y \in B$, *then the following formulas are equivalent*:

(1) $x < y;$ (2) $x+y = y;$ (3) $x.y = x;$

(4) $x'+y = 1;$ (5) $x.y' = 0.$

If, however, the symbol '$<$' is retained in the list of primitive expressions, then the identity sign '$=$' can be deleted from it, because Post. $\mathfrak{A}_2$ can be adopted as its definition.

The symbol '$=$' is frequently treated not as a specific symbol of Boolean algebra, but as the sign of logical identity. From now on we shall adopt this standpoint. It results in important simplifications in the system of primitive expressions and postulates of Boolean algebra. It is especially helpful in reducing the list of primitive expressions. In addition to 'B', it suffices to

---

† This result can be improved. In deriving any three of the postulates listed above from the remaining postulates of the system we can avoid the use of $\mathfrak{A}_8$ and $\mathfrak{A}_9$ so that the derivation of $\mathfrak{A}_{10}^a$ and $\mathfrak{A}_{10}^b$ can essentially be carried out within the ordinary, and not extended, system of Boolean algebra. To this end, however, we must supplement the system $[\mathfrak{A}_1 - \mathfrak{A}_7]$ with appropriate definitions of 'Σ' and 'Π'. Moreover, Posts. $\mathfrak{A}_{10}^a$ and $\mathfrak{A}_{10}^b$ must be put in a somewhat weaker form, and in fact provided with certain existential assumptions: $\mathfrak{A}_{10}^a$ with the assumption that $\sum\limits_{y \in X} y$ exists, and $\mathfrak{A}_{10}^b$ with the assumption that $\prod\limits_{y \in X} y$ exists. Cf. A. Tarski (77), p. 211; also B. Jónsson and A. Tarski, (36c), part I, p. 905.

include in this list one of the following expressions: '$x < y$', '$x+y$', '$x.y$', '$\sum_{y \in X} y$', or '$\prod_{y \in X} y$', in order to be able to define all the remaining expressions. For each of the expressions just listed a corresponding system of postulates can be constructed which is equivalent to the system $[\mathfrak{A}_1 - \mathfrak{A}_{10}]$ but is much simpler.[1]

As an example we here introduce a simple postulate system based on the two fundamental expressions 'B' and '$x < y$', and we prove its equivalence with the system $[\mathfrak{A}_1 - \mathfrak{A}_{10}]$. The system consists of four sentences:

POSTULATE $\mathfrak{B}_1$. *If* $x, y \in B$, $x < y$, *and* $y < x$, *then* $x = y$.

POSTULATE $\mathfrak{B}_2$. *If* $x, y, z \in B$, $x < y$, *and* $y < z$, *then* $x < z$.

POSTULATE $\mathfrak{B}_3$. *If* $x, y \in B$ *and* x *non-* $< y$, *then there is an element* $z \in B$ *such that* $z < x$, *and* z *non-* $< y$, *and the formulas*: $u \in B$, $u < y$, *and* $u < z$ *always imply* $u < v$, *for every* $v \in B$.

POSTULATE $\mathfrak{B}_4$. *If* $X \subseteq B$, *then there exists an element* $x \in B$ *which satisfies the following conditions*: (1) $y < x$ *for every* $y \in X$; (2) *if* $z \in B$, $z < x$, *and if, for every* $y \in X$, *the formulas* $u \in B$, $u < y$, *and* $u < z$ *always imply* $u < v$ *for every* $v \in B$, *then also* $z < v$ *for every* $v \in B$.

The content of the last two postulates will be clear if it is noticed that, on the basis of the earlier postulate system the condition '*the formulas* $u \in B$, $u < y$, *and* $u < z$ *always imply* $u < v$ *for every* $v \in B$' is equivalent to the formula '$y.z = 0$'.

Similarly, the condition '$z < v$ *for every* $v \in B$' can be replaced by the equation '$z = 0$'.

The remaining expressions of Boolean algebra are defined in the following way:

DEFINITION $\mathfrak{B}_5$. (a) *For all* $x, y, z \in B$, $z = x+y$ *if and only if* z *is the only element which satisfies the conditions*:

(1) $x < z$ *and* $y < z$,
(2) *for every* $u \in B$, *if* $x < u$ *and* $y < u$, *then* $z < u$,

---

[1] E. V. Huntington has constructed such postulate systems for the ordinary system of Boolean algebra; cf. Huntington, E. V. (32) and (34).

(b) *for all* $x, y, z \in B$, $z = x \cdot y$ *if and only if* z *is the only element which satisfies the conditions*:

(1) $z < x$ *and* $z < y$,

(2) *for every* $u \in B$, *if* $u < x$ *and* $u < y$, *then* $u < z$;

(c) *for every* $x \in B$, $x = 0$ *if and only if* x *is the only element which satisfies the condition*: $x < v$ *for every* $v \in B$;

(d) *for every* $x \in B$, $x = 1$ *if and only if* x *is the only element which satisfies the condition*: $v < x$ *for every* $v \in B$;

(e) *for every* $x, y \in B$, $y = x'$ *if and only if* y *is the only element which satisfies the conditions*:

(1) *the formulas* $u \in B$, $u < x$, *and* $u < y$ *always imply* $u < v$ *for every* $v \in B$;

(2) *the formulas* $u \in B$, $x < u$, *and* $y < u$ *always imply* $v < u$ *for every* $v \in B$.

DEFINITION $\mathfrak{B}_6$. *If* $x \in B$ *and* $X \subseteq B$, *then*

(a) $x = \sum\limits_{y \in X} y$ *if and only if* x *is the only element which satisfies the conditions*:

(1) $y < x$ *for every* $y \in X$;

(2)   *if* $z \in B$ *and* $y < z$ *for every* $y \in X$, *then* $x < z$;

(b) $x = \prod\limits_{x \in X} y$ *if and only if* x *is the only element which satisfies the conditions*:

(1) $x < y$ *for every* $y \in X$;

(2)   *if* $z \in B$ *and* $z < y$ *for every* $y \in X$, *then* $z < x$.

THEOREM 1. *The postulate system* $[\mathfrak{A}_1 – \mathfrak{A}_{10}]$ *is equivalent to the system of postulates and definitions* $[\mathfrak{B}_1 – \mathfrak{B}_6]$.

Proof I. We shall show how the sentences $\mathfrak{B}_1 – \mathfrak{B}_6$ can be derived from Posts. $\mathfrak{A}_1 – \mathfrak{A}_{10}$.

$\mathfrak{B}_1$ trivially follows from Post. $\mathfrak{A}_2$; $\mathfrak{B}_2$ is identical with $\mathfrak{A}_1^b$. We pass on to $\mathfrak{B}_3$. Let x, $y \in B$ and x *non-* $< y$. We put $z = x \cdot y'$. We see at once that $z \in B$ and $z < x$. If we had $z < y$, i.e. $x \cdot y' < y$, we should have (in view of the formula $x \cdot y < y$) $x \cdot y + x \cdot y' < y$, whence $x \cdot (y + y') < y$, $x \cdot 1 = y$, and $x < y$, contrary to the hypothesis; hence z *non-* $< y$. Finally if

$u \in B$, $u < y$, and $u < z$, then $u < y.z = y.(x.y')$; from this by easy transformations it follows that

$$u < x.(y.y') = x.0 = 0,$$

and since $0 < v$ for every $v \in B$, so also $u < v$ for every $v \in B$. Thus the element z satisfies all conditions of Post. $\mathfrak{B}_3$.

In order to derive $\mathfrak{B}_4$ we take any set $X \subseteq B$ and put $x = \sum_{y \in X} y$. According to $\mathfrak{A}_8^{a,b}$ we have $x \in B$ and $y < x$ for every $y \in X$. Further, we consider any element $z \in B$ which satisfies the premisses of condition (2) of Post. $\mathfrak{B}_4$, i.e. (α) $z < x$, and (β) for every $y \in X$ the formulas $u \in B$, $u < y$, and $u < z$ always imply $u < v$ for every $v \in B$. If in (β) 'u' is replaced by '$z.y$', then it follows that $z.y < v$ for all $y \in X$ and $v \in B$ whence, by $\mathfrak{A}_8^c$, $\sum_{y \in X}(z.y) < v$, and since, by $\mathfrak{A}_{10}^a$, $z.x = z.\sum_{y \in X} y = \sum_{y \in X}(z.y)$, we have further $z.x < v$ for every $v \in B$. On the other hand (α) gives $z = z.x$; thus finally $z < v$ for every $v \in B$. The element x consequently satisfies all conditions of the Post. $\mathfrak{B}_4$.

$\mathfrak{B}_5^a$ is immediately derivable from $\mathfrak{A}_3$; we use $\mathfrak{A}_2$ in order to prove that $x+y$ is the only element which satisfies conditions (1) and (2) of Def. $\mathfrak{B}_5^a$. In a similar way $\mathfrak{B}_5^b$ is derived from $\mathfrak{A}_4$, $\mathfrak{B}_5^{c,d}$ from $\mathfrak{A}_6$; $\mathfrak{B}_6$ from $\mathfrak{A}_8$ and $\mathfrak{A}_9$.

There remains $\mathfrak{B}_5^e$. Let $x, y \in B$ and $y = x'$. If $u \in B$, $u < x$, and $u < y$, then $u < x.y = x.x' = 0$, and since $0 < v$, we also have $u < v$ for every $v \in B$; hence y satisfies condition (1). In a similar way it is established that y satisfies condition (2).

If any other element $y_1 \in B$ likewise satisfies these two conditions, it is easy to prove that we have $x.y_1 = 0$ and $x+y_1 = 1$; with the help of the usual methods of Boolean algebra, we obtain from these formulas $y_1 = x' = y$. Consequently y is the only element which satisfies conditions (1) and (2). Since, according to $\mathfrak{A}_7^a$, there corresponds an element $x' \in B$ to every element $x \in B$, we easily conclude that, conversely, if y is the only element which satisfies (1) and (2), then $y = x'$. $\mathfrak{B}_5^e$ is in this way derived.

We have thus shown that the sentences $\mathfrak{B}_1$–$\mathfrak{B}_6$ actually follow from the Posts. $\mathfrak{A}_1$–$\mathfrak{A}_{10}$, q.e.d.

II. The proof in the opposite direction likewise presents no great difficulties.

$\mathfrak{A}_1^a$ is derived from $\mathfrak{B}_3$, by substituting 'x' for 'y' in $\mathfrak{B}_3$ and so convincing ourselves that the assumption: x *non*-$<$ x leads to a contradiction. $\mathfrak{A}_1^b$ coincides with $\mathfrak{B}_2$. $\mathfrak{A}_2$ is an equivalence, thus a conjunction of two implications; one of these implications coincides with $\mathfrak{B}_1$, the other follows directly from the Post. $\mathfrak{A}_1^a$ already derived (it must not be forgotten that the symbol '$=$' is always to be regarded as the symbol of logical identity).

Now we come to Post. $\mathfrak{A}_3$. Let X be any subset of B. According to $\mathfrak{B}_4$ there is an element x which satisfies the following conditions:

(1) $$x \in B;$$

(2) $$y < x \text{ for every } y \in X;$$

(3) *if* $z \in B$, $z < x$ *and if, for every* $y \in X$, *the formulas* $u \in B$, $u < y$, *and* $u < z$ *always imply* $u < v$ *for every* $v \in B$, *then* $z < v$ *for every* $v \in B$.

We shall prove that the element x also satisfies the following condition:

(4)   *if* $z \in B$ *and if* $y < z$ *for every* $y \in X$, *then* $x < z$.

In fact let us suppose that this is not the case; let z be an element of the set B such that

(5) $$y < z \text{ for every } y \in X,$$

and that nevertheless x *non*-$<$ z. By Post. $\mathfrak{B}_3$ ('z' and 't' being substituted for 'y' and 'z' respectively) there is an element t which satisfies the following conditions:

(6) $$t \in B \text{ and } t < x;$$

(7) $$t \text{ non-} < z;$$

(8) *the formulas* $u \in B$, $u < z$, *and* $u < t$ *always imply* $u < v$ *for every* $v \in B$.

Let us now consider any element $y \in X$ and suppose that the formulas $u \in B$, $u < y$, and $u < t$ hold. By formula (5) we have

$y < z$; by virtue of $\mathfrak{B}_2$ the inclusions $u < y$ and $y < z$ give $u < z$. With the help of (8) we infer from this that $u < v$ for every $v \in B$. Consequently,

(9) *the formulas $u \in B$, $u < y$, and $u < t$ always imply $u < v$, for every $y \in X$ and $v \in B$.*

In view of (6) and (9) we can apply (3), replacing 'z' by 't'; we thus obtain $t < v$ for every $v \in B$ and in particular $t < z$, which obviously contradicts formula (7). We must therefore assume that condition (4) is satisfied.

According to (2) and (4) the element $z \in B$ satisfies both conditions of Def. $\mathfrak{B}_6^a$. From $\mathfrak{B}_1$ it is easy to infer that x is the only element which satisfies both these conditions; hence $x = \sum_{y \in X} y$. If in (1), (2), and (4) we replace 'x' by '$\sum_{y \in X} y$', we obtain the complete Post. $\mathfrak{A}_8$.

From Defs. $\mathfrak{B}_5^a$ and $\mathfrak{B}_6^a$ together it follows that the formulas $x = \sum_{t \in X} t$ and $x = y + z$ are equivalent in case the set $X \subseteq B$ consists of only the two elements y and z, $X = \{y, z\}$; Post. $\mathfrak{A}_3$ thus represents a special case of Post. $\mathfrak{A}_8$. $\mathfrak{A}_6$ is similarly derived; we apply $\mathfrak{A}_8$ twice, it being assumed the first time that X is the empty set, and the second time that $X = B$. Finally by means of an analogous method we prove $\mathfrak{A}_4$ and $\mathfrak{A}_9$. For example, to obtain $\mathfrak{A}_9$ we consider any set $X \subseteq B$ and apply $\mathfrak{A}_8$, not to this set X, but to the set Y of all those elements $y \in B$ which satisfy the condition: $y < x$ for every $x \in X$; from Def. $\mathfrak{B}_6$ it results that $\prod_{x \in X} x = \sum_{y \in Y} y$. $\mathfrak{A}_4$ can be derived from $\mathfrak{A}_9$ as a special case by putting $X = \{y, z\}$.

Before we proceed, it may be remarked that on the basis of the Posts. $\mathfrak{A}_1$–$\mathfrak{A}_4$ and $\mathfrak{A}_6$ already obtained, Post. $\mathfrak{B}_3$ can easily be transformed in the following way:

(10) *if $x, y \in B$ and x non- $< y$, then there exists an element $z \in B$ such that $z < x$, $z \neq 0$, and $z . y = 0$.*

The condition (2) of Post. $\mathfrak{B}_4$ may be similarly modified: *if $z \in B$, $z < x$ and if $z . y = 0$ for every $y \in X$, then $z = 0$.* Moreover, we remind ourselves that in establishing $\mathfrak{A}_8$ it was proved that the element x which satisfies the conclusion of Post. $\mathfrak{B}_4$ is identical with $\sum\limits_{y \in X} y$. We can thus state the following:

(11) *if* $X \subseteq B$, $z \in B$, $z < \sum\limits_{y \in X} y$ *and if* $z.y = 0$ *for every* $y \in X$, *then* $z = 0$.

With the help of (10) and (11) we prove $\mathfrak{A}_{10}$ without difficulty. Let $x \in B$ and $X \subseteq B$. In view of $\mathfrak{A}_8^b$ we have $z < \sum\limits_{y \in X} y$ for every $z \in X$, whence by $\mathfrak{A}_4$ and $\mathfrak{B}_2$ it is easy to obtain $x.z < x. \sum\limits_{y \in X} y$ for every $z \in X$, and thus by $\mathfrak{A}_8^c$

(12) $$\sum\limits_{y \in X} (x.y) < x. \sum\limits_{y \in X} y.$$

Should the opposite inclusion not hold, then by (10) there would be an element $z \in B$ such that

(13) $$z < x. \sum\limits_{y \in X} y \quad and \quad z. \sum\limits_{y \in X} (x.y) = 0 ;$$

(14) $$z \neq 0.$$

As a consequence of Post. $\mathfrak{A}_8^b$ we have $x.y < \sum\limits_{y \in X} (x.y)$ for every $y \in X$, whence $z.(x.y) < z. \sum\limits_{y \in X} (x.y)$ and, in view of (13), $z.(x.y) = (z.x).y = 0$ for every $y \in B$; from the formula (13) it follows that $z.x < \sum\limits_{y \in X} y$. Hence by applying condition (11) we obtain $z.x = 0$; but on account of (13) $z < x$, and this implies $z = z.x$, so that at last we have $z = 0$, in contradiction to formula (14).

In this way it is established that the inclusion (12) actually holds in the other direction; hence, by virtue of Post. $\mathfrak{B}_1$, we immediately derive $\mathfrak{A}_{10}^a$.

As a special case of $\mathfrak{A}_{10}^a$ we obtain $\mathfrak{A}_5^a$. From $\mathfrak{A}_5^a$ we can immediately derive $\mathfrak{A}_5^b$. On the other hand $\mathfrak{A}_{10}^b$ must be proved with the help of a method analogous to that used for Post. $\mathfrak{A}_{10}^a$. (The inclusion $x + \prod\limits_{y \in X} y < \prod\limits_{y \in X} (x+y)$ follows easily from $\mathfrak{A}_9$, $\mathfrak{A}_3$, and $\mathfrak{B}_2$; to prove the opposite inclusion we employ an indirect mode of argument making essential use of (10) as well as Post. $\mathfrak{A}_5^a$, which has already been derived.)

$\mathfrak{A}_7$ still remains to be considered. On the basis of the remaining postulates we first transform Def. $\mathfrak{B}_5^e$ in the following way:

(15) *if* $x, y \in B$, *then* $y = x'$ *if and only if both* $x.y = 0$ *and* $x+y = 1$.

(It is superfluous to add here that y is the only element which satisfies both the equalities mentioned, for from Posts. $\mathfrak{A}_1$–$\mathfrak{A}_6$ it can easily be concluded that two elements of this kind must be identical.)

Let us consider further any element $x \in B$. We denote by the symbol 'X' the set of all elements $z \in B$ such that $x.z = 0$, $X = E[z \in B \; and \; x.z = 0]$, and finally put $y = \sum_{z \in X} z$. By Post. $\mathfrak{A}_{10}^a$ we have the formula $x.y = x. \sum_{z \in X} z = \sum_{z \in X} (x.z)$; and since all terms of the sum $\sum_{z \in X} (x.z)$ are equal to 0, we also have $\sum_{z \in X} (x.z) = 0$, as can easily be seen from $\mathfrak{A}_8^c$. Thus

(16) $$x.y = 0.$$

Let us assume then that $x+y \neq 1$ and that consequently $1 \, non\text{-} < x+y$. Hence it follows by (10) that there exists an element $z \in B$ such that $z \neq 0$ and $z.(x+y) = 0$, whence $z.x + z.y = 0$ and $z.x = 0 = z.y$. If $z.x = 0$, then $z \in X$, from which, by virtue of $\mathfrak{A}_8^b$, we have $z < \sum_{z \in X} z = y$ and consequently $z = z.y$; this equality, together with the formula $z.y = 0$, gives $z = 0$, contrary to the inequality previously stated. We must thus assume that

(17) $$x+y = 1.$$

From (15)–(17) it is seen at once that $y = x'$, whence $\mathfrak{A}_7$ follows by suitable substitution.

In this way all postulates of the system $[\mathfrak{A}_1$–$\mathfrak{A}_{10}]$ can be derived from the postulates and definitions $\mathfrak{B}_1$–$\mathfrak{B}_6$; the two systems are thus finally seen to be equivalent, q.e.d.

The postulate system $[\mathfrak{B}_1$–$\mathfrak{B}_4]$ can be formally simplified:

THEOREM 2. *The postulate system* $[\mathfrak{B}_1$–$\mathfrak{B}_4]$ *is equivalent to the system which consists of* $\mathfrak{B}_2$ *and the following sentence*:

POSTULATE $\mathfrak{B}_4^*$. *If $X \subseteq B$, then there is exactly one element $x \in B$ which satisfies the conditions* (1) *and* (2) *of Post.* $\mathfrak{B}_4$.

Proof. $\mathfrak{B}_4^*$ was implicitly derived from the system $[\mathfrak{B}_1\text{–}\mathfrak{B}_4]$ in the proof of Th. 1 (Part II). In fact, it was shown there that every element $x \in B$ which satisfies the conditions (1) and (2) of Post. $\mathfrak{B}_4$ also satisfies the formula $x = \sum_{y \in X} y$; hence there can be only one such element. It remains to be proved that the Posts. $\mathfrak{B}_1$ and $\mathfrak{B}_3$ can be derived from $\mathfrak{B}_2$ and $\mathfrak{B}_4^*$.

In order to simplify the exposition we shall define the auxillary symbol ')(' (the formula 'x)(y' may be read *'the elements x and y are disjoint'*).

(1) *If x, $y \in B$, then x)(y if and only if the formulas $u \in B$, $u < x$, and $u < y$ always imply $u < v$ for every $v \in B$.*

We next prove some lemmas.

(2)   *If $x, y, z \in B$, $x < y$ and y)(z, then x)(z.*

This follows directly from (1) and Post. $\mathfrak{B}_2$.

(3)   *If $y \in B$ and y)(y, then $y < v$ for every $v \in B$.*

To prove this we use $\mathfrak{B}_4^*$, putting $X = \{y\}$. If we consider (1), we see that there is an element $x \in B$ which satisfies the conditions: (α) $y < x$, and (β) if $x \in B$, $z < x$ and z)(y, then $z < v$ for every $v \in B$. If in (β) 'z' is replaced by 'y', then, using (α), we at once obtain (3).

(4)   *If $y, z \in B$, $y < z$ and y)(z, then $y < v$ for every $v \in B$.*

Consider any element $u \in B$ such that $u < y$; by applying the premiss $y < z$, we obtain $u < z$ with the help of $\mathfrak{B}_2$. By definition (1) it follows from the formulas y)(z, $u < y$, and $u < z$ that $u < v$ for every $v \in B$. Thus, if $u \in B$ and $u < y$, then $u < v$ for every $v \in B$; if we put in (1) 'y' for 'x', it follows that y)(y and hence, according to (3), that $y < v$ for every $v \in B$, which was to be proved.

(5)   *If $x \in B$, then $x < x$.*

As in the proof of (3) we put, $X = \{x\}$ in $\mathfrak{B}_4^*$, and obtain an element x_1 which satisfies the two conditions: (α) $x < x_1$, and (β) if $z \in B$, $z < x_1$ and z)(x, then $z < v$ for every $v \in B$.

Further, we consider the set $X_1 = \underset{y}{E}[y \in B \text{ and } y < x]$. We have
(γ) $y < x$ for every $y \in X_1$, and (δ) if $z \in B$, $z < x$, and $z)(y$ for
every $y \in X_1$, then $z < v$ for every $v \in B$ (for from the hypothesis of (δ) it follows that $z)(z$, whence, by (3), $z < v$ for every
$v \in B$). If we now apply $\mathfrak{B}_2$, from (α) and (γ) we obtain (ϵ) $y < x_1$
for every $y \in X_1$. Finally the following condition is satisfied:
(ζ) if $z \in B$, $z < x_1$ and $z)(y$ for every $y \in X_1$, then $z < v$ for every
$v \in B$. In fact if $u \in B$, $u < z$, and $u < x$, then $u \in X_1$, whence,
by the hypothesis of condition (ζ), $z)(u$, or what, according to (1),
amounts to the same thing, $u)(z$; from (4) and the formulas
$u < z$ and $u)(z$ it follows that $u < v$ for every $v \in B$; by (1)
we thus have $z)(x$; thus the element z satisfies the hypothesis
of condition (β), and consequently $z < v$ for every $v \in B$, which
was to be proved. By (1) it follows from (γ)–(ζ) that not only x
but also x_1 satisfies both conditions of the Post. $\mathfrak{B}_4$ as applied
to the set X_1. From this by $\mathfrak{B}_4^*$ we have $x = x_1$, and this identity
together with (α) at once gives (5).

(6) *If* $x, y \in B$, $y < x$ *and if for every* $z \in B$ *the formulas* $z < x$
    and $z)(y$ *imply* $z < y$, *then* $x = y$.

In order to prove this we again put $X = \{y\}$. In view of (5)
we have (α) $t < y$ for every $t \in X$ and, moreover, by virtue of
(4), (β) if $z \in B$, $z < y$, and $z)(t$ for $t \in X$, then $z < v$ for every
$v \in B$. On the other hand the hypothesis of condition (6) im-
plies (γ) $t < x$ for every $t \in X$, as well as (δ) if $z \in B$, $z < x$, and
$z)(t$ for $t \in X$, then $z < v$ for every $v \in B$ [for if $z < x$ and $z)(y$,
then by hypothesis we have $z < y$ and hence by (4), $z < v$ for
every $v \in B$]. From (1) and (α)–(δ) it follows that y as well as x
satisfies both conditions of Post. $\mathfrak{B}_4$; by $\mathfrak{B}_4^*$ the elements x and
y are therefore identical, q.e.d.

Now we turn to the proof of Posts. $\mathfrak{B}_1$ and $\mathfrak{B}_3$.

$\mathfrak{B}_1$ follows immediately from (6), for if $x < y$ and $y < x$, then
by $\mathfrak{B}_2$ the formula $z < x$ always implies $z < y$; the hypothesis
of (6) is thus satisfied, and consequently $x = y$. It is also not
difficult to derive $\mathfrak{B}_1$ from $\mathfrak{B}_4^*$ directly with the help of $\mathfrak{B}_2$, by
putting $X = \underset{z}{E}[z \in B \text{ and } z < x]$ and arguing as in the proof
of Lemma 5.

To prove $\mathfrak{B}_3$ we proceed indirectly: we assume that this postulate is not satisfied, thus that (α) x *non-*$<$ y and that nevertheless, in view of (1), (β) for every $z \in B$ the formulas $z < x$ and $z)(y$ always imply $z < y$. Now $\mathfrak{B}_4^*$ is applied to the set $X = \{x, y\}$; we obtain an element $t \in B$ such that (γ) $x < t$, (δ) $y < t$, and (ϵ) for every $z \in B$, the formulas $z < t$, $z)(x$ and $z)(y$ imply $z < v$ for every $v \in B$. It is proved that the element t likewise satisfies the following condition: (ζ) if $z \in B$, $z < t$, and $z)(y$ then $z < y$. In order to prove this we consider any element $u \in B$ such that $u < z$ and $u < x$; according to (2) the formulas $u < z$ and $z)(y$ always imply $u)(y$; with the help of (β), from the formulas $u < x$ and $u)(y$ we obtain the inclusion $u < y$, whence, by virtue of (4), $u < v$ for every $v \in B$. Hence it follows by (1) that $z)(x$. The element z which satisfies the hypothesis of condition (ζ) thus also satisfies the hypothesis of condition (ϵ); consequently $z < v$ for every $v \in B$ and in particular $z < y$. If now in (6) we replace 'x' by 't' and take note of (δ) and (ζ), we obtain $t = y$, whence, by (γ), $x < y$. In this way we come to a contradiction with assumption (α), and hence we must assume that Post. $\mathfrak{B}_3$ is satisfied.

It is thus proved that the postulate systems $[\mathfrak{B}_1-\mathfrak{B}_4]$ and $[\mathfrak{B}_2, \mathfrak{B}_4^*]$ are actually equivalent.[1]

[1] The formulation of Posts. $\mathfrak{B}_4$ and $\mathfrak{B}_4^*$ as well as some fragments of the proofs of Ths. 1 and 2 have been influenced by the researches of S. Leśniewski. The extended system of Boolean algebra is closely related to the deductive theory developed by S. Leśniewski and called by him *mereology*. The foundations of mereology have been briefly discussed in article II, where bibliographical references to the relevant works of Leśniewski will also be found. The relation of the part to the whole, which can be regarded as the only primitive notion of mereology, is the correlate of Boolean-algebraic inclusion. The postulate system $[\mathfrak{B}_2, \mathfrak{B}_4^*]$ has been obtained by a slight modification of the postulate system for mereology (Posts. I and II) suggested in II; regarding the relation of the latter system to the original postulate system of Leśniewski see II, p. 25, footnote 2.

The formal difference between mereology and the extended system of Boolean algebra reduces to one point: the axioms of mereology imply (under the assumption of the existence of at least two different individuals) that there is no individual corresponding to the Boolean-algebraic zero, i.e. an individual which is a part of every other individual. If a set B of elements (together with the relation of inclusion) constitutes a model of the extended system of Boolean algebra, then, by removing the zero element from B, we obtain a model for mereology; if, conversely, a set C is a model for mereology, then, by adding a new element to C and by postulating that this element is in the relation of

§ 2. The Atomistic System of Boolean Algebra

There are many problems which, although they are formulated exclusively in the terms of Boolean algebra, yet they cannot be resolved either in the ordinary or in the extended system of this theory. Two groups of questions especially deserve attention: those of the first group are connected with the concept of cardinal number and are especially concerned with the cardinal number of the set B; those of the second group involve the concept of an *atom* or *indecomposable element*.[1] We shall deal mainly with the problems of the second group and proceed at once to define the concept of an atom, introducing a special symbol, '*At*', to denote the set of all atoms.

Definition ℭ. $x \in At$ (1) *if and only if* $x \in B$ *and* $x \neq 0$, *and* (2), *for every element* $y \in B$, *the formulas* $y < x$ *and* $y \neq 0$ *imply* $y = x$.

Various equivalent transformations of this definition are known:

Theorem 3. *On the basis of Posts.* $\mathfrak{A}_1$–$\mathfrak{A}_{10}$ *and Def.* ℭ *the following conditions are equivalent for all* $x \in B$; (1) $x \in At$; (2) *every element* $y \in B$ *satisfies one and only one of the two formulas*: $x < y$ *or* $x < y'$; (3) $x \neq 0$ *and the formula* $x = y + z$, *for all elements* $y, z \in B$, *implies* $x = y$ *or* $x = z$; (4) $x \neq 0$ *and the formula* $x < y + z$, *for all elements* $y, z \in B$, *implies* $x < y$ *or* $x < z$; (5) *for every set* $X \subseteq B$ *the formula* $x = \sum_{y \in X} y$ *implies* $x \in X$; (6) *for every set* $X \subseteq B$, *the formula* $x < \sum_{y \in X} y$ *implies the existence of an element* $y \in X$ *such that* $x < y$.

The proof presents no difficulties.

inclusion to every element of C, we obtain a model for the extended system of Boolean algebra. Apart from these formal differences and similarities, it should be emphasized that mereology, as it was conceived by its author, is not to be regarded as a formal theory where primitive notions may admit many different interpretations. Regarding Ths. 1 and 2 see Tarski, A. (78).

[1] This concept is treated in Schröder, E. (62) vol. 2, Part 1, pp. 318, 349. There will be found a definition of the atom which is equivalent to the Def. ℭ given here, as well as a proof of several properties of this concept, including those formulated here in Th. 3. Instead of the expression 'atom' Schröder uses the term 'Individual' (in German).

In connexion with Def. $\mathfrak{C}$ a whole series of questions arises: Do atoms exist at all? If so, are atoms included in every element which belongs to the universe of discourse and is distinct from 0? Further, is every element the sum of the atoms included in it? These questions can be neither affirmed nor denied on the basis of the previous assumptions. But all the above problems are solved in a positive sense the moment we enlarge the postulate system $[\mathfrak{A}_1-\mathfrak{A}_{10}]$ or $[\mathfrak{B}_1-\mathfrak{B}_4]$ by adding the following sentence:

POSTULATE $\mathfrak{D}$. *If* $x \in B$ *and* $x \neq 0$, *then there is an element* $y \in At$ *such that* $y < x$.

The totality of postulates $[\mathfrak{A}_1-\mathfrak{A}_{10}, \mathfrak{D}]$ and the theorems following from them we call the *atomistic system of Boolean algebra*.

The Post. $\mathfrak{D}$, although seeming to be logically weak, yet brings with it a whole series of far-reaching consequences. In what follows we shall discuss some of these, showing at the same time that they are equivalent to the new postulate. The first group of these consequences is formed by sentences which in one way or another express the possibility of representing any element as the sum of atoms:

POSTULATE $\mathfrak{D}_1$. (*a*) $1 = \sum\limits_{y \in At} y$;

(*b*) *if* $x \in B$, *then* $x = \sum\limits_{y \in At \text{ and } y < x} y$.

POSTULATE $\mathfrak{D}_2$. *Let* $x, y \in B$; *then*

(*a*) *if for every element* $z \in At$ *the formula* $z < x$ *implies* $z < y$, *then* $x < y$;

(*b*) *if the formulas* $z < x$ *and* $z < y$ *are equivalent for every element* $z \in At$, *then* $x = y$.

THEOREM 4. *On the basis of Posts.* $\mathfrak{A}_1-\mathfrak{A}_{10}$ *and Def.* $\mathfrak{C}$, *Post.* $\mathfrak{D}$ *is equivalent to each of the Posts.* $\mathfrak{D}_1^a$, $\mathfrak{D}_1^b$, $\mathfrak{D}_2^a$, *and* $\mathfrak{D}_2^b$.[1]

Proof. First $\mathfrak{D}_1^a$ is derived from $\mathfrak{D}$. If $\sum\limits_{y \in At} y \neq 1$, then $\left(\sum\limits_{y \in At} y\right)' \neq 0$; hence, according to Post. $\mathfrak{D}$, there would be a $z \in At$ such that $z < \left(\sum\limits_{y \in At} y\right)'$. But at the same time by virtue

---

[1] Schröder in (62), vol. 2, part 1, pp. 337–8, expresses the conjecture that Post. $\mathfrak{D}_1$ cannot be derived from $\mathfrak{D}$.

of $\mathfrak{A}_8^b$ we should have $z < \sum\limits_{y \in At} y$. From the condition (2) of Th. 3 $\left(\text{where '}x\text{' is replaced by '}z\text{' and '}y\text{' by '} \sum\limits_{y \in At} y\text{'}\right)$ it follows, however, that these last two given inclusions contradict one another. It must thus be assumed that $1 = \sum\limits_{y \in At} y$, q.e.d.

If $\mathfrak{D}_1^a$ is now assumed, $\mathfrak{D}_1^b$ is proved in the following way. Let $x \in B$; then by virtue of $\mathfrak{D}_1^a$ and $\mathfrak{A}_{10}^a$ we have

$$x = x \cdot 1 = x \cdot \sum_{y \in At} y = \sum_{y \in At} (x \cdot y).$$

On the other hand we infer from the condition (2) of Th. 3 that for any $y \in At$ either $y \cdot x = y$ or $y \cdot x = 0$, according to whether the inclusion $y < x$ holds or not. Consequently the set

$$X = \underset{y}{E}[y < x \text{ and } y \in At]$$

differs from the set of all elements of the form $y \cdot x$, with $y \in At$, at most by the absence of the null element. Hence with the help of $\mathfrak{A}_8$ we easily conclude that

$$\sum_{y < x \text{ and } y \in At} y = \sum_{y \in At} (x \cdot y)$$

and finally that

$$x = \sum_{y < x \text{ and } y \in At} y,$$

as was to be shown.

From $\mathfrak{D}_1^b$ we obtain $\mathfrak{D}_1^a$ by the use of Post. $\mathfrak{A}_8^c$ in which we put $X = \underset{z}{E}[z \in At \text{ and } z < x]$ and $z = y$. From $\mathfrak{D}_2^a$ by means of $\mathfrak{A}_2$ we can at once derive $\mathfrak{D}_2^b$.

In order, finally, to prove $\mathfrak{D}$ on the basis of $\mathfrak{D}_2^b$, we proceed as follows: If $x \in B$ and $x \neq 0$, then according to $\mathfrak{D}_2^b$ the formulas $z < x$ and $z < 0$ cannot be equivalent for every $z \in At$; since by Def. $\mathfrak{C}$ we always have $z \neq 0$ and consequently $z \, non\text{-}< 0$ for every $z \in At$, there must be an element $z \in At$ such that $z < x$, which was to be proved.

Thus Posts. $\mathfrak{D}$–$\mathfrak{D}_2^b$ are in fact equivalent.

By Th. 1 we can replace in Th. 4 the postulate system $[\mathfrak{A}_1$–$\mathfrak{A}_{10}]$ by the system of postulates and definitions $[\mathfrak{B}_1$–$\mathfrak{B}_6]$. In this connexion we note that in the system $[\mathfrak{B}_1$–$\mathfrak{B}_6]$ enlarged by Def. $\mathfrak{C}$ and Post. $\mathfrak{D}_2^a$, Post. $\mathfrak{B}_3$ is superfluous, since it can easily be derived from $\mathfrak{D}_2^a$ (the formula '$x \neq 0$' in $\mathfrak{C}$ must however first be replaced by the condition '*there is an element $z \in B$ such that x non-$< z$*', and similarly for the formula '$y \neq 0$'). Post. $\mathfrak{B}_2$ can also be eliminated from the system if we strengthen $\mathfrak{D}_2^a$ in the following way:

POSTULATE $\mathfrak{D}_2^*$. *If $x, y \in B$, then $x < y$ if and only if for every $z \in At$ the formula $z < x$ implies $z < y$.*

The sentences $\mathfrak{B}_1$, $\mathfrak{B}_4$, and $\mathfrak{D}_2^*$ thus form a postulate system which suffices for the establishment of the atomistic system of Boolean algebra.

It should be pointed out that, on the basis of the usual system of Boolean algebra, i.e. Posts. $\mathfrak{A}_1$–$\mathfrak{A}_7$, Th. 4 does not lose its validity: the sentences $\mathfrak{D}$–$\mathfrak{D}_2^b$ remain equivalent (if the symbol '$\sum$' is understood in the sense of Def. $\mathfrak{B}_6^a$).†

From Ths. 3 and 4 it follows that in the Boolean algebra enriched by Post. $\mathfrak{D}$ a certain method of reasoning can be used which is familiar from the calculus of classes. If in the calculus of classes the inclusion '$X \subseteq Y$', or the equality '$X = Y$' is to be proved, it is customary to show that every element of the set X is also an element of the set Y, or that the two sets consist of the same elements. According to postulates $\mathfrak{D}_2^a$ and $\mathfrak{D}_2^b$ we can proceed in a completely analogous manner in the atomistic system of Boolean algebra by operating with atoms and making use of those of their properties that are established in conditions (2), (4), and (6) of Th. 3.

By means of the method of reasoning just described two sentences can be derived from $\mathfrak{D}$ which are generalizations of Posts. $\mathfrak{A}_5$ and $\mathfrak{A}_{10}$, these are the *general distributive laws for addition and multiplication*:

POSTULATE $\mathfrak{D}_3$. *Let $\mathfrak{R}$ be any class of subsets X of the set B $\left(\sum_{X \in \mathfrak{R}} X \subseteq B \right)$; and let $\mathfrak{L}$ be the class of all those sets Y which are included in the sum of all sets of the class $\mathfrak{R}$ and have at least one*

---

† In the original of this article it was erroneously stated that on the basis of the ordinary system of Boolean algebra Post. $\mathfrak{D}$ is essentially weaker than Posts. $\mathfrak{D}_1^a$–$\mathfrak{D}_2^b$. Actually, however, the proof that all these postulates are equivalent can be carried out along the same lines as the proof outlined above for Th. 4. The fact that Posts. $\mathfrak{D}_1^a$–$\mathfrak{D}_2^b$ are derivable from Post. $\mathfrak{D}$ is also a direct consequence of Th. 24 in article VIII, § 3.

element in common with every set of the class $\Re$ $\left(Y \subseteq \sum_{X \in \Re} X; \ if \right.$
$X \in \Re, \ then \ X.Y \neq 0 \right)$. *Then*:

$$(a) \qquad \prod_{X \in \Re} \sum_{z \in X} z = \sum_{Y \in \Omega} \prod_{z \in Y} z;$$

$$(b) \qquad \sum_{X \in \Re} \prod_{z \in X} z = \prod_{Y \in \Omega} \sum_{z \in Y} z.$$

The question now arises whether Post. $\mathfrak{D}_3$ can be derived from Posts. $\mathfrak{A}_1$–$\mathfrak{A}_{10}$ without the help of $\mathfrak{D}$. We have shown in collaboration with A. Lindenbaum that this is not possible: $\mathfrak{D}$ not only implies $\mathfrak{D}_3$, but is actually equivalent to it.

THEOREM 5. *On the basis of Posts.* $\mathfrak{A}_1$–$\mathfrak{A}_{10}$ *and Def.* $\mathfrak{C}$ *Posts.* $\mathfrak{D}, \mathfrak{D}_3^a$, *and* $\mathfrak{D}_3^b$ *are equivalent.*

Proof. $\mathfrak{D}_3^a$ is derived from $\mathfrak{D}$ with the help of the method discussed above: it is shown that for every $u \in At$ the formulas
(α) $u < \prod_{X \in \Re} \sum_{z \in X} z$ and (β) $u < \sum_{Y \in \Omega} \prod_{z \in Y} z$ are equivalent. In fact, by $\mathfrak{A}_9^b$ and $\mathfrak{A}_1^b$ it follows from (α) that $u < \sum_{z \in X} z$ for every $X \in \Re$. Since $u \in At$, we infer, with the help of Th. 5, that for every set $X \in \Re$ there is a corresponding element $z \in X$ such that $u < z$. Thus, if Y contains all elements z which belong to at least one set $X \in \Re$ and satisfy the inclusion $u < z$, then Y has elements in common with every set X and therefore belongs to the class Ω. On the other hand, by $\mathfrak{A}_9^a$, the formula $u < \prod_{z \in Y} z$ holds. Hence with the help of $\mathfrak{A}_8^b$ and $\mathfrak{A}_1^b$ we obtain the formula (β) without difficulty. The proof of the implication in the opposite direction is still easier (actually the inclusion $\sum_{Y \in \Omega} \prod_{z \in Y} z < \prod_{X \in \Re} \sum_{z \in X} z$ can be derived from Posts. $\mathfrak{A}_1$–$\mathfrak{A}_{10}$ alone). Since Post. $\mathfrak{D}_2^b$ follows from $\mathfrak{D}$, the equivalence of the formulas (α) and (β) for every $u \in At$ leads directly to the equation $\mathfrak{D}_3^a$.

In turn we shall show how Post. $\mathfrak{D}$ can be derived from $\mathfrak{D}_3^a$. Let x be any element of the set B which is distinct from 0. By the symbol '$\Re$' we denote the class of all sets which consist of only two elements of the form $x.y$ and $x.y'$, where y is some element of the set B. For every set $X \in \Re$ we have

$$\sum_{z \in X} z = x.y + x.y' = x.(y + y') = x.1 = x,$$

whence we obtain at once $\prod_{X \in \Re} \sum_{z \in X} z = x$. Next let us form for the class $\Re$ the class $\mathfrak{L}$ in the way prescribed by the hypothesis of Post. $\mathfrak{D}_3$. By $\mathfrak{D}_3^a$ we have

$$\prod_{X \in \Re} \sum_{z \in X} z = \sum_{Y \in \mathfrak{L}} \prod_{z \in Y} z,$$

whence

$$\sum_{Y \in \mathfrak{L}} \prod_{z \in Y} z = x \neq 0.$$

From this it is easily inferred that there exists a set $Y \in \mathfrak{L}$, such that $\prod_{z \in Y} z \neq 0$. We consider any element $y \in B$. Since $X = \{x.y, \ x.y'\} \in \Re$, X has at least one element in common with the set Y, i.e. either $x.y \in Y$ or $x.y' \in Y$. Accordingly, by $\mathfrak{A}_9^b$, either $\prod_{z \in Y} z < x.y$ or $\prod_{z \in Y} z < x.y'$ holds; hence $\prod_{z \in Y} z < x$ and either $\prod_{z \in Y} z < y$ or $\prod_{z \in Y} z < y'$. However, the last two inclusions cannot both hold, since $\prod_{z \in Y} z \neq 0$. It is thus seen that the element $\prod_{z \in Y} z$ satisfies the condition (2) of Th. 3, whence $\prod_{z \in Y} z \in At$. In this way we have derived $\mathfrak{D}$, in fact for any given element $x \neq 0$ we have constructed an element $y \in At$ such that $y < x$. As regards Post. $\mathfrak{D}_3^b$ it is not difficult to prove that it is equivalent to $\mathfrak{D}_3^a$ (for this purpose we can use the generalized De Morgan laws, which can easily be derived from Posts. $\mathfrak{A}_1$–$\mathfrak{A}_{10}$). Thus the sentences $\mathfrak{D}$, $\mathfrak{D}_3^a$ and $\mathfrak{D}_3^b$ are finally seen to be equivalent, as was to be proved.

It should be emphasized that the distributive laws are not given the sharpest possible form in Post. $\mathfrak{D}_3$ since the class $\mathfrak{L}$ contains in general many superfluous sets. Only in Post. $\mathfrak{D}_3^*$ will these laws be formulated in a completely adequate way.

POSTULATE $\mathfrak{D}_3^*$. *Let $\Re$ be any class of non-empty subsets of the set B, and let $\mathfrak{F}$ be the class of all those functions f which correlate some element $f(X) \in X$ with every set $X \in \Re$. Then:*

(a) $$\prod_{X \in \Re} \sum_{y \in X} y = \sum_{f \in \mathfrak{F}} \prod_{X \in \Re} f(X);$$

(b) $$\sum_{X \in \Re} \prod_{y \in X} y = \prod_{f \in \mathfrak{F}} \sum_{X \in \Re} f(X).$$

Th. 5 remains valid if $\mathfrak{D}_3$ in it is replaced by $\mathfrak{D}_3^*$; but when deriving $\mathfrak{D}_3^*$ from $\mathfrak{D}$, we have to make use of the axiom of of choice.†

We now formulate two sentences which are equivalent, not only to Post. $\mathfrak{D}$ (on the basis of Posts. $\mathfrak{A}_1$–$\mathfrak{A}_{10}$ or $\mathfrak{B}_1$–$\mathfrak{B}_4$), but to the whole atomistic postulate system of Boolean algebra, i.e. [$\mathfrak{B}_1$–$\mathfrak{B}_4$, $\mathfrak{D}$]. These sentences are quite different in character from the postulates so far considered.

POSTULATE $\mathfrak{E}_1$. *There is a set E as well as a function F, which satisfies the following conditions*: (1) *if $x \in B$, then $F(x) \subseteq E$*; (2) *if $X \subseteq E$, then there exists exactly one element $x \in B$ such that $X = F(x)$*; (3) *if $x, y \in B$, then $x < y$ if and only if $F(x) \subseteq F(y)$.*

POSTULATE $\mathfrak{E}_2$. *There is a class $\mathfrak{R}$ of sets and a function F which satisfy the following conditions*: (1) *if $X \in \mathfrak{R}$, then also* $\sum_{Y \in \mathfrak{R}} Y - X \in \mathfrak{R}$; (2) *if $\mathfrak{L} \subseteq \mathfrak{R}$, then also* $\sum_{Y \in \mathfrak{L}} Y \in \mathfrak{R}$; (3) *if $x \in B$, then $F(x) \in \mathfrak{R}$*; (4) *if $X \in \mathfrak{R}$, then there is exactly one element $x \in B$ such that $X = F(x)$*; (5) *if $x, y \in B$, then $x < y$ if and only if $F(x) \subseteq F(y)$.*

THEOREM 6. *On the basis of Defs. $\mathfrak{B}_5$, $\mathfrak{B}_6$, and $\mathfrak{C}$, the postulate system* [$\mathfrak{B}_1$–$\mathfrak{B}_4$, $\mathfrak{D}$] (*or, what amounts to the same thing,* [$\mathfrak{A}_1$–$\mathfrak{A}_{10}$, $\mathfrak{D}$]) *is equivalent to each of the Posts. $\mathfrak{E}_1$ and $\mathfrak{E}_2$.*

Proof. In order to derive Post. $\mathfrak{E}_1$ from the system [$\mathfrak{B}_1$–$\mathfrak{B}_4$, $\mathfrak{D}$] we put $E = At$ and $F(x) = E[y \in At \text{ and } y < x]$ for $x \in B$.

The proof that the set E and the function F determined by these formulas satisfy the conditions (1)–(3) is based on Ths. 3,

---

† The result stated in Th. 5 can be improved in the following way. Instead of $\mathfrak{D}_3$ we consider a somewhat weaker Post. $\mathfrak{D}_4$. The only difference between $\mathfrak{D}_3$ and $\mathfrak{D}_4$ consists in the fact that in $\mathfrak{D}_4$ the formulas (*a*) and (*b*) of $\mathfrak{D}_3$ are assumed to hold under certain existential hypotheses. For example, formula (*a*) is provided with the hypothesis that the following sums and products exist (and hence are elements of B): $\sum_{z \in X} z$ for every $X \in \mathfrak{R}$, $\prod_{X \in \mathfrak{R}} \sum_{z \in X} z$, and $\prod_{z \in Y} z$ for every $Y \in \mathfrak{L}$. $\mathfrak{D}_3$ and $\mathfrak{D}_4$ are, of course, equivalent within the extended system of Boolean algebra. On the other hand, by somewhat modifying the proof of Th. 5, we can show that $\mathfrak{D}$, $\mathfrak{D}_4^a$, and $\mathfrak{D}_4^b$ are equivalent within the ordinary system of Boolean algebra, i.e. on the basis of Posts. $\mathfrak{A}_1$–$\mathfrak{A}_7$ and Defs. $\mathfrak{B}_6$ and $\mathfrak{C}$. These remarks still hold if we consider, instead of $\mathfrak{D}_4$, Post. $\mathfrak{D}_4^*$ obtained in an analogous manner from $\mathfrak{D}_3^*$.

5, and 6, and presents no difficulty. If by $\Re$ we understand the class of all subsets of the set E, we at once obtain $\mathfrak{E}_2$ from $\mathfrak{E}_1$. The derivation of the Posts. $\mathfrak{B}_1$–$\mathfrak{B}_4$ and $\mathfrak{D}$ from $\mathfrak{E}_2$ is somewhat more complicated. However, only Post. $\mathfrak{D}$ causes some difficulty. The problem reduces to showing that every set $X \in \Re$ distinct from 0 contains an atomic subset, i.e. another set $Y \in \Re$, likewise distinct from 0, which includes no set of the class $\Re$ other than 0 and itself. For this purpose we consider any element $x \in X$ and put $Y = \prod_{x \in Z \text{ and } Z \in \Re} Z$; with the help of (1) and (2) it is not difficult to prove that the set Y is just such an atomic set.[1] The theorem is in this way proved.

The sense of Th. 6 can be expressed in the following way: *the atomistic system of Boolean algebra is isomorphic with the ordinary calculus of classes*. Moreover, in this theorem is contained a rather interesting fact belonging to the theory of sets: every class $\Re$ of sets which satisfies conditions (1) and (2) of Post. $\mathfrak{E}_2$ (i.e. is closed under the operations of complementation and unrestricted addition) is isomorphic with the class of all subsets of a certain set E with respect to the relation of inclusion.[2]

---

[1] Cf. Tarski (75), pp. 235 ff.

[2] This article was conceived as the first part of a more comprehensive paper. The second part (which was never published) included, among other things, a discussion of the so-called atomless system of Boolean algebra, i.e. the system in which Post. $\mathfrak{D}$ is replaced by a postulate stating that there are no atoms. The atomless system of Boolean algebra can, of course, be constructed as a system in which 'B' and '$x < y$' occur as the only primitive expressions. A model for such a system is provided by the family of all regular open sets of a Euclidean space and the relation of set-theoretical inclusion. In view of the remarks made above in p. 333, footnote, this follows immediately from the discussion in II, in particular, from Th. B.

XII

FOUNDATIONS OF THE CALCULUS
OF SYSTEMS†

THE results here communicated belong to *general metamathematics* (which I formerly called the *general methodology of the deductive sciences*), and thus to a discipline whose task it is to define the meaning of general metamathematical concepts which appear in the discussion of the most diverse deductive theories and to establish the basic properties of these concepts. Only those theories will be considered here, however, whose construction presupposes a logical basis of a greater or lesser extent, and at least the whole sentential calculus.

In a previous communication (article III of the present volume) I have sketched a system of foundations for the general metamathematics of theories with the above characteristics. Here in § 1 a modification will be undertaken in this system whereby, I believe, greater naturalness and simplicity is reached. In § 2 I sketch a calculus—I call it the *calculus of deductive systems*, or simply the *calculus of systems*—which essentially facilitates metamathematical investigations of the kind considered here; at the same time an interesting analogy will be apparent between this calculus and intuitionistic logic. In § 3 I extend the calculus of systems by introducing the concept of an axiomatizable system. In § 4 two further kinds of deductive systems are considered, namely irreducible and complete systems, and some rather deep theorems are stated which can be obtained within the calculus of systems and which characterize certain classes of deductive systems with respect to cardinality and structure. In § 5 the general results will be

† BIBLIOGRAPHICAL NOTE. This article originally appeared in two parts under the title 'Grundzüge des Systemenkalkül, Erster Teil', *Fundamenta Mathematicae*, vol. 25 (1935), pp. 503–26, and 'Grundzüge des Systemenkalkül, Zweiter Teil', ibid., vol. 26 (1936), pp. 283–301. The first part comprised the Introduction and §§ 1–3, the second part §§ 4–5 and the Appendix. Earlier publications of the author in the same directions are III and V; see also the historical remarks at the end of the article.

illustrated by applying them to some special, and in fact very elementary, deductive theories.[1]

Proofs will not be given here. With few exceptions they are very simple.

§ 1. PRIMITIVE CONCEPTS AND AXIOMS

In the postulate system given in III four primitive expressions appeared: 'S'—'*the set of all meaningful sentences*', '$Cn(X)$' —'*the set of consequences of the set X of sentences*', '$\bar{x}$'—'*the negation of the sentence x*', '$x \to y$'—'*the implication with the antecedent x and the consequent y*'.[2] With the help of these terms, the concept of *deductive system* was there defined, i.e. the concept of a set of sentences which contains as elements all its consequences; the class of all deductive systems was denoted by the symbol '$\mathfrak{S}$'. It was there remarked that among deductive systems a smallest exists, i.e. a system which is a sub-system of all other deductive systems. It is the system $Cn(0)$, the set of consequences of the empty set. This system, which here for brevity will be denoted by 'L', can be interpreted as the set of all logically valid sentences (or, more generally, as the set of all those sentences which from the start we recognize as true when undertaking the construction of the deductive theory that is the object of our metamathematical investigation).[3]

It now appears that the postulate system mentioned takes on a

---

[1] In order to avoid misunderstandings it is important to note that I use the expressions 'deductive theory' and 'deductive system' in quite distinct senses. By deductive theories I understand here the models (realizations) of the axiom system which is given in § 1. Since in this system four primitive concepts appear, every quadruple of concepts which satisfies all the axioms of the system is its model. In order not to depart too much from the usual meaning of the term 'deductive theory', I have in mind only those models of the axiom system which are constituted by certain sets of expressions and operations on expressions. On the other hand, deductive systems (in the domain of a particular deductive theory) are certain special sets of expressions which I shall characterize precisely at the beginning of § 1 as well as in Def. 5 of § 2.

[2] The symbolism employed here differs slightly from that of III; in fact '$\bar{x}$' is used instead of '$n(x)$', and '$x \to y$' instead of '$c(x, y)$'. Otherwise, just as in III, I employ the customary set-theoretical notation; in particular I use formulas of the type '$x, y,... \in X$' as abbreviations of the expressions '*the elements x, y,... belong to the set X*'.

[3] Cf. III, p. 33 and V, Th. 9 (*b*), p. 70, of the present work.

simpler and, as it seems to me, also more natural form if, instead of '$Cn(X)$', the symbol 'L' is included in the list of primitive expressions. For the foundation of general metamathematics (to the extent to which this task was attempted in III) the following five axioms then suffice:

AXIOM 1. $0 < \bar{\bar{S}} \leqslant \aleph_0$.

AXIOM 2. *If $x, y \in S$, then $\bar{x} \in S$ and $x \to y \in S$.*

AXIOM 3. $L \subseteq S$.

AXIOM 4. *If $x, y, z \in S$, then $(\bar{x} \to x) \to x \in L$, $x \to (\bar{x} \to y) \in L$ and $(x \to y) \to ((y \to z) \to (x \to z)) \in L$.*

AXIOM 5. *If $x, x \to y \in L$ (where $y \in S$), then $y \in L$.*

The content of these axioms is so clear that it scarcely requires an explanation. From the intuitive standpoint they may nevertheless admit of certain doubts which I shall not here discuss.[1]

The Axs. 2, 4, and 5 are adapted from the system of primitive expressions, axioms and rules of inference of Łukasiewicz, which forms a sufficient basis for the development of the sentential calculus.[2] It is clear that we could equally well use any other formal system of the sentential calculus as a starting point. In particular, if we took Nicod's sentential calculus as our model, then the number of the primitive expressions appearing in the axioms would be smaller, and Axs. 2 and 4 would be somewhat simpler.[3]

We will show how, with the help of the primitive concepts adopted, the concept of consequence can be defined. For this purpose we put in the first place:

DEFINITION 1. $x+y = \bar{x} \to y$, $x.y = \overline{x \to \bar{y}}$ *for all $x, y \in S$.*

The concepts of the sum and the product of sentences can be extended to an arbitrary finite number of summands and factors by recursion:

---

[1] Cf. III, p. 31, note 3; IX, p. 282, note 2; and VIII, footnote, p. 156.
[2] Cf. IV, pp. 39 and 43; III, p. 33, the remarks in connexion with Th. 3*.
[3] Cf. Nicod, J. (55).

DEFINITION 2. $\sum\limits_{i=1}^{n} x_i = \prod\limits_{i=1}^{n} x_i = x_1$, if $n = 1$ and $x_1 \in S$;

$\sum\limits_{i=1}^{n} x_i = \sum\limits_{i=1}^{n-1} x_i + x_n$ and $\prod\limits_{i=1}^{n} x_i = \prod\limits_{i=1}^{n-1} x_i . x_n$, if n is an arbitrary natural number > 1 and $x_1, x_2, ..., x_n \in S$.

The definition of the concept of *consequence* runs as follows:

DEFINITION 3. *For an arbitrary set* $X \subseteq S$ *the set* $Cn(X)$ *consists of those, and only those, sentences* $y \in S$, *which satisfy the following condition: either* $y \in L$, *or there exist sentences*

$$x_1, x_2, ..., x_n \in X,$$

such that $\left(\prod\limits_{i=1}^{n} x_i \right) \to y \in L.$

This definition can be transformed in the following way:

THEOREM 1. *For an arbitrary set* $X \subseteq S$ *the set* $Cn(X)$ *is the intersection of all sets* Y *which satisfy the following two conditions:* (1) $L + X \subseteq Y$; (2) *if* $x, x \to y \in Y$ (*where* $y \in S$), *then* $y \in Y$.

It is easy to show that the set $Cn(X)$ is one of the sets Y which satisfy the conditions (1) and (2) of Th. 1; it is thus the smallest of these sets. Since the operation which correlates the sentence y with two given sentences x and $x \to y$ is commonly called the operation of *detachment*, we can, on the basis of Th. 1, characterize the set $Cn(X)$ as the smallest set which includes the sets L and X and is closed under the operation of detachment.†

It might appear that the concept of consequence here introduced is essentially narrower than that which is generally used in the construction of deductive theories; in the definition of this concept we have made use of only one of the rules of inference, the rule of detachment, as though we had totally forgotten the existence of other rules of this kind, e.g. the rule of substitution and the rules of universal quantification.[1] All this, however, depends essentially on the interpretation of the symbol 'L'.

---

[1] Cf. for instance VIII, p. 180.

† In connexion with Def. 3, cf. the papers of K. Ajdukiewicz, (2) and (4). Keeping in mind the meaning of $Cn(X)$ as determined by Def. 3 we recognize in Th. 1 one of the formulations of the so-called deduction theorem. In fact this is just the form in which the deduction theorem was first established by the author in 1921; see III, p. 32, footnote †.

If the system of logic on which we base deductive theories is so constituted that, whenever a given sentence y can be obtained with the help of any of the rules of inference from another sentence x, or from other sentences $x_1,..., x_n$, then the implication, $x \to y$, or $\left(\prod_{i=1}^{n} x_i \right) \to y$, belongs to the set L of all logically valid sentences (actually all known and sufficiently developed systems of logic satisfy this condition), then the concept of consequence is in no way narrowed. In fact, the sentence y will under these conditions be a consequence of every set of sentences which contains the sentence x or the sentences $x_1,..., x_n$, not only in the usual sense, but also in the sense of Def. 3 or of Th. 1. We can thus say that the rule of detachment (even in the special form in which it appears in Def. 3) suffices as the only rule of inference in constructing deductive theories, provided that these theories stand on a sufficiently rich logical basis.

The following two theorems can now be proved:

THEOREM 2. (a) If $X \subseteq S$, then $X \subseteq Cn(X) \subseteq S$;

(b) if $X \subseteq S$, then $Cn(Cn(X)) = Cn(X)$;

(c) if $X \subseteq S$, then $Cn(X) = \sum\limits_{Y \subseteq X \text{ and } \overline{Y} < \aleph_0} Cn(Y)$;

(d) there exists a sentence $x \in S$ such that $Cn(\{x\}) = S$;

(e) if $X \subseteq S$, $y, z \in S$ and $y \to z \in Cn(X)$, then $z \in Cn(X+\{y\})$;

(f) if $X \subseteq S$, $y, z \in S$ and $z \in Cn(X+\{y\})$, then $y \to z \in Cn(X)$;

(g) if $x \in S$, then $Cn(\{x, \bar{x}\}) = S$;

(h) if $x \in S$, then $Cn(\{x\}) . Cn(\{\bar{x}\}) = Cn(0)$.

THEOREM 3. $L = Cn(0)$.

Th. 2 shows that from the Axs. 1–5 given above and from Def. 3 all Axs. 1–10* which were proposed in III can be derived (we have neglected the Axs. 1 and 6* of III in the formulation of Th. 2, because the first is contained in the new Ax. 1, and the second coincides with the new Ax. 2). It is easy to show that the converse of this result also holds: if we regard Th. 3 as the definition of the symbol 'L', then we can derive from this definition and the axiom system of III the new Axs. 1–5 and Def. 3 as theorems. Finally we can thus assert that the two systems of axioms and primitive notions are equivalent.

§ 2. The Calculus of Systems

On the basis outlined in the previous paragraph two calculi can be constructed which are very useful in metamathematical investigations, namely the *calculus of sentences* and the *calculus of deductive systems*; the first is a complete and the second a partial interpretation of the formal system which is usually called the *algebra of logic* or *Boolean algebra*.

In order to make the further considerations more specific we shall here formulate *in extenso* a well-known system of postulates which suffices for the construction of the algebra of logic. In this system eight primitive concepts appear: 'B'—'the universe of discourse', '$x < y$'—'the element x stands in the relation of inclusion to the element y', '$x = y$'—'the element x stands in the relation of equality to the element y', '$x+y$'—'the sum of the elements x and y', '$x.y$'—'the product of the elements x and y', '0'—'the zero element' ('the empty element'), '1'—'the unit element' ('the universal element'), '$\bar{x}$'—'the complement of the element x'.

The system contains seven postulates:

POSTULATE I. *(a) If $x \in B$, then $x < x$; (b) if $x, y, z \in B$, $x < y$ and $y < z$, then $x < z$.*

POSTULATE II. *If $x, y \in B$, then $x = y$ if and only if both $x < y$ and $y < x$.*

POSTULATE III. *If $x, y \in B$, then (a) $x+y \in B$; (b) $x < x+y$ and $y < x+y$; (c) if $z \in B$, $x < z$ and $y < z$, then $x+y < z$.*

POSTULATE IV. *If $x, y \in B$, then (a) $x.y \in B$; (b) $x.y < x$ and $x.y < y$; (c) if $z \in B$, $z < x$ and $z < y$, then $z < x.y$.*

POSTULATE V. *If $x, y, z \in B$, then (a) $x.(y+z) = x.y+x.z$ and (b) $x+(y.z) = (x+y).(x+z)$.*

POSTULATE VI. *(a) 0, 1 $\in B$; (b) if $x \in B$, then $0 < x$ and $x < 1$.*

POSTULATE VII. *If $x \in B$, then (a) $\bar{x} \in B$, (b) $x.\bar{x} = 0$, and (c) $x+\bar{x} = 1$.*

The system of the algebra of logic can be extended by introducing two infinite operations: '$\sum_{y \in X} y$'—'the sum of all elements

of the set X' and '$\prod_{y \in X} y$'—'*the product of all elements of the set X'*.
It is then necessary to add the following postulates:

POSTULATE VIII. *If $X \subseteq B$, then* (a) $\sum_{y \in X} y \in B$ *and* (b) $x < \sum_{y \in X} y$
for every $x \in X$; (c) *if moreover $z \in B$, and $x < z$ for every $x \in X$,*
then $\sum_{y \in X} y < z$.

POSTULATE IX. *If $X \subseteq B$, then* (a) $\prod_{y \in X} y \in B$ *and* (b) $\prod_{y \in X} y < x$
for every $x \in X$; (c) *if moreover $z \in B$, and $z < x$ for every $x \in X$,*
then $z < \prod_{y \in X} y$.

POSTULATE X. *If $x \in B$ and $X \subseteq B$, then* (a) $x . \sum_{y \in X} y = \sum_{y \in X} (x.y)$
and (b) $x + \prod_{y \in X} y = \prod_{y \in X} (x+y)$.

Posts. I–VII together with all theorems which can be derived
from them, form the *ordinary*, and Posts. I–X the *extended* (or
complete) *system of the algebra of logic*.[1]

We pass now to the *calculus of sentences*, which we shall only
describe briefly. In order to avoid terminological misunderstanding we shall call it the *sentential algorithm* instead of simply
sentential calculus; the term 'sentential calculus' already has a
definite meaning, other than that intended here.

The set S forms the universe of discourse of the sentential
algorithm. Between the elements of this set we shall define two
relations: '$x \supset y$' ('*x implies y*') and '$x \equiv y$' ('*x is equivalent
to y*').

DEFINITION 4. (a) $x \supset y$ *if and only if $x, y \in S$ and $x \rightarrow y \in L$;*
(b) $x \equiv y$ *if and only if both $x \supset y$ and $y \supset x$.*

The following can now be proved:

THEOREM 4. *If in the ordinary system of the algebra of logic we
replace in all the postulates the symbols 'B', '$<$', and '$=$' respectively by 'S', '$\supset$', and '$\equiv$'; if further '0' is replaced by the variable
'u' and '1' by the variable 'v', making everywhere in the corresponding postulates the assumptions that $u \in S$ and $\bar{u} \in L$ as well as
$v \in L$; and if the remaining symbols are left unchanged; then Posts.
I–VII are satisfied.*

[1] Cf. XI, pp. 320–33.

It is to be noted in connexion with this theorem that the symbols '0' and '1' of the algebra of logic cannot be 'effectively' interpreted on our foundations, for we cannot define a single constant which designates an individual sentence.[1]

Th. 4 shows that all postulates of the ordinary system of the algebra of logic can be derived from Axs. 1–5 proposed in § 1 on the basis of suitably chosen definitions (Defs. 1 and 4). It is also easily seen that Ax. 1 plays no part in the proof of this theorem. This result can be reversed: from Posts. I–VII of the algebra of logic (when the replacements indicated in Th. 4 are carried out) one can derive Axs. 2–5 provided that '$x \to y$' is suitably defined, for example by putting $x \to y = \bar{x}+y$. We can thus assert that Axs. 2–5 form a system of statements which is equivalent to the system of postulates for the ordinary algebra of logic. (The problem of the equivalence of the systems of primitive concepts which appear in the two systems of statements would require a more exhaustive discussion.[2])

Since we have chosen as foundation for this work the system of axioms of § 1 we have restricted ourselves in essence to

---

[1] The interpretation of the algebra of logic in sentential algorithm can clearly be modified so as to avoid the replacement of logical identity by another equivalence relation. To obtain this modification we consider, instead of sentences $x \in S$, the equivalence classes $X \subseteq S$ each of which consists of all sentences y which are equivalent, in the sense of Def. 4 b, to some sentence x ($y \equiv x$). For these equivalence classes we define in an appropriate way the relations and operations $\supset$, $+$, $.$, etc.; for instance, the formula '$X \supset Y$' will express the fact that $x \supset y$ (in the sense of Def. 4 a), for all $x \in X$ and $y \in Y$. The elements 0 and 1 are now interpreted 'effectively'—in fact 0 as the set of all sentences $x \in S$ such that $\bar{x} \in L$, and 1 simply as the set L.

[2] As I have already mentioned in § 1, any system of axioms and rules of inference which suffices for the formal construction of the sentential calculus can be used as a model in formulating an axiom system for the domain in which we are interested in this paper. If we compare this remark with the facts just discussed we reach the conviction that there exists a general method which enables us to obtain a system of postulates for the algebra of logic from every system of axioms and rules of inference for the sentential calculus. It would be superfluous to describe here more exactly in what this method consists, especially as it has been developed in recent years in the articles of Bernstein and Huntington; cf. Bernstein, B. A. (6), Huntington, E. V. (34), (35). I take this opportunity of mentioning that the system of Axs. 1–5 of § 1 and its connexion with the algebra of logic were already known to me in the year 1930. (I mention this system implicitly in article V of the present work, p. 62, as the basis upon which the unpublished second part of V has been constructed.)

characterizing the set S as a non-empty, at most denumerable, set, whose elements with respect to the operations defined on them satisfy the postulates of the algebra of logic. Thus, from a purely formal point of view, the investigations which rest on this foundation do not transcend the boundary of the algebra of logic, but they differ from the discussions previously carried out in that field by the fact that they concern concepts which have not been dealt with hitherto. It is not the elements of the system S itself, the relations holding between them, nor the operations performed on them, which form the chief subject of the investigation, but certain selected sets of these elements, namely deductive systems.[1]

Let us remind ourselves of the definition of *deductive system*:

DEFINITION 5. $X \in \mathfrak{S}$ *if and only if* $Cn(X) \subseteq X \subseteq S$.

With the help of the primitive concepts which we have adopted, one may also characterize this concept as follows:

THEOREM 5. $X \in \mathfrak{S}$ *holds if and only if* $L \subseteq X \subseteq S$ *holds and if the formulas* $x, x \to y \in X$ *and* $y \in S$ *always imply* $y \in X$.

The calculus of deductive systems, which we shall call for short the *calculus of systems*, represents (as will be seen below) a very essential extension of the sentential algorithm. The correlates of all primitive concepts of the algebra of logic occur in the calculus of systems. The class $\mathfrak{S}$ forms the universe of discourse. L is regarded as the zero system, S as the unit system. Inclusion and equality between systems as well as the product (the intersection) of systems preserve their usual sense as laid down in the calculus of classes (which, as is well known, presents one of the commonest interpretations of the algebra of logic); equality thus coincides with logical identity. On the other hand the addition of systems, which we call *logical addition* and denote by the symbol '$\dotplus$', does not coincide with set-theoretical addi-

---

[1] In fact the calculus of deductive systems outlined in this paper proves to coincide with what was somewhat later developed as the calculus of Boolean-algebraic ideals; in particular axiomatizable deductive systems (discussed in § 3) coincide with principal ideals, and complete deductive systems (§ 4) with prime ideals. Compare the note of the author, (80). See also the papers of Stone, M. H. (67) and (69).

tion, for this last operation when applied to systems, does not in general yield a new deductive system;[1] the same holds for the operation of complementation.

The definition of the *logical sum* of the systems X and Y is as follows:

DEFINITION 6. $X \dotplus Y = Cn(X+Y)$ *for any given* $X, Y \in \mathfrak{S}$.

The *logical complement* or the *negation of the system* X, symbolically $\overline{X}$, we define as follows:

DEFINITION 7. $\overline{X} = \prod\limits_{x \in X} Cn(\{\overline{x}\})$ *for any given* $X \in \mathfrak{S}$.

We shall learn below (Th. 12) other equivalent formulations of this definition.

On the basis of the definitions just stated the following theorem can be proved:

THEOREM 6. *If in all postulates of the ordinary system of the algebra of logic we replace the variables* 'x', 'y', 'z' *by the variables* 'X', 'Y', 'Z' *which denote deductive systems; if, further, we replace the constants* 'B', '$<$', '$+$', '0', *and* '1' *respectively by the symbols* '$\mathfrak{S}$', '$\subseteq$', '$\dotplus$', 'L', *and* 'S'; *then all postulates are satisfied with the exception of Post. VII c, which fails; but the following consequence of Post. VII c preserves its validity:*

POSTULATE VII (d). *If* $X, Y \in \mathfrak{S}$ *and* $X \cdot Y = L$, *then* $Y \subseteq \overline{X}$.

The essential difference between the calculus of systems and, e.g., the calculus of classes thus lies in the fact that, in the calculus of systems, instead of the law of excluded middle $X \dotplus \overline{X} = S$, only a certain weaker consequence of this law holds. As will be seen from Ths. 17 and 37 given below, the law of excluded middle holds in the case when the class $\mathfrak{S}$ is finite (or, what amounts to the same thing, when the set S does not contain infinitely many sentences such that no two of them are equivalent). If, however, the class $\mathfrak{S}$ is infinite—and we encounter this case as a rule when we consider special deductive sciences—then there exist systems which do not satisfy this law.

The breakdown of the law of excluded middle has further

---

[1] Cf. III, p. 33, Th. 7*; V, p. 71, Th. 12 and the remarks added to it.

consequences. The law of double negation holds but only in one direction, and only the law of triple negation holds in both directions:

THEOREM 7. *If $X \in \mathfrak{S}$, then $X \subseteq \bar{\bar{X}}$ and $\bar{X} = \bar{\bar{\bar{X}}}$.*

Of the two De Morgan laws only one is true, of the four laws of contraposition two drop out and two remain valid:

THEOREM 8. *If $X, Y \in \mathfrak{S}$, then*

$$(a) \quad \overline{X \dotplus Y} = \bar{X}.\bar{Y}; \qquad (b) \quad \overline{\overline{X.Y}} = \bar{\bar{X}}.\bar{\bar{Y}};$$

    (c)   *the formula $X \subseteq Y$ implies $\bar{Y} \subseteq \bar{X}$;*

    (d)   *the formula $X \subseteq \bar{Y}$ implies $Y \subseteq \bar{X}$.*

Ths. 7 and 8 can be derived from the system of postulates described in Th. 6 by purely algebraic methods from the domain of the algebra of logic. The same applies to other theorems of the calculus of systems which are known to us, provided they do not have an existential character.

The formal resemblance of the calculus of systems to the intuitionistic sentential calculus of Heyting is striking:[1] we might say that the formal relation of the calculus of systems to the ordinary calculus of classes is exactly the same as the relation of Heyting's sentential calculus to the ordinary sentential calculus. In other words, if we construct the calculus of classes on the basis of intuitionistic logic, then this calculus, it seems, will not formally differ from the calculus of systems. We can also express this as follows: the system of postulates described in Th. 6 forms (if we abstract from the specific sense of the terms appearing in these postulates) a sufficient basis for a system of the algebra of logic which has the intuitionistic calculus of classes as one of its interpretations. It is clear that the last remarks would require a more exact and detailed elaboration.†

The calculus of systems can be extended by the introduction of infinite operations. The product (the intersection) of systems of a class $\mathfrak{K}$ is understood exactly as in the calculus of classes

---

[1] Cf. Heyting, A. (28).

† For such a more exact and detailed formulation see X and VII, in particular § 5; cf. also Stone, M. H. (70).

(the set S is regarded as the product of the empty class). On the other hand the definition of the *logical sum of systems of the class* $\Re$, symbolically $\overset{\cdot}{\underset{Y\in\Re}{\sum}} Y$, forms the natural generalization of Def. 6:

Definition 8. $\overset{\cdot}{\underset{Y\in\Re}{\sum}} Y = Cn\Big(\underset{Y\in\Re}{\sum} Y \Big)$ *for an arbitrary class* $\Re \subseteq \mathfrak{S}$.

In the case when the class $\Re$ consists of sets $Y_1, Y_2, Y_3,...$ which form a finite sequence with n terms, or an infinite sequence, we write respectively: '$Y_1 \dotplus Y_2 \dotplus ... \dotplus Y_n$', or '$Y_1 \dotplus Y_2 \dotplus ... \dotplus Y_n \dotplus ...$' instead of '$\overset{\cdot}{\underset{Y\in\Re}{\sum}} Y$'.

Theorem 9. *If in all postulates of the extended system of the algebra of logic—in addition to the changes described in Th. 6—we everywhere replace the variables 'X' and 'Y' by the variables '$\Re$' and '$\mathfrak{L}$', denoting classes of systems, and the constants '$\sum$' and '$\prod$' by the constants '$\overset{\cdot}{\sum}$' and '$\prod$'; then in addition to Posts. I–VI and VII a, b, also Posts. VIII, IX, and X a are satisfied, on the other hand Post. X b together with Post. VII c fails.*

According to this theorem, finite multiplication is distributive under infinite addition but the distributive law for finite addition under infinite multiplication no longer holds in the calculus of systems; *a fortiori* the general distributive laws for one of the infinite operations under the other fails.[1] By comparing Ths. 17 and 24 given below we see that the distributive law of finite addition under infinite multiplication preserves its validity in the same cases as does the law of excluded middle.

From the postulates described in Ths. 6 and 9 other theorems of the calculus of systems which involve infinite operations are derivable, e.g. that one of the two De Morgan laws which was valid for finite operations:

Theorem 10. *If $\Re \subseteq \mathfrak{S}$, then* $\overline{\overset{\cdot}{\underset{Y\in\Re}{\sum}} Y} = \underset{Y\in\Re}{\prod} \overline{Y}$.

Theorems of the calculus of systems are also known which

---

[1] These laws are not even valid in the extended system of the algebra of logic; cf. article XI of the present work, § 2, especially pp. 337–9.

cannot be derived from the postulates described in Ths. 6 and 9; as an example we give

THEOREM 11. *If $\Re$ is an arbitrary class of deductive systems such that $S = \sum\limits_{Y \in \Re} Y$, then there is a finite subclass $\mathfrak{L}$ of $\Re$ such that*

$$S = \sum_{Y \in \mathfrak{L}} Y.$$

We now turn our attention to the following theorem:

THEOREM 12. *If $X \in \mathfrak{S}$, then*

(a) $$\bar{X} = \sum_{X.Y=L} Y$$

and

(b) $$\bar{X} = \prod_{X+Y=S} Y.$$

The formula (a) of this theorem is easily derivable from those postulates of the algebra of logic which are satisfied in the calculus of systems, in particular from Posts. VII b, d; the formula (b) cannot be derived in this way, and in its proof we must directly refer to the adopted axioms and to the definitions of the symbols appearing in the formula.

Th. 12 shows how the concept of the logical complement can be characterized with the help of other concepts belonging to the calculus of systems. We see that the logical complement of the system X, as in the ordinary calculus of classes, is the sum of all systems *disjoint* with X, i.e. those for which $X.Y = L$ holds, and likewise it is the intersection of all systems Y which *logically supplement* X: $X \dot{+} Y = S$. Moreover, by virtue of the law of contradiction, $\bar{X}$ is the largest system which is logically disjoint with X; since, however, the law of the excluded middle fails, $\bar{X}$ is in general not the smallest system which logically supplements X (from Th. 12 it only follows that, if such a smallest system exists at all then it coincides with $\bar{X}$).

§ 3. AXIOMATIZABLE SYSTEMS

Since the calculus of systems deviates from the ordinary calculus of classes, a series of problems of the following kind arises: one considers an arbitrary theorem of the calculus of classes which loses its general validity when carried over into the calculus of systems; the problem then is to characterize in the simplest possible manner the class of those systems for which the theorem in question preserves its validity. The most important of these problems concerns the law of excluded middle and leads to the concept of the *axiomatizable system*, already discussed in III.[1]

A deductive system X is called axiomatizable if it has at least one axiom system, i.e. a finite set Y of sentences such that $X = Cn(Y)$. Since, however, the finite set Y can be replaced by a set which contains a single sentence (namely the logical product of the sentences of the set Y ordered in any manner), this definition can be somewhat simplified; we denote the class of all axiomatizable systems by the symbol '$\mathfrak{A}$' and set down:

DEFINITION 9. $X \in \mathfrak{A}$ *if and only if there is an $x \in S$ such that* $X = Cn(\{x\})$.

Hence by Ax. 1 it follows at once:

THEOREM 13. $\overline{\overline{\mathfrak{A}}} \leqslant \aleph_0$.

The following theorems express implicitly elementary properties of axiomatizable systems:

THEOREM 14. *If $x \in S$, then (a) $Cn(\{x\}) = L$ if and only if $x \in L$; (b) $Cn(\{x\}) = S$ if and only if $\bar{x} \in L$; (c) $Cn(\{\bar{x}\}) = \overline{Cn(\{x\})}$.*

THEOREM 15. *If $x, y \in S$, then*

(a) $Cn(\{x\}) \subseteq Cn(\{y\})$ *if and only if $y \supset x$;*

(b) $Cn(\{x\}) = Cn(\{y\})$ *if and only if $x \equiv y$;*

(c) $Cn(\{x+y\}) = Cn(\{x\}) \cdot Cn(\{y\}))$;

(d) $Cn(\{x \cdot y\}) = Cn(\{x, y\}) = Cn(\{x\}) \dotplus Cn(\{y\})$.

From Ths. 14 and 15 it follows that the algorithm of sentences is homomorphic to that segment of the calculus of systems in which we restrict ourselves to the discussion of axiomatizable systems. (This homomorphism is not an isomorphism for the reason that a single axiomatizable system is correlated with all equivalent

---

[1] Cf. III, p. 35, Def. 9; also V, p. 76, § 4, especially Def. 3 b. (I use the concept of axiomatizability in these two articles in a wider sense than here, since I relate it not only to deductive systems but to arbitrary sets of sentences.)

sentences.)† This also justifies the remark made above that the calculus of arbitrary systems—axiomatizable and non-axiomatizable—forms an essential extension of the sentential algorithm. Every theorem of the calculus of systems involving axiomatizable systems can be translated completely or partially (according to whether in this theorem we are dealing exclusively with axiomatizable systems or not) into the language of the sentential algorithm, and vice versa. It must be noted, however, that the two calculi are dual with respect to one another: to the sum of systems corresponds the product of sentences, to the product of systems the sum of sentences, and so on.

In this connexion the following theorem, which is a translation of Th. 4, can be stated:

THEOREM 16. *If in the postulates of the ordinary system of the algebra of logic we carry out all transformations which were described in the hypothesis of Th. 6, but with the difference that 'B' is replaced not by 'S' but by 'A', then all the Posts. I–VII are satisfied.*

From this it follows in particular that the law of excluded middle holds in the calculus of axiomatizable systems (as the correlate of the law of contradiction in the sentential algorithm). Far more interesting, however, is the fact that the axiomatizable systems are the only systems which satisfy the law of excluded middle:

THEOREM 17. $X \in \mathfrak{A}$ *if and only if* $X \in \mathfrak{S}$ *and* $X \dotplus \bar{X} = S$.

Th. 17 is of interest if only on account of the analogy discussed above between the calculus of systems and intuitionistic logic; as is to be expected, the law of excluded middle holds when applied to systems which in a certain sense have a finitistic character, while losing its validity for the remaining systems. Th. 17 in view of its form could be adopted as the definition of the concept of axiomatizability; this definition would be formulated entirely in terms of the calculus of systems. The problem

† If the algorithm of sentences is modified in the way indicated in the first foot-note on p. 349, it obviously becomes isomorphic to the calculus of axiomatizable systems.

of a simple characterization of those systems which obey the
law of excluded middle is completely solved in Th. 17. Below
(in Ths. 23, 24, and 26) we shall find answers to analogous
questions concerning some other theorems of the calculus of
classes.

In proving Th. 17 it is convenient to make use of the following theorem, which is the converse of Posts. III a and IV a from
Th. 16:

THEOREM 18. *If* $X, Y \in \mathfrak{S}$, $X \dotplus Y \in \mathfrak{A}$ *and* $X . Y \in \mathfrak{A}$, *then*
$X, Y \in \mathfrak{A}$.[1] *More generally, if* $X, Y \in \mathfrak{S}$ *and* $X \dotplus Y \in \mathfrak{A}$, *then there
exist systems* $X_1, Y_1 \in \mathfrak{A}$ *such that* $X_1 \subseteq X$, $Y_1 \subseteq Y$ *and*

$$X \dotplus Y = X_1 \dotplus Y_1.$$

In order to complete the foundations of the calculus of
systems we next give a theorem which establishes a certain
connexion between axiomatizable systems and systems of an
arbitrary character, and which does not follow from the theorems
of the calculus of systems hitherto stated:

THEOREM 19. *If* $X, Y \in \mathfrak{S}$, *and if for every system* $Z \in \mathfrak{A}$ *the
formula* $Z \subseteq X$ *implies* $Z \subseteq Y$, *then* $X \subseteq Y$.

We shall now regard the postulates described in Ths. 6 and 9
(obtained by transforming and partly by weakening the postulates of the algebra of logic) together with Ths. 11, 13, 17, and
19 as a *system of postulates for the calculus of systems*, which is
enriched through infinite operations and the concept of axiomatizability. It can be shown that every theorem of the calculus
of systems which can be proved on the basis of the axioms of
§ 1 can also be derived from the postulates just described. We
include in the calculus of systems only those theorems which
contain exclusively symbols and terms of the following three
kinds: (1) terms of a general logical character, (2) constants
which appear in the postulates, as well as symbols which can be
defined with their help, and finally (3) variables which denote
deductive systems, classes of systems and relations between

---

[1] An analogous theorem also holds in the sentential calculus of Heyting, A.
(28) (cf. p. 514, note 1):

$$\vdash : a \vee b . \vee \neg (a \vee b) : \wedge : a \wedge b . \vee \neg (a \wedge b) : \supset . a \vee \neg a . \wedge . b \vee \neg b.$$

systems, classes of those classes and relations, and so on). This remark applies in particular to Ths. 12, 16, and 18.[1]

The postulates of the calculus of systems listed above are not independent of one another (e.g. Posts. V a, b can be derived from the remaining sentences); similarly the primitive concepts of the calculus appearing in these postulates are not independent of one another, i.e. some of these concepts can be defined with the help of the remaining ones. By reducing the number of the primitive concepts a significant simplification of the system of postulates can be reached. E.g., suppose we take as primitive concepts only the two symbols, '$\mathfrak{S}$' and the inclusion sign '$\subseteq$' (we abstract from the general logical meaning of this sign). We can then define the remaining terms of the calculus in the following way:

$\mathbf{D_1}$. *For any given class* $\mathfrak{R}$ *of systems* $\displaystyle\sum_{Y \in \mathfrak{R}} Y$ *is the unique system* $X \in \mathfrak{S}$ *which satisfies the conditions*: (1) $Y \subseteq X$ *for every* $Y \in \mathfrak{R}$; (2) *if* $Z \in \mathfrak{S}$ *and* $Y \subseteq Z$ *for every* $Y \in \mathfrak{R}$, *then also* $X \subseteq Z$.

$\mathbf{D_2}$. *For any given class* $\mathfrak{R}$ *of systems,* $\displaystyle\prod_{Y \in \mathfrak{R}} Y = \sum_{Z \in \mathfrak{L}} Z$, *where* $\mathfrak{L}$ *is the class of all systems* Z *which satisfy the condition* $Z \subseteq Y$ *for every* $Y \in \mathfrak{R}$.

$\mathbf{D_3}$. *For any given systems* X *and* Y,

$$X \dotplus Y = \sum_{Z \in \mathfrak{R}} Z \quad and \quad X.Y = \prod_{Z \in \mathfrak{R}} Z,$$

where $\mathfrak{R} = \{X, Y\}$.

$\mathbf{D_4}$. $\displaystyle S = \sum_{Y \in \mathfrak{S}} Y, \ L = \prod_{Y \in \mathfrak{S}} Y$.

$\mathbf{D_5}$. *For any given system* X, $\displaystyle \bar{X} = \sum_{Y \in \mathfrak{S} \ and \ X.Y = L} Y$.

Finally we can take Th. 17 as the definition of the symbol '$\mathfrak{A}$'. It will be seen that with the adoption of the above definitions

[1] For a simpler, although equivalent, formulation of the system of postulates for the calculus of systems, and for a clarification of the part played by these postulates in the development of this calculus, see Tarski (80).

the following postulates suffice for the foundation of the whole calculus of systems: Posts. Ia, VIIIa, and Xa, which were mentioned in Ths. 6 and 9, and Ths. 11, 13, and 19. (Post. VIIIa expresses the fact that for any class $\Re$ of systems there exists exactly *one* system X which satisfies the two conditions given in Def. D_1).

It is worth mentioning that within the theory of sets a variety of relatively simple interpretations for the calculus of systems can be obtained in the following way. We consider any class $\mathfrak{A}$ of sets which satisfies the following conditions: (1) $\mathfrak{A}$ is not empty and is at most denumerable; (2) $\mathfrak{A}$ is a field of sets, i.e. the formula $X, Y \in \mathfrak{A}$ always implies $X + Y \in \mathfrak{A}$ and $X - Y \in \mathfrak{A}$;[1] (3) if $\Re$ is any subclass of $\mathfrak{A}$ and $\sum_{Y \in \mathfrak{A}} Y = \sum_{Y \in \Re} Y$, then there is a finite subclass $\mathfrak{L}$ of $\Re$ such that $\sum_{Y \in \mathfrak{A}} Y = \sum_{Y \in \mathfrak{L}} Y$. Then we form the class $\mathfrak{S}$ of all sets which are sums of an arbitrary number of sets of the class $\mathfrak{A}$. In turn we interpret further terms of the calculus of systems as follows. To the symbolic expressions '$X \subseteq Y$' and '$X \cdot Y$' we ascribe the usual set-theoretical meaning. Similarly we regard the sums $X \dotplus Y$ and $\sum_{Y \in \Re} Y$ as the usual sums of sets (in the sense of set theory). We put, further, $S = \sum_{Y \in \mathfrak{A}} Y$ and $L = 0$; finally we interpret the symbols '$\prod_{Y \in \Re} Y$' and '$\overline{X}$' (to which the usual meaning cannot be ascribed) by following the lines of Defs. D_2 and D_5. It turns out that the system of concepts obtained in this way satisfies all the postulates of the calculus of systems.

From the point of view of the problems in which we are interested the fact that the calculus of systems can be interpreted in various ways within the calculus of systems itself is of great importance. In order to formulate the relevant theorems conveniently we introduce the following notation:

DEFINITION 10. (a) $\mathfrak{S}^X$ is the class of all systems $Y \in \mathfrak{S}$ such that $Y \subseteq X$;

   (b) $\mathfrak{S}_X$ is the class of all systems $Y \in \mathfrak{S}$ such that $X \subseteq Y$.

[1] Cf. Hausdorff, F. (25), pp. 78 ff.

THEOREM 20. *Let A be any deductive system; let $\overline{X}^A = A \cdot \overline{X}$
for every system $X \in \mathfrak{S}$; finally let $\mathfrak{A}^A$ be the class of all systems
$X \in \mathfrak{S}^A$, such that $A \subseteq X \dotplus \overline{X}$ (and in particular let $\mathfrak{A}^A = \mathfrak{S}^A \cdot \mathfrak{A}$
in case $A \in \mathfrak{A}$). In the postulates of the calculus of systems replace
everywhere the symbols 'S', '$\overline{X}$', '$\mathfrak{S}$', and '$\mathfrak{A}$' respectively by 'A',
'$\overline{X}^A$', '$\mathfrak{S}^A$', and '$\mathfrak{A}^A$', and leave the remaining symbols unchanged.
Then all postulates of the calculus of systems remain valid with
the exception of Ths. 11 and 13, either of which holds if and only if
$A \in \mathfrak{A}$.*

THEOREM 21. *Let B be any deductive system; further let*

$$\overline{X}_B = \sum_{Y \in \mathfrak{S} \text{ and } X \cdot Y \subseteq B}^{\cdot} Y$$

*for every system $X \in \mathfrak{S}$ (and in particular let $\overline{X}_B = B \dotplus \overline{X}$ in case
$X \in \mathfrak{A}$); finally let $\mathfrak{A}_B$ be the class of all systems of the form $B \dotplus Y$
where $Y \in \mathfrak{A}$ (and in particular let $\mathfrak{A}_B = \mathfrak{S}_B \cdot \mathfrak{A}$, in case $B \in \mathfrak{A}$).
In the postulates of the calculus of systems replace everywhere the
symbols 'L', '$\overline{X}$', '$\mathfrak{S}$', and '$\mathfrak{A}$' respectively by 'B', '$\overline{X}_B$', '$\mathfrak{S}_B$',
and '$\mathfrak{A}_B$', leaving the remaining symbols unchanged. Then all
postulates of the calculus of systems are satisfied.*

The theorems just stated show that all results derived from
the postulates of the calculus of systems can be relativized, and
in fact in two directions: (1) We can narrow the set S of all meaningful sentences by replacing it by any axiomatizable system A
and restricting the discussion to those deductive systems which
are included in A (when we are concerned with results which
can be proved without the help of Ths. 11 and 13, the assumption of the axiomatizability of the system A is superfluous).
(2) We can enlarge the set L of all logical theorems by replacing
it by any deductive system B and restricting the discussion to
those systems which include the system B.[1] We can, of course,
subject every result to the twofold relativization—first in one
direction and then in the other. It is not difficult to formulate
a general theorem which embraces Ths. 20 and 21 as special
cases and makes possible the simultaneous relativization in both
directions.

---

[1] Cf. an analogous result in article V of the present work, pp. 67 f., Th. 6
and the accompanying remarks.

All further theorems of the present article are consequences of the postulates of the calculus of systems.

THEOREM 22. *If* $X \in \mathfrak{S}$, *then*

(a)
$$X = \sum_{Y \in \mathfrak{S}^X . \mathfrak{A}}^{\cdot} Y$$

and

(b)
$$\overline{\overline{X}} = \prod_{Y \in \mathfrak{S}_X . \mathfrak{A}} Y.$$

Th. 22 a stands in close connexion with Th. 19; in the system of postulates for the calculus of systems Th. 19 could be replaced by Th. 22 a. With the help of Th. 22 b we can set up a series of necessary and sufficient conditions for those systems which satisfy the law of double negation:

THEOREM 23. *For an arbitrary system* $X \in \mathfrak{S}$ *the following conditions are equivalent*: (1) $X = \overline{\overline{X}}$; (2) *there exists a system* $Y \in \mathfrak{S}$ *such that* $X = \overline{Y}$; (3) *the formula* $\overline{X} \subseteq Y$ *always implies the formula* $\overline{Y} \subseteq X$ *for* $Y \in \mathfrak{S}$; (4) *the formula* $\overline{X} \subseteq \overline{Y}$ *always implies the formula* $Y \subseteq X$ *for* $Y \in \mathfrak{S}$; (5) $X = \prod_{Y \in \mathfrak{S}_X . \mathfrak{A}} Y$; (6) *there exists a class* $\mathfrak{R} \subseteq \mathfrak{A}$, *such that* $X = \prod_{Y \in \mathfrak{R}} Y$.

Consequently the systems which satisfy the law of double negation coincide with the intersections of axiomatizable systems.

In Th. 17 we already recognized a characteristic property of axiomatizable systems which was formulated entirely in terms of the calculus of systems; we shall now give two other properties of this kind:

THEOREM 24. *For an arbitrary system* $X \in \mathfrak{S}$ *the following conditions are equivalent*: (1) $X \in \mathfrak{A}$; (2) *every class* $\mathfrak{R} \subseteq \mathfrak{S}$ *satisfies the formula* $X \dotplus \prod_{Y \in \mathfrak{R}} Y = \prod_{Y \in \mathfrak{R}} (X \dotplus Y)$; (3) *for every class* $\mathfrak{R} \subseteq \mathfrak{S}$ *the formula* $X = \sum_{Y \in \mathfrak{R}}^{\cdot} Y$ *implies the existence of a finite class* $\mathfrak{L} \subseteq \mathfrak{R}$ *such that* $X = \sum_{Y \in \mathfrak{L}}^{\cdot} Y$.

In condition (3) of this theorem both identity signs can be replaced by inclusion signs.

Non-axiomatizable systems can be characterized in the following way:

THEOREM 25. *The following conditions are equivalent*:
(1) $X \in \mathfrak{S}-\mathfrak{A}$; (2) *there is an infinite class* $\mathfrak{K} \subseteq \mathfrak{S}$ *such that* $X = \sum_{Y \in \mathfrak{K}} \dot{} \, Y$, *where* $Y \neq L$ *for any* $Y \in \mathfrak{K}$ *and* $Y.Z = L$ *for any two distinct systems* $Y, Z \in \mathfrak{K}$; (3) *there exists an infinite sequence of systems* $Y_n \in \mathfrak{S}$, *such that* $X = Y_1 \dot{+} Y_2 \dot{+} ... \dot{+} Y_n \dot{+} ...$ *where* $Y_n \subseteq Y_{n+1}$ *and* $Y_n \neq Y_{n+1}$ *for every natural number* n. *It can, moreover, be assumed in conditions* (2) *and* (3) *that all systems of the class* $\mathfrak{K}$ *or all systems* Y_n *are axiomatizable.*

The non-axiomatizable systems are thus sums of infinite classes of logically disjoint systems and also sums of infinite series of strictly increasing systems.

If we translate the condition (2) of the above theorem into the language of the sentential algorithm (after having replaced '$\mathfrak{S}$' by '$\mathfrak{A}$'), we come to the conclusion that for every system $X \in \mathfrak{S}$ a set $Y \subseteq X$ can be constructed such that $(\alpha) \; X = Cn(Y)$, $(\beta) \; Y.L = 0$, and $(\gamma) \; x+y \in L$ for all $x, y \in Y$ (or, what amounts to the same thing, (β') if $x, y \in Y$, $x \supset z$, and $y \supset z$, then $z \in L$); if, moreover, $X \in \mathfrak{S}-\mathfrak{A}$, then (δ) the set Y is infinite. The conditions (β) and (γ) express the fact that Y is a set of *maximal independent sentences* in the sense of Sheffer; hence it results *a fortiori* that Y is a set of independent sentences in the usual sense and therefore a basis of the system X.[1]

In an analogous manner we easily infer, either from condition (3) of Th. 25, or directly, that with every system $X \in \mathfrak{S}$ an infinite sequence of sentences $y_n \in X$ can be correlated which satisfy the conditions: (α) for every sentence $x \in X$ there exists a sentence y_n such that $y_n \supset x$ (thus, if Y is the set of all members of our sequence, then $Cn(Y) = X$); $(\beta) \; y_{n+1} \supset y_n$ for every natural number n; if, moreover, $X \in \mathfrak{S}-\mathfrak{A}$, then (γ) the formula $y_n \equiv y_{n+1}$ does not hold for any natural n. The sequence of sentences which satisfy the condition (β) could be called a *logically increasing sequence of sentences*; if, moreover, the condi-

---

[1] Cf. Sheffer, H. M. (63), p. 32; for the concept of basis cf. III, pp. 35 f., Def. 7 and Th. 17* and also V, pp. 88 f., especially Def. 5.

tion (γ) is satisfied, then we speak of a *strictly increasing sequence of sentences*.

By a *logical product of an infinite sequence of sentences* y_n we understand a sentence x such that (α) $x \supset y_n$ for every natural number n, and (β) if z is any sentence such that $z \supset y_n$ for every natural number n, then $z \supset x$ (there may be many such sentences x but they are mutually equivalent); in an analogous manner we define the logical product of an arbitrary set of sentences. It seems natural to call an increasing sequence of sentences y_n *convergent* when a logical product of this sequence exists. Since such a logical product is also the product of the whole system X with which the sequence considered was correlated (in the way indicated in the preceding paragraph),we shall in an analogous manner also call the system X itself *convergent*. When the logical product x of the system X itself belongs to this system, the formula $X = Cn(\{x\})$ is easily seen to hold, and the system X is simply axiomatizable.

The class of convergent systems we shall denote by the symbol '$\mathfrak{C}$'; translating the definition of convergence into the language of the calculus of systems we obtain

DEFINITION 11. $X \in \mathfrak{C}$ *if and only if* $X \in \mathfrak{S}$ *and* $\prod\limits_{Y \in \mathfrak{S}_X . \mathfrak{A}} Y \in \mathfrak{A}$;

in other words, if there exists a system Y *such that* (1) $Y \in \mathfrak{A}$ *and* $X \subseteq Y$, *as well as* (2) *for every system* Z *the formulas* $Z \in \mathfrak{A}$ *and* $X \subseteq Z$ *imply* $Y \subseteq Z$.

Various transformations of this definition are known:

THEOREM 26. *For any given system* $X \in \mathfrak{S}$ *the following conditions are equivalent*: (1) $X \in \mathfrak{C}$, (2) $\bar{X} \in \mathfrak{A}$, (3) $\bar{\bar{X}} \in \mathfrak{A}$, (4) $\bar{X} \dotplus \bar{\bar{X}} = S$; (5) $\overline{X \cdot Y} = \bar{X} \dotplus \bar{Y}$ *for every system* $Y \in \mathfrak{S}$; (6) $\overline{\overline{X \dotplus Y}} = \bar{\bar{X}} \dotplus \bar{\bar{Y}}$ *for every system* $Y \in \mathfrak{S}$.

We can thus characterize convergent systems most simply as those systems whose logical complements are axiomatizable and which therefore satisfy the law of excluded middle in a weaker form: $\bar{X} \dotplus \bar{\bar{X}} = S$.

It follows from Ths. 23, 24, and 26 that some theorems of the calculus of classes which are not generally valid in the calculus of systems regain their validity when we assume for at least one of the systems to which these theorems refer that it is axiomatizable (e.g. the laws of contraposition, the distributive law under infinite multiplication), or at least that it is convergent (De Morgan's law).

As examples of other theorems which concern convergent systems we give

THEOREM 27. (*a*) $\mathfrak{A} \subseteq \mathfrak{C}$;

(*b*) $X \in \mathfrak{C}$ *if and only if* $X \in \mathfrak{S}$ *and* $\bar{X} \in \mathfrak{C}$;

(*c*) *if* $X, Y \in \mathfrak{C}$, *then* $X \dotplus Y \in \mathfrak{C}$ *and* $X.Y \in \mathfrak{C}$.

THEOREM 28. *If* $X, Y \in \mathfrak{S}$, $X \dotplus Y \in \mathfrak{A}$ *and* $X.Y \in \mathfrak{C}$, *then* $X, Y \in \mathfrak{C}$.

In this theorem, which is an analogue of Th. 18, one cannot replace '$\mathfrak{A}$' by '$\mathfrak{C}$'.

The question arises whether there exist convergent non-axiomatizable systems, in other words whether there exist non-axiomatizable systems whose complements are axiomatizable. We shall see later (in Th. 38) that the answer is in the affirmative assuming that non-axiomatizable systems exist at all, that is, assuming that the class $\mathfrak{S}$ is infinite. On the other hand it will turn out that on the same assumption there are also non-convergent systems. We have thus three classes of systems: (1) axiomatizable systems, which are correlated with single sentences; (2) non-axiomatizable convergent systems, which correspond to infinite convergent strictly increasing sequences of sentences, and finally (3) divergent systems, which correspond to infinite divergent sequences. The systems of the first two classes have in general similar properties; one of the principal differences consists in the fact that no system of the second class satisfies the law of double negation.

§ 4. IRREDUCIBLE AND COMPLETE SYSTEMS

On the basis of the calculus of systems certain theorems of a deeper nature can be established, concerning the cardinality and structure of various kinds of deductive systems; in the formulations and proofs of these theorems two concepts occur which have not yet been used in this article, namely the concepts of an *irreducible* and of a *complete* system. We denote the class of irreducible systems by the symbol '$\mathfrak{I}$', that of complete systems by the symbol '$\mathfrak{B}$'.[1] The definitions of these two concepts are quite analogous:

DEFINITION 12. *$X \in \mathfrak{I}$ if and only if $X \neq L$; and if for every system $Y \in \mathfrak{S}$ for which $Y \subseteq X$, we have $Y = L$ or $Y = X$ (in other words, if $\mathfrak{S}^X = \{L, X\}$).*

DEFINITION 13. *$X \in \mathfrak{B}$ if and only if $X \in \mathfrak{S}$, $X \neq S$, and if, for every system $Y \in \mathfrak{S}$ for which $X \subseteq Y$, we have $Y = X$ or $Y = S$ (in other words, if $\mathfrak{S}_X = \{S, X\}$).*

The following are simple consequences of these definitions:

THEOREM 29. *The following conditions are mutually equivalent:* (1) $X \in \mathfrak{I}$; (2) $X \in \mathfrak{S}$, $X \neq L$, *and the formulas Y, $Z \in \mathfrak{S}$ and $X = Y \dotplus Z$ always imply $X = Y$ or $X = Z$;* (3) $X \in \mathfrak{S}$, $X \neq S$, *and for every system $Y \in \mathfrak{S}$ we have $X \subseteq Y$ or $X \subseteq \overline{Y}$.*

THEOREM 30. *The following conditions are mutually equivalent:* (1) $X \in \mathfrak{B}$; (2) $X \in \mathfrak{S}$, $X \neq S$, *and the formulas Y, $Z \in \mathfrak{S}$ and $X = Y.Z$ always imply $X = Y$ or $X = Z$;* (3) $X \in \mathfrak{S}$, $X \neq S$, *and for every system $Y \in \mathfrak{S}$ we have $Y \subseteq X$ or $\overline{Y} \subseteq X$.*

In the conditions (3) of these two theorems the formula '$Y \in \mathfrak{S}$' can be replaced by '$Y \in \mathfrak{A}$'.

THEOREM 31. (a) *If $X, Y \in \mathfrak{I}$ and $X \neq Y$, then $X.Y = L$;*
(b) *if $X, Y \in \mathfrak{B}$ and $X \neq Y$, then $X \dotplus Y = S$.*

THEOREM 32. (a) *If $\mathfrak{R} \subseteq \mathfrak{I}$, $\mathfrak{L} \subseteq \mathfrak{I}$ and $\sum\limits_{X \in \mathfrak{R}} X \subseteq \sum\limits_{X \in \mathfrak{L}} X$, then $\mathfrak{R} \subseteq \mathfrak{L}$;*
(b) *if $\mathfrak{R} \subseteq \mathfrak{B}$, $\mathfrak{L} \subseteq \mathfrak{B}$ and $\prod\limits_{X \in \mathfrak{R}} X \subseteq \prod\limits_{X \in \mathfrak{L}} X$, then $\mathfrak{L}.\mathfrak{A} \subseteq \mathfrak{R}$.*

---

[1] I have already dealt with the concept of completeness in my earlier works (cf. III, p. 34, Def. 4; also V, pp. 93 ff., especially Def. 7); the concept is used here in a somewhat different sense, because (1) it is applied exclusively to deductive systems and not to arbitrary sets of sentences, and (2) only those deductive systems are called complete which—in the old terminology—are both complete and consistent. The concept of an irreducible system is here introduced for the first time.

Contrary to what might be supposed on the ground of Defs. 12 and 13 as well as Ths. 29 and 30, there exists between the properties of the classes $\mathfrak{J}$ and $\mathfrak{V}$ no exact dual correspondence. A small difference can already be seen in Th. 32; the divergence appears clearly in the further theorems.

THEOREM 33. *$X \in \mathfrak{J}$ if and only if $X \in \mathfrak{S}$ and $\bar{X} \in \mathfrak{V}$.*

In this theorem '$\mathfrak{J}$' cannot simply be replaced by '$\mathfrak{V}$' and vice versa; on the other hand we have:

THEOREM 34. *If $X \in \mathfrak{V}.\mathfrak{A}$, then $\bar{X} \in \mathfrak{J}$; if $X \in \mathfrak{V}-\mathfrak{A}$, then $\bar{X} = L$.*

One of the essential differences between the properties of the classes $\mathfrak{J}$ and $\mathfrak{V}$ consists in the fact that the class $\mathfrak{J}$ consists exclusively of axiomatizable systems, whilst the class $\mathfrak{V}$ can contain both axiomatizable and non-axiomatizable systems; from Th. 34 it follows, however, that complete systems in all cases possess axiomatizable complements and are thus convergent:

THEOREM 35. $\mathfrak{J} \subseteq \mathfrak{A},\ \mathfrak{V} \subseteq \mathfrak{C}$.

The theorem of Lindenbaum, according to which every system distinct from S can be extended to a complete system[1] is of a deeper character. This result can be improved in the following way:

THEOREM 36. *If $X \in \mathfrak{S}$, then $X = \displaystyle\prod_{Y \in \mathfrak{S}_X.\mathfrak{V}} Y$.*

Thus every system can be represented as an intersection of complete systems. The corresponding dual theorem for the class $\mathfrak{J}$ fails; if the class $\mathfrak{S}$ is infinite, then, as can easily be seen from Th. 37 below, there certainly are systems which are not sums of irreducible systems.

We now come to the theorems announced at the beginning of this section.

---

[1] Cf. III, p. 34, Th. 12; V, p. 98, Th. 56.

THEOREM 37. *The following conditions are equivalent*:
(1) $\overline{\overline{\mathfrak{S}}} < \aleph_0$; (2) $\overline{\overline{\mathfrak{V}}} < \aleph_0$; (3) *there exists a natural number n such that* $\overline{\overline{\mathfrak{J}}} = n$ *and* $\overline{\overline{\mathfrak{S}}} = 2^n$; (4) *there exists a natural number n such that* $\overline{\overline{\mathfrak{V}}} = n$ *and* $\overline{\overline{\mathfrak{S}}} = 2^n$; (5) $\mathfrak{S} \subseteq \mathfrak{A}$; (6) $\mathfrak{V} \subseteq \mathfrak{A}$; (7) $S = \sum_{Y \in \mathfrak{J}} Y$;
(8) *for every* $X \in \mathfrak{S}$, $X = \sum_{Y \in \mathfrak{S}^X . \mathfrak{J}} Y$; (9) *for every* $X \in \mathfrak{S}$,
$X = \prod_{Y \in \mathfrak{S}_X . \mathfrak{V}. \mathfrak{A}} Y$.

THEOREM 38. *The following conditions are equivalent*:
(1) $\overline{\overline{\mathfrak{S}}} \geqslant \aleph_0$; (2) $\overline{\overline{\mathfrak{S}}} = 2^{\aleph_0}$; (3) $\overline{\overline{\mathfrak{A}}} = \aleph_0$; (4) $\overline{\overline{\mathfrak{C} - \mathfrak{A}}} \geqslant \aleph_0$; (5) $\overline{\overline{\mathfrak{S} - \mathfrak{C}}} = 2^{\aleph_0}$.

By applying the relativization theorems—20 and 21—we can derive consequences of a more general nature from the last two theorems. These consequences concern those deductive systems which are included in a given axiomatizable system A or which include a given (not necessarily axiomatizable) system B. Nevertheless it must be noticed that in generalizing Ths. 37 and 38, all kinds of systems involved in these theorems undergo relativization with respect to the set A or the set B; not only do the classes $\mathfrak{S}$ and $\mathfrak{A}$ go over respectively into $\mathfrak{S}^A$ and $\mathfrak{A}^A$, or into $\mathfrak{S}_B$ and $\mathfrak{A}_B$ (cf. Ths. 20 and 21), but an analogous transformation must also be performed on classes $\mathfrak{C}$, $\mathfrak{J}$, and $\mathfrak{V}$. By relativizing the definitions of these classes we easily realize that (under the assumption $A \in \mathfrak{A}$) $\mathfrak{C}^A = \mathfrak{S}^A . \mathfrak{C}$ and $\mathfrak{J}^A = \mathfrak{S}^A . \mathfrak{J}$, and that the class $\mathfrak{V}^A$ consists of those and only those systems X which satisfy the formulas $X \neq A$ and $\mathfrak{S}^A . \mathfrak{S}_X = \{A, X\}$ ($\mathfrak{V}^A$ can also be characterized as the class of all systems of the form $A . Y$, where Y is an arbitrary complete system which does not include A). Moreover, the class $\mathfrak{C}_B$ consists of all systems $X \in \mathfrak{S}_B$ which satisfy the condition: among the systems of the class $\mathfrak{A}_B$ which include X there is a smallest one (if $B \in \mathfrak{A}$, then $\mathfrak{C}_B = \mathfrak{S}_B . \mathfrak{C}$); the class $\mathfrak{J}_B$ consists of all systems X which satisfy the formulas $X \neq B$ and $\mathfrak{S}_B . \mathfrak{S}^X = \{B, X\}$ (the formulas $X \in \mathfrak{J}_Y$ and $Y \in \mathfrak{V}^X$ are thus equivalent); finally we have $\mathfrak{V}_A = \mathfrak{S}_A . \mathfrak{V}$. In the case of relativizations based on Th. 20 the assumption $A \in \mathfrak{A}$ is essentially necessary. Nevertheless, it

can be shown that the equivalence of the relativized conditions
(1), (3), and (5) of Th. 37 does not depend on this assumption,
since from each of these conditions it follows that $A \in \mathfrak{A}$; the
same applies to the conditions (7) and (8) if we supplement each
of them with the words '*and* $\overline{\overline{\mathfrak{J}}}{}^A < \aleph_0$'. Similarly the relativized
formulas (1), (2), (3), and (5) of Th. 38 are equivalent indepen-
dently of the assumption that the system A is axiomatizable.
(All this holds on the condition that we interpret the symbol
'$\mathfrak{A}^A$' as '$\mathfrak{S}^A . \mathfrak{A}$', and neglect the more general interpretation
indicated in Th. 20.)

Th. 37 can be extended by including in it various theorems
of the ordinary calculus of classes which are not generally valid
in the calculus of systems as conditions equivalent to the formulas
(1)–(9), e.g. the law of excluded middle (the equivalence of which
to formula (5) can be inferred from Th. 17), the law of double
negation, as well as the laws of contraposition stated in Th. 23,
further, the weaker law of excluded middle and De Morgan's
law which appear in Th. 26. Some of the theorems of the calculus
of classes just mentioned can be used to supplement the relativizations of Th. 37 described above if the range of applications
of these theorems is restricted to systems of the class $\mathfrak{S}^A$ or $\mathfrak{S}_B$;
thus, for example, under the assumption that $A \in \mathfrak{A}$, the conditions (1)–(9) of Th. 37 relativized with respect to the system A
are equivalent to the condition: $X = \overline{\overline{X}}$ *for every system* $X \in \mathfrak{S}^A$.

The power of the class $\mathfrak{S}^A$ can be called the *cardinal degree
of capacity of the system* A, the power of the class $\mathfrak{S}_B$ can be
called the *cardinal degree of completeness of the system* B.[1] Since
$\mathfrak{S} = \mathfrak{S}^S = \mathfrak{S}_L$, the power of the class $\mathfrak{S}$ is both the degree of
capacity of the system S and the degree of completeness of the
system L; this cardinal number can be called the *cardinal degree
of capacity*, or the *degree of completeness, of the deductive theory
discussed*. From Ths. 37 and 38 it follows that the power of
the class $\mathfrak{S}$ is either a natural number of the form 2^n or equals
$2^{\aleph_0}$. By virtue of the relativized theorems this applies to the
degree of capacity and the degree of completeness of an arbi-

---

[1] The second of these concepts I have previously introduced; cf. V of the
present volume, pp. 100 ff., especially Def. 8.

trary deductive system as well. In case the degree of completeness of a theory is finite, the structure of the theory is very simple; as Th. 37 shows, all systems are then axiomatizable, and hence the calculus of systems is not formally distinct from the calculus of classes; every system can be represented—and in fact in only one way—as the sum of a finite number of irreducible systems or as the intersection of a finite number of axiomatizable complete systems. The situation is almost equally clear in the case when non-axiomatizable systems exist but only one of them is complete. This case is treated in the following theorem:

THEOREM 39. *The formulas* $(1)\ \overline{\overline{\mathfrak{B}-\mathfrak{A}}} = 1$ *and* $(2)\ \sum_{Y \in \mathfrak{J}}^{\cdot} Y \in \mathfrak{B}$ *are mutually equivalent and imply the following consequences*:
$(3)\ \overline{\overline{\mathfrak{A}}} = \overline{\overline{\mathfrak{C}-\mathfrak{A}}} = \overline{\overline{\mathfrak{J}}} = \overline{\overline{\mathfrak{B}}} = \overline{\overline{\mathfrak{B}.\mathfrak{A}}} = \aleph_0;\ (4)\ if\ \overline{\overline{\mathfrak{S}^X.\mathfrak{J}}} = n < \aleph_0,$
then $X \in \mathfrak{A},\ X = \sum_{Y \in \mathfrak{S}^X.\mathfrak{J}}^{\cdot} Y,\ \overline{X} = \prod_{Y \in \mathfrak{S}^X.\mathfrak{J}} \overline{Y},\ \overline{\overline{\mathfrak{S}^X}} = 2^n\ and\ \overline{\overline{\mathfrak{S}_X}} = 2^{\aleph_0};$
$(5)\ if\ \overline{\overline{\mathfrak{J}-\mathfrak{S}^X}} = n < \aleph_0,\ then\ either\ X \in \mathfrak{A},\ X = \prod_{Y \in \mathfrak{J}-\mathfrak{S}^X} \overline{Y},\ and$
$\overline{\overline{\mathfrak{S}_X}} = 2^n,\ or\ X \in \mathfrak{C}-\mathfrak{A},\ X = \sum_{Y \in \mathfrak{S}^X.\mathfrak{J}}^{\cdot} Y\ and\ \overline{\overline{\mathfrak{S}_X}} = 2^{n+1},\ while\ in$
both cases $\overline{X} = \sum_{Y \in \mathfrak{J}-\mathfrak{S}^X}^{\cdot} Y\ and\ \overline{\overline{\mathfrak{S}^X}} = 2^{\aleph_0};\ (6)\ if\ \overline{\overline{\mathfrak{S}^X.\mathfrak{J}}} = \overline{\overline{\mathfrak{J}-\mathfrak{S}^X}} = \aleph_0,$
then $X \in \mathfrak{S}-\mathfrak{C},\ X = \sum_{Y \in \mathfrak{S}^X.\mathfrak{J}}^{\cdot} Y,\ \overline{X} = \sum_{Y \in \mathfrak{J}-\mathfrak{S}^X}^{\cdot} \overline{Y},\ and\ \overline{\overline{\mathfrak{S}^X}} = \overline{\overline{\mathfrak{S}_X}} = 2^{\aleph_0}.$

The cases described in Ths. 37 and 39 possess many points of contact. Characteristic for both cases is the role of irreducible systems as basic elements out of which arbitrary deductive systems can be constructed, either with the help of the operation of addition alone or with the help of two operations—addition and complementation. The logical alternative of the two cases considered can be characterized as follows:

THEOREM 40. *The following conditions are equivalent*:
$(1)\ \overline{\overline{\mathfrak{B}-\mathfrak{A}}} \leqslant 1;\ (2)\ if\ X = \sum_{Y \in \mathfrak{J}}^{\cdot} Y,\ then\ \overline{\overline{\mathfrak{S}^X}} \leqslant 2;\ (3)\ for\ every\ system$
$X \in \mathfrak{A}\ either\ \overline{\overline{\mathfrak{S}^X}} < \aleph_0\ or\ \overline{\overline{\mathfrak{S}_X}} < \aleph_0;\ (4)\ for\ every\ system\ X \in \mathfrak{S}$
either $X = \sum_{Y \in \mathfrak{S}^X.\mathfrak{J}}^{\cdot} Y\ or\ X = \prod_{Y \in \mathfrak{J}-\mathfrak{S}^X} \overline{Y}.$

B b

To conclude this section we will discuss the concept of the *structural type* of a theory.[1] We say that two deductive theories (or, more generally, two arbitrary models of the system of axioms and primitive concepts listed in § 1) possess the same structural type if the class of all deductive systems of the first theory is isomorphic with the class of all systems of the second theory with respect to the relation of inclusion, i.e. if between the systems of the two theories a one-one correlation can be set up which satisfies the following condition: if X_1 and Y_1 are two arbitrary systems of the first theory, and X_2 and Y_2 are the correlated systems of the second theory, then $X_1 \subseteq Y_1$ if and only if $X_2 \subseteq Y_2$. This can also be expressed otherwise: between the sentences of the two theories a many-many correlation can be set up so that if x_1 and y_1 are two arbitrary sentences of the first theory whilst x_2 and y_2 are any two correlated sentences of the second theory, then the formulas $x_1 \supset y_1$ and $x_2 \supset y_2$ are equivalent.

It turns out that two cardinal numbers exert an essential influence on the structural type of a theory: the power $\mathfrak{a}$ of the class of all complete axiomatizable systems and the power $\mathfrak{n}$ of the class of all complete non-axiomatizable systems. The ordered pair of these numbers $(\mathfrak{a}, \mathfrak{n})$ we call the *characteristic pair of a deductive theory*. The numbers $\mathfrak{a}$ and $\mathfrak{n}$ cannot be quite arbitrary; from Ths. 37 and 38 it follows that $\mathfrak{a} \leqslant \aleph_0$ and $\mathfrak{n} \leqslant 2^{\aleph_0}$, it also follows that the case $\mathfrak{a} < \aleph_0, 0 < \mathfrak{n} < \aleph_0$ and the case $\mathfrak{a} = \aleph_0, \mathfrak{n} = 0$ are excluded. Other relations between the numbers $\mathfrak{a}$ and $\mathfrak{n}$ have been established by A. Mostowski[2]; he has proved that only the following pairs of numbers can appear as characteristic pairs of deductive theories: (1) the pair $(\mathfrak{a}, 0)$, where $\mathfrak{a}$ is any natural number distinct from 0; (2) the pair $(\aleph_0, \mathfrak{n})$, where $\mathfrak{n}$ is any number $\leqslant \aleph_0$; (3) the pair $(\mathfrak{a}, 2^{\aleph_0})$, where $\mathfrak{a}$ is any number $\leqslant \aleph_0$ (hence it follows in particular that the number $\mathfrak{n}$ as well as the number $\mathfrak{v} = \mathfrak{a} + \mathfrak{n} = \overline{\overline{\mathfrak{B}}}$ satisfies the continuum hypothesis). Moreover, it turns out that the num-

---

[1] The importance of this concept has been recognized, independently of me, by Lindenbaum.

[2] The results of Mostowski referred to in this section are published in Mostowski, A. (53c), pp. 34–53.

bers in question are subject to no further restrictions: given an arbitrary pair $(\mathfrak{a}, \mathfrak{n})$ of one of the three types just listed we can always construct a deductive theory such that $\overline{\overline{\mathfrak{B} \cdot \mathfrak{A}}} = \mathfrak{a}$ and $\overline{\overline{\mathfrak{B} - \mathfrak{A}}} = \mathfrak{n}$.

Theories with the characteristic pair $(\mathfrak{a}, 0)$ where $\mathfrak{a} < \aleph_0$ possess the simplest structure; following them are the theories with the pair $(\aleph_0, 1)$. From Ths. 37 and 39, in which we have studied theories of this kind, it follows that in these simplest cases the characteristic pair completely determines the structural type of a theory: two theories with the same characteristic pair possess the same structural type. Mostowski has shown that this result can be extended to all characteristic pairs $(\aleph_0, \mathfrak{n})$ so long as $\mathfrak{n} < \aleph_0$, but that in general the characteristic pair does not determine the structural type of the theory; among the theories with the characteristic pair $(\aleph_0, \aleph_0)$ non-denumerably many different structural types appear.

The results concerning the structural type of an entire theory can be relativized with respect to individual systems of the theory. In that case it is necessary to distinguish between the *inner* and the *outer structural types of the system*: two systems X and Y (of one theory or even of two different deductive theories) possess the same inner (or outer) structural types if the classes $\mathfrak{S}^X$ and $\mathfrak{S}^Y$ (or $\mathfrak{S}_X$ and $\mathfrak{S}_Y$) are isomorphic with respect to the relation of inclusion. The structural type of the whole theory is both the inner structural type of the system S and the outer of the system L. The method of relativization enables us, in the investigation of structural types, to restrict ourselves to one single deductive theory without the scope of our discussion being thereby essentially restricted: for every theory with the characteristic pair $(\mathfrak{a}, 2^{\aleph_0})$ is, as Mostowski has shown, universal in the sense that among the structural types of individual systems of this theory the structural types of all possible theories are already represented.

§ 5. Applications to Special Deductive Theories

We shall now illustrate our general discussion by examples of some special deductive theories in which the mutual relations between various classes of deductive systems are ordered in a particularly clear manner.

The first of the theories which we have in mind can be called the *elementary theory of dense order*. In the theorems of this theory denumerably many variables of different form appear; let these be, for example, the signs 'a'', 'a''', 'a'''', and so on. Moreover we have here four constants: the *symbol of the ordering relation* 'R', the *negation sign* 'N', the *implication sign* 'C', and the *universal quantifier* 'Π'.[1]

An expression which is composed of the symbol 'R' and two variables following it, e.g. '$Ra'a''$' (read 'a' *precedes* a''' or 'a'' *follows* a''), we call an *elementary formula*. The expression consisting of the negation sign and an arbitrary expression x following it is called the *negation of the expression x*, symbolically $\bar{x}$; thus e.g. '$NRa'a''$' (read 'a' *does not precede* a''' or '*it is false that* a' *precedes* a''') is the negation of the elementary formula '$Ra'a''$'. The formula which is formed by the implication sign and two succeeding expressions x and y is called the *implication with the antecedent x and the consequent y*, symbolically $x \to y$; thus, e.g., '$CRa'a''NRa''a'$' (read '*If* a' *precedes* a'' *then* a'' *does not precede* a'') is an implication with the antecedent '$Ra'a''$' and the consequent '$NRa''a'$'. By putting in front of an arbitrary expression x the universal quantifier followed by a variable we obtain the *universal quantification of the expression x with respect to the given variable*; e.g. '$\Pi a'NRa'a'$' (read: '*for every* a' *it is false that* a' *precedes itself*') is the universal quantification of the expression '$NRa'a'$' with respect to the variable 'a''.

Further, we define recursively a certain category of expressions called *sentential functions*: sentential functions of the 1st order are elementary formulas, functions of higher order arise out of functions of lower order by applying one of the three operations: negation, implication, and universal quantification. Among the variables which occur in a given sentential function we distinguish by a well-known method *free* and *bound variables*.[2] A sentential function containing no free variables is called a

---

[1] For symbolism see IV, pp. 39 and 54; VIII, p. 168.
[2] Cf. article IX, Def. 6, and article VIII, Def. 11.

(*meaningful*) *sentence*; the set of all sentences is denoted by the symbol 'S'.

Finally we define recursively the concept of *logical theorem*. Among the logical theorems of the 1st order, i.e. among the *logical axioms*, we include sentential functions of two kinds: (1) all expressions of the type $(\bar{x} \to x) \to x$, $x \to (\bar{x} \to y)$, and $(x \to y) \to ((y \to z) \to (x \to z))$, where x, y, and z are arbitrary sentential functions; (2) the following four sentences:

$$c_1 = \text{'}N\Pi a'\Pi a''N Ra'a'''\text{'},$$

$$c_2 = \text{'}\Pi a'\Pi a''C Ra'a''N Ra''a'\text{'},$$

$$c_3 = \text{'}\Pi a'\Pi a''\Pi a'''C Ra'a''C N Ra'''a''Ra'a'''\text{'},$$

$$c_4 = \text{'}\Pi a'\Pi a''C Ra'a''N\Pi a'''C Ra'a'''N Ra'''a''\text{'}.$$

As is seen, the axioms of the 1st kind comprise all sentential functions which can be obtained by substitution from the axioms of the sentential calculus. The first of the axioms of the 2nd kind expresses the fact that the relation R is not empty: there exist two elements of which one precedes the other. The next two characterize R as a relation which orders all the elements of the universe of discourse.[1] Finally, the last asserts that the order established by the relation R is dense: if a' precedes a'', then there is an element a''' which succeeds a' and precedes a''.[2]

Logical theorems of higher orders are obtained from the theorems of lower orders with the help of one of the four well-known operations: substitution, detachment, and the insertion and deletion of the universal quantifier in the consequent.[3] The set of all logical theorems which contain no free variables is denoted by the symbol 'L'.

We have thus shown how all primitive expressions of general

---

[1] The axioms c_2 and c_3 deviate somewhat from the usual postulates of ordering. The reason is that among the signs of the theory discussed the identity sign does not appear; cf. Łukasiewicz, J. (50), pp. 35 ff.

[2] The attention of the reader, who may be surprised at the presence of the sentences c_1–c_4 among the logical axioms and consequently among the elements of the set L, should be directed to the fact that the set L can be interpreted as the totality of sentences which are accepted as true when we begin to develop the deductive theory under consideration (cf. § 1).

[3] Cf. IX, p. 285; VIII, p. 181.

metamathematics are to be interpreted in their application to the special deductive theory which has been chosen as the subject of investigation. It can easily be established that the *Axs.* 1–5 *of* § 1 *preserve their validity* in this interpretation. Thus all consequences of these axioms are also valid; in particular the calculus of systems can be applied in our discussion.

Let us now consider two special sentences:

$$d_1 = \text{`}\Pi a' N \Pi a'' N R a' a'''\text{'}  \quad \text{and} \quad d_2 = \text{`}\Pi a' N \Pi a'' N R a'' a'\text{'}.$$

As can easily be realized these sentences express the fact that for every element there is another element which follows it or which precedes it: in other words, in the ordering of the elements which is established by the relation R there is no last and no first element.

With the help of a frequently used method, which consists in reducing the sentences to normal form and successively eliminating the quantifiers,[1] the following theorem can be established (we formulate it in the terms of the sentential algorithm which was introduced in §§ 1 and 2):

Every sentence $x \in S$ satisfies exactly one of the following sixteen formulas: (1) $x \in L$, (2) $\bar{x} \in L$, (3) $x \equiv d_1$, (4) $x \equiv \overline{d_1}$, (5) $x \equiv d_2$, (6) $x \equiv \overline{d_2}$, (7) $x \equiv d_1 + d_2$, (8) $x \equiv d_1 + \overline{d_2}$, (9) $x \equiv \overline{d_1} + d_2$, (10) $x \equiv \overline{d_1} + \overline{d_2}$, (11) $x \equiv d_1 . d_2$, (12) $x \equiv d_1 . \overline{d_2}$, (13) $x \equiv \overline{d_1} . d_2$, (14) $x \equiv \overline{d_1} . \overline{d_2}$, (15) $x \equiv d_1 . d_2 + \overline{d_1} . \overline{d_2}$, (16) $x \equiv d_1 . \overline{d_2} + \overline{d_1} . d_2$.

This theorem implicitly contains the solution of all the more important metamathematical problems relating to the theory of dense order. Translated into the language of the calculus of systems it states that the class $\mathfrak{S}$ contains sixteen different systems; if we put $D_1 = Cn(\{d_1\})$ and $D_2 = Cn(\{d_2\})$, then these systems are L, S, D_1, $\overline{D_1}$, D_2, $\overline{D_2}$, $D_1 . D_2$, $D_1 . \overline{D_2}$, $\overline{D_1} . D_2$, $\overline{D_1} . \overline{D_2}$, $D_1 \dotplus D_2$, $D_1 \dotplus \overline{D_2}$, $\overline{D_1} \dotplus D_2$, $\overline{D_1} \dotplus \overline{D_2}$, $D_1 . \overline{D_2} \dotplus \overline{D_1} . D_2$, and $D_1 . D_2 \dotplus \overline{D_1} . \overline{D_2}$. All these systems are axiomatizable; the degree of content and the degree of completeness of each of them is obviously finite. Four of the systems are irreducible: $D_1 . D_2$,

---

[1] So far as I know this method was first used in Skolem, Th. (64). It was applied to the theory of dense order by Langford, C. H. (45), pp. 16 ff. The facts given below concerning this theory supplement Langford's results.

$D_1 . \overline{D_2}, \overline{D_1} . D_2,$ and $\overline{D_1} . \overline{D_2}.$ Four others, the logical complements of the preceding systems, are complete: $\overline{D_1 . D_2} = \overline{D_1} \dotplus \overline{D_2}$, $\overline{D_1 . \overline{D_2}} = \overline{D_1} \dotplus D_2$, $\overline{\overline{D_1} . D_2} = D_1 \dotplus \overline{D_2}$, and $\overline{\overline{D_1} . \overline{D_2}} = D_1 \dotplus D_2$; the characteristic pair of the theory is thus $(4, 0)$. Every system is the sum of all the irreducible systems included in it and also the intersection of all complete systems which include it. In a word, the relations between the systems are exactly of the kind anticipated in Th. 37.

The second theory with which we shall deal—the *elementary theory of isolated order*—has an infinitistic character, and the variety of its systems is much larger. On a superficial view this theory differs only slightly from the theory of dense order: the same constants and variables appear in it, the same expressions are called sentential functions and sentences, the same operations are used in the recursive definition of the concept of logical theorem. The sole distinction consists in the fact that we omit c_4 from the list of logical axioms and replace it by two sentences:

$$c_5 = \text{`}\Pi a' \Pi a'' C R a' a'' N \Pi a''' C R a' a''' N \Pi a'''' C R a' a'''' N R a'''' a'''\text{'},$$

$$c_6 = \text{`}\Pi a' \Pi a'' C R a'' a' N \Pi a''' C R a''' a' N \Pi a'''' C R a'''' a' N R a''' a''''\text{'}.$$

These two new axioms state that every element which, in the order established by the relation R, is not a last or a first element has an immediate successor or an immediate predecessor.[1]

In order to study the structure of deductive systems in the theory of isolated order we single out a particular sequence of sentences e_n; translated into English the sentence e_n expresses the fact that there exist at most $n+1$ different elements. We put

$$e_1 = \text{`}\Pi a' \Pi a'' \Pi a''' C R a' a'' N R a'' a'''\text{'},$$

$$e_2 = \text{`}\Pi a' \Pi a'' \Pi a''' \Pi a'''' C R a' a'' C R a'' a''' N R a''' a''''\text{'},$$

and so on; it is easily seen how the recursive definition of e_n should look. With the help of this sequence and of the sentences d_1 and d_2 formulated in connexion with the theory of dense order, we define a new sequence of sentences i_n, namely:

---

[1] Langford, C. H. (45), pp. 459 ff., is concerned with a similar but more special theory, namely with the elementary theory of the order type ω.

$i_1 = d_1 + \overline{d_2}, i_2 = \overline{d_1} + d_2, i_3 = \overline{d_1} + \overline{d_2}, i_4 = \overline{e_1}$, and $i_n = e_{n-4} + \overline{e_{n-3}}$
for every natural $n \geqslant 5$. The content of the sentences i_1, i_2,
and i_3 is clear (the sentence i_3, for example, states that in the
order established by the relation R either a first or a last element
occurs); the sentences i_4, i_5, i_6,... state respectively that the
number of all elements is not exactly equal to 2, 3, 4,.... .

With the help of the above-mentioned method (that of the
successive elimination of quantifiers)[1] we can prove the following theorem, which has a decisive significance for the metamathematical discussion of the theory of isolated order:

*Every sentence $x \in S$ satisfies one and only one of the four con-
ditions*: (1) $x \in L$; (2) $\bar{x} \in L$; (3) *there exists a finite increasing
sequence of natural numbers n_1, n_2,..., n_p such that $x \equiv \prod_{k=1}^{p} i_{n_k}$;
(4) there exists a finite increasing sequence of natural numbers
n_1, n_2,..., n_p such that $x \equiv \sum_{k=1}^{p} \overline{i_{n_k}}$. The sequences n_1, n_2,..., n_p
which occur in (3) and (4) are uniquely determined by the sen-
tence x.*

We state here the most important consequences of this theorem in the field of the calculus of systems.

Let us put $I_n = Cn(\{i_n\})$ for every natural number n. It can
be shown that the class of all systems I_n is identical with the
class $\mathfrak{I}$; since for any two different natural numbers n and p the
corresponding systems I_n and I_p are always different, we have
$\overline{\overline{\mathfrak{I}}} = \aleph_0$.

The class $\mathfrak{B}$ consists of all systems $\overline{I_n}$, where n is an arbitrary
natural number, and of the system $I_1 \dotplus I_2 \dotplus ... \dotplus I_n \dotplus ...$; this
latter system is not axiomatizable, whilst the remaining ones are
axiomatizable. We have therefore $\overline{\overline{\mathfrak{B}}} = \overline{\overline{\mathfrak{B} . \mathfrak{A}}} = \aleph_0$, $\overline{\overline{\mathfrak{B} - \mathfrak{A}}} = 1$;
the deductive theory discussed thus has the characteristic pair
$(\aleph_0, 1)$.

The class $\mathfrak{A}$ contains the systems L and S as well as all systems
of the form $I_{n_1} \dotplus I_{n_2} \dotplus ... \dotplus I_{n_p}$ and $\overline{I_{n_1}} . \overline{I_{n_2}} ..... \overline{I_{n_p}}$, where the

---

[1] Cf. p. 374, footnote (the application of this method is however not so simple
in the case now discussed as it was in the previous one).

natural numbers n_1, n_2,..., n_p form an arbitrary finite increasing sequence; hence it follows that $\overline{\overline{\mathfrak{A}}} = \aleph_0$. Systems of the form $I_{n_1} \dot{+} I_{n_2} \dot{+} ... \dot{+} I_{n_p}$ have a finite degree of capacity equal to 2^p, but an infinite degree of completeness equal to $2^{\aleph_0}$; it is the other way round in the case of the systems of the form $\overline{I_{n_1}} . \overline{I_{n_2}} . .... \overline{I_{n_p}}$. It is to be noted that the system S and the systems of the form $\overline{I_{n_1}} . \overline{I_{n_2}} . .... \overline{I_{n_p}}$ are the only systems of the theory of isolated order which cannot be represented as sums of the irreducible systems included in them.

The class $\mathfrak{S} - \mathfrak{A}$ consists of systems of the form

$$I_{n_1} \dot{+} I_{n_2} \dot{+} ... \dot{+} I_{n_p} \dot{+} ...,$$

where the numbers n_1, n_2,..., n_p,... form an arbitrary infinite increasing sequence; these systems are convergent or divergent according to whether the number of natural numbers which are not terms of the sequence n_1, n_2,..., n_p,... is finite or not. Hence $\overline{\overline{\mathfrak{C}}} = \overline{\overline{\mathfrak{C} - \mathfrak{A}}} = \aleph_0$, while $\overline{\overline{\mathfrak{S}}} = \overline{\overline{\mathfrak{S} - \mathfrak{A}}} = \overline{\overline{\mathfrak{S} - \mathfrak{C}}} = 2^{\aleph_0}$. The degree of capacity of every system of the class $\mathfrak{S} - \mathfrak{A}$ is equal to $2^{\aleph_0}$. If $X = I_{n_1} \dot{+} I_{n_2} \dot{+} ... \dot{+} I_{n_p} \dot{+} ...$ is a convergent system, then $\overline{X} = I_{m_1} \dot{+} I_{m_2} \dot{+} ... \dot{+} I_{m_q}$ and $\overline{\overline{X}} = \overline{I_{m_1}} . \overline{I_{m_2}} . .... \overline{I_{m_q}}$, where m_1, m_2,..., m_q are all the distinct natural numbers which do not occur in the sequence n_1, n_2,..., n_p,.... The degree of completeness of such a system X is equal to 2^{q+1}. If, however,

$$X = I_{n_1} \dot{+} I_{n_2} \dot{+} ... \dot{+} I_{n_p} \dot{+} ...$$

is a divergent system, then

$$\overline{X} = I_{m_1} \dot{+} I_{m_2} \dot{+} ... \dot{+} I_{m_q} \dot{+} ...,$$

where m_1, m_2,..., m_q,... is the infinite increasing sequence of all those natural numbers which do not occur in the sequence n_1, n_2,..., n_p,...; moreover, $\overline{\overline{X}} = I_{n_1} \dot{+} I_{n_2} \dot{+} ... \dot{+} I_{n_p} \dot{+} ...$, so that the system X satisfies the law of double negation. The degree of completeness of a divergent system X is infinite, equal to $2^{\aleph_0}$.

In a word, we have here just the case which was anticipated and described in Th. 39.

Many other examples of deductive theories are known whose characteristic pair is $(\aleph_0, 1)$ and which—in the sense of the concluding remarks of § 4—have the same structural type as the elementary theory of isolated order. An especially simple theory of this kind is the *elementary theory of identity*. Its construction is very similar to both the previous theories; the difference consists only in that in the system of logical axioms, instead of the sentences c_1, c_2,..., c_6, only two sentences are included, '$\Pi a' R a' a''$' and '$\Pi a' \Pi a'' \Pi a''' C R a' a'' C R a' a''' R a'' a'''$', which together characterize R as a reflexive, symmetrical, and transitive relation. The symbol 'R' we can here call the *identity sign*, the formula '$R a' a''$' is read 'a' *is identical with* a'''. All results concerning the theory of isolated order can also be extended to the theory of identity, provided that the sequence of sentences i_n is suitably defined. It is not difficult to formulate the definition of this sequence: the sentences i_1, i_2, i_3,... are to express respectively the idea that the number of all elements is not exactly equal to 1, 2, 3,...; we can for example put:

$$i_1 = \text{'}N\Pi a' \Pi a'' R a' a''\text{'},$$

$$i_2 = \text{'}\Pi a' \Pi a'' C \Pi a''' C N R a' a''' R a'' a''' R a' a''\text{'},$$

and so on. We may mention yet another theory of the same structural type, namely the *atomistic algebra of logic*.[1] Again, it differs from the previously considered theories in only one respect, i.e. in having a different list of logical axioms. We call the symbol 'R' in this theory the *inclusion sign*: we now read the formula '$R a' a''$' as '*the element a' stands in the relation of inclusion to the element a''*' or '*the element a' is included in the element a''*'. As logical axioms of the theory, apart from the sentential functions which can be obtained by substitution into the axioms of the sentential calculus, we adopt sentences which suffice for the development of the ordinary system of the algebra of logic (which, however, are formulated with the help of the

[1] Cf. article XI, p. 334 of the present work. (Nevertheless, we have in mind here an atomistic theory of the algebra of logic in which infinite operations do not occur.) The structural properties of the theory in question can be derived from the results in Skolem (64). In article VIII, pp. 199–208 I have given the most important metamathematical theorems concerning the atomistic algebra of logic (although there the problem is treated from quite a different point of view).

inclusion sign as the only primitive term)[1] and in addition the
following sentence:

$$l = \text{`}\Pi a' C \Pi a'' C R a'' a' N \beta \Pi a'' R a' a'' \text{'}$$

where β is an abbreviation of the sentential function

$$N C \Pi a''' C N \Pi a'' R a''' a'' C R a''' a'' R a'' a''' \Pi a''' R a'' a''';$$

'β' expresses the fact that the element a'' is an atom (an indecomposable element), i.e. a non-empty element which includes
no non-empty element distinct from itself, and the sentence l
asserts that every non-empty element includes at least one atom.
The definition of the sequence i_n for the atomistic algebra of
logic resembles the corresponding definition for the elementary
theory of identity; the sentence i_n states that the number of all
atoms is not exactly equal to n.

In conclusion we give examples of deductive theories of higher
structural types. If in the postulate system of the atomistic
algebra of logic we replace the sentence l by the logically weaker
sentence

$$l_1 = \text{`}N \Pi a' C \Pi a'' C \beta R a'' a' N \Pi a''' C \Pi a'' C \beta R a'' a''' R a' a''' \text{'},$$

(where 'β' has its previous meaning), which states only that
among the elements which include all atoms there is a smallest
one, we obtain an example of a theory with the characteristic
pair $(\aleph_0, 2)$. If, on the other hand, we simply delete the sentence
l, then the resulting theory, which is clearly the general algebra
of logic, proves to have $(\aleph_0, \aleph_0)$ as its characteristic pair. If we
strike out the sentence c_4 from the axiom system of the theory
of dense order (or, what amounts to the same thing, the sentences c_5 and c_6 from the axiom system of the theory of isolated
order) and pass in this way to the *general elementary theory of
order*, then the structure of the theory becomes enormously more
complicated: while the theory of dense order has one of the
simplest structural types, the general theory of order has a very
composite type, and its characteristic pair is the greatest pos-
sible one, namely $(\aleph_0, 2^{\aleph_0})$. If we drop the sentences c_1, c_2, and
c_3 from the axiom system of the general theory of order, we

---

[1] Cf. Huntington, E. V. (32), and VIII, p. 179.

obtain a new theory with the characteristic pair $(\aleph_0, 2^{\aleph_0})$, namely the *general elementary theory of a binary relation*. This theory embraces in a certain sense all previously considered theories. Within the discussion of this theory all results sketched in this section find their place. Thus, for example, all results concerning arbitrary deductive systems of the theory of dense order can be regarded as results relating to those systems of the general theory of a binary relation which include the system

$$A = Cn(\{c_1, c_2, c_3, c_4\}).$$

APPENDIX

In spite of their elementary character, the results stated in § 5 have a series of interesting consequences extending far beyond the limits of the present work. I will here discuss, although very briefly and sketchily, some of these consequences.

Let α be an arbitrary order type.[1] Let us consider an arbitrary set X and an arbitrary relation R which orders this set according to the type α. Let us assume that the variables 'a''', 'a'''', 'a''''',..., which appear in the elementary theory of order, denote the elements of the set X exclusively, and confine our attention to the set of all those sentences of the theory of order which are true with the given interpretation of the variables. It is not difficult to see that this set depends neither on the set X nor on the relation R, but exclusively on the order type α; on that account we denote it by the symbol '$T(\alpha)$'. It can be shown that—for any given order type α—the set $T(\alpha)$ is a complete deductive system.

The formula $T(\alpha) = T(\beta)$ states that all properties of the order types α and β which can be formulated in the language of the elementary theory of order are identical; for that reason we call the types α and β which satisfy this formula *elementarily indistinguishable* or *elementarily equivalent*.

---

[1] The remarks given below refer to ordering relations, but they can be extended to arbitrary relations. We must speak, then, not of types of order, but of relation numbers in the sense of Whitehead, A. N., and Russell, B. (90), vol. II, pp. 30 ff.

It follows from the well-known theorem of Skolem–Löwenheim[1] that every order type is elementarily equivalent to an order type which is at most denumerable. From this it follows that every class of order types of which no two are elementarily equivalent can have at most the power $2^{\aleph_0}$, for this is the power of the class of all denumerable types (this can also be proved more simply without the theorem of Skolem–Löwenheim). On the other hand, it is not difficult to construct a class of elementarily distinguishable (non-equivalent) order types the power of which is actually $2^{\aleph_0}$ (in the construction we imitate the proof of the set-theoretical theorem just mentioned, according to which the class of denumerable order types has the power $2^{\aleph_0}$).[2]

If we restrict ourselves to dense order types, we reach the result that only four such elementarily non-equivalent order types exist; in fact, the set $T(\alpha)$ is always a complete system, and in the elementary theory of dense order only four complete systems appear. As an example of four dense elementarily non-equivalent order types, the types η, $1+\eta$, $\eta+1$, $1+\eta+1$ may serve, where 'η' denotes as usual the type of the natural ordering of the set of all rational numbers. By using the symbols which were introduced in the discussion of the theory of dense order it can be shown that

$$T(\eta) = D_1 \dotplus D_2, \qquad T(1+\eta) = D_1 \dotplus \overline{D_2},$$
$$T(\eta+1) = \overline{D_1} \dotplus D_2, \quad \text{and} \quad T(1+\eta+1) = \overline{D_1} \dotplus \overline{D_2}.$$

Every other type of dense order is elementarily equivalent to one of these four types. Thus, for example, for the type λ of the natural order of the set of all real numbers we have $T(\lambda) = T(\eta)$. Nevertheless, the types η and λ are distinguished by a whole series of properties: η is a denumerable type, λ is not; λ is a continuous type, η is not. From this it follows that properties of order types such as *denumerability and continuity are not elementary properties*, in the sense that *they cannot be expressed in the language of the elementary theory of order* (as far as denumerability is concerned this also follows from the theorem of Skolem–Löwenheim).

---

[1] Cf. Skolem, Th. (66), pp. 23 ff.
[2] Cf. Hausdorff, F. (25), pp. 49 f.

In a similar manner we reach the result at there can be only a denumerable number of isolated order types (i.e. order types having no proper cuts, but only jumps and gaps) of which no two are elementarily equivalent. Let us, in fact, consider the class consisting of all finite types as well as the types ω, ω^*, $\omega^*+\omega$, and $\omega+\omega^*$, where 'ω', as usual, denotes the order type of the less-than relation in the domain of natural numbers. It is easy to see that no two types of this class are elementarily equivalent. It can also be shown that

$$T(\omega^*) = \overline{I_1}, \quad T(\omega) = \overline{I_2}, \quad T(\omega^*+\omega) = \overline{I_3}, \quad T(n) = \overline{I_{n+2}}$$

for every natural $n \geqslant 2$ and finally that

$$T(\omega+\omega^*) = I_1 \dotplus I_2 \dotplus ... \dotplus I_n \dotplus ... .$$

Since in this way all complete systems of the theory of isolated order are exhausted (cf. § 5), every other isolated order type must be elementarily equivalent to one of the types considered. Thus, e.g., $T(\omega) = T(\omega+\omega^*+\omega)$; on the other hand, ω is the type of a well-ordered set while $\omega+\omega^*+\omega$ is not. Hence it follows that *well-ordering is not an elementary property of the order types, it cannot be expressed in the elementary theory of order.*

It seems that a new, wide realm of investigation is here opened up. It is perhaps of interest that these investigations can be carried out within the framework of mathematics itself (e.g. set theory) and that the concepts and methods of metamathematics are essentially superfluous: all concepts which occur in these investigations (e.g. the concept of an elementary property of an order type or the concept of elementarily indistinguishable order types) can be defined in purely mathematical terms.[1]

NOTE. In the present article I have brought together results which have been obtained at various times.

The results in §§ 1–4 belong, for the most part, to the years 1930–2. I reported on them in 1933 to the Warsaw Section of the Polish Mathematical Society in a lecture entitled 'On

---

[1] For this purpose it suffices to apply the method which I have developed in VI. For precise mathematical definitions of the notions involved and for an outline of the general theory of these notions see Tarski, A. (84a).

the foundations of the methodology of deductive sciences' (cf. Mazurkiewicz, S. (53) where—making use of the calculus of systems—a new method for the foundation of probability theory is developed).

The results of § 5 had already been obtained in the years 1926–8, naturally without applying the apparatus of the calculus of systems, which was not at my disposal at that time. (Cf. VIII, p. 205, note 2, as well as Presburger, M. (61)). During the same period I became aware of the facts discussed in the appendix. But I was able to state these facts in a correct and precise form only with the help of the methods which were developed in VI and VIII, during the years 1929–30. The facts discussed in the appendix, together with other results of a related kind, were communicated by me in 1932 in a lecture at the Warsaw Section of the Polish Mathematical Society.

XIII

ON THE LIMITATIONS OF THE MEANS OF EXPRESSION OF DEDUCTIVE THEORIES†

THE metamathematical results presented in this article are applicable to various systems of mathematical logic and to various deductive theories based upon these systems of logic,[1] but in order to make the following remarks more specific, we adopt as *logic* an arbitrary system which satisfies the following two conditions:

(*a*) The system includes as a subsystem the logic of *Principia Mathematica* which is assumed to be strictly formalized and to be modified as follows:

(1) a simple theory of types and the axiom of extensionality are assumed;

(2) all defined operators, e.g. '$\hat{x}(\phi x)$' or '$(\imath x)(\phi x)$', are eliminated;

(3) only those sentential functions which contain no free variables are regarded as sentences.

(*b*) The system possibly contains still further logical sentences (which do not occur as axioms or theorems in *Principia Mathematica*), as, for example, the axioms of infinity and of choice, etc., but no new undefined constants appear in it.

---

[1] I announced Th. 9 to the Second Polish Philosophical Congress in Warsaw in 1927 in a lecture on the concepts of consistency and completeness; the result is mentioned in Fränkel, A. (16), p. 352, note 3. Other results of the present communication were first presented at the sessions of the Logical Section of the Warsaw Philosophical Society on 15 June 1932 (Lindenbaum: *On some methodological questions connected with the foundations of Geometry*, I) and on 1 February 1933 (Lindenbaum and Tarski, *On some metalogical theorems*). To the same circle of ideas belong Tarski, A. and Lindenbaum, A. (85) and article X of the present work.

---

† BIBLIOGRAPHICAL NOTE. This article is the text of a lecture given by the author on 12 June 1935 to the Mathematical Colloquium conducted by Karl Menger in the University of Vienna. The article first appeared under the title 'Über die Beschränktheit der Ausdrucksmittel deduktiver Theorien' in *Ergebnisse eines mathematischen Kolloquiums*, fascicule 7 (1934–5), pp. 15–22. For further historical references see footnote 1.

For other concepts and symbols used in the following discussion see X, pp. 297–8. In particular, we shall employ symbolic expressions of the form

$$R \frac{x', y', z', \dots}{x'', y'', z'', \dots}$$

(explained in X, p. 310), which express the fact that the relation R maps the class of all individuals onto itself in one-one fashion, so that the individuals, classes, relations, etc., x', y', z',... are mapped on x'', y'', z'',..., respectively. As a general scheme of a sentential function the symbolic expression '$\sigma(a, b, c, \dots; x, y, z, \dots)$' is used; here '$a$', '$b$', '$c$',... are all extra-logical constants and 'x', 'y', 'z',... are all free variables of the given sentential function. The conjunction of all axioms of a deductive axiomatizable system,[1] in which 'a', 'b', 'c',... appear as primitive extra-logical constants, is represented in the schematic form '$\alpha(a, b, c, \dots)$'.

§ 1

THEOREM 1. *Every sentence of the form*

$$(x', x'', y', y'', z', z'', \dots, R) :. R \frac{x', y', z', \dots}{x'', y'', z'', \dots} . \supset \ : \sigma(x', y', z', \dots).$$
$$\equiv . \sigma(x'', y'', z'', \dots)$$

is logically provable.

Roughly speaking, Th. 1 states that every relation between objects (individuals, classes, relations, etc.) which can be expressed by purely logical means is invariant with respect to every one-one mapping of the 'world' (i.e. the class of all individuals) onto itself and this invariance is logically provable. The theorem is certainly plausible and had already been used as a premiss in certain intuitive considerations. Nevertheless it had never before been precisely formulated and exactly proved. The proof of Th. 1 is carried out by induction with respect to the form of the sentential function '$\sigma(x', y', z', \dots)$'.[2]

It should be noted that the symbol 'σ' which occurs in the

[1] See article V of the present work.
[2] A detailed proof of Th. 1 has recently been published in Mostowski, A. (53d), pp. 200 ff. The formulation of the theorem in Mostowski's work is, however, weaker than that given in the present article.

schemata of sentential functions must not be regarded as an independent symbol (say as a variable predicate). For example, in contrast to Th. 1, sentences of the form

$$(x', x'', y', y'', z', z'', ..., R, f) : . R \frac{x', y', z', ...}{x'', y'', z'', ...} . \supset : f(x', y', z', ...).$$
$$\equiv . f(x'', y'', z'', ...)$$

are, in general, not provable but refutable in logic. On the other hand, by virtue of Th. 1 sentences of the following form are provable:

$$(x', x'', y', y'', z', z'', ..., R, f', f'') : . R \frac{x', y', z', ..., f'}{x'', y'', z'', ..., f''} .$$
$$. \supset : f'(x', y', z', ...). \equiv . f''(x'', y'', z'', ...).$$

If we apply Th. 1 to sentential functions with a single free variable, and in fact an individual variable, we obtain as an immediate corollary

THEOREM 2. *If 'x'' and 'x''' are individual variables, then every sentence of the form*

$$(x', x'') : \sigma(x'). \equiv . \sigma(x'')$$

is provable in logic.

Intuitively interpreted: no two individuals are distinguishable by purely logical means. By an easy transformation we obtain from Th. 2

THEOREM 3. *If 'x' is an individual variable, then every sentence of the form*

$$(x). \sigma(x). \vee . (x). \sim \sigma(x)$$

is logically provable.

In other words, apart from the universal class and the empty class, no class of individuals can be defined by purely logical means.[1] It should be noted that the following sharper formulation of Th. 3 in general fails: *If 'x' is an individual variable, then for every sentential function 'σ(x)' either the sentence '(x). σ(x)' or*

---

[1] The words 'define', 'definable', etc., are used in two distinct senses: in the first sense it is a question of a *formal* relation of certain expressions to other expressions of a theory (cf. article X of the present work); in the second sense of a *semantical* relation between objects and expressions (cf. article VI). Here these words are used in the second sense.

the sentence '(x). $\sim$ σ(x)' is provable in logic. This holds if and only if the basic system of logic is complete.

If we now apply Th. 1 to sentential functions with two free variables, we obtain

THEOREM 4. *If 'x' and 'y' are individual variables, then every sentence of the form*

$$(x,y).\sigma(x,y). \vee .(x,y). \sim \sigma(x,y). \vee : (x,y) : \sigma(x,y).$$
$$\equiv .x = y : \vee : (x,y) : \sigma(x,y). \equiv .x \neq y$$

is logically provable.

Consequently only four binary relations between individuals are definable by purely logical means, namely the universal relation, the empty relation, identity, and diversity (thus the four moduli of Schröder's algebra of relations).

This theorem can be extended to many-termed relations: for every natural number n only a specifiable finite number of n-termed relations between individuals can be defined by purely logical means, and each of these relations can be expressed by means of identity and the concepts of the sentential calculus.

Yet another special case of Th. 1 will be stated here concerning sentential functions with one free variable denoting a class of individuals; '$Nc(x)$' will denote the cardinal number (in the sense of *Principia Mathematica*) of the class x and '$\bar{x}$' will denote the complement of x.

THEOREM 5. *If 'x'' and 'x''' are variables ranging over classes of individuals, then every sentence of the form*

$$(x',x'') :. Nc(x') = Nc(x''). Nc(\bar{x}') = Nc(\bar{x}'') : \supset : \sigma(x'). \equiv .\sigma(x'')$$

is logically provable.

Thus, no two classes of individuals which have equal cardinal numbers and whose complements also have equal cardinal numbers, are distinguishable by purely logical means.[1]

It is customary to say that our logic is a logic of extensions and not of intensions, since two concepts with different intensions but identical extensions are logically indistinguishable.

---

[1] That is the solution of a problem which was first formulated in Tarski, A. (75), p. 182.

In the light of Th. 5 this assertion can be sharpened: our logic is not even a logic of extensions, but merely a logic of cardinality, since two concepts with different extensions are still logically indistinguishable if only the cardinal numbers of their extensions are equal and the cardinal numbers of the extensions of the complementary concepts are also equal.

We pass now to the applications of Th. 1 to special deductive theories, and to make the discussion more specific we concentrate upon Euclidean geometry. It is well known that Euclidean geometry can be constructed by means of a single primitive concept, e.g. by means of the four-termed relation of the congruence of two pairs of points. We shall denote this relation by 'a'. The expression '$a(x, y, z, t)$' will thus be read: '*the point x is the same distance from the point y as the point z is from the point t*'. We also assume that the axiom system of Euclidean geometry has been supplemented by a sentence to the effect that every individual is a point (so that space is identified with the class of all individuals).[1] As a special case of Th. 1 we obtain at once

THEOREM 6. *Every sentence of the form*

$$(x', x'', y', y'', z', z'', ..., R) :. R\frac{a, x', y', z', ...}{a, x'', y'', z'', ...} . \supset : \sigma(a; x', y', z', ...).$$
$$\equiv . \sigma(a; x'', y'', z'', ...)$$

is provable in Euclidean geometry.

Intuitively speaking, every relation between objects (points, sets of points, classes of sets of points, etc.) which can be expressed in terms of logic and geometry is invariant with respect to every one-one mapping of space onto itself in which the relation a is preserved, i.e. with respect to every similarity transformation (and not only with respect to every isometrical transformation).

Using the specific properties of the axioms of Euclidean geometry, it is easy from Th. 6 to derive analogues of Ths. 2–4. Thus, by means of logic and Euclidean geometry no two points can be distinguished, neither can any specific set of points,

---

[1] Cf. the concluding remarks of this article.

apart from the whole space and the empty set, be defined, nor any two-termed relation between points, apart from the universal relation, the empty relation, identity, and diversity (in particular, in geometry no direction on a straight line can be singled out). The distinction between logic and geometry comes to light, however, in the discussion of three-termed relations; for, as is well known, there are infinitely many three- (and many-) termed relations which are definable in Euclidean geometry.

In the two following corollaries of Th. 6 the formulas '$x' \sim x'''$' ('*the point sets x' and x'' are geometrically similar*'), '$x' \cong x'''$' ('*the point sets x' and x'' are congruent*'), and '$S(x)$' ('*x is a line segment*') appear, which are known to be definable by means of 'a'.

THEOREM 7. *If 'x'' and 'x''' are variables ranging over point sets, then every sentence of the form*

$$(x', x'') :. x' \sim x'' . \supset : \sigma(a; x') . \equiv . \sigma(a; x'')$$

is provable in Euclidean geometry.

THEOREM 8. *Under the same hypothesis, every sentence of the form*

$$(x') : S(x') . \sigma(a; x') . \supset . (\exists x'') . S(x'') . \sigma(a; x'') . \sim (x' \cong x'')$$

is provable in Euclidean geometry.

Thus, no two similar point sets can be distinguished by means of logic and Euclidean geometry, and no absolute unit of length is definable within Euclidean geometry.

Various results of this kind can also be reached by investigating other deductive theories, e.g. *projective geometry* and *topology*. *Th. 6 holds generally for every theory based on logic*, because no use is made in its proof of specific properties of the axioms of geometry. Not only Ths. 2–4, but also the analogue of Th. 4 for three-termed relations, can be carried over to projective geometry; in fact there are only finitely many three-termed (although infinitely many four-termed) relations between points which can be defined in this geometry (in particular the relation of lying between is not definable in projective geometry). In topology only those two- and many-termed relations between points are at all definable which can be defined by purely logical means.

§ 2

We shall here state two general metamathematical theorems as a further application of Th. 1.

As is well known, an axiom system '$\alpha(a,b,c,...)$' is said to be *categorical* or *monomorphic* if the sentence

$$(x',x'',y',y'',z',z'',...) : \alpha(x',y',z',...) . \alpha(x'',y'',z'',...).$$

$$. \supset . (\exists R) . R \frac{x',\, y',\, z',...}{x'',\, y'',\, z'',...}$$

is logically provable. The axiom system is said to be *non-ramifiable* if, for every sentential function '$\sigma(x,y,z,...)$', the disjunction

$$(x,y,z,...) : \alpha(x,y,z,...) . \supset . \sigma(x,y,z,...) :$$

$$: \vee : (x,y,z,...) : \alpha(x,y,z,...) . \supset . \sim \sigma(x,y,z,...)$$

is logically provable. If, on the other hand, for every sentential function a member of this disjunction is logically provable, then the system is *decision-definite* or *complete*.[1] By the application of Th. 1 we can now prove

THEOREM 9. *Every categorical axiom system is non-ramifiable.*

For example, the axiomatically constructed arithmetic of natural numbers is categorical and thus non-ramifiable. It is also incomplete, i.e. there are arithmetical sentences which can be neither proved nor disproved by means of the axioms of arithmetic.[2] On account of the non-ramifiability the independence proof for such sentences cannot be carried out in any direction by an interpretation in logic.

The converse of Th. 9 proves to hold under a supplementary assumption. We define an axiom system '$\alpha(a,b,c,...)$' to be

---

[1] For these concepts cf. Fraenkel, A. (16), pp. 347–54 (especially p. 353). There are certain points of contact, although of a superficial kind, between § 2 of this article and Carnap, R. (9).

[2] Cf. Gödel, K. (22), pp. 174 f. and 182 ff.

effectively interpretable in logic if there is in logic a sentential function '$\tau(x, y, z, ...)$' such that the following three formulas are logically provable:

$$(x, y, z, ...): \tau(x, y, z, ...). \supset . \alpha(x, y, z, ...);$$

$$(\exists x, y, z, ...) . \tau(x, y, z),$$

$$(x', x'', y', y'', z', z'', ...): \tau(x', y', z', ...) . \tau(x'', y'', z'', ...).$$

$$\supset . x' = x''. y' = y''. z' = z''... .$$

If we take into consideration the rules of definition usually adopted in logic, this formulation can be considerably simplified: the axiom system '$\alpha(a, b, c, ...)$' is effectively interpretable in logic if and only if there are logical constants 'a''', 'b''', 'c''', ... (undefined or defined) such that the sentence '$\alpha(a', b', c', ...)$' obtained by substituting these logical constants for the primitive terms in the axiom system discussed is logically provable.

THEOREM 10. *Every non-ramifiable axiom system which is effectively interpretable in logic is categorical.*

It sometimes happens that an effective interpretation of an axiom system in logic can be found only at a higher type (this applies, for example, to the axiom system of the arithmetic of real numbers). It is then also possible to infer categoricity from non-ramifiability, under certain conditions which are satisfied by the familiar logical systems. We do not know, however, whether the converse of Th. 9 holds quite generally.

In all such considerations it is important to bear in mind the relativity of the three concepts—categoricity, non-ramifiability, and completeness, with respect to the adopted system of logic. *If*, for example, *the logic is complete, then the concepts of non-ramifiability and completeness have the same extension for every axiom system, so that every categorical system is* (not only non-ramifiable but also) *complete*. The three concepts discussed were here introduced as a-concepts in the sense of Carnap. If we consider the corresponding f-concepts,[1] then we find that f-completeness and f-non-ramifiability have the same extension;

---

[1] Concerning a- and f-concepts, see Carnap, R. (10), pp. 123 ff.

the analogue of Th. 9 proves to hold, but the problem of its converse is open in this case as well.

In conclusion we note the following: when applying Th. 1 to special deductive theories and to general metamathematics we must restrict ourselves to those axiom systems from which it follows that there are no individuals outside the domain of discourse of the theory discussed. We could remove this restriction by means of an appropriate generalization of Th. 1. To obtain this generalization we should introduce a new notion, in fact the notion of a *sentential function intrinsic under a given variable (or constant) 'a'* or, in other words, the notion of a *property intrinsic for a given class a*. Roughly speaking, a property is intrinsic for a class a if it involves exclusively elements of a, subsets of a, relations between elements of a, etc. (and not, for example, individuals outside of a). We do not give here a precise definition of the new notion, nor do we formulate the generalization of Th. 1 in which this notion is essentially involved. We should like to mention only that, with the help of this notion, various new results—of the same general character as those previously discussed—can be formulated and established. Thus, for instance, it can be shown that every property of a point set x, which is expressed in terms of logic and topology, and is intrinsic for x, is invariant under any homomorphic transformation (of the set x into another set x').[1]

---

[1] Cf. Kuratowski, C. (41), p. 13.

XIV

ON EXTENSIONS OF INCOMPLETE SYSTEMS OF THE SENTENTIAL CALCULUS†

ACCORDING to a well-known theorem of A. Lindenbaum,[1] every consistent set of sentences of an arbitrary deductive theory can be extended to form a consistent and complete system. The question arises how many such extensions are there; in other words, how great is the number of all the consistent and complete systems which include a given set of sentences? I shall here prove a general theorem from the meta-sentential calculus by which under certain assumptions only a single extension exists in the domain of the sentential calculus. I shall show further that most of the incomplete systems of the sentential calculus so far treated satisfy these assumptions, so that the ordinary (two-valued) system of the sentential calculus constitutes the only consistent and complete extension for all these systems.

It is well known that in the expressions of the sentential calculus special variables, the so-called *sentential variables* $p, q, r, \ldots$, occur, and in addition the *implication sign* C and possibly still other constants, e.g. the *negation sign* N, the *signs of disjunction* A, and *conjunction* K, and so on.[2] Correlated with each constant is a *fundamental operation on expressions* by which compound expressions can be formed from simpler ones. For example, corresponding to the implication sign is the operation

---

[1] See Article V, Th. 56, p. 98 of the present book.

[2] As regards symbolism cf. IV. Concerning terminology and the concepts involved see the foregoing and article V of the present book. It should be noted that the signs 'p', 'q',..., 'C', 'N',... and the expressions composed of these signs are not used here as expressions of the sentential calculus, but as designations of the corresponding expressions.

---

† BIBLIOGRAPHICAL NOTE. This article contains the text of a lecture given by the author in 1935 to the Mathematical Colloquium conducted by Karl Menger in the University of Vienna. The article first appeared under the title 'Über die Erweiterungen der unvollständigen Systeme des Aussagenkalküls' in *Ergebnisse eines mathematischen Kolloquiums*, fascicule 7 (1934–5), pp. 51–57.

of *forming an implication*, by means of which from two given expressions a and b the implication Cab is formed. The intersection of all sets of expressions which contain all the sentential variables and are closed under each of the fundamental operations is called the *set of (meaningful) sentences* and denoted by S.

Two operations are further distinguished which correlate certain sentences with other sentences, namely the operations of *substitution* and *detachment*. It would be somewhat tedious and superfluous here to explain more precisely the intuitively obvious meaning of the concept of substitution. The set of all sentences which can be obtained from the sentences of a given set X by substitution is denoted by '$Sb(X)$'. More generally, we denote by '$Sb_Y(X)$' the set of all sentences which are obtained by replacing all variables in the sentences of the set X by sentences of the set Y (in such a way that variables of the same shape which occur in a given sentence are replaced by sentences of the same shape). The operation of detachment correlates the sentence b with two given sentences of the form a and Cab. If X is any set of sentences, then the intersection of all sets of sentences which include the set $Sb(X)$ and are closed under the operation of detachment is called the *set of consequences of X*, in symbols $Cn(X)$.

By means of the concepts of sentence and set of consequences other metalogical concepts can be defined. Thus, for example, a set X of sentences is called a *deductive system*, or simply a *system*, if $Cn(X) = X$. The set X of sentences is called *inconsistent* or *consistent* according to whether the formula $Cn(X) = S$ holds or not. The set X of sentences is called *complete* if every sentence a either is a consequence of the set X or, when added to X, yields an inconsistent set $X+\{a\}$. More generally we call the set X of sentences *complete with respect to a set Y of sentences* if for every sentence b of the set Y either $b \in Cn(X)$ holds or the set $X+\{b\}$ is inconsistent.

Various elementary properties of the above concepts are assumed to be known.[1] In addition to the symbols listed above

---

[1] Several such properties are established in article V of the present book.

I shall use the customary symbolism of set theory. The general theorem mentioned at the outset is as follows:

THEOREM 1. *Let X be a consistent set of sentences, which satisfies the following condition: there is a set Z of sentences such that X is complete with respect to the set $Sb_Z(S)$ and the set $X + Sb_Z(S)$ is inconsistent. Then there exists exactly one consistent and complete deductive system Y, which includes the set X.*

For the proof of this theorem we require certain lemmas.

LEMMA 1. *For every set X of sentences $Sb_X(S)$ is the smallest set of sentences which includes X and is closed under every fundamental operation.*

The proof—based on an exact definition of the concept of substitution—presents no difficulties. As a corollary we easily obtain:

LEMMA 2. *For every set X of sentences*

$$X \subseteq Sb_X(S) = Sb_{Sb_X(S)}(S).$$

LEMMA 3. *For all sets X and Y of sentences*

$$Sb_Y(Cn(X)) \subseteq Cn(Sb_Y(Sb(X))).$$

Proof. Let Z be the set of all sentences a which satisfy the formula $Sb_Y(\{a\}) \subseteq Cn(Sb_Y(Sb(X)))$. It is easily shown that $Sb(X) \subseteq Z$ and that Z is closed under the operation of detachment (i.e. if $a \in Z$ and $Cab \in Z$, then $b \in Z$). From this it results immediately that $Cn(X) \subseteq Z$, in other words that $Sb_Y(\{a\}) \subseteq Cn(Sb_Y(Sb(X)))$ for every sentence $a \in Cn(X)$; and thus finally $Sb_Y(Cn(X)) \subseteq Cn(Sb_Y(Sb(X)))$, q.e.d.

If in Lemma 3 we put $Y = S$, then by a simple transformation we obtain the formula $Sb(Cn(X)) \subseteq Cn(X)$ for every set X of sentences. By virtue of this formula the definition of the set of consequences given above can be replaced by the following: $Cn(X)$ is the intersection of all sets of sentences which include X and are closed under the operations of substitution and detachment.

LEMMA 4. *If the sets $X + Y$ and $X + Z$ of sentences are inconsistent, then the set $X + Sb_Z(Sb(Y))$ is also inconsistent.*

Proof. If, in Lemma 3, we replace 'X' by '$X+Y$' and 'Y' by 'Z' we obtain $Sb_Z(Cn(X+Y)) \subseteq Cn(Sb_Z(Sb(X+Y)))$. Since, by hypothesis, $Cn(X+Y) = S$, it follows that

$$Sb_Z(S) \subseteq Cn(Sb_Z(Sb(X+Y))),$$

whence, by Lemma 2, $Z \subseteq Cn(Sb_Z(Sb(X+Y)))$ and further

(1) $X+Z \subseteq X+Cn(Sb_Z(Sb(X+Y)))$.

Clearly we have

$$Sb_Z(Sb(X+Y)) = Sb_Z(Sb(X))+Sb_Z(Sb(Y))$$
$$\subseteq Sb(X)+Sb_Z(Sb(Y)) \subseteq Cn(X)+Sb_Z(Sb(Y));$$

hence by applying the formula $Cn(Cn(X)+Y) = Cn(X+Y)$[1] we derive

$$Cn(Sb_Z(Sb(X+Y)) \subseteq Cn(Cn(X)+Sb_Z(Sb(Y)))$$
$$= Cn(X+Sb_Z(Sb(Y))).$$

Since in addition $X \subseteq X+Sb_Z(Sb(Y)) \subseteq Cn(X+Sb_Z(Sb(Y)))$, we have

(2) $X+Cn(Sb_Z(Sb(X+Y))) \subseteq Cn(X+Sb_Z(Sb(Y)))$.

The inclusions (1) and (2) at once give

$$X+Z \subseteq Cn(X+Sb_Z(Sb(Y)));$$

since the set $X+Z$ is inconsistent, $Cn(X+Sb_Z)(Sb(Y)))$ is also inconsistent.[2] From this it follows immediately that $X+Sb_Z(Sb(Y))$ is inconsistent, q.e.d.

LEMMA 5. *If X and Y are consistent deductive systems and if $X.Y$ is complete with respect to a set Z of sentences, then we have $X.Z = Y.Z$.*

Proof. Let a be any sentence of the set $X.Z$. Then we have $X.Y+\{a\} \subseteq X$, and the set $X.Y+\{a\}$ is thus consistent. Since $X.Y$ is complete with respect to Z and $a \in Z$, it follows that $a \in Cn(X.Y)$. The set $X.Y$, as the intersection of two systems, is itself a system, i.e. $Cn(X.Y) = X.Y$[3]; consequently $a \in X.Y$. Thus the formula $a \in X.Z$ always implies $a \in X.Y$. Accordingly, $X.Z \subseteq X.Y$, $X.Z = X.Y.Z$. In an analogous manner it can be proved that $Y.Z = X.Y.Z$, whence finally we have

$$X.Z = Y.Z, \text{ q.e.d.}$$

[1] Cf. V, Th. 2, p. 65.   [2] Cf. V, Th. 46, p. 91.   [3] Cf. V, Th. 11, p. 71.

We turn now to the proof of Th. 1. According to the theorem of Lindenbaum already mentioned, the consistent set X of sentences can be extended to at least one consistent and complete system Y. Let us suppose there is another such system Y_1 distinct from Y. Since Y and Y_1 are complete, the set $Y+Y_1$ of sentences must be inconsistent.[1] But since

$$X+Sb_Z(S) \subseteq Y+Sb_Z(S)$$

and $X+Sb_Z(S)$ is inconsistent, $Y+Sb_Z(S)$ must also be inconsistent. Hence it follows by Lemma 4 that

(1) *the set $Y+Sb_{Sb_Z(S)}(Y_1)$ is inconsistent.*

Moreover, we have

$$Sb_{Sb_Z(S)}(Y_1) \subseteq Sb(Y_1) \subseteq Cn(Y_1) = Y_1$$

since Y_1 is a deductive system. By Lemma 2,

$$Sb_{Sb_Z(S)}(Y_1) \subseteq Sb_{Sb_Z(S)}(S) = Sb_Z(S),$$

whence

$$Sb_{Sb_Z(S)}(Y_1) \subseteq Y_1.Sb_Z(S) \quad \text{and} \quad Y+Sb_{Sb_Z(S)}(Y_1) \subseteq Y+Y_1.Sb_Z(S).$$

The last inclusion, together with (1), shows that

(2) *the set $Y+Y_1.Sb_Z(S)$ is inconsistent.*

We also have $X \subseteq Y$ and $X \subseteq Y_1$; consequently $X \subseteq Y.Y_1$. Hence the set $Y.Y_1$, like the set X, is complete with respect to $Sb_Z(S)$. By means of Lemma 5 we conclude from this that $Y.Sb_Z(S) = Y_1.Sb_Z(S)$, and that consequently

(3) $\qquad\qquad Y = Y+Y_1.Sb_Z(S).$

From (2) and (3) it follows at once that Y is inconsistent, contrary to our assumption. Hence the system Y_1 cannot be distinct from Y. Th. 1 is thus proved.

In order to apply this theorem to special systems, let us consider first certain systems of the so-called restricted sentential calculus, i.e. the theory in which the implication sign is the only constant.[2]

THEOREM 2. *If X is a consistent set of sentences which contains the three sentences $CpCqp$, $CpCCpqq$, and $CCqrCCpqCpr$ as elements, then the ordinary (two-valued) system of the restricted sentential calculus is the only consistent and complete deductive system which includes the set X.*

[1] Cf. V, Th. 53, p. 97.   [2] Cf. IV, pp. 52–4.

LEMMA 6. *If* $X_0 = \{CpCqp,\ CpCCpqq,\ CCqrCCpqCpr\}$ *and* $Z = \{p\}$, *then for every sentence* $a \in Sb_Z(S)$ *we have either* $a \in Cn(X_0)$ *or both* $Cap \in Cn(X_0)$ *and* $Cpa \in Cn(X_0)$.

Proof. Let Y be the set of all sentences which satisfy the conclusion of the lemma. By applying the usual methods of the sentential calculus it can be shown without difficulty that the sentence Cpp is a consequence of the set X_0. Hence it follows immediately that for $a \in Z$, i.e. for $a \rightleftharpoons p$, both $Cap \in Cn(X_0)$ and $Cpa \in Cn(X_0)$, whence $a \in Y$. Consequently $Z \subseteq Y$. Next consider any two sentences a and b of the set Y; we shall show that $Cab \in Y$ also holds. Here three cases are to be distinguished: (1) $b \in Cn(X_0)$; (2) $Cap \in Cn(X_0)$, $Cpa \in Cn(X_0)$, $Cbp \in Cn(X_0)$, and $Cpb \in Cn(X_0)$; (3) $a \in Cn(X_0)$, $Cbp \in Cn(X_0)$, and $Cpb \in Cn(X_0)$.

For the sake of brevity only the first case will be investigated in detail. Since $CpCqp \in X_0$, we have $CbCab \in Sb(X_0)$ and hence $CbCab \in Cn(X_0)$; since moreover $b \in Cn(X_0)$ and the set $Cn(X_0)$ is closed under the operation of detachment, we obtain $Cab \in Cn(X_0)$. Analogously it can be shown that in case (2) we also have $Cab \in Cn(X_0)$, whilst in case (3) we have both

$$CCabp \in Cn(X_0) \quad \text{and} \quad CpCab \in Cn(X_0).$$

Thus in each case $Cab \in Y$. Consequently we have established that Y includes the set Z and is closed under the operation of implication formation, i.e. the only fundamental operation of the restricted sentential calculus. According to Lemma 1 it follows that $Sb_Z(S) \subseteq Y$; and thus for every sentence $a \in Sb_Z(S)$ we have either $a \in Cn(X_0)$ or $Cap \in Cn(X_0)$ and $Cpa \in Cn(X_0)$, q.e.d.

LEMMA 7. *If* $X_0 = \{CpCqp,\ CpCCpqq,\ CCqrCCpqCpr\}$ *and* $Z = \{p\}$, *then* (a) *the set* X_0 *is complete with respect to the set* $Sb_Z(S)$ *and* (b) *the set* $X_0 + Sb_Z(S)$ *is inconsistent.*

Proof. (a) easily results from the previous lemma. In fact for every sentence $a \in Sb_Z(S)$ we have either $a \in Cn(X_0)$ or $Cap \in Cn(X_0)$; thus, if a does not belong to the set $Cn(X_0)$, then the set $X_0 + \{a\}$ is inconsistent, since its set of consequences contains both the sentences a and Cap, hence also the sentential variable p, and consequently all possible sentences as elements (for every sentence can be obtained by substitution from the variable p). For analogous reasons (b) also holds.

Proof of Theorem 2. It is a well-known fact that the ordinary system L of the sentential calculus (which can be defined, for example, by means of the familiar two-valued matrix) is consistent and complete[1] and contains the sentences $CpCqp$, $CpCCpqq$, and $CCqrCCpqCpr$ as elements. Hence it follows that the set X_0 which consists of these three sentences is consistent. This fact, together with Lemma 7, permits us to apply Th. 1 and in this way to conclude that L is the only consistent and complete system which includes X_0. If now X is any consistent set of sentences such that $X_0 \subseteq X$, then, according to Lindenbaum's theorem, X can be extended to a consistent and complete system Y. Since $X_0 \subseteq Y$, Y cannot be distinct from L. Consequently L is the only consistent and complete system which includes X as a sub-set, q.e.d.

From Th. 2 we immediately obtain the following

COROLLARY. *The ordinary system of the restricted sentential calculus is the only consistent and complete system which includes the sentences* $CpCqp$, $CpCCpqq$, *and* $CCqrCCpqCpr$ *as elements.*

Analogous corollaries can be derived from Ths. 3 and 4 given below.

It should be noted in connexion with Th. 2 that the class of all systems of the sentential calculus which contains the sentences $CpCqp$, $CpCCpqq$, and $CCqrCCpqCpr$ is fairly comprehensive. For example, to this class belong all *many-valued systems* of the sentential calculus which originate with J. Łukasiewicz,[2] as well as the *positive system* of the sentential calculus originated by D. Hilbert.[3] Each of these systems can therefore be extended in only one way to a consistent and complete system.

---

[1] Cf. IV, pp. 43 ff.

[2] Cf. IV, pp. 47 ff.

[3] The set of all consequences of those axioms of Hilbert for the sentential calculus which contain no negation sign is called the positive system of the sentential calculus, see (29a), p. 153.

We turn now to those systems of the sentential calculus in which two constants occur, the implication sign and the negation sign. For these we have

THEOREM 3. *If X is a consistent set of sentences which contains the four sentences $CpNNp$, $CqCpq$, $CNpCpq$, and $CpCNqNCpq$ as elements, then the ordinary (two-valued) system of the sentential calculus is the only consistent and complete system which includes the set X.*

The proof is completely analogous to that of Th. 2 and is based on the following lemma which corresponds exactly to Lemmas 6 and 7:

LEMMA 8. *If $X_0 = \{CpNNp, CpCqp, CNpCpq, CpCNqNCpq\}$ and $Z = \{CpNNp\}$, then: (a) for every sentence $a \in Sb_Z(S)$ either $a \in Cn(X_0)$ or $Na \in Cn(X_0)$, (b) X_0 is complete with respect to the set $Sb_Z(S)$, and (c) the set $X_0 + Sb_Z(S)$ is inconsistent.*

Finally we consider systems of the sentential calculus in which, in addition to the signs of implication and negation, two other constants occur, the signs of disjunction and conjunction:

THEOREM 4. *If X is a consistent set of sentences which contains the following ten sentences as elements: $CpNNp$, $CqCpq$, $CNpCpq$, $CpCNqNCpq$, $CpApq$, $CqApq$, $CNpCNqNApq$, $CpCqKpq$, $CNpNKpq$, and $CNqNKpq$, then the ordinary (two-valued) system of the sentential calculus is the only consistent and complete system which includes the set X.*

To the systems X which satisfy the hypothesis of Th. 4 (or 3) belong again all many-valued systems of Łukasiewicz and also, among others, the *intuitionistic system* of the sentential calculus of A. Heyting.[1] For each of these systems the ordinary two-valued system constitutes the only consistent and complete extension.[2]

---

[1] Heyting, A. (28).

[2] This result was previously obtained for the many-valued systems of the sentential calculus by another method, cf. IV, p. 49, Th. 21 and the accompanying footnote.

XV

THE ESTABLISHMENT OF SCIENTIFIC SEMANTICS†

THE word 'semantics' is used here in a narrower sense than usual. We shall understand by semantics the totality of considerations concerning those concepts which, roughly speaking, express certain connexions between the expressions of a language and the objects and states of affairs referred to by these expressions. As typical examples of semantical concepts we may mention the concepts of *denotation, satisfaction,* and *definition,* which appear, for example, in the following statements:

> *The expression 'the victor of Jena' denotes Napoleon; snow satisfies the condition 'x is white'; the equation 'x³ = 2' defines (determines uniquely) the cube root of the number 2.*

The concept of *truth* also—and this is not commonly recognized—is to be included here, at least in its classical interpretation, according to which 'true' signifies the same as 'corresponding with reality'.

Concepts from the domain of semantics have traditionally played a prominent part in the discussions of philosophers, logicians, and philologists. Nevertheless they have long been regarded with a certain scepticism. From the historical point of view this scepticism is well founded; for, although the content of the semantical concepts, as they occur in colloquial language, is clear enough, yet all attempts to characterize this content more precisely have failed, and various discussions in which these concepts appeared and which were based on quite plausible and seemingly evident premisses, have often led to paradoxes and antinomies. It suffices to mention here the anti-

† BIBLIOGRAPHICAL NOTE. This article is a summary of an address which was given at the International Congress of Scientific Philosophy in Paris, 1935. The article first appeared in Polish under the title 'O ugruntowaniu naukowej semantyki', in *Przegląd Filozoficzny,* vol. 39 (1936), pp. 50–57, and later in German under the title 'Grundlegung der wissenschaftlichen Semantik', in *Actes du Congrès International de Philosophie Scientifique,* vol. 3 (Actualités Scientifiques et Industrielles, vol. 390), Paris, 1936, pp. 1–8.

nomy of the liar, the Grelling–Nelson antinomy of heterological terms and the Richard antinomy of definability.

The main source of the difficulties met with seems to lie in the following: it has not always been kept in mind that the semantical concepts have a relative character, that they must always be related to a particular language. People have not been aware that the language *about which* we speak need by no means coincide with the language *in which* we speak. They have carried out the semantics of a language in that language itself and, generally speaking, they have proceeded as though there was only one language in the world. The analysis of the antinomies mentioned shows, on the contrary, that the semantical concepts simply have no place in the language to which they relate, that the language which contains its own semantics, and within which the usual logical laws hold, must inevitably be inconsistent. Only in recent years has attention been given to all these facts (as far as I know Leśniewski was the first to become fully aware of them).

As soon as we give full recognition to the above circumstances, and carefully avoid the errors so far committed, the task of laying the foundations of a scientific semantics, i.e. of characterizing precisely the semantical concepts and of setting up a logically unobjectionable and materially adequate way of using these concepts, presents no further insuperable difficulties. Naturally, in doing this we must proceed cautiously, making full use of the apparatus which modern logic provides and carefully attending to the requirements of present-day methodology.

In the solution of this problem we can distinguish several steps. We must begin with the description of the language whose semantics we wish to construct. In particular we must enumerate the primitive terms of the language and give the' rules of definition by which new terms distinct from the primitive ones can be introduced into the language. Next we must distinguish those expressions of the language which are called sentences, separate the axioms from the totality of sentences, and finally formulate the rules of inference by means of which theorems can be derived from these axioms. The description of a language is exact and clear only if it is purely structural, that is to say, if we employ in it only those concepts which relate to the form and arrangement of the signs and compound expressions of the language. Not every language can be described in this purely structural manner. The languages for which such a description can be given are called *formalized languages*. Now, since the degree of exactitude of all further investigations depends essentially on the clarity and precision of this description, *it is only the semantics of formalized languages which can be constructed by exact methods*.

The next step is the construction of the language on the basis of which the semantics of the given language is to be developed and which, for the sake of brevity, we shall call the *metalanguage*. The most important point in this construction is the problem of equipping the metalanguage with a sufficiently rich vocabulary. But the solution of this problem is prescribed by the particular nature of the semantical concepts. In fact, semantical concepts express certain relations between objects (and states of affairs) referred to in the language discussed and expressions of the language referring to these objects. Hence the statements which establish the essential properties of semantical concepts must contain both the designation of the objects referred to (thus the expressions of the language itself), and the terms which are used in the structural description of the language. The latter terms belong to the domain of the so-called *morphology of language* and are the designations of individual expressions of the language, of structural properties of expressions, of structural relations between expressions, and so on. The metalanguage which is to form the basis for semantical investigations must thus contain both kinds of expression: the expressions of the original language, and the expressions of the morphology of language. In addition to these, the metalanguage, like every other language, must contain a larger or smaller stock of purely logical expressions. The question now arises whether a metalanguage which is equipped exclusively with expressions of the kinds mentioned forms a sufficient basis for semantical investigations. We shall return to this question later.

Our next task is to determine the conditions under which we should be inclined to regard a way of using semantical concepts as materially adequate and in accordance with ordinary usage. We shall explain this more exactly by reference to the concept of truth. We regard the truth of a sentence as its 'correspondence with reality'. This rather vague phrase, which can certainly lead to various misunderstandings and has often done so in the past, is interpreted as follows. We shall regard as valid all such statements as:

> the sentence 'it is snowing' is true if and only if it is snowing;
> the sentence 'the world war will begin in the year 1963' is true if
> and only if the world war will begin in 1963.

Quite generally we shall accept as valid every sentence of the form

> the sentence x is true if and only if p

where 'p' is to be replaced by any sentence of the language under investigation and 'x' by any individual name of that sentence provided this name occurs in the metalanguage. (In colloquial language such names are usually formed by means of quotation marks.) Statements of this form can be regarded as partial definitions of the concept of truth. They explain in a precise way, and in conformity with common usage, the sense of all special expressions of the type: the sentence x is true. Now, if we succeed in introducing the term 'true' into the metalanguage in such a way that every statement of the form discussed can be proved on the basis of the axioms and rules of inference of the metalanguage, then we shall say that the way of using the concept of truth which has thus been established is *materially adequate*. In particular, if we succeed in introducing such a concept of truth by means of a definition, then we shall also say that the corresponding definition is materially adequate. We can apply an analogous method to any other semantical concepts as well. For each of these concepts we formulate a system of statements, which are expressed in the form of equivalences and have the character of partial definitions; as regards their contents, these statements determine the sense of the concept concerned with respect to all concrete, structurally described expressions of the language being investigated. We then agree to regard a way of using (or a definition of) the semantical concept in question as materially adequate if it enables us to prove in the metalanguage all the partial definitions just mentioned. By way of illustration we give here such a partial definition of the concept of satisfaction:

John and Peter satisfy the sentential function 'X and Y are brothers' if and only if John and Peter are brothers.

It should also be noted that, strictly speaking, the described conventions (regarding the material adequacy of the usage of semantical concepts) are formulated in the metametalanguage and not in the metalanguage itself.

All the problems so far have been of a preparatory and auxiliary nature. Only after these preliminaries can we approach our chief problem. This is the establishment of a materially correct way of using the semantical concepts in the metalanguage. At this point two procedures come into consideration. In the first the semantical concepts (or at least some of them) are introduced into the metalanguage as new primitive concepts and their basic properties are established by means of axioms. Among these axioms are included all statements which secure the materially adequate use of the concepts in question. In this way semantics becomes an independent deductive theory based upon the morphology of language. But when this method, which seems easy and simple, is worked out in detail various objections present themselves. The setting up of an axiom system sufficient for the development of the whole of semantics offers considerable difficulties. For certain reasons, which we shall not go into here, the choice of axioms always has a rather accidental character, depending on inessential factors (such as e.g. the actual state of our knowledge). Various criteria which we should like to use in this connexion prove to be inapplicable. Moreover, the question arises whether the axiomatically constructed semantics is consistent. The problem of consistency arises, of course, whenever the axiomatic method is applied, but here it acquires a special importance, as we see from the sad experiences we have had with the semantical concepts in colloquial language. Apart from the problem of consistency, a method of constructing a theory does not seem to be very natural from the psychological point of view if in this method the role of primitive concepts—thus of concepts whose meaning should appear evident—is played by concepts which have led to various misunderstandings in the past. Finally, should this method prove to be the only possible one and not be regarded as merely a transitory stage, it would arouse certain doubts from a general philosophical point of view. It seems to me that it would then be difficult to bring this method into harmony with the postulates of the unity of science and of physicalism (since the concepts of semantics would be neither logical nor physical concepts).

In the second procedure, which has none of the above disadvantages, the semantical concepts are defined in terms of the usual concepts of the metalanguage and are thus reduced to purely logical concepts, the concepts of the language being investigated and the specific concepts of the morphology of language. In this way semantics becomes a part of the morphology of language if the latter is understood in a sufficiently wide sense. The question arises whether this method is applicable at all. It seems to me that this problem can now be regarded as definitely solved. It proves to be closely connected with the theory of logical types. The chief result relevant to this question can be formulated as follows:

It is possible to construct in the metalanguage methodologically correct and materially adequate definitions of the semantical concepts if and only if the metalanguage is equipped with variables of higher logical type than all the variables of the language which is the subject of investigation.

It would be impossible to establish here, even in general outline, the thesis just formulated. It can only be pointed out that it has been found useful, in defining the semantical concepts, to deal first with the concept of satisfaction; both because a definition of this concept presents relatively few difficulties, and because the remaining semantical concepts are easily reducible to it. It is also to be noted that from the definitions of semantical concepts various important theorems of a general nature concerning the concepts defined can be derived. For example, on the basis of the definition of truth we can prove the laws of contradiction and of excluded middle (in their metalogical formulation).

With this the problem of establishing semantics on a scientific basis is completely solved. Only the future can definitely say whether further investigations in this field will prove to be fruitful for philosophy and the special sciences, and what place semantics will win for itself in the totality of knowledge. But it does seem that the results hitherto reached justify a certain optimism in this respect. The very fact that it has been possible to define the semantical concepts, at least for formalized languages, in a correct and adequate manner seems to be not entirely without importance from the philosophical standpoint. The problem of the definition of truth, for example, has often been emphasized as one of the fundamental problems of the theory of knowledge. Also certain applications of semantics, and especially of the theory of truth, in the domain of the methodology of the deductive sciences, or metamathematics, seem to me to deserve attention. Using the definition of truth we are in a position to carry out the proof of consistency for deductive theories in which only (materially) true sentences are (formally) provable; this can be done under the condition that the metalanguage in which the proof of consistency is carried out is equipped with variables of higher type than all the variables which occur in the sentences of the theory discussed. Admittedly such a proof has no great cognitive value since it rests on logically stronger premises than the assumptions of the theory whose consistency is proved. Nevertheless, the result seems to be of some interest for the reason that it cannot be improved. For it follows from the investigations of Gödel that the proof of consistency cannot be carried out if the metalanguage contains no variables of higher type. The definition of truth has yet another consequence which is likewise connected with the investigations of Gödel. As is well known, Gödel has developed a method which makes it possible, in every theory which includes the arithmetic of natural numbers as a part, to construct sentences which can be neither proved nor disproved in this theory. But he has also pointed out that the undecidable sentences constructed by this method become decidable if the theory under investigation is enriched by the addition of variables of higher type. The proof that the sentences involved actually in this way become decidable again rests on the definition of truth. Similarly—as I have shown by means of the methods used in developing semantics—for any given deductive theory it is possible to indicate concepts which cannot be defined in this theory, although in their content they belong to the theory, and become definable in it if the theory is enriched by the introduction of higher types. Summarizing, we can say that the establishment of scientific semantics, and in particular the definition of truth, enables us to match the negative results in the field of metamathematics with corresponding positive ones, and in that way to fill to some extent the gaps which have been revealed in the deductive method and in the very structure of deductive science.[1]

[1] More detailed information about many of the problems discussed in this article can be found in VIII. Attention should also be called to my later paper, Tarski, A. (82). While the first part of that paper is close in its content to the present article, the second part contains polemical remarks regarding various objections which have been raised against my investigations in the field of semantics. It also includes some observations about the applicability of semantics to empirical sciences and their methodology.

XVI

ON THE CONCEPT OF LOGICAL CONSEQUENCE†

THE concept of *logical consequence* is one of those whose introduction into the field of strict formal investigation was not a
matter of arbitrary decision on the part of this or that investi-
gator; in defining this concept, efforts were made to adhere to
the common usage of the language of everyday life. But these
efforts have been confronted with the difficulties which usually
present themselves in such cases. With respect to the clarity
of its content the common concept of consequence is in no way
superior to other concepts of everyday language. Its extension
is not sharply bounded and its usage fluctuates. Any attempt
to bring into harmony all possible vague, sometimes contradictory, tendencies which are connected with the use of this
concept, is certainly doomed to failure. We must reconcile ourselves from the start to the fact that every precise definition of
this concept will show arbitrary features to a greater or less
degree.

Even until recently many logicians believed that they had
succeeded, by means of a relatively meagre stock of concepts,
in grasping almost exactly the content of the common concept
of consequence, or rather in defining a new concept which coincided in extent with the common one. Such a belief could easily
arise amidst the new achievements of the methodology of deductive science. Thanks to the progress of mathematical logic
we have learnt, during the course of recent decades, how to
present mathematical disciplines in the shape of formalized
deductive theories. In these theories, as is well known, the

† BIBLIOGRAPHICAL NOTE. This is a summary of an address given at the
International Congress of Scientific Philosophy in Paris, 1935. The article
first appeared in print in Polish under the title 'O pojciu wynikania logicz-
nego' in *Przegląd Filozoficzny*, vol. 39 (1936), pp. 58–68, and then in German
under the title 'Über den Begriff der logischen Folgerung', *Actes du Congrès
International de Philosophie Scientifique*, vol. 7 (Actualités Scientifiques et
Industrielles, vol. 394), Paris, 1936, pp. 1–11.

proof of every theorem reduces to single or repeated application of some simple rules of inference—such as the rules of substitution and detachment. These rules tell us what transformations of a purely structural kind (i.e. transformations in which only the external structure of sentences is involved) are to be performed upon the axioms or theorems already proved in the theory, in order that the sentences obtained as a result of such transformations may themselves be regarded as proved. Logicians thought that these few rules of inference exhausted the content of the concept of consequence. Whenever a sentence follows from others, it can be obtained from them—so it was thought—in more or less complicated ways by means of the transformations prescribed by the rules. In order to defend this view against sceptics who doubted whether the concept of consequence when formalized in this way really coincided in extent with the common one, the logicians were able to bring forward a weighty argument: the fact that they had actually succeeded in reproducing in the shape of formalized proofs all the exact reasonings which had ever been carried out in mathematics.

Nevertheless we know today that the scepticism was quite justified and that the view sketched above cannot be maintained. Some years ago I gave a quite elementary example of a theory which shows the following peculiarity: among its theorems there occur such sentences as:

A_0. 0 *possesses the given property P*,

A_1. 1 *possesses the given property P*,

and, in general, all particular sentences of the form

A_n. *n possesses the given property P*,

where 'n' represents any symbol which denotes a natural number in a given (e.g. decimal) number system. On the other hand the universal sentence:

A. *Every natural number possesses the given property P*,

cannot be proved on the basis of the theory in question by means of the normal rules of inference.[1] This fact seems to me to speak

---

[1] For a detailed description of a theory with this peculiarity see IX; for the discussion of the closely related rule of infinite induction see VIII, pp. 258 ff.

for itself. It shows that the formalized concept of consequence, as it is generally used by mathematical logicians, by no means coincides with the common concept. Yet intuitively it seems certain that the universal sentence A follows in the usual sense from the totality of particular sentences $A_0, A_1,..., A_n,...$. Provided all these sentences are true, the sentence A must also be true.

In connexion with situations of the kind just described it has proved to be possible to formulate new rules of inference which do not differ from the old ones in their logical structure, are intuitively equally infallible, i.e. always lead from true sentences to true sentences, but cannot be reduced to the old rules. An example of such a rule is the so-called rule of infinite induction according to which the sentence A can be regarded as proved provided all the sentences $A_0, A_1,..., A_n,...$ have been proved (the symbols 'A_0', 'A_1', etc., being used in the same sense as previously). But this rule, on account of its infinitistic nature, is in essential respects different from the old rules. It can only be applied in the construction of a theory if we have first succeeded in proving infinitely many sentences of this theory—a state of affairs which is never realized in practice. But this defect can easily be overcome by means of a certain modification of the new rule. For this purpose we consider the sentence B which asserts that all the sentences $A_0, A_1,..., A_n,...$ are *provable* on the basis of the rules of inference hitherto used (not that they have actually been proved). We then set up the following rule: if the sentence B is proved, then the corresponding sentence A can be accepted as proved. But here it might still be objected that the sentence B is not at all a sentence of the theory under construction, but belongs to the so-called metatheory (i.e. the theory *of* the theory discussed) and that in consequence a practical application of the rule in question will always require a transition from the theory to the metatheory.[1] In order to avoid this objection we shall restrict

---

[1] For the concept of metatheory and the problem of the interpretation of a metatheory in the corresponding theory see article VIII, pp. 167 ff., 184, and 247 ff.

consideration to those deductive theories in which the arithmetic of natural numbers can be developed, and observe that
in every such theory all the concepts and sentences of the
corresponding metatheory can be interpreted (since a one-one
correspondence can be established between expressions of a
language and natural numbers).[1] We can replace in the rule
discussed the sentence B by the sentence B', which is the arithmetical interpretation of B. In this way we reach a rule which
does not deviate essentially from the rules of inference, either
in the conditions of its applicability or in the nature of the
concepts involved in its formulation or, finally, in its intuitive
infallibility (although it is considerably more complicated).

Now it is possible to state other rules of like nature, and even
as many of them as we please. Actually it suffices in fact to
notice that the rule last formulated is essentially dependent
upon the extension of the concept 'sentence provable on the
basis of the rules hitherto used'. But when we adopt this rule
we thereby widen the extension of this concept. Then, for the
widened extension we can set up a new, analogous rule, and
so on ad infinitum. It would be interesting to investigate
whether there are any objective reasons for assigning a special
position to the rules ordinarily used.

The conjecture now suggests itself that we can finally succeed
in grasping the full intuitive content of the concept of consequence by the method sketched above, i.e. by supplementing
the rules of inference used in the construction of deductive
theories. By making use of the results of K. Gödel[2] we can
show that this conjecture is untenable. In every deductive
theory (apart from certain theories of a particularly elementary
nature), however much we supplement the ordinary rules of
inference by new purely structural rules, it is possible to construct sentences which follow, in the usual sense, from the
theorems of this theory, but which nevertheless cannot be
proved in this theory on the basis of the accepted rules of

---

[1] For the concept of metatheory and the problem of the interpretation of a
metatheory in the corresponding theory see article VIII, pp. 167 ff., 184, and
247 ff.

[2] Cf. Gödel, K. (22), especially pp. 190 f.

inference.[1] In order to obtain the proper concept of consequence, which is close in essentials to the common concept, we must resort to quite different methods and apply quite different conceptual apparatus in defining it. It is perhaps not superfluous to point out in advance that—in comparison with the new—the old concept of consequence as commonly used by mathematical logicians in no way loses its importance. This concept will probably always have a decisive significance for the practical construction of deductive theories, as an instrument which allows us to prove or disprove particular sentences of these theories. It seems, however, that in considerations of a general theoretical nature the proper concept of consequence must be placed in the foreground.[2]

The first attempt to formulate a precise definition of the proper concept of consequence was that of R. Carnap.[3] But this

[1] In order to anticipate possible objections the range of application of the result just formulated should be determined more exactly and the logical nature of the rules of inference exhibited more clearly; in particular it should be exactly explained what is meant by the structural character of these rules.

[2] An opposition between the two concepts in question is clearly pointed out in article IX, pp. 293 ff. Nevertheless, in contrast to my present standpoint, I have there expressed myself in a decidedly negative manner about the possibility of setting up an exact formal definition for the proper concept of consequence. My position at that time is explained by the fact that, when I was writing the article mentioned, I wished to avoid any means of construction which went beyond the theory of logical types in any of its classical forms; but it can be shown that it is impossible to define the proper concept of consequence adequately whilst using exclusively the means admissible in the classical theory of types; unless we should thus limit our considerations solely to formalized languages of an elementary and fragmentary character (to be exact, to the so-called languages of finite order; cf. article VIII, especially pp. 268 ff.). In his extremely interesting book, Carnap, R. (10), the term (*logical*) *derivation* or *derivability* is applied to the old concept of consequence as commonly used in the construction of deductive theories, in order to distinguish it from the concept of *consequence* as the proper concept. The opposition between the two concepts is extended by Carnap to the most diverse derived concepts ('f-*concepts*' and 'a-*concepts*', cf. pp. 88 ff., and 124 ff.); he also emphasizes—to my mind correctly—the importance of the proper concept of consequence and the concepts derived from it, for general theoretical discussions (cf. e.g. p. 128).

[3] Cf. Carnap, R. (10), pp. 88 f., and Carnap, R. (11) especially p. 181. In the first of these works there is yet another definition of consequence which is adapted to a formalized language of an elementary character. This definition is not considered here because it cannot be applied to languages of a more complicated logical structure. Carnap attempts to define the concept of logical consequence not only for special languages, but also within the framework of what he calls '*general syntax*'. We shall have more to say about this on p. 416, note 1.

attempt is connected rather closely with the particular properties of the formalized language which was chosen as the subject
of investigation. The definition proposed by Carnap can be
formulated as follows:

*The sentence X follows logically from the sentences of the class
K if and only if the class consisting of all the sentences of K and
of the negation of X is contradictory.*

The decisive element of the above definition obviously is the
concept 'contradictory'. Carnap's definition of this concept is
too complicated and special to be reproduced here without long
and troublesome explanations.[1]

I should like to sketch here a general method which, it seems
to me, enables us to construct an adequate definition of the
concept of consequence for a comprehensive class of formalized
languages. I emphasize, however, that the proposed treatment
of the concept of consequence makes no very high claim to
complete originality. The ideas involved in this treatment will
certainly seem to be something well known, or even something
of his own, to many a logician who has given close attention to
the concept of consequence and has tried to characterize it
more precisely. Nevertheless it seems to me that only the
methods which have been developed in recent years for the
establishment of scientific semantics, and the concepts defined
with their aid, allow us to present these ideas in an exact form.[2]

Certain considerations of an intuitive nature will form our
starting-point. Consider any class K of sentences and a sentence
X which follows from the sentences of this class. From an intuitive standpoint it can never happen that both the class K
consists only of true sentences and the sentence X is false.
Moreover, since we are concerned here with the concept of
logical, i.e. *formal*, consequence, and thus with a relation which is
to be uniquely determined by the form of the sentences between
which it holds, this relation cannot be influenced in any way by
empirical knowledge, and in particular by knowledge of the

---

[1] See footnote 3 on p. 413.
[2] The methods and concepts of semantics and especially the concepts of
truth and satisfaction are discussed in detail in article VIII; see also article XV.

objects to which the sentence X or the sentences of the class K refer. The consequence relation cannot be affected by replacing the designations of the objects referred to in these sentences by the designations of any other objects. The two circumstances just indicated, which seem to be very characteristic and essential for the proper concept of consequence, may be jointly expressed in the following statement:

(F) *If, in the sentences of the class* K *and in the sentence* X, *the constants—apart from purely logical constants—are replaced by any other constants (like signs being everywhere replaced by like signs), and if we denote the class of sentences thus obtained from* K *by* 'K'', *and the sentence obtained from* X *by* 'X'', *then the sentence* X' *must be true provided only that all sentences of the class* K' *are true.*

[For the sake of simplifying the discussion certain incidental complications are disregarded, both here and in what follows. They are connected partly with the theory of logical types, and partly with the necessity of eliminating any defined signs which may possibly occur in the sentences concerned, i.e. of replacing them by primitive signs.]

In the statement (F) we have obtained a necessary condition for the sentence X to be a consequence of the class K. The question now arises whether this condition is also sufficient. If this question were to be answered in the affirmative, the problem of formulating an adequate definition of the concept of consequence would be solved affirmatively. The only difficulty would be connected with the term 'true' which occurs in the condition (F). But this term can be exactly and adequately defined in semantics.[1]

Unfortunately the situation is not so favourable. It may, and it does, happen—it is not difficult to show this by considering special formalized languages—that the sentence X does not follow in the ordinary sense from the sentences of the class K although the condition (F) is satisfied. This condition may in fact be satisfied only because the language with which we are

---

[1] See footnote 2 on p. 414.

dealing does not possess a sufficient stock of extra-logical constants. The condition (F) could be regarded as sufficient for the sentence X to follow from the class K only if the designations of all possible objects occurred in the language in question. This assumption, however, is fictitious and can never be realized.[1] We must therefore look for some means of expressing the intentions of the condition (F) which will be completely independent of that fictitious assumption.

Such a means is provided by semantics. Among the fundamental concepts of semantics we have the concept of the *satisfaction of a sentential function* by single objects or by a sequence of objects. It would be superfluous to give here a precise explanation of the content of this concept. The intuitive meaning of such phrases as: *John and Peter satisfy the condition 'X and Y are brothers'*, or *the triple of numbers 2, 3, and 5 satisfies the equation 'x+y = z'*, can give rise to no doubts. The concept of satisfaction—like other semantical concepts—must always be relativized to some particular language. The details of its precise definition depend on the structure of this language. Nevertheless, a general method can be developed which enables us to construct such definitions for a comprehensive class of formalized languages. Unfortunately, for technical reasons, it would be impossible to sketch this method here even in its general outlines.[2]

One of the concepts which can be defined in terms of the concept of satisfaction is the concept of *model*. Let us assume that in the language we are considering certain variables correspond to every extra-logical constant, and in such a way that every sentence becomes a sentential function if the constants in it are replaced by the corresponding variables. Let L be any class of sentences. We replace all extra-logical constants which

---

[1] These last remarks constitute a criticism of some earlier attempts to define the concept of formal consequence. They concern, in particular, Carnap's definitions of logical consequence and a series of derivative concepts (L-consequences and L-concepts, cf. Carnap, R. (10), pp. 137 ff.). These definitions, in so far as they are set up on the basis of 'general syntax', seem to me to be materially inadequate, just because the defined concepts depend essentially, in their extension, on the richness of the language investigated.

[2] See footnote 2 on p. 414.

occur in the sentences belonging to L by corresponding variables, like constants being replaced by like variables, and unlike by unlike. In this way we obtain a class L' of sentential functions. An arbitrary sequence of objects which satisfies every sentential function of the class L' will be called a *model* or *realization of the class L of sentences* (in just this sense one usually speaks of models of an axiom system of a deductive theory). If, in particular, the class L consists of a single sentence X, we shall also call the model of the class L the *model of the sentence X*.

In terms of these concepts we can define the concept of logical consequence as follows:

The sentence X follows logically from the sentences of the class K if and only if every model of the class K is also a model of the sentence X.†

It seems to me that everyone who understands the content of the above definition must admit that it agrees quite well with common usage. This becomes still clearer from its various consequences. In particular, it can be proved, on the basis of this definition, that every consequence of true sentences must be true, and also that the consequence relation which holds between given sentences is completely independent of the sense of the extra-logical constants which occur in these sentences. In brief, it can be shown that the condition (F) formulated above is necessary if the sentence X is to follow from the sentences of the class K. On the other hand, this condition is in general not sufficient, since the concept of consequence here defined (in agreement with the standpoint we have taken) is independent of the richness in concepts of the language being investigated.

Finally, it is not difficult to reconcile the proposed definition with that of Carnap. For we can agree to call a class of sentences

† After the original of this paper had appeared in print, H. Scholz in his article 'Die Wissenschaftslehre Bolzanos, Eine Jahrhundert-Betrachtung', *Abhandlungen der Fries'schen Schule*, new series, vol. 6, pp. 399–472 (see in particular p. 472, footnote 58) pointed out a far-reaching analogy between this definition of consequence and the one suggested by B. Bolzano about a hundred years earlier.

contradictory if it possesses no model. Analogously, a class of sentences can be called *analytical* if every sequence of objects is a model of it. Both of these concepts can be related not only to classes of sentences but also to single sentences. Let us assume further that, in the language with which we are dealing, for every sentence X there exists a negation of this sentence, i.e. a sentence Y which has as a model those and only those sequences of objects which are not models of the sentence X (this assumption is rather essential for Carnap's construction). On the basis of all these conventions and assumptions it is easy to prove the *equivalence of the two definitions*. We can also show —just as does Carnap—that those and only those sentences are analytical which follow from every class of sentences (in particular from the empty class), and those and only those are contradictory from which every sentence follows.[1]

I am not at all of the opinion that in the result of the above discussion the problem of a materially adequate definition of the concept of consequence has been completely solved. On the contrary, I still see several open questions, only one of which— perhaps the most important—I shall point out here.

Underlying our whole construction is the division of all terms of the language discussed into logical and extra-logical. This division is certainly not quite arbitrary. If, for example, we were to include among the extra-logical signs the implication sign, or the universal quantifier, then our definition of the concept of consequence would lead to results which obviously contradict ordinary usage. On the other hand, no objective grounds are known to me which permit us to draw a sharp

---

[1] Cf. Carnap, R. (10), pp. 135 ff., especially Ths. 52.7 and 52.8; Carnap, R. (11), p. 182, Ths. 10 and 11. Incidentally I should like to remark that the definition of the concept of consequence here proposed does not exceed the limits of syntax in Carnap's conception (cf. Carnap, R. (10), pp. 6 ff.). Admittedly the general concept of satisfaction (or of model) does not belong to syntax; but we use only a special case of this concept—the satisfaction of sentential functions which contain no extra-logical constants, and this special case can be characterized using only general logical and specific syntactical concepts. Between the general concept of satisfaction and the special case of this concept used here approximately the same relation holds as between the semantical concept of true sentence and the syntactical concept of analytical sentence.

boundary between the two groups of terms. It seems to be possible to include among logical terms some which are usually regarded by logicians as extra-logical without running into consequences which stand in sharp contrast to ordinary usage. In the extreme case we could regard all terms of the language as logical. The concept of *formal* consequence would then coincide with that of *material* consequence. The sentence X would in this case follow from the class K of sentences if either X were true or at least one sentence of the class K were false.[1]

In order to see the importance of this problem for certain general philosophical views it suffices to note that the division of terms into logical and extra-logical also plays an essential part in clarifying the concept 'analytical'. But according to many logicians this last concept is to be regarded as the exact formal correlate of the concept of *tautology* (i.e. of a statement

---

[1] It will perhaps be instructive to juxtapose the three concepts: 'derivability' (cf. p. 413, note 2), 'formal consequence', and 'material consequence', for the special case when the class K, from which the given sentence X follows, consists of only a finite number of sentences: $Y_1, Y_2,..., Y_n$. Let us denote by the symbol 'Z' the conditional sentence (the implication) whose antecedent is the conjunction of the sentences $Y_1, Y_2,..., Y_n$ and whose consequent is the sentence X. The following equivalences can then be established:

the sentence X is (logically) derivable from the sentences of the class K if and only if the sentence Z is logically provable (i.e. derivable from the axioms of logic);

the sentence X follows formally from the sentences of the class K if and only if the sentence Z is analytical;

the sentence X follows materially from the sentences of the class K if and only if the sentence Z is true.

Of the three equivalences only the first can arouse certain objections; cf. article XII, pp. 342-64, especially 346. In connexion with these equivalences cf. also Ajdukiewicz, K. (2), p. 19, and (4), pp. 14 and 42.

In view of the analogy indicated between the several variants of the concept of consequence, the question presents itself whether it would not be useful to introduce, in addition to the special concepts, a general concept of a relative character, and indeed the concept of *consequence with respect to a class L of sentences*. If we make use again of the previous notation (limiting ourselves to the case when the class K is finite), we can define this concept as follows:

the sentence X follows *from the sentences of the class K* with respect to the class L of sentences *if and only if the sentence Z belongs to the class L.*

On the basis of this definition, derivability would coincide with consequence with respect to the class of all logically provable sentences, formal consequences would be consequences with respect to the class of all analytical sentences, and material consequences those with respect to the class of all true sentences.

which 'says nothing about reality'), a concept which to me personally seems rather vague, but which has been of fundamental importance for the philosophical discussions of L. Wittgenstein and the whole Vienna Circle.[1]

Further research will doubtless greatly clarify the problem which interests us. Perhaps it will be possible to find important objective arguments which will enable us to justify the traditional boundary between logical and extra-logical expressions. But I also consider it to be quite possible that investigations will bring no positive results in this direction, so that we shall be compelled to regard such concepts as 'logical consequence', 'analytical statement', 'and 'tautology' as relative concepts which must, on each occasion, be related to a definite, although in greater or less degree arbitrary, division of terms into logical and extra-logical. The fluctuation in the common usage of the concept of consequence would—in part at least—be quite naturally reflected in such a compulsory situation.

[1] Cf. Wittgenstein, L. (91), Carnap, R. (10), pp. 37–40.

XVII

SENTENTIAL CALCULUS AND TOPOLOGY†

In this article I shall point out certain formal connexions between the sentential calculus and topology (as well as some other mathematical theories). I am concerned in the first place with a topological interpretation of two systems of the sentential calculus, namely the ordinary (two-valued) and the intuitionistic (Brouwer–Heyting) system. With every sentence $\mathfrak{A}$ of the sentential calculus we correlate, in one-one fashion, a sentence $\mathfrak{A}_1$ of topology in such a way that $\mathfrak{A}$ is provable in the two-valued calculus if and only if $\mathfrak{A}_1$ holds in every topological space. An analogous correlation is set up for the intuitionistic calculus. The present discussion seems to me to have a certain interest not only from the purely formal point of view; it also throws an interesting light on the content relations between the two systems of the sentential calculus and the intuitions underlying these systems.

In order to avoid possible misunderstandings I should like to emphasize that I have not attempted to adapt the methods of reasoning used in this article to the requirements of intuitionistic logic.[1] For valuable help in completing this work I am indebted to Professor A. Mostowski.

---

[1] Most results of this article were obtained in the year 1935. The connexion between the intuitionistic calculus and Boolean algebra (or the theory of deductive systems, see § 5) was discovered by me still earlier, namely in 1931. Some remarks to this effect can be found in article XII of the present book and in Tarski, A. (80). Only after completing this paper did I become acquainted with the work, then newly published, of Stone, M. H. (70). In spite of an entirely different view of Brouwerian logic there is certainly some connexion between particular results of the two works, as can easily be seen comparing Stone's Th. 7, p. 22, and my Th. 4.11. In their mathematical content these two theorems are closely related. But this does not at all apply to the two works as wholes. In particular, Th. 4.24, in which I see the kernel of this paper, tends in quite a different direction from Stone's considerations.

---

† BIBLIOGRAPHICAL NOTE. This article is the text of an address given by the author on 30 September 1937 to the Third Polish Mathematical Congress in Warsaw (see *Annales de la Société polonaise de mathématique*, vol. 16 (1937), p. 192). The article first appeared under the title 'Der Aussagenkalkül und die Topologie', in *Fundamenta Mathematicae*, vol. 31 (1938), pp. 103–34.

§ 1. The Two-valued and the Intuitionistic Sentential Calculus

As is well known, the most elementary part of mathematical logic is the *sentential calculus*. In the expressions of the sentential calculus variables of only one kind occur, namely *sentential variables*, which represent whole sentences. As sentential variables the letters 'X', 'Y', 'Z',... will be used. In addition to the variables, four constants occur in the sentential calculus: the *implication* sign '$\rightarrow$', the *disjunction* sign '$\vee$', the *conjunction* sign '$\wedge$', and the *negation* sign '$\sim$' (a fifth constant, the *equivalence* sign '$\Leftrightarrow$', will not be involved here).

Arbitrary *expressions* which are composed of sentential variables, the four constants mentioned, and possibly brackets, will be denoted by the letters '$\mathfrak{A}$', '$\mathfrak{B}$', '$\mathfrak{C}$',.... . It is assumed that the sentential variables are ordered in an (infinite) sequence with distinct terms: $\mathfrak{B}_1, \mathfrak{B}_2,..., \mathfrak{B}_n,....$. The symbols '$\mathfrak{A} \rightarrow \mathfrak{B}$', '$\mathfrak{A} \vee \mathfrak{B}$', and '$\mathfrak{A} \wedge \mathfrak{B}$' denote the *implication*, the *disjunction*, and the *conjunction* of $\mathfrak{A}$ and $\mathfrak{B}$ respectively, that is the compound expressions which are formed when $\mathfrak{A}$ and $\mathfrak{B}$ are combined by the corresponding constant '$\rightarrow$', '$\vee$', or '$\wedge$'. Analogously the *negation* of $\mathfrak{A}$ is denoted by '$\sim \mathfrak{A}$'.[1]

In addition to the signs just listed the usual set-theoretical symbolism will be used.

The only expressions of the sentential calculus with which we shall deal will be sentences:

DEFINITION 1.1. *An expression $\mathfrak{A}$ is called a* sentential function, *or (for short) a* sentence, *if $\mathfrak{A}$ belongs to every system which* (i) *contains all sentential variables among its elements,* (ii) *is closed under the operations of forming implications, disjunctions, conjunctions, and negations; in other words, the system of all sentences is the smallest system which has the properties* (i) *and* (ii).

---

[1] In principle we thus use the signs '$\rightarrow$', '$\vee$', '$\wedge$', and '$\sim$' in two senses: in the logical and in the metalogical sense. In practice only the second sense is involved. In §§ 4 and 5 yet another, quite different sense will be given to these signs.

In the system of all sentences two subsystems are distinguished: the provable sentences of the two-valued calculus and those of the intuitionistic calculus.

DEFINITION 1.2. *The sentence* $\mathfrak{A}$ *is called an* axiom *of the two-valued calculus, or of the intuitionistic calculus, if there are sentences* $\mathfrak{B}$, $\mathfrak{C}$, *and* $\mathfrak{D}$ *such that* $\mathfrak{A}$ *satisfies one of the following formulas* (i)–(x) *in the first case, or one of the formulas* (i)–(ix) *and* (xi) *in the second case:*

(i) $\mathfrak{A} = \mathfrak{B} \rightarrow (\mathfrak{C} \rightarrow \mathfrak{B})$,

(ii) $\mathfrak{A} = [\mathfrak{B} \rightarrow (\mathfrak{C} \rightarrow \mathfrak{D})] \rightarrow [(\mathfrak{B} \rightarrow \mathfrak{C}) \rightarrow (\mathfrak{B} \rightarrow \mathfrak{D})]$,

(iii) $\mathfrak{A} = \mathfrak{B} \rightarrow (\mathfrak{B} \vee \mathfrak{C})$,

(iv) $\mathfrak{A} = \mathfrak{C} \rightarrow (\mathfrak{B} \vee \mathfrak{C})$,

(v) $\mathfrak{A} = (\mathfrak{B} \rightarrow \mathfrak{D}) \rightarrow \{(\mathfrak{C} \rightarrow \mathfrak{D}) \rightarrow [(\mathfrak{B} \vee \mathfrak{C}) \rightarrow \mathfrak{D}]\}$,

(vi) $\mathfrak{A} = (\mathfrak{B} \wedge \mathfrak{C}) \rightarrow \mathfrak{B}$,

(vii) $\mathfrak{A} = (\mathfrak{B} \wedge \mathfrak{C}) \rightarrow \mathfrak{C}$,

(viii) $\mathfrak{A} = (\mathfrak{D} \rightarrow \mathfrak{B}) \rightarrow \{(\mathfrak{D} \rightarrow \mathfrak{C}) \rightarrow [\mathfrak{D} \rightarrow (\mathfrak{B} \wedge \mathfrak{C})]\}$,

(ix) $\mathfrak{A} = \sim\mathfrak{B} \rightarrow (\mathfrak{B} \rightarrow \mathfrak{C})$,

(x) $\mathfrak{A} = (\sim\mathfrak{B} \rightarrow \mathfrak{B}) \rightarrow \mathfrak{B}$,

(xi) $\mathfrak{A} = (\mathfrak{B} \rightarrow \sim\mathfrak{B}) \rightarrow \sim\mathfrak{B}$.[1]

DEFINITION 1.3. *If* $\mathfrak{A}$, $\mathfrak{B}$, *and* $\mathfrak{C}$ *are three sentences and if* $\mathfrak{A} = \mathfrak{B} \rightarrow \mathfrak{C}$, *then* $\mathfrak{C}$ *is said to be the* result of the detachment *of the sentence* $\mathfrak{B}$ *from the sentence* $\mathfrak{A}$.

DEFINITION 1.4. *The system* $\mathfrak{W}$ *of provable sentences of the two-valued calculus is the smallest system of sentences which contains all the axioms of the two-valued calculus and is closed under the operation of detachment. The system* $\mathfrak{W}$ *of provable sentences of the intuitionistic calculus is the smallest system which contains all the axioms of the intuitionistic calculus and is closed under the same operation.*[2]

It is known that from these definitions the following two theorems are derivable:

---

[1] Cf. similar axiom systems: for the two-valued calculus in Hilbert, D. and Bernays, P. (31) and for the intuitionist calculus in Gentzen, G. (18). An axiom system for the intuitionistic calculus was first given in Heyting, A. (28). The problem of the mutual independence of the axioms (or, more exactly, axiom-schemata) of 1.2 is not here discussed.

[2] The operation of substitution need not be considered in 1.4 since the axiom system defined in 1.2 is already closed under this operation.

THEOREM 1.5. $\mathfrak{IR} \subseteq \mathfrak{ZR}$ (*but not conversely*: *if, e.g.,* $\mathfrak{B}$ *is a sentential variable, then* $\mathfrak{B} \vee \sim \mathfrak{B}$ *and* $\sim \mathfrak{B} \vee \sim \sim \mathfrak{B}$ *belong to* $\mathfrak{ZR}$ *but not to* $\mathfrak{IR}$).

THEOREM 1.6. *For every sentence* $\mathfrak{A}$ *the following conditions are equivalent*: (i) $\mathfrak{A} \in \mathfrak{ZR}$, (ii) $\sim \sim \mathfrak{A} \in \mathfrak{ZR}$, (iii) $\sim \sim \mathfrak{A} \in \mathfrak{IR}$.[1]

An exact proof of these theorems (as well as of all the results given below) must of course be based upon a suitable axiom-system for the meta-sentential calculus and not only upon the definitions of the notions involved. But it would be superfluous here to formulate such an axiom system explicitly.[2]

§ 2. The Matrix Method

The definitions in § 1 afford no criterion that would enable us to decide in each particular case whether a given sentence $\mathfrak{A}$ is provable in the two-valued or in the intuitionistic calculus. Such a criterion is provided by the so-called *matrix method*.[3]

DEFINITION 2.1. *Let there be given a set* $\mathscr{W}$ *of arbitrary elements, an element* $A \in \mathscr{W}$, *three binary operations* $\twoheadrightarrow$, $\curlyvee$, *and* $\curlywedge$, *and a unary operation* $\sim$. *It is assumed that* $\mathscr{W}$ *is closed under these operations and that the following holds*:

if $Y \in \mathscr{W}$ *and* $A \twoheadrightarrow Y = A$, *then* $Y = A$.

Under these assumptions, the ordered sextuple

$$\mathsf{M} = [\mathscr{W}, A, \twoheadrightarrow, \curlyvee, \curlywedge, \sim]$$

is called a (normal logical) *matrix*.

Note 2.2. If $\mathsf{M} = [\mathscr{W}, A, \twoheadrightarrow, \curlyvee, \curlywedge, \sim]$ is a matrix, then $\mathscr{W}$ is sometimes called the *value system*, A the *designated element*, and $\twoheadrightarrow$, $\curlyvee$, $\curlywedge$, $\sim$ the *fundamental operations* (*first, second*, etc.) of M.

---

[1] The first part of Th. 1.5 follows easily from 1.2–1.4 (it suffices to show that every sentence $\mathfrak{A}$ of the form 1.2 (xi) belongs to $\mathfrak{ZR}$. For the second part cf. Heyting, A. (28), p. 56. Th. 1.6 was stated in Glivenko, T. (19).

[2] In this connexion see VIII, p. 173, IX, p. 282.

[3] Cf. IV, p. 41. The concept of matrix is there rather more widely conceived than here, since matrices with more than one designated element are also considered (cf. 2.1, 2.2).

DEFINITION 2.3. *Two matrices* $\mathsf{M}_1 = [\,\mathscr{W}_1, A_1, \twoheadrightarrow_1, \Upsilon_1, \curlywedge_1, \sim_1]$ *and* $\mathsf{M}_2 = [\,\mathscr{W}_2, A_2, \twoheadrightarrow_2, \Upsilon_2, \curlywedge_2, \sim_2]$ *are said to be* isomorphic *if there is a function* F *which maps* $\mathscr{W}_1$ *on* $\mathscr{W}_2$ *in a one-one fashion and satisfies the following formulas:* $F(A_1) = A_2$, $F(X \twoheadrightarrow_1 Y) = F(X) \twoheadrightarrow_2 F(Y)$, $\quad F(X \,\Upsilon_1 Y) = F(X) \,\Upsilon_2 F(Y)$, $F(X \curlywedge_1 Y) = F(X) \curlywedge_2 F(Y)$, *and* $F(\sim_1 X) = \sim_2 F(X)$ *for all* $X, Y \in \mathscr{W}_1$.

COROLLARY 2.4. *Every matrix* M *is isomorphic with itself; if* M_1 *is isomorphic with* M_2, *then* M_2 *is isomorphic with* M_1; *if* M_1 *is isomorphic with* M_2 *and* M_2 *is isomorphic with* M_3, *then* M_1 *is isomorphic with* M_3.   [By 2.3]

DEFINITION 2.5. *Let* $\mathsf{M} = [\,\mathscr{W}, A, \twoheadrightarrow, \Upsilon, \curlywedge, \sim]$ *be a matrix and* $\mathfrak{A}$ *a sentence. The following formulas define (recursively) a function* $F_{\mathfrak{A},\mathsf{M}}$ *which correlates an element*

$$F_{\mathfrak{A},\mathsf{M}}(X_1,...,X_n,...) \in \mathscr{W}$$

with every infinite sequence of elements $X_1,..., X_n,... \in \mathscr{W}$:

(i) $F_{\mathfrak{A},\mathsf{M}}(X_1,...,X_n,...) = X_p$ *if* $\mathfrak{A} = \mathfrak{B}_p$ $(p = 1,2,...)$.

(ii) $F_{\mathfrak{A},\mathsf{M}}(X_1,...,X_n,...)$
$$= F_{\mathfrak{B},\mathsf{M}}(X_1,...,X_n,...) \twoheadrightarrow F_{\mathfrak{C},\mathsf{M}}(X_1,...,X_n,...),$$

if $\mathfrak{A} = \mathfrak{B} \to \mathfrak{C}$ *(where* $\mathfrak{B}$ *and* $\mathfrak{C}$ *are any sentences).*

(iii), (iv) *analogously for the operations* Υ *and* $\vee$, *or* $\curlywedge$ *and* $\wedge$.

(v) $F_{\mathfrak{A},\mathsf{M}}(X_1,...,X_n,...) = \sim F_{\mathfrak{B},\mathsf{M}}(X_1,...,X_n,...)$ *if* $\mathfrak{A} = \sim \mathfrak{B}$ *(where* $\mathfrak{B}$ *is any sentence).*

We say that the sentence $\mathfrak{A}$ *is* satisfied by *the matrix* M, *in symbols* $\mathfrak{A} \in \mathfrak{E}(\mathsf{M})$, *if* $F_{\mathfrak{A},\mathsf{M}}(X_1,..., X_n,...) = A$ *for all* $X_1,..., X_n,... \in \mathscr{W}$.[1]

Note 2.6. It is sometimes said that a matrix M is *adequate* for the system $\mathfrak{S}$ of sentences if $\mathfrak{E}(\mathsf{M}) = \mathfrak{S}$.

COROLLARY 2.7. *If the matrices* M_1 *and* M_2 *are isomorphic, then* $\mathfrak{E}(\mathsf{M}_1) = \mathfrak{E}(\mathsf{M}_2)$.   [By 2.1, 2.3, 2.5]

---

[1] In the formulation of 2.5, we could use functions with finitely many variables, but that would create certain technical difficulties in our further considerations. Another, although equivalent, definition of the set $\mathfrak{E}(\mathsf{M})$ was given in IV.

DEFINITION 2.8. *If*

$M = [\mathscr{W}, A, \twoheadrightarrow, \curlyvee, \curlywedge, \sim]$ *and* $M_1 = [\mathscr{W}_1, A, \twoheadrightarrow, \curlyvee, \curlywedge, \sim]$ *are two matrices and if* $\mathscr{W}_1 \subseteq \mathscr{W}$, *then* M_1 *is called a* submatrix *of* M.

COROLLARY 2.9. *If* M *is any matrix and* M_1 *is a submatrix of* M, *then* $\mathfrak{E}(M) \subseteq \mathfrak{E}(M_1)$.   [By 2.1, 2.5, 2.8]

DEFINITION 2.10. *We denote by* ZK *the ordered sextuple* $[\mathscr{W}, 1, \twoheadrightarrow, \curlyvee, \curlywedge, \sim]$, *where* $\mathscr{W} = \{0, 1\}$, $x \twoheadrightarrow y = 1-x+x.y$, $x \curlyvee y = x+y-x.y$, $x \curlywedge y = x.y$, *and* $\sim x = 1-x$ *for all* $x, y \in \mathscr{W}$.

The following result is well known:

THEOREM 2.11. ZK *is a matrix and* $\mathfrak{E}(ZK) = 3\mathfrak{R}$.[1]

For the system $3\mathfrak{R}$ there is, in contrast to $3\mathfrak{R}$, no adequate matrix with a finite value system.[2] We can, however, construct an infinite sequence of matrices $IK_1, \ldots, IK_n, \ldots$ with finite value systems such that $\prod_{n=1}^{\infty} \mathfrak{E}(IK_n) = 3\mathfrak{R}$. We shall now describe the construction of this sequence.[3]

DEFINITION 2.12. *Let* $M = [\mathscr{W}, B, \twoheadrightarrow, \curlyvee, \curlywedge, \sim]$ *be a matrix and* A *any element which does not belong to* $\mathscr{W}$. *We put:*

(i) $\mathscr{W}^* = \mathscr{W} + \{A\}$;

(ii) $X \twoheadrightarrow^* Y = X \twoheadrightarrow Y$ *if* $X, Y \in \mathscr{W}$ *and* $X \twoheadrightarrow Y \neq B$; $X \twoheadrightarrow^* Y = A$ *if* $X, Y \in \mathscr{W}$ *and* $X \twoheadrightarrow Y = B$; $X \twoheadrightarrow^* A = A$ *for* $X \in \mathscr{W}^*$; $A \twoheadrightarrow^* Y = Y$ *for* $Y \in \mathscr{W}^*$;

(iii) $X \curlyvee^* Y = X \curlyvee Y$ *for* $X, Y \in \mathscr{W}$; $Z \curlyvee^* A = A \curlyvee^* Z = A$ *for* $Z \in \mathscr{W}^*$;

(iv) $X \curlywedge^* Y = X \curlywedge Y$ *for* $X, Y \in \mathscr{W}$; $Z \curlywedge^* A = A \curlywedge^* Z = Z$ *for* $Z \in \mathscr{W}^*$;

(v) $\sim^* X = \sim X$ *if* $X \in \mathscr{W}$ *and* $\sim X \neq B$; $\sim^* X = A$ *if* $X \in \mathscr{W}$ *and* $\sim X = B$; $\sim^* A = \sim B$.

The ordered sextuple $[\mathscr{W}^*, A, \twoheadrightarrow^*, \curlyvee^*, \curlywedge^*, \sim^*]$ *is denoted by* M*.

---

[1] 2.11 easily results from the well-known theorem according to which the system $3\mathfrak{R}$ is complete. The first completeness proof for $3\mathfrak{R}$ is due to Post, E. (60), pp. 180 ff.

[2] Cf. Gödel, K. (23), p. 40.

[3] This result is due to Jaśkowski, S. (36). (Our account deviates only in inessentials from that of Jaśkowski. The operation Γ of Jaśkowski is replaced by the operation * defined in 2.12, which serves the same purpose.)

Note 2.13. The operation on matrices just defined is not unambiguous because its result depends on the choice of the element A. This fact does not affect the subsequent considerations because all matrices which can be obtained from the given matrix M by means of the operation $*$ are isomorphic with one another. The ambiguity of 'M$*$' is, moreover, avoidable by constructing a set-theoretical function F which correlates an element $F(\mathscr{W})$ *non*-$\in\mathscr{W}$ with every system $\mathscr{W}$, and then replacing 'A' in 2.12 by '$F(\mathscr{W})$' (certain difficulties which may arise in connexion with the theory of logical types will here be ignored).

COROLLARY 2.14. *If* M *is a matrix, then* M$*$ *is also a matrix and* $\mathfrak{E}(M*) \subseteq \mathfrak{E}(M)$; *if the matrices* M_1 *and* M_2 *are isomorphic, then* M_1^* *and* M_2^* *are also isomorphic.*   [By 2.1, 2.3, 2.5, 2.12]

DEFINITION 2.15. *Let n be a natural number and*

$$M = [\mathscr{W}, A, \twoheadrightarrow, \curlyvee, \curlywedge, \sim]$$

a matrix. We put:

(i) $\mathscr{W}^n =$ *the system of ordered n-tuples* $[X_1,..., X_n]$ *with*

$$X_1,..., X_n \in \mathscr{W};$$

(ii) $A^n = [X_1,..., X_n]$, *where* $X_1 = ... = X_n = A$;

(iii) $[X_1,..., X_n] \twoheadrightarrow^n [Y_1,..., Y_n] = [X_1 \twoheadrightarrow Y_1,..., X_n \twoheadrightarrow Y_n]$ *for*

$$X_1,..., X_n, Y_1,..., Y_n \in \mathscr{W};$$

(iv), (v) *analogously for* $\curlyvee$ *and* $\curlywedge$;

(vi) $\sim^n [X_1,..., X_n] = [\sim X_1,..., \sim X_n]$ *for* $X_1,..., X_n \in \mathscr{W}$.

$[\mathscr{W}^n, A^n, \twoheadrightarrow^n, \curlyvee^n, \curlywedge^n, \sim^n]$ *is called the nth* power *of the matrix* M *and is denoted by* 'M^n'.

COROLLARY 2.16. *If n is a natural number and* M *is a matrix, then* M^n *is also a matrix and we have* $\mathfrak{E}(M^n) = \mathfrak{E}(M)$; *if* M_1 *and* M_2 *are isomorphic matrices, then* M_1^n *and* M_2^n *are also isomorphic.*
                                        [By 2.1, 2.3, 2.5, 2.15]

DEFINITION 2.17. $IK_1 = ZK$, $IK_{n+1} = ((IK_n)^n)*$ *for every natural number n.*

On the basis of this definition the following theorem can be proved:

THEOREM 2.18. *In order that* $\mathfrak{A} \in \mathfrak{IR}$, *it is necessary and sufficient that* $\mathfrak{A} \in \mathfrak{E}(\mathsf{IK}_n)$ *for every natural number n; in other words*

$$\prod_{n=1}^{\infty} \mathfrak{E}(\mathsf{IK}_n) = \mathfrak{IR}.^1$$

Note 2.19. It is known that this theorem can be improved: with every sentence $\mathfrak{A}$ a well-determined natural number can be correlated (depending exclusively on the structure of this sentence) such that the formulas $\mathfrak{A} \in \mathfrak{E}(\mathsf{IK}_n)$ and $\mathfrak{A} \in \mathfrak{IR}$ are equivalent.[1]

The decision criterion mentioned at the beginning of this section is provided by 2.11 for the system $\mathfrak{IR}$ and by 2.18 in its improved form just mentioned for the system $\mathfrak{IR}$.[2]

§ 3. TOPOLOGICAL SPACES[3]

We first recall some familiar topological concepts:

DEFINITION 3.1. *A non-empty set S is called a* topological space (*with fundamental operation* $^-$), *if the following conditions are satisfied:*

(i) *if* $X \subseteq S$, *then* $\bar{\bar{X}} = \bar{X} \subseteq S$;

(ii) *if* $X \subseteq S$ *and X consists of at most one element, then* $\bar{X} = X$;

(iii) *if* $X \subseteq S$ *and* $Y \subseteq S$, *then* $\overline{X+Y} = \bar{X}+\bar{Y}$.

Note 3.2. Let S be a topological space and A a non-empty subset of S. We now define an operation $^{-(A)}$ relative to A by the formula $\bar{X}^{(A)} = A . \bar{X}$ for every $X \subseteq A$. On the basis of 3.1 it can then easily be shown that A is a topological space with the fundamental operation $^{-(A)}$; such a space is called a *subspace* of S.

DEFINITION 3.3. *If S is a topological space, a set X is said to be* open (*in S*), *in symbols* $X \in \mathcal{O}(S)$, *if* $X = S - \overline{S - X}$.

DEFINITION 3.4. *A subset X of a topological space S is said to be* dense (*in S*) *if* $\bar{X} = S$.

---

[1] See p. 426, note 3.

[2] Another decision criterion for the intuitionistic calculus was given in Gentzen, G. (18).

[3] For what follows cf. Kuratowski, C. (41), in particular pp. 15 ff., 38, 40, 82 f., 95, and 101 ff.

DEFINITION 3.5. *The topological space S is said to be* isolated *if x non-$\in \overline{S-\{x\}}$, and* dense-in-itself *if $x \in \overline{S-\{x\}}$ for every $x \in S$.*

COROLLARY 3.6. *A topological space S is isolated if and only if $\overline{X} = X$ for every $X \subseteq S$.*   [By 3.1, 3.5]

DEFINITION 3.7. *A topological space S is called* normal *if, for any two sets $X_1 \subseteq S$ and $X_2 \subseteq S$ such that $\overline{X}_1 . \overline{X}_2 = 0$, there exist two disjoint sets $Y_1, Y_2 \in \mathcal{O}(S)$ such that $\overline{X}_1 \subseteq Y_1$ and $\overline{X}_2 \subseteq Y_2$.*

DEFINITION 3.8. *We say that the topological space S is a* space with a countable basis *if there is an infinite sequence of non-empty sets $X_1,..., X_n,... \in \mathcal{O}(S)$ such that every non-empty set $Y \in \mathcal{O}(S)$ can be represented in the form $Y = X_{i_1}+...+X_{i_n}+...$ (where $i_1,..., i_n,...$ is a sequence of natural numbers).*

For later applications we shall distinguish another special class of topological spaces:

DEFINITION 3.9. *A topological space S is called an E-space if it satisfies the following condition:*

For every natural number n and every non-empty set $A \in \mathcal{O}(S)$, there exist non-empty, pair-wise disjoint sets

$$B_1,..., B_n \in \mathcal{O}(S)$$

for which

(i) $B_1+...+B_n \subseteq A$ *and* $B_1+...+B_n \neq A$,

(ii) $\overline{A-(B_1+...+B_n)} \supseteq \overline{A}-A$,

(iii) $\overline{B}_1. \,.....\, \overline{B}_n \supseteq A-(B_1+...+B_n)$.

THEOREM 3.10. *Every normal and dense-in-itself topological space S with a countable basis is an E-space.*

Proof.[1] According to 3.8 there exists an infinite sequence of sets $C_1,..., C_k,...$ with the following properties:

(1) *the sets $C_1,..., C_k,...$ are open and non-empty;*

(2) *every open and non-empty set X can be represented in the form*

$$X = C_{i_1}+...+C_{i_k}+... \,.$$

Since, by hypothesis, the space S is normal and dense-in-itself,

---

[1] Originally I had proved the theorem for the Euclidean straight line (and its subspaces dense-in-themselves). I am indebted to Professor S. Eilenberg for the general proof.

we easily obtain from (2), by means of 3.5 and 3.7,

(3) *for every non-empty set* $X \in \mathcal{O}(S)$ *there exists a set* C_k *such that*
$$\bar{C}_k \subseteq X \quad and \quad \bar{C}_k \neq X.$$

We now consider any natural number n and a set A for which

(4) $\qquad\qquad A \in \mathcal{O}(S)$ *and* $A \neq 0$;

we shall construct sets $B_1,..., B_n$ which satisfy the following conditions:

(5) $B_1,..., B_n$ *are open, non-empty, and pair-wise disjoint*;

(6) $\qquad B_1+...+B_n \subseteq A$ *and* $B_1+...+B_n \neq A$;

(7) $\qquad \overline{A-(B_1+...+B_n)} \supseteq \bar{A}-A$;

(8) $\qquad \bar{B}_1. ... . \bar{B}_n \supseteq A-(B_1+...+B_n).$

We carry this out first for the case $n = 1$; that is to say we construct a set B such that

(9) $\qquad B \in \mathcal{O}(S),\ B \neq 0,\ B \subseteq A,$ *and* $B \neq A$,

(10) $\qquad \overline{A-B} \supseteq \bar{A}-A$ *and* $\bar{B} \supseteq A-B.$

A set B of this kind is most easily obtained in the following way. We consider the sets $A_k = C_k.(\bar{A}-A)$ and choose from each such set (if it is not empty) a point a_k; let $D = \{a_1,..., a_k,...\}$. It is easy to see that $\bar{D} \supseteq \bar{A}-A$ (for otherwise, according to (2), there would be for $X = S-\bar{D}$ a set $C_k \subseteq S-\bar{D} \subseteq S-D$, having an element in common with $\bar{A}-A$, and this would contradict the definition of D). The set D is at most denumerable and not *empty* (apart from the trivial case $\bar{A}-A = 0$). We order the elements of D in an infinite sequence $d_1,..., d_k,...$ in such a way that in this sequence every element $x \in D$ occurs either only once or infinitely many times, according to whether $x \in \overline{D-\{x\}}$ or not. Moreover, since S is a normal space with a countable basis it is *metricizable*: with every pair of points $x, y \in S$ a real number $|x-y|$ can be correlated, the so-called *distance between x and y*, in such a way that the following conditions are satisfied:

(i) *for all* $x, y \in S$ *the formulas* $|x-y| = 0$ *and* $x = y$ *are equivalent*;

(ii) $|x-y|+|x-z| \geqslant |y-z|$ *for* $x, y, z \in S$;

(iii) *if* $X \subseteq S$, *then* $x \in \bar{X}$ *if and only if for every* $r > 0$ *there is an element* $y \in X$ *with* $|x-y| < r$.

Now, by (iii), since $D \subseteq \bar{A}$, for every $x \in D$ there are points $y \in A$ which lie as near to x as we please. Accordingly, we can choose for every natural k a point $e_k \in A$ with

$$|d_k - e_k| < 1/k,$$

and we put $B = A - \{e_1, ..., e_k, ...\}$; it can be shown without difficulty that B satisfies formulas (9) and (10) (the fact that S is dense-in-itself plays an essential part in this).

The case $n = 1$ has thus been dealt with. We turn now to the general case, making use of the set B just defined.

We first put, for every $X \subseteq S$,

(11) $$X^+ = S - \overline{S - \bar{X}}.$$

From (11), 3.1, and 3.3, we easily obtain the following rules for calculating with the operation $^+$:

(12) $\qquad (X^+)^+ = X^+ \in \mathcal{O}(S)$ *for every* $X \subseteq S$;

(13) $\qquad X \subseteq X^+$ *and* $\bar{X} = \overline{X^+}$ *for every* $X \in \mathcal{O}(S)$;

(14) $\qquad$ *if* $X \in \mathcal{O}(S)$, $Y \subseteq S$ *and* $X \subseteq \bar{Y}$, *then* $X \subseteq Y^+$;

(15) *if* $X_1 + ... + X_n \subseteq S$ *and the sets* $\bar{X}_1, ..., \bar{X}_n$ *are pair-wise disjoint, then* $(X_1 + ... + X_n)^+ = X_1^+ + ... + X_n^+$.

Further, we define by recursion an infinite sequence of sets $G_1, ..., G_l, ...$:

(16) $G_1 = C_k^+$, *where k is the smallest natural number for which*
$$\bar{C}_k \subseteq B \quad and \quad \bar{C}_k \neq B;$$

(17) $G_{l+1} = C_k^+$ *where k is the smallest natural number for which*
$$\bar{C}_k \subseteq B - \overline{G_1 + ... + G_l} \quad and \quad \bar{C}_k \neq B - \overline{G_1 + ... + G_l}.$$

On the basis of (1), (3), (9), (12), and (13) it is next shown (by means of an easy induction) that by (16) and (17) a set G_l is actually correlated with every natural number l and also that these sets G_1 satisfy the following conditions:

(18) G_l *is open, non-empty, and such that* $G_l^+ = G_l \subseteq \bar{G}_l \subseteq B$ *for every* l;

(19) *the sets* $\bar{G}_1, ..., \bar{G}_l, ...$ *(and hence also the sets* $G_1, ..., G_l, ...$*) are pair-wise disjoint.*

Let $C = B - \overline{G_1 + ... + G_l + ...}$. In view of (9) we have $C \in \mathcal{O}(S)$. If C were not empty, then by (9) there would be a number

k for which $\bar{C}_k \subseteq C$ and $\bar{C}_k \neq C$. We should then *a fortiori* have $\bar{C}_k \subseteq B - \overline{G_1 + ... + G_l}$ and $\bar{C}_k \neq B - \overline{G_1 + ... + G_l}$ for every l, and we could infer from that by (16) and (17) that C_k is identical with one of the G_l, whence, in consequence of (1) and (13),

$$\bar{C}_k = \bar{G}_l \subseteq B - \bar{G}_l.$$

The set G_l must thus be empty, in contradiction to (18). Consequently $C = 0$, i.e. $B \subseteq \overline{G_1 + ... + G_l + ...}$. Accordingly, by (10) we have $\bar{A} \subseteq \overline{A - B + \bar{B}} \subseteq \bar{B} \subseteq \overline{G_1 + ... + G_l + ...}$. On the other hand, from (18) and (9) we get

$$G_1 + ... + G_l + ... \subseteq B \subseteq A,$$

so that finally

(20) $$\bar{A} = \overline{G_1 + ... + G_l + ...}\,.$$

We shall now show the following:

(21) *If $X \in \mathcal{O}(S)$, $X \subseteq A$, and X non-$\subseteq G_1 + ... + G_l + ...$, then there exist infinitely many numbers l for which $X . G_l \neq 0$.*

For this purpose we consider an open set $X \subseteq A$ which has elements in common with only a finite number of sets G_l. There is thus an l_0 such that $X . G_l = 0$ for every $l > l_0$. The set $Y = X - \overline{G_1 + ... + G_{l_0}}$ has then no element in common with the whole sum $G_1 + ... + G_{l_0} + ...$. Since in addition $Y \in \mathcal{O}(S)$, we also have $Y . \overline{G_1 + ... + G_{l_0} + ...} = 0$; thus, on account of (20), $Y . \bar{A} = 0$ and *a fortiori* $Y . A = 0$. On the other hand,

$$Y \subseteq X \subseteq A\,;$$

consequently $Y = 0$, i.e. $X \subseteq \overline{G_1 + ... + G_{l_0}}$. From this, by means of (14), (15), and (19), we obtain

$$X \subseteq (G_1 + ... + G_{l_0})^+ = G_1^+ + ... + G_{l_0}^+,$$

whence, in view of (18),

$$X \subseteq G_1 + ... + G_{l_0} \subseteq G_1 + ... + G_{l_0} + ...\,.$$

We have thus shown that every open set $X \subseteq A$, which has elements in common with only finitely many sets G_l, is included in $G_1 + ... + G_l + ...$. Hence by contraposition we obtain (21).

Let

(22) $C_{p_1}, ..., C_{p_k}, ...$ *be those sets of the sequence $C_1, ..., C_k, ...$ which are included in A but not in $G_1 + ... + G_l + ...$.*

The set $G_1,..., G_l,...$ may be divided into n systems of sets $\mathscr{M}_1,..., \mathscr{M}_n$ in such a way that the following conditions are satisfied:

(23) $$\mathscr{M}_1+...+\mathscr{M}_n = \{G_1,..., G_l,...\};$$

(24) *the systems* $\mathscr{M}_1,..., \mathscr{M}_n$ *are non-empty and pair-wise disjoint;*

(25) *for every set* C_{p_k}, $k = 1, 2,...,$ *and for every number* j, $1 \leqslant j \leqslant n$, *there is a set* $X \in \mathscr{M}_j$, *such that*

$$C_{p_k}.X \neq 0.$$

To prove this we apply the following procedure. In view of (1), (22), and (21), there certainly exist n sets $G_{l_1},..., G_{l_n}$ which have no elements in common with C_{p_1}. We include in the system $\mathscr{M}_j$ the sets G_{l_j}, $1 \leqslant j \leqslant n$. Similarly, there exist n sets $G_{l_{n+1}},..., G_{l_{2.n}}$, which are distinct from $G_{l_1},..., G_{l_n}$ and have no element in common with C_{p_2}; the set $G_{l_{n+j}}$, $1 \leqslant j \leqslant n$, is again included in $\mathscr{M}_j$. This procedure can obviously be continued without end. The sets G_l possibly then remaining are subsequently arbitrarily included in the systems $\mathscr{M}_1,..., \mathscr{M}_n$ (e.g. all may be included in the system $\mathscr{M}_1$).

We now put

(26) $$B_j = \sum_{X \in \mathscr{M}_j} X \quad for \quad j = 1, 2,..., n.$$

From (18), (19), (23), (24), and (26) it is seen at once that the sets $B_1,..., B_n$ just defined satisfy the condition (5). By (18), (23), and (26) we have $B_1+...+B_n = G_1+...+G_l+... \subseteq B$ and consequently $A-(B_1+...+B_n) \supseteq A-B$. Hence by means of (9) and (10) we obtain (6) and (7). Finally, let us suppose that the formula $\bar{B}_j \supseteq A-(B_1+...+B_n)$ does not hold for a given j, $1 \leqslant j \leqslant n$. We thus have

$$A-\bar{B}_j \, non\text{-} \subseteq B_1+...+B_n = G_1+...+G_l+...,$$

whence, by virtue of (4), $A-\bar{B}_j$ is open. By means of (2) and (22) we infer from this the existence of a set C_{p_k} which is included in $A-\bar{B}_j$ and has no element in common with B_j. But this is in obvious contradiction to (25) and (26). Accordingly, our supposition is refuted and (8) holds.

We have thus constructed (for every natural n and every non-empty open set A) sets $B_1,..., B_n$ which satisfy the conditions (5)–(8). Hence, by 3.9, S is an E-space, which was to be proved.

Note 3.11. It is clear from 3.10 that the Euclidean spaces (of any number of dimensions) belong to the E-spaces. Also, every subspace of a Euclidean space which is dense-in-itself is an E-space.

§ 4. TOPOLOGICAL INTERPRETATION OF THE TWO-VALUED AND OF THE INTUITIONISTIC SENTENTIAL CALCULUS

We now define, for subsets of an arbitrary topological space, four operations which we denote by the same symbols as were used for the operations on sentences which were discussed at the beginning of § 1.

DEFINITION 4.1. *If S is a topological space, we put, for all sets $X \subseteq S$ and $Y \subseteq S$,*

(i) $X \to Y = S - \overline{X - Y}$,

(ii) $X \vee Y = X + Y$ *(the ordinary set-theoretical sum)*,

(iii) $X \wedge Y = X.Y$ *(the ordinary set-theoretical product)*,

(iv) $\backsim X = X \to 0 \ (= S - \overline{X})$.

COROLLARY 4.2. (i) *If S is a topological space, $X \subseteq S$ and $Y \subseteq S$, then $X \to Y$, $\backsim X \in \mathcal{O}(S)$, and in fact $X \to Y$ is the largest open set Z for which $X.Z \subseteq Y$, and $\backsim X$ is the largest open set disjoint from X.*

(ii) *If in addition $X, Y \in \mathcal{O}(S)$ then we also have $X \vee Y \in \mathcal{O}(S)$ and $X \wedge Y \in \mathcal{O}(S)$, and in fact $X \vee Y$ is the smallest open set which includes X and Y, and $X \wedge Y$ is the largest open set which is included in X and Y.*   [By 3.1, 3.3, 4.1]

COROLLARY 4.3. *If S is a topological space, $X \subseteq S$ and $Y \subseteq S$, then $X \to Y = S$ if and only if $X \subseteq Y$; in particular $S \to Y = S$ if and only if $Y = S$.*   [By 3.1, 4.1 (i)]

COROLLARY 4.4. *If S is a topological space and $X \subseteq S$, then X is dense in S if and only if $\backsim \backsim X = S$.*   [By 3.1, 3.4, 4.1 (iv)]

DEFINITION 4.5. *The ordered sextuple $[\mathcal{O}(S), S, \to, \vee, \wedge, \backsim]$ where S is a topological space, is denoted by $\mathsf{O}(S)$.*

THEOREM 4.6. *For every topological space S, $\mathsf{O}(S)$ is a matrix.*
[By 2.1, 3.1, 3.3, 4.2, 4.3, 4.5]

The matrix $\mathsf{O}(S)$, like every other matrix, uniquely determines a system of sentences, namely $\mathfrak{E}(\mathsf{O}(S))$. We shall investigate the relation of this system to the systems $3\mathfrak{K}$ and $\mathfrak{IK}$ in detail.

LEMMA 4.7. *If S is a topological space and $\mathscr{S} = \{S, 0\}$, then $\mathsf{M} = [\mathscr{S}, S, \rightarrow, \vee, \wedge, \sim]$ is a submatrix of $\mathsf{O}(S)$ which is isomorphic with* ZK.

The proof (by 2.1, 2.3, 2.8, 2.10, 3.1, 3.3, and 4.1) presents no difficulties.

THEOREM 4.8. *For every topological space S we have*

$$\mathfrak{E}(\mathsf{O}(S)) \subseteq 3\mathfrak{K}.$$

[By 2.7, 2.9, 2.11, 4.7]

LEMMA 4.9. *If S is any topological space and $\mathfrak{A}$ an axiom of the intuitionistic sentential calculus, then $\mathfrak{A} \in \mathfrak{E}(\mathsf{O}(S))$.*

Proof. In accordance with 1.2 it is necessary to distinguish in the proof ten cases according to the form of the axiom $\mathfrak{A}$. Since the method of arguing is in all cases nearly the same, we shall consider only one case in detail, say 1.2 (ix).

Let then

(1) $$\mathfrak{A} = \sim\mathfrak{B} \rightarrow (\mathfrak{B} \rightarrow \mathfrak{C}),$$

where $\mathfrak{B}$ and $\mathfrak{C}$ are any sentences. We construct in accordance with 2.5 (and with the help of 4.5, 4.6) the functions $F_{\mathfrak{A},\mathsf{O}(S)}$, $F_{\mathfrak{B},\mathsf{O}(S)}$, $F_{\mathfrak{C},\mathsf{O}(S)}$. We consider further an arbitrary sequence of sets $X_1,..., X_n,... \in \mathcal{O}(S)$ and put

(2) $$F_{\mathfrak{A},\mathsf{O}(S)}(X_1,..., X_n,...) = X, \qquad F_{\mathfrak{B},\mathsf{O}(S)}(X_1,..., X_n,...) = Y,$$
$$F_{\mathfrak{C},\mathsf{O}(S)}(X_1,..., X_n,...) = Z.$$

By 2.5 (ii), (v), from (1) and (2) we obtain $X = \sim Y \rightarrow (Y \rightarrow Z)$, whence, by 4.1 (i), (iv),

(3) $$X = (S-\overline{Y}) \rightarrow (Y \rightarrow Z) = S-\overline{(S-\overline{Y})-(S-\overline{Y-Z})}.$$

Since, by virtue of 3.1 (iii), $\overline{Y-Z} \subseteq \overline{Y}$ and consequently

$S - \overline{Y} \subseteq S - \overline{\overline{Y} - Z}$, it results from (3) and 3.1 (ii) that $X = S$. By (2) we thus have

$$F_{\mathfrak{A}, \mathcal{O}(S)}(X_1, ..., X_n, ...) = S \text{ for all } X_1, ..., X_n, ... \in \mathcal{O}(S).$$

Hence, by 2.5, we obtain $\mathfrak{A} \in \mathfrak{C}(\mathcal{O}(S))$, q.e.d.

LEMMA 4.10. *Let S be any topological space and $\mathfrak{A}$, $\mathfrak{B}$, $\mathfrak{C}$ three sentences such that $\mathfrak{A} = \mathfrak{B} \to \mathfrak{C}$. If $\mathfrak{A}$, $\mathfrak{B} \in \mathfrak{C}(\mathcal{O}(S))$, then also $\mathfrak{C} \in \mathfrak{C}(\mathcal{O}(S))$; in other words, the system $\mathfrak{C}(\mathcal{O}(S))$ is closed under the operation of detachment.*

Proof. In accordance with 2.5 (and with the help of 4.5, 4.6) we construct the functions $F_{\mathfrak{A}, \mathcal{O}(S)}$, $F_{\mathfrak{B}, \mathcal{O}(S)}$, and $F_{\mathfrak{C}, \mathcal{O}(S)}$; we then have, for all sets $X_1, ..., X_n, ... \in \mathcal{O}(S)$,

(1) $F_{\mathfrak{A}, \mathcal{O}(S)}(X_1, ..., X_n, ...)$
$$= F_{\mathfrak{B}, \mathcal{O}(S)}(X_1, ..., X_n, ...) \to F_{\mathfrak{C}, \mathcal{O}(S)}(X_1, ..., X_n, ...),$$

and, since $\mathfrak{A}$, $\mathfrak{B} \in \mathfrak{C}(\mathcal{O}(S))$,

(2) $F_{\mathfrak{A}, \mathcal{O}(S)}(X_1, ..., X_n, ...) = S = F_{\mathfrak{B}, \mathcal{O}(S)}(X_1, ..., X_n, ...).$

By virtue of 4.3 the formulas (1) and (2) yield

$$F_{\mathfrak{C}, \mathcal{O}(S)}(X_1, ..., X_n, ...) = S \text{ for all } X_1, ..., X_n, ... \in \mathcal{O}(S),$$

whence, by 2.5, $\mathfrak{C} \in \mathfrak{C}(\mathcal{O}(S))$. The system $\mathfrak{C}(\mathcal{O}(S))$ is thus closed under the operation of detachment (cf. 1.3), q.e.d.

THEOREM 4.11. *For every topological space S we have*

$$\mathfrak{IR} \subseteq \mathfrak{C}(\mathcal{O}(S)).$$

[By 1.4, 4.9, 4.10]

THEOREM 4.12. *For every topological space S and every sentence $\mathfrak{A}$ the conditions $\mathfrak{A} \in \mathfrak{IR}$ and $\sim \sim \mathfrak{A} \in \mathfrak{C}(\mathcal{O}(S))$ are equivalent.*

[By 1.6, 4.8, 4.11]

In view of Ths. 4.8 and 4.11, the double inclusion $\mathfrak{IR} \subseteq \mathfrak{C}(\mathcal{O}(S)) \subseteq \mathfrak{IR}$ holds for every topological space S. We shall now show that there exist spaces S for which $\mathfrak{C}(\mathcal{O}(S)) = \mathfrak{IR}$ and also spaces S for which $\mathfrak{C}(\mathcal{O}(S)) = \mathfrak{IR}$. In fact the first equality holds if and only if S is an isolated (and thus, so to speak, a degenerate) space. The second holds for all E-spaces, and thus in particular for all normal spaces which are dense-in-themselves and have a countable basis (cf. 3.9–3.11).

LEMMA 4.13. *Let S be a topological space. In order that every sentence $\mathfrak{A}$ of the form $\mathfrak{A} = (\sim\mathfrak{B} \to \mathfrak{B}) \to \mathfrak{B}$ (where $\mathfrak{B}$ is any sentence) should belong to $\mathfrak{C}(O(S))$, it is necessary and sufficient that S be isolated.*

Proof. If S is an isolated space, we easily obtain from 3.6 and 4.1 (i), (iv) the formula $(\sim X \to X) \to X = S$ for every $X \subseteq S$, and hence by means of 2.5, 4.5, and 4.6 (just as in the proof of 4.9) we conclude that every sentence $\mathfrak{A} = (\sim\mathfrak{B} \to \mathfrak{B}) \to \mathfrak{B}$ belongs to $\mathfrak{C}(O(S))$.

Suppose now, conversely, that $\mathfrak{C}(O(S))$ contains all sentences

$$\mathfrak{A} = (\sim\mathfrak{B} \to \mathfrak{B}) \to \mathfrak{B}.$$

By 1.1 we can, in particular, assume that $\mathfrak{B}$ is a sentential variable, say $\mathfrak{B} = \mathfrak{B}_1$. According to 2.5 and on account of 4.5, 4.6 we then have for every sequence of open sets $X_1, ..., X_n ...$

$$F_{\mathfrak{B}, O(S)}(X_1, ..., X_n, ...) = X_1$$

and $\quad F_{\mathfrak{A}, O(S)}(X_1, ..., X_n, ...) = (\sim X_1 \to X_1) \to X_1.$

Since $\mathfrak{A} \in \mathfrak{C}(O(S))$, we obtain

$$(\sim X_1 \to X_1) \to X_1 = S,$$

whence, on account of 4.3, $(\sim X_1 \to X_1) \subseteq X_1$ and further, by virtue of 3.1 and 4.1, $S - \overline{S - \overline{X}_1} \subseteq X_1$ for every $X_1 \in O(S)$. If, in particular, x is an element of S, we conclude from 3.1 and 3.3 that $S - \{x\} \in O(S)$; consequently

$$S - \overline{S - \overline{S - \{x\}}} \subseteq S - \{x\},$$

and therefore $\overline{S - \overline{S - \{x\}}} \supseteq \{x\}$; from this we see at once that $\overline{S - \{x\}} \neq S$ and $\overline{S - \{x\}} = S - \{x\}$. We thus have

$$x \, non\text{-}\in \overline{S - \{x\}} \, for \, every \, x \in S;$$

but this means that the space S is isolated (cf. 3.5). Lemma 4.13 thus holds in both directions.

THEOREM 4.14. *Let S be a topological space. In order that $\mathfrak{C}(O(S)) = 3\mathfrak{N}$, it is necessary and sufficient that S be isolated.*

Proof. Let S be an isolated space. Then, according to 1.2, 4.9, and 4.13, $\mathfrak{C}(O(S))$ contains all axioms of the two-valued sentential calculus and, by 4.10, is closed under the operation of detachment. Consequently, by 1.4 we have

$$\mathfrak{3R} \subseteq \mathfrak{C}(\mathsf{O}(S)),$$

whence, on account of 4.8, $\mathfrak{C}(\mathsf{O}(S)) = \mathfrak{3R}$. If, conversely, this latter equality is satisfied, then the system $\mathfrak{C}(\mathsf{O}(S))$ contains in particular all sentences $\mathfrak{A}$ of the form $\mathfrak{A} = (\smallsmile \mathfrak{B} \to \mathfrak{B}) \to \mathfrak{B}$ (cf. 1.2 (x) and 1.4); in view of 4.13 the space S is therefore isolated, q.e.d.

Before continuing we shall subject the operations $\to$ and $\smallsmile$ to a relativization (cf. 3.2).

DEFINITION 4.15. *If S is a topological space and $A \subseteq S$, then for all sets $X \subseteq S$ and $Y \subseteq S$ we put:*

(i) $X \underset{A}{\to} Y = A - \overline{X - Y}$,

(ii) $\underset{A}{\smallsmile} X = X \underset{A}{\to} 0 \ (= A - \overline{X})$.

COROLLARY 4.16. *If S is a topological space, $A \subseteq S$, $X \subseteq S$, and $Y \subseteq S$, then $X \underset{A}{\to} Y \subseteq A$ and $\underset{A}{\smallsmile} X \subseteq A$; if moreover $A \in \mathcal{O}(S)$, then $X \underset{A}{\to} Y$, $\underset{A}{\smallsmile} X \in \mathcal{O}(S)$.*   [By 3.1, 3.3, 4.15]

COROLLARY 4.17. *Let S be a topological space and $A \subseteq S$.*

(i) *If $X \subseteq A$ and $Y \subseteq S$, then we have $X \underset{A}{\to} Y = A$ if and only if $X \subseteq Y$.*

(ii) *If $Y \subseteq A$, then $A \underset{A}{\to} Y = A$ if and only if $Y = A$.*

(iii) *$X \underset{A}{\to} A = A$ for every $X \subseteq A$.*

(iv) *$A \underset{A}{\to} Y = Y$ for every open set $Y \subseteq A$.*

(v) *$\underset{A}{\smallsmile} A = A \underset{A}{\to} 0 = 0$ and $\underset{A}{\smallsmile} 0 = 0 \underset{A}{\to} 0 = A$.*

[By 3.1, 3.3, 4.15]

COROLLARY 4.18. *If S is a topological space, then*

$$X \underset{S}{\to} Y = X \to Y \quad and \quad \underset{S}{\smallsmile} X = \smallsmile X$$

for all sets $X \subseteq S$ and $Y \subseteq S$.   [By 4.1 (i), (iv), 4.15]

The following lemma is a generalization of 4.7:

LEMMA 4.19. *If S is a topological space, $A \subseteq S$, $A \neq 0$, and $\mathscr{T} = \{A, 0\}$, then $\mathsf{N} = [\mathscr{T}, A, \underset{A}{\rightarrow}, \vee, \wedge, \underset{A}{\backsim}]$ is a matrix which is isomorphic with* ZK.

[By 2.1, 2.3, 2.10, 3.1, 4.1 (ii), (iii), 4.17 (iii), (v)]

LEMMA 4.20. *Premisses*:

(α) *S is a topological space;*

(β) *$A, B \in \mathcal{O}(S)$, $B \subseteq A$, $B \neq A$, $\overline{A-B} \supseteq \bar{A} - A$ and $\bar{B} \supseteq A - B$;*

(γ) *$\mathscr{S}$ is a system of open sets $X \subseteq B$;*

(δ) *if $X, Y \in \mathscr{S}$ and $X - Y \neq 0$, then $\overline{X-Y} \supseteq \bar{B} - B$;*

(ϵ) *$\mathsf{M} = [\mathscr{S}, B, \underset{B}{\rightarrow}, \vee, \wedge, \underset{B}{\backsim}]$ is a matrix;*

(ζ) *$\mathscr{T} = \mathscr{S} + \{A\}$ and $\mathsf{N} = [\mathscr{T}, A, \underset{A}{\rightarrow}, \vee, \wedge, \underset{A}{\backsim}]$.*

Conclusions:

(i) *$\mathscr{T}$ is a system of open sets $X \subseteq A$;*

(ii) *if $X, Y \in \mathscr{T}$ and $X - Y \neq 0$, then $\overline{X-Y} \supseteq \bar{A} - A$;*

(iii) *N is a matrix and $\mathsf{N} = \mathsf{M}^*$.*

Proof. According to the premisses (α) and (β), and in view of 3.1, we have $\bar{B} \supseteq B$ and $\bar{B} \supseteq A - B$, whence $\bar{B} \supseteq A$; and since $A \supseteq B$, we get

(1) $$\bar{A} = \bar{B} \quad and \quad \bar{A} - A \subseteq \bar{B} - B.$$

From (γ) and (ζ) we obtain at once the conclusion (i):

(2) $$\mathscr{T} \text{ is a system of open sets } X \subseteq A.$$

We shall next prove the conclusion (ii):

(3) $$\text{if } X, Y \in \mathscr{T} \text{ and } X - Y \neq 0, \text{ then } \overline{X-Y} \supseteq \bar{A} - A.$$

In fact, if $X - Y \neq 0$, then $Y \neq A$ by (2); thus $Y \in \mathscr{S}$ on account of (ζ). If also $X \in \mathscr{S}$, then it follows from (δ) and (1) that

$$\overline{X-Y} \supseteq \bar{B} - B \supseteq \bar{A} - A.$$

But, if X *non-*$\in \mathscr{S}$, then $X = A$, and, if $Y \neq B$, then by (γ) we have $B - Y \neq 0$ and hence by (δ) and (1) we obtain

$$\overline{X-Y} \supseteq \overline{B-Y} \supseteq \bar{B} - B \supseteq \bar{A} - A$$

(since $B \in \mathscr{S}$ by virtue of (ϵ) and 2.1). For $X = A$ and $Y = B$ the formula $\overline{X-Y} \supseteq \bar{A} - A$ follows directly from (β). (3) thus holds in all cases.

The operation $\underset{A}{\to}$ has the following properties:

(4) *if* $X, Y \in \mathscr{S}$, *then either* $X \underset{A}{\to} Y = A$ *or* $= X \underset{B}{\to} Y$, *according to whether* $X \underset{B}{\to} Y = B$ *or* $\neq B$;

(5) $\qquad\qquad X \underset{A}{\to} A = A$ *for every* $X \in \mathscr{T}$;

(6) $\qquad\qquad A \underset{A}{\to} Y = Y$ *for every* $Y \in \mathscr{T}$.

If, in fact, $X, Y \in \mathscr{S}$ and $X \underset{B}{\to} Y = B$, then by 4.17 (i) we have $X \subseteq Y$ and $X \underset{A}{\to} Y = A$ (since $X \subseteq B \subseteq A$ on account of (β) and (γ)). But if $X \underset{B}{\to} Y = B - \overline{X - Y} \neq B$, then it follows from (δ) that $\overline{X - Y} \supseteq \overline{B} - B$, and thus, in view of (1), we have

$$A - \overline{X - Y} \subseteq A - (\overline{B} - B) = (A - \overline{B}) + B = B$$

and therefore $A - \overline{X - Y} \subseteq B - \overline{X - Y}$; on the other hand, since $A \supseteq B$ and hence $A - \overline{X - Y} \supseteq B - \overline{X - Y}$, we finally get

$$A - \overline{X - Y} = B - \overline{X - Y}, \qquad \text{i.e. } X \underset{A}{\to} Y = X \underset{B}{\to} Y.$$

In this way (4) is proved. (5) and (6) result immediately from 4.17 (iii), (iv), and (2).

From (2), by 4.1 (ii), (iii), we also obtain

(7) $Z \vee A = A \vee Z = A$ *and* $Z \wedge A = A \wedge Z = Z$ *for every* $Z \in \mathscr{T}$.

The operation $\underset{A}{\backsim}$ satisfies the following conditions:

(8) $\qquad\qquad \underset{A}{\backsim} A = \underset{B}{\backsim} B = 0$;

(9) *if* $X \in \mathscr{S}$, *then* $\underset{A}{\backsim} X = A$ *or* $= \underset{B}{\backsim} X$, *according to whether* $\underset{B}{\backsim} X = B$ *or* $\neq B$.

(8) follows directly from 4.17 (v). With the help of 2.1 and (ϵ) we infer from (8) that $0 \in \mathscr{S}$; in view of this we put in (4) $Y = 0$ and by means of 4.15 (ii) we immediately obtain (9).

If now we compare the premisses (ϵ), (ζ) and the formulas (4)–(9) with 2.12 (i)–(v), we see at once that

(10) $\qquad\qquad \mathsf{N} = \mathsf{M}^*$,

whence, by 2.14,

(11) $\qquad\qquad \mathsf{N}$ *is a matrix*.

With (2), (3), (10), and (11) the proof is complete.

LEMMA 4.21. *Premisses*:

(α) S *is a topological space*;

(β) $B_1,..., B_n \in \mathcal{O}(S)$, $B_1,..., B_n$ *are non-empty, pair-wise disjoint sets*, $B_1+...+B_n = B$ *and* $\bar{B}_1. ... . \bar{B}_n \supseteq \bar{B}-B$;

(γ) *for* $p = 1, 2,..., n$, $\mathcal{S}_p$ *is a system of open sets* $X \subseteq B_p$;

(δ) *if* $X, Y \in \mathcal{S}_p$ $(p = 1, 2,..., n)$ *and* $X-Y \neq 0$, *then*

$$\overline{X-Y} \supseteq \bar{B}_p-B_p;$$

(ϵ) $\mathsf{M} = [\mathscr{W}, A, \twoheadrightarrow, \curlyvee, \curlywedge, \sim]$ *is a matrix*;

(ζ) *for* $p = 1, 2,..., n$, $\mathsf{M}_p = [\mathcal{S}_p, B_p, \underset{B_p}{\rightarrow}, \vee, \wedge, \underset{B_p}{\sim}]$ *is a matrix and is isomorphic with* M;

(η) $\mathcal{S}$ *is the system of sets* $X = X_1+...+X_n$, *where*

$$X_1 \in \mathcal{S}_1,..., X_n \in \mathcal{S}_n, \quad \text{and} \quad \mathsf{P} = [\mathcal{S}, B, \underset{B}{\rightarrow}, \wedge, \vee, \underset{B}{\sim}].$$

Conclusions:

(i) $\mathcal{S}$ *is a system of open sets* $X \subseteq B$;

(ii) *if* $X, Y \in \mathcal{S}$ *and* $X-Y \neq 0$, *then* $\overline{X-Y} \supseteq \bar{B}-B$;

(iii) P *is a matrix and is isomorphic with* M^n.

Proof. The assertion (i), i.e.

(1) $\qquad\qquad \mathcal{S}$ *is a system of open sets* $X \subseteq B$,

results easily from (β), (γ), and (η); since, by 3.1 and 3.3, every sum of open sets is itself an open set.

In order to prove (ii) we consider two arbitrary sets $X, Y \in \mathcal{S}$ such that $X-Y \neq 0$. By (η),

$$X = X_1+...+X_n \quad \text{and} \quad .Y = Y_1+...+Y_n,$$

where $X_p, Y_p \in \mathcal{S}_p$ for $p = 1, 2,..., n$. Since, in view of (γ) and (β), we have $X_p \subseteq B_p, Y_p \subseteq B_p$ and the sets $B_1,..., B_n$ are pair-wise disjoint we obviously have

$$X-Y = (X_1+...+X_n)-(Y_1+...+Y_n)$$
$$= (X_1-Y_1)+...+(X_n-Y_n);$$

if, therefore, $X-Y \neq 0$, there must be a p, $1 \leqslant p \leqslant n$, such that $X_p-Y_p \neq 0$. Hence by (δ) we get $\overline{X_p-Y_p} \supseteq \bar{B}_p-B_p$ and consequently $\overline{X-Y} \supseteq \bar{B}_p-B_p$, because by ($\alpha$) and 3.1 (iii)

$$\overline{X-Y} = \overline{X_1-Y_1}+...+\overline{X_n-Y_n}.$$

But according to (β)

$$\bar{B}_p - B_p \supseteq (\bar{B}-B)-B_p = \bar{B}-(B+B_p) = \bar{B}-B,$$

so that finally $\overline{X-Y} \supseteq \bar{B}-B$. Thus we have

(2)   *if $X, Y \in \mathscr{S}$ and $X-Y \neq 0$ then $\overline{X-Y} \supseteq \bar{B}-B$.*

We turn now to (iii) and in accordance with 2.15 (and making use of 2.16 and (ϵ)) we construct the matrix

$$\mathsf{M}^n = [\mathscr{W}^n, A^n, \twoheadrightarrow^n, \varphi^n, \downarrow^n, \sim^n].$$

By (ζ) (and 2.4), M is isomorphic with $\mathsf{M}_1,..., \mathsf{M}_n$; hence by 2.3 there exist functions $F_1,..., F_n$, which produce this isomorphism and which in particular map $\mathscr{W}$ onto $S_1,..., S_n$ in one-one fashion. We now put

(3) $F(U) = F_1(U_1)+...+F_n(U_n)$ *for* $U = [U_1,..., U_n] \in \mathscr{W}^n$.

Since here $F_p(U_p) \in \mathscr{S}_p$ $(p = 1, 2,..., n)$, whence by (iii) $F_p(U_p) \subset B_p$, and the sets $B_1,..., B_n$ are pair-wise disjoint we obtain, with the help of (η),

(4)   *the function F maps $\mathscr{W}^n$ onto $\mathscr{S}$ in one-one fashion.*

By means of 2.3, 2.15, and (β) we conclude that

(5) $F(A^n) = F_1(A)+...+F_n(A) = B_1+...+B_n = B.$

Further, let

(6) $U = [U_1,..., U_n] \in \mathscr{W}^n$ and $V = [V_1,..., V_n] \in \mathscr{W}^n$.

By 2.15, (6), (3), 2.3, (ζ), and 4.15 (i) we get

(7) $F(U \twoheadrightarrow^n V) = F([U_1 \twoheadrightarrow V_1,..., U_n \twoheadrightarrow V_n])$

$\qquad = F_1(U_1 \twoheadrightarrow V_1)+...+F_n(U_n \twoheadrightarrow V_n)$

$\qquad = (F_1(U_1) \underset{B_1}{\rightarrow} F_1(V_1))+...+(F_n(U_n) \underset{B_n}{\rightarrow} F_n(V_n))$

$\qquad = (B_1-\overline{F_1(U_1)-F_1(V_1)})+...+(B_n-\overline{F_n(U_n)-F_n(V_n)}).$

For $p, q = 1,..., n$ and $p \neq q$, $F_p(U_p) \in \mathscr{S}_p$, $F_q(V_q) \in \mathscr{S}_q$, and thus by (iii) and (ii) we have

$$F_p(U_p) \subseteq B_p, \qquad F_q(V_q) \subseteq B_q, \qquad F_p(U_p).B_q = F_p(U_p).F_q(V_q) = 0;$$

since in addition $B_q \in \mathscr{O}(S)$, we also have

$$\overline{F_p(U_p)-F_p(V_p)}.B_q = 0.$$

From this with the help of (7) and 3.1 (iii) we obtain

(8) $F(U \twoheadrightarrow^n V) = (B_1 + ... + B_n) -$
$$- (\overline{F_1(U_1) - F_1(V_1)} + ... + \overline{F_n(U_n) - F_n(V_n)})$$
$$= B - (\overline{F_1(U_1) - F_1(V_1)} + ... + \overline{(F_n(U_n) - F_n(V_n))}$$
$$= B - (\overline{F_1(U_1) + ... + F_n(U_n)}) - (F_1(V_1) + ... + F_n(V_n)));$$

by virtue of 4.15 (i), formulas (3), (6), and (8) yield

(9) $F(U \twoheadrightarrow {}^n V) = B - \overline{F(U) - F(V)} = F(U) \underset{B}{\to} F(V)$ for   all
$U, V \in \mathscr{W}^n$.

In an analogous but more simple manner we obtain the formulas

(10) $F(U \curlyvee^n V) = F(U) \vee F(V)$ and $F(U \curlywedge {}^n V) = F(U) \wedge F(V)$
for $U, V \in \mathscr{W}^n$;

(11) $\qquad\qquad F(\sim^n U) = \underset{B}{\sim} F(U)$ for $U \in \mathscr{W}^n$.

By means of 2.1, 2.15, and 2.16 we infer from (4), (5), (9)–(11) that $B \in \mathscr{S}$ and $\mathscr{S}$ is closed under $\underset{B}{\to}$, $\vee$, $\wedge$, $\underset{B}{\sim}$; in view of (η), (1), and 4.17 (ii), P is thus a matrix. We also conclude— again from (4), (5), (9), (10), (11)—that the matrices P and M^n satisfy the conditions of the definition 2.3; consequently we have

(12) $\qquad\qquad$ P *is a matrix isomorphic with* M^n.

By (1), (11), and (12), all the conclusions of the lemma are satisfied.

LEMMA 4.22. *Premisses*:

(α) *S is an E-space*;

(β) M *is a matrix*;

(γ) *for every non-empty set $B \in \mathcal{O}(S)$ there is a system $\mathscr{S}$ of open sets $X \subseteq A$ with the following properties*:

(γ_1) *if $X, Y \in \mathscr{S}$ and $X - Y \neq 0$, then $\overline{X - Y} \supseteq \bar{B} - B$*;

(γ_2) $\mathsf{M}' = [\mathscr{S}, B, \underset{B}{\to}, \vee, \wedge, \underset{B}{\sim}]$ *is a matrix which is isomorphic with* M.

Conclusion: for every natural number n and every non-empty set $A \in \mathcal{O}(S)$ there is a system $\mathcal{T}$ of open sets $X \subseteq A$ with the following properties:

(i) *if $X, Y \in \mathcal{T}$ and $X-Y \neq 0$, then $\overline{X-Y} \supseteq \overline{A}-A$;*

(ii) $\mathsf{N} = [\mathcal{T}, A, \underset{A}{\to}, \vee, \wedge, \underset{A}{\sim}]$ *is a matrix isomorphic with* $(\mathsf{M}^n)^*$.

Proof. Let n be a natural number and A a non-empty open set. By virtue of 3.9 and in view of (α) there exist sets $B_1,..., B_n$, which satisfy the following conditions:

(1) $B_1,..., B_n$ *are non-empty, pair-wise disjoint open sets;*

(2) $$B_1 + ... + B_n = B \subseteq A \text{ and } B \neq A;$$

(3) $$\overline{A-B} \supseteq \overline{A}-A;$$

(4) $$\overline{B}_1 ..... \overline{B}_n \supseteq A-B.$$

From (2)–(4) we easily obtain

(5) $$\overline{B}_1 ..... \overline{B}_n \supseteq (\overline{A}-A)+(A-B) = \overline{A}-B \supseteq \overline{B}-B.$$

By (β) and (γ) there exist systems of sets $\mathcal{S}_1,..., \mathcal{S}_n$ with the following properties:

(6)   *for $p = 1, 2,..., n$, $\mathcal{S}_p$ is a system of open sets $X \subseteq B_p$;*

(7)   *if $X, Y \in \mathcal{S}_p$ ($p = 1, 2,..., n$) and $X-Y \neq 0$, then*
$$\overline{X-Y} \supseteq \overline{B}_p - B_p;$$

(8)   *for $p = 1, 2,..., n$, $\mathsf{M}_p = [\mathcal{S}_p, B_p, \underset{B_p}{\to}, \vee, \wedge, \underset{B_p}{\sim}]$ is a matrix which is isomorphic with M.*

Let us now put

(9) $\mathcal{S} = \{$*the system of sets* $X = X_1 + ... + X_n$, *where*
$$X_1 \in \mathcal{S}_1,..., X_n \in \mathcal{S}_n\};$$
$$\mathsf{P} = [\mathcal{S}, B, \underset{B}{\to}, \vee, \wedge, \underset{B}{\sim}].$$

By (α), (β), (1), (2), and (5)–(9) the premises of 4.21 are satisfied. Consequently we have

(10) $\mathcal{S}$ *is a system of open sets* $X \subseteq A$;

(11)   *if $X, Y \in \mathcal{S}$ and $X-Y \neq 0$, then $\overline{X-Y} \supseteq \overline{B}-B$;*

(12) P *is a matrix which is isomorphic with* M^n.

Let

(13) $\mathscr{T} = \mathscr{S} + \{A\}$ and $\mathsf{N} = [\mathscr{T}, A, \underset{A}{\rightarrow}, \vee, \wedge, \underset{A}{\curlyvee}].$

By (α), (1)–(4), and (9)–(13), the premisses of 4.20 hold. Consequently,

(14) $\mathscr{T}$ is a system of open sets $X \subseteq A$;

(15)   if $X, Y \in \mathscr{T}$ and $X - Y \neq 0$, then $\overline{X - Y} \supseteq \overline{A} - A$;

(16) N is a matrix and $\mathsf{N} = \mathsf{P}^*$.

From (12) and (16) by means of 2.14 we obtain

(17) N is a matrix which is isomorphic with $(\mathsf{M}^n)^*$.

In view of (14), (15), and (17) the proof of Lemma 4.22 is complete.

LEMMA 4.23. *If S is an E-space, then for every natural number n the matrix $\mathsf{O}(S)$ contains a submatrix which is isomorphic with IK_n.*

Proof. By means of an inductive procedure we shall establish a logically stronger conclusion, namely

(1)   *for every non-empty set $A \in \mathcal{O}(S)$ there is a system $\mathscr{T}$ of open sets $X \subseteq A$ with the following properties:*

   (i) *if $X, Y \in \mathscr{T}$ and $X - Y \neq 0$, then $\overline{X - Y} \supseteq \overline{A} - A$;*

   (ii) $\mathsf{N} = [\mathscr{T}, A, \underset{A}{\rightarrow}, \vee, \wedge, \underset{A}{\curlyvee}]$ *is a matrix isomorphic with* IK_n.

In fact by 2.17 and 4.19, (1) holds for $n = 1$. Assuming that (1) is satisfied for a given natural n we apply 4.22 (with $\mathsf{M} = \mathsf{IK}_n$) and with the help of 2.17 we easily see that (1) also holds for $n+1$.

If we now put $A = S$ in (1) we at once obtain the conclusion of the lemma from 2.8, 3.1, 3.2, 4.5, and 4.18.

THEOREM 4.24. *If S is an E-space, then $\mathfrak{E}(\mathsf{O}(S)) = \mathfrak{IR}$.*

Proof. By 4.23 there is for every natural n a submatrix N_n which is isomorphic with IK_n. Now, if $\mathfrak{A} \in \mathfrak{E}(\mathsf{O}(S))$, then by 2.9 and 2.7 we have $\mathfrak{A} \in \mathfrak{E}(\mathsf{N}_n) = \mathfrak{E}(\mathsf{IK}_n)$ for $n = 1, 2,...$ and consequently, by 2.18, $\mathfrak{A} \in \mathfrak{IR}$. Accordingly, $\mathfrak{E}(\mathsf{O}(S)) \subseteq \mathfrak{IR}$; hence by 4.9 we at once obtain $\mathfrak{E}(\mathsf{O}(S)) = \mathfrak{IR}$, q.e.d.

Note 4.25. The following theorem is easily established:

Let S be a topological space, A a non-empty open set $\subseteq S$, $\mathcal{T}$ the system of open sets $X \subseteq A$, and $\mathsf{N} = [\mathcal{T}, A, \underset{A}{\rightarrow}, \vee, \wedge, \underset{A}{\sim}]$. Then N is a matrix and we have $\mathfrak{C}(\mathsf{O}(S)) \subseteq \mathfrak{C}(\mathsf{N})$. If, in particular, $\mathfrak{C}(\mathsf{N}) = \mathfrak{JR}$, then $\mathfrak{C}(\mathsf{O}(S)) = \mathfrak{JR}$.

From this we see that the converse of Th. 4.24 does not hold; the formula $\mathfrak{C}(\mathsf{O}(S)) = \mathfrak{JR}$ applies, for example, to all normal spaces with a countable basis, which include a non-empty open set that is dense-in-itself, and, more generally, to every space S which includes an open E-space (cf. 3.2), independently of whether S itself is an E-space.

On the other hand there are also spaces S such that the matrix $\mathsf{O}(S)$ is adequate neither for the system $\mathfrak{JR}$, nor for the system $\mathfrak{JR}$, but for an intermediate system. Examples can be found both among normal spaces with a countable basis and among those which have no countable basis. For instance, normal spaces are known which are dense in themselves and satisfy the condition

$$\text{if } X \in \mathcal{O}(S), \text{ then } \bar{X} \in \mathcal{O}(S).$$

For every such space S the system $\mathfrak{C}(\mathsf{O}(S))$ is, by 4.14, distinct from $\mathfrak{JR}$. It is also easily shown that this system contains all sentences $\mathfrak{A}$ of the form $\mathfrak{A} = \smallsmile \mathfrak{B} \vee \smallsmile \smallsmile \mathfrak{B}$ and consequently cannot coincide with $\mathfrak{JR}$ (cf. 1.5). The problem of setting up an exact correlation between the topological properties of a space S and the logical (or rather metalogical) properties of the corresponding system $\mathfrak{C}(\mathsf{O}(S))$ is still by no means completely solved.

We shall now put some of the results obtained into a more intuitive and more lucid form.

Let $\mathfrak{A}$ be a sentential function of the sentential calculus in which, in addition to the constants '$\rightarrow$', '$\vee$', etc., the sentential variables 'X', 'Y', 'Z',... occur. We give the expression $\mathfrak{A}$ the following schematic form:

$$\mathfrak{A} = \text{'}\phi(X, Y, Z, ...)\text{'}.$$

Now let us suppose that the variables 'X', 'Y', 'Z',... do not represent sentences but denote sets of points of a topological space S. We give to the constants '$\rightarrow$', '$\vee$', etc., the meaning explained in 4.1. With this interpretation $\mathfrak{A}$ is no longer a sentential function, but a designatory function, which (exactly like 'X', 'Y',...) denotes a set of the space S. In view of this we can construct the following sentences $\mathfrak{A}_1$ and $\mathfrak{A}_2$:

$\mathfrak{A}_1 =$ '*For any open sets* X, Y, Z,... *of the space* S, *the set* $\phi(X, Y, Z,...)$ *is dense in* S.'

$\mathfrak{A}_2 =$ '*For any open sets* X, Y, Z,... *of the space* S,

$$\phi(X, Y, Z,...) = S\text{'}.$$

[Strictly speaking, $\mathfrak{A}_1$ and $\mathfrak{A}_2$ are not sentences but sentential functions, because free variables, e.g. 'S', occur in them.]

We now consider two expressions: '$\mathfrak{A}_1$ *holds* (or *is valid* or *is satisfied*) *in the space* S' and '$\mathfrak{A}_2$ *holds in the space* S.' The intuitive meaning of these expressions seems to be completely clear. Nevertheless, certain difficulties are encountered when one tries to explain their meaning in a strictly formal way.[1] In connexion with 2.5 the easiest way is to interpret the second expression as being synonymous with the expression: '$\mathfrak{A}$ *is satisfied by the matrix* $O(S)$'. In order to construct a definition for the first expression, we note that the sentence $\mathfrak{A}_1$, by virtue of 4.4, admits of the following equivalent transformation:

'*For any open sets* X, Y, Z,... *of the space* S,

$$\sim \sim \phi(X, Y, Z,...) = S\text{'}.$$

Hence we can say that the expression '$\mathfrak{A}_1$ *holds in the space* S' means the same as '$\sim \sim \mathfrak{A}$ *is satisfied by the matrix* $O(S)$'.

On the basis of these stipulations we obtain from 4.11, 4.12, and 4.24 the following formulations:

FIRST PRINCIPAL THEOREM. *Let* $\mathfrak{A}$ *be a sentence of the sentential calculus and* S *any topological space. The following conditions are then equivalent*:

    (i) $\mathfrak{A}$ *is provable in the two-valued calculus*;

    (ii) $\mathfrak{A}_1$ *holds in the space* S;

    (iii) $\mathfrak{A}_1$ *holds in every topological space*.

---

[1] The concept of validity (and of satisfiability) of a sentence belongs to semantics. For the problem of an exact definition of this concept cf. article VIII of the present work, in particular pp. 189 and 199.

SECOND PRINCIPAL THEOREM. *Let $\mathfrak{A}$ be a sentence of the sentential calculus and S any E-space. The following conditions are then equivalent*:

(i) $\mathfrak{A}$ *is provable in the intuitionistic calculus*;

(ii) $\mathfrak{A}_2$ *holds in the space S*;

(iii) $\mathfrak{A}_2$ *holds in every topological space.*

Ths. 4.8 and 4.14 can also be brought into an analogous form.

In connexion with the second principal theorem it is worth remembering that, in particular, all Euclidean spaces are E-spaces (cf. 3.11).

With the theorems just stated, the decision criteria which were mentioned in § 2 (cf. 2.19) can now be applied to topological sentences of the form $\mathfrak{A}_1$ or $\mathfrak{A}_2$ (and even to somewhat more extensive classes of topological sentences). We are in a position to decide, in each particular case, whether a sentence of this form is generally valid in topology.

In conclusion it should be noted that the sentential calculus can be interpreted in topology in various ways; the interpretation discussed above is obviously not the only possible one. If we are dealing, for example, with the two-valued calculus, we derive a quite trivial and in fact a general set-theoretical (not especially topological) interpretation from 4.14; every set S can in fact be made into an isolated topological space by putting $\bar{X} = X$ for every $X \subseteq S$ (cf. 3.6). A less trivial interpretation of this calculus is obtained in the following way. We consider the so-called *regular open sets* of a topological space S, i.e. sets $X \subseteq S$ for which $X = S - \overline{S - \bar{X}}$; let $\mathcal{O}'(S)$ be the system of all these sets. We define for the sets $X, Y \in \mathcal{O}'(S)$ the operations $\rightarrow$, $\wedge$, and $\sim$ exactly as in 4.1, but we put

$$X \vee 'Y = S - \overline{S - \bar{X} - \bar{Y}}$$

($X \vee 'Y$ is thus the smallest regular open set which includes X and Y). It can then be shown that the matrix

$$\mathsf{O}'(S) = [\mathcal{O}'(S),\ S,\ \rightarrow,\ \vee',\ \wedge,\ \sim]$$

is adequate for the system $\mathbf{3\Re}$.[1] In view of this we correlate with the sentence

$$\mathfrak{A} = `\phi(X, Y, Z, ...)'$$

the following sentence:

$\mathfrak{A}'_1 = `For\ any\ regular\ open\ sets\ X,\ Y,\ Z, ...\ of\ the\ space\ S,$ $\phi'(X, Y, Z, ...) = S'$,

where $`\phi'(X, Y, Z, ...)'$ is obtained from $`\phi(X, Y, Z, ...)'$ by replacing the sign $`\vee\,'$ by the sign $`\vee\,''$. It then appears that the first principal theorem remains valid if $`\mathfrak{A}_1'$ in it is replaced by $`\mathfrak{A}'_1'$.

§ 5. Interpretation of the Sentential Calculus in Boolean Algebra and in Related Mathematical Theories[2]

Generalized Boolean algebra is here regarded as a part of abstract algebra, namely as the theory of Boolean rings:

DEFINITION 5.1. *A set R with at least two elements is called a* Boolean ring *(with the fundamental operations $+$ and $.$) if the following formulas hold for all elements $x,\ y,\ z \in R$:*

(i) $x+y,\ x.y \in R$,

(ii) $x+(y+z) = (x+y)+z$,

(iii) $x = y+(x+y)$,

(iv) $x.(y.z) = (x.y).z$,

(v) $x.(y+z) = x.y+x.z$,

(vi) $x.y = y.x$,

(vii) $x.x = x$.

---

[1] This follows from the fact that the family $\mathcal{O}'(R)$ together with the operations $\vee\,'$, $\wedge$, and $\frown$ satisfy the postulates of Boolean algebra. This fact was noticed by me as far back as 1927, and was implicitly stated in Th. B of II, where, however, a different terminology was used; cf. XI, p. 341, footnote 2. Compare also Tarski, A. (81), Th. 7.23, p. 178, and footnote 25, p. 181, where a reference to an earlier paper of von Neumann is given.

[2] For various topics discussed in this section see the following papers: article XI of this book (the foundations of Boolean algebra, the concept of atom); article XII of this book (the theory of deductive systems); Stone, M. H. (67) (the relation of Boolean algebra to general abstract algebra and to the theory of fields of sets); Stone, M. H. (68) (the relation of Boolean algebra to general topology); Tarski, A. (80) (operations on Boolean-algebraic ideals).

Note 5.2. It can be shown that formulas (iii) and (vi) in 5.1 can be replaced by the following equivalent conditions:

(iii') *there is an element* $u \in R$ *such that*

$$x = y + u = u + y;$$

(vi') $(y + z) \cdot x = y \cdot x + z \cdot x.$

Hence we see that Boolean rings coincide with those rings (in the sense of abstract algebra) in which every element x satisfies 5.1 (vii).

DEFINITION 5.3. *Let R be a Boolean ring.*

(i) *We shall say that x is* divisible *by y, or that y is a* divisor *of x, in symbols $y \mid x$ (or $x < y$), if $x, y \in R$ and if there is a $z \in R$ such that $x = y \cdot z$.*

(ii) *We denote by* 0 *that element $x \in R$ which is divisible by every $y \in R$.*

The two symbols '$\mid$' and '0' can be defined in various other (equivalent) ways.

DEFINITION 5.4. *The Boolean ring R is called* atomistic *or* atomless *according to whether, for every element $y \in R$ distinct from 0, there are finitely or infinitely many elements x, which are divisible by y.*

Note 5.5. Every element y of a Boolean ring R is called an *atom* if there exist exactly two elements x which are divisible by y (in fact $x = 0$ and $x = y$). A ring R is atomistic if and only if every element $y \neq 0$ of R can be represented as a sum of a finite number of atoms[1]; R is atomless if and only if there are no atoms in R.

DEFINITION 5.6. *If R is a Boolean ring, then a non-empty set $I \subseteq R$ is called an* ideal *(in R), in symbols $I \in \mathscr{I}(R)$, if for any two elements x and y of I their sum $x + y$ is in I, and for every element y in I all the elements x such that $y \mid x$ are in I.*

---

[1] In article XI of the present work the term 'atomistic' is used in a wider sense.

DEFINITION 5.7. *If R is a Boolean ring, then for any $I, J \in \mathscr{I}(R)$ we put*

(i) $\quad I \to J = \sum\limits_{X \in \mathscr{I}(R), I.X \subseteq J} X;$

(ii) $\quad I \vee J = \prod\limits_{X \in \mathscr{I}(R), I + J \subseteq X} X;$

(iii) $\quad I \wedge J = I.J$ *(the intersection of I and J)*;

(iv) $\quad \sim I = I \to \{0\} \quad \left(= \sum\limits_{X \in \mathscr{I}(R), I.X = \{0\}} X \right).$

From 5.1, 5.6, and 5.7 we easily obtain

COROLLARY 5.8. *For every Boolean ring R we have*

(i) *$\{0\}$, $R \in \mathscr{I}(R)$, and in fact $\{0\}$ is the smallest and R the largest ideal in R;*

(ii) *if $I, J \in \mathscr{I}(R)$ then $I \to J, I \vee J, I \wedge J, \sim I$ are ideals in R, in fact $I \to J$ is the largest ideal X for which $I.X \subseteq J$ and $\sim I$ the largest ideal X for which $I.X = \{0\}$, moreover, $I \vee J$ is the smallest ideal which includes I and J, finally $I \wedge J$ is the largest ideal which is included in I and J.*

DEFINITION 5.9. *The ordered sextuple $[\mathscr{I}(R), R, \to, \vee, \wedge, \sim]$, where R is a Boolean ring, is denoted by $\mathfrak{I}(R)$.*

The following well-known theorem exhibits a close formal connexion between Boolean algebra and topology[1]:

THEOREM 5.10. *A normal topological space $R^{\times}$, which satisfies the following conditions, can be correlated with every Boolean ring R:*

(i) *there is a function F which maps the system $\mathscr{I}(R)$ onto the system $\mathcal{O}(R^{\times})$ in such a way that the formulas*

$$I \subseteq J \quad \text{and} \quad F(I) \subseteq F(J)$$

are equivalent for all $I, J \in \mathscr{I}(R)$ (in particular $F(\{0\}) = 0$ and $F(R) = R^{\times}$);

(ii) *$R^{\times}$ is isolated if and only if R is atomistic;*

(iii) *$R^{\times}$ is dense in itself if and only if R is atomless;*

(iv) *$R^{\times}$ is a space with a countable basis if and only if R is denumerable.*

[1] See Stone, M. H. (68).

THEOREM 5.11. *For every Boolean ring R, $\mathsf{I}(R)$ is a matrix;*
if $R^\times$ is the topological space correlated with the ring R according to
5.10, *then the matrices $\mathsf{I}(R)$ and $\mathsf{O}(R^\times)$ are isomorphic.*

Proof. We first show by means of 4.2 and 5.8 that the function
F, which by 5.10 (i) maps the system $\mathsf{I}(R)$ on to $\mathcal{O}(R^\times)$ in one-
one fashion, satisfies the following formulas:

$$F(I \to J) = F(I) \to F(J), \qquad F(I \vee J) = F(I) \vee F(J),$$

$$F(I \wedge J) = F(I) \wedge F(J), \quad \text{and} \quad F(\sim I) = \sim F(I)$$

for all I, $J \in \mathscr{I}(R)$ (where the signs '$\to$', '$\vee$', etc., are to be
interpreted on the left-hand side of each formula in the sense of
Boolean algebra, and on the right-hand side in the topological
sense). Hence we easily conclude, with the help of 2.1, 2.3,
4.5, 4.6, and 5.9, that $\mathsf{I}(R)$ is a matrix and is in fact isomorphic
with $\mathsf{O}(R^\times)$, q.e.d.

With the help of the last two theorems all the results of § 4
can be carried over to Boolean algebra:

THEOREM 5.12. *For every Boolean ring R we have*

$$\mathfrak{I}\mathfrak{R} \subseteq \mathfrak{E}(\mathsf{I}(R)) \subseteq \mathfrak{I}\mathfrak{R}.$$

[By 2.7, 4.8, 4.11, 5.11]

THEOREM 5.13. *For every Boolean ring R and every sentence $\mathfrak{A}$*
the conditions $\mathfrak{A} \in \mathfrak{I}\mathfrak{R}$ and $\sim\sim\mathfrak{A} \in \mathfrak{E}(\mathsf{I}(R))$ are equivalent.

[By 2.7, 4.12, 5.11 or 1.6, 5.12]

THEOREM 5.14. *Let R be a Boolean ring. In order that*

$$\mathfrak{E}(\mathsf{I}(R)) = \mathfrak{I}\mathfrak{R}$$

it is necessary and sufficient that R be atomistic.

[By 2.7, 4.14, 5.10 (ii), 5.11]

THEOREM 5.15. *If R is an atomless Boolean ring, then*

$$\mathfrak{E}(\mathsf{I}(R)) = \mathfrak{I}\mathfrak{R}.$$

[By 2.7, 3.10, 4.24, 5.10 (iii), (iv), 5.11]

Note 5.16. Note 4.25 can *mutatis mutandis* be applied to
Th. 5.15. The converse of this theorem does not hold. The
formula $\mathfrak{E}(\mathsf{I}(R)) = \mathfrak{I}\mathfrak{R}$, applies for example, to every denumerable Boolean ring R which, although not atomless itself, in-
cludes an atomless subring (i.e. an atomless ring $R_1 \subseteq R$ with

the same fundamental operations as R). On the other hand, many examples of *Boolean* rings can be given for which neither $\mathfrak{E}(\mathsf{I}(R)) = \mathfrak{ZR}$ nor $\mathfrak{E}(\mathsf{I}(R)) = \mathfrak{ZR}$ holds. To such rings belong, in particular, all infinite *completely additive* rings, i.e. Boolean rings R which satisfy the following condition:

for every set $X \subseteq R$ there exists the greatest common divisor of all elements $x \in X$ (i.e. an element y which is a common divisor of all $x \in X$ and is divisible by every other common divisor of these elements).

For the completely additive rings R the following property is characteristic: $\backsim I \vee \backsim \backsim I = R$ holds for every $I \in \mathscr{I}(R)$; in other words, the system $\mathfrak{E}(\mathsf{I}(R))$ contains all sentences $\mathfrak{A}$ of the form $\mathfrak{A} = \backsim \mathfrak{B} \vee \backsim \backsim \mathfrak{B}$ and cannot therefore be identical with $\mathfrak{ZR}$ (cf. 1.5). On the other hand, as is easily shown, no infinite completely additive ring R is atomistic, whence, by virtue of 5.14, $\mathfrak{E}(\mathsf{I}(R)) \neq \mathfrak{ZR}$.

Ths. 5.12–5.15 can be expressed in a form analogous to that of the first and second principal theorems of § 4. In particular, two sentences $\mathfrak{A}_1$ and $\mathfrak{A}_2$ of Boolean algebra can be correlated with every sentence $\mathfrak{A}$ of the sentential calculus in such a way that $\mathfrak{A}$ is respectively provable in the two-valued or in the intuitionistic calculus if and only if $\mathfrak{A}_1$ or $\mathfrak{A}_2$ holds for every Boolean ring. From this we obtain a decision criterion for sentences of the form $\mathfrak{A}_1$ or $\mathfrak{A}_2$.

The remarks at the end of § 4 regarding other possible interpretations of the sentential calculus can also be extended to Boolean algebra. The regular open sets in that case are to be replaced by the ideals I which satisfy the formula $\backsim \backsim I = I$.

All these results hold not only for the formal system of Boolean algebra, but also for every realization of this system. The best known of these realizations is the theory of fields of sets, i.e. of systems of sets which are closed under the operations of addition and subtraction. Every field of sets with at least two elements is, as is easily seen, a Boolean ring with the so-called symmetric subtraction $Y \ominus Y = (X-Y)+(Y-X)$ and the ordinary set-theoretical multiplication as the fundamental operations. The simplest examples of special Boolean rings, which were mentioned in 5.14–5.16, can be drawn directly from the theory of fields of sets. Every field of sets, for instance, which consists of all finite subsets of a given (finite or infinite) set is an example of an atomistic ring. In order to obtain a denumerable atomless ring, we consider the set X of positive rational numbers $x \leqslant 1$ and form the field of sets which consists of all sums of finitely many intervals $\subseteq X$ (the right-hand endpoint being included in the interval, but not the left-hand one). As examples of completely additive rings (in the sense of 5.16) the fields of sets which consist of all subsets of a given set may serve.

Another important realization of Boolean algebra is *general metamathematics*, i.e. the *theory of deductive systems*.[1]

Finally, some of the results obtained can be extended to a more general theory, namely to the *theory of lattices*. In fact we consider lattices in which the operation S of infinite addition (infinite join operation) is always performable and in which finite multiplication is distributive under both finite and infinite addition.[2] As examples of such lattices the system of all open sets of a topological space and the system of all ideals of a Boolean ring may be mentioned. Let L be a lattice of this type with at least two different elements; we put

$$1 = \underset{z \in L}{\mathsf{S}}\, z, \qquad 0 = \underset{z \in 0}{\mathsf{S}}\, z \quad \text{(0 denoting the empty set)},$$

$$x \vee y = x + y, \qquad x \wedge y = x.y,$$

$$x \to y = \underset{(x \wedge z) \vee y = y}{\mathsf{S}}\, z, \qquad \sim x = x \to 0$$

for all $x, y \in L$. It then results that $\mathsf{M}(L) = [L, 1, \to, \vee, \wedge, \sim]$ is a matrix, and that Ths. 5.12 and 5.13 remain valid if '$\mathsf{M}(L)$' is substituted for '$\mathsf{I}(R)$' in them. Ths. 5.14 and 5.15 can also be carried over to the theory of lattices. In this way we obtain an interpretation of the two-valued and of the intuitionistic calculus in the theory of lattices.

---

[1] See XII.

[2] More generally, we could consider here arbitrary distributive lattices (without infinite addition) in which for any elements $x,\ y \in L$ there is an element $z \in L$ such that (i) $x.z + y = y$, and (ii) $u + z = z$ whenever $u \in L$ and $x.u + y = y$. Such lattices are intimately related to the so-called Brouwerian algebras (or Brouwerian logics). Cf. Birkhoff, G. (7a) and McKinsey, J. C. C. and Tarski, A. (53a).

ABBREVIATIONS

Akad. der Wiss.	*Akademie der Wissenschaften*
Amer. J. Math.	*American Journal of Mathematics*
Amer. math. Soc.	*American Mathematical Society*
Ann. Soc. polon. Math.	*Annales de la Société Polonaise de Mathématique*
Bull.	*Bulletin de la* or *Bulletin of the*
C.R.	*Comptes Rendus du* or *Comptes Rendus des Séances de la*
Cambridge phil. Soc.	*Cambridge Philosophical Society*
ed.	edition
Erg. math. Koll.	*Ergebnisse eines mathematischen Kolloquiums*
Fund. Math.	*Fundamenta Mathematicae*
J.	*Journal of* or *Journal of the*
London math. Soc.	*London Mathematical Society*
Mat.-nat. klasse	*Matematisk-naturvidenskapelig klasse*
Math. Ann.	*Mathematische Annalen*
Math. Z.	*Mathematische Zeitschrift*
Mh. Math. Phys.	*Monatshefte für Mathematik und Physik*
Nat. Acad. Sci.	*National Academy of Sciences*
Phys.-math. Klasse	*Physikalisch-mathematische Klasse*
Proc.	*Proceedings of the*
Roy. Irish Acad.	*Royal Irish Academy*
Skrift. Videnskapsselskapet	*Skrifter utgit av Videnskapsselskapet i Kristiania*
Soc. Sci. Lett.	*Société des Sciences et des Lettres de*
Trans.	*Transactions of*
Trav.	*Travaux de la*

BIBLIOGRAPHY

(1) W. ACKERMANN, 'Über die Erfüllbarkeit gewisser Zählaus-
drücke', *Math. Ann.* c (1928), 638–49.

(2) K. AJDUKIEWICZ, *Z metodologji nauk dedukcyjnych* (From the
methodology of the deductive sciences) (Lwów, 1921).

(3) K. AJDUKIEWICZ, *Główne zasady metodologji nauk i logiki formalnej*
(Main principles of the methodology of the sciences and of formal
logic) (Warszawa, 1928).

(4) K. AJDUKIEWICZ, Logiczne podstawy nauczania (The logical
foundations of teaching), *Encyclopidja Wychowania*, ii (Warszawa, 1934), 1–75.

(5) P. BERNAYS, 'Axiomatische Untersuchung des Aussagenkalküls
der *Principia mathematica*', *Math. Z.*, xxv (1926), 305–20.

(5a) P. BERNAYS and M. SCHÖNFINKEL, 'Zum Entscheidungsproblem
der mathematischen Logik', *Math. Ann.*, xcix (1928), 342–
72.

(6) B. A. BERNSTEIN, 'Whitehead and Russell's theory of deduction
as a mathematical science', *Bull. Amer. math. Soc.*, xxxvii
(1931), 480–8.

(6a) E. W. BETH, 'On Padoa's method in the theory of definition',
Indagationes Mathematicae, xv (1953), 330–9.

(7) G. BIRKHOFF, 'On the combination of subalgebras', *Proc. Cambridge phil. Soc.*, xxix (1933), 441–6.

(7a) G. BIRKHOFF, *Lattice Theory*, Amer. Math. Soc. Colloquium Publications XXV (revised ed.). (New York, 1948).

(8) R. CARNAP, *Abriß der Logistik* (Wien, 1929).

(9) R. CARNAP, 'Bericht über Untersuchungen zur allgemeinen Axiomatik', *Erkenntnis*, i (1930), 303–7.

(10) R. CARNAP, *Logische Syntax der Sprache* (Wien, 1934).

(11) R. CARNAP, 'Ein Gültigkeitskriterium für die Sätze der klassischen Mathematik', *Mh. Math. Phys.*, xlii (1935), 163–90.

(11a) A. CHURCH, 'Special cases of the decision problem', *Revue Philo-
sophique de Louvain*, xlix (1951), 203–21.

(12) L. CHWISTEK, 'The theory of constructive types (Principles of
logic and mathematics)', Part I, *Ann. Soc. polon. Math.*,
ii (1924), 9–48.

(13) L. CHWISTEK, 'Neue Grundlagen der Logik und Mathematik',
Math. Z., xxx (1929), 704–24.

(14) L. CHWISTEK, 'Über die Hypothesen der Mengenlehre', *Math.
Z.*, xxv (1926), 439–73.

(14a) L. COUTURAT, *L'algèbre de la logique* (Paris, 1905).

(15) R. DEDEKIND, *Was sind und was sollen die Zahlen?* 5th. ed.
(Braunschweig, 1923).

(15a) A. H. DIAMOND and J. C. C. McKINSEY, 'Algebras and their
subalgebras', *Bull. Amer. math. Soc.*, liii (1947), 959–62.

(16) A. FRAENKEL, *Einleitung in die Mengenlehre*, 3rd. ed. (Berlin, 1928).

(17) M. FRÉCHET, 'Des familles et fonctions additives d'ensembles abstraits', *Fund. Math.*, iv (1923), 328–65.

(18) G. GENTZEN, 'Untersuchungen über das logische Schließen', *Math. Z.*, xxxix (1934), 176–210 and 405–31.

(19) T. GLIVENKO, 'Sur quelques points de la logique de M. Brouwer', *Academie Royale de Belgique, Bull. Classe des Sciences*, xv (1929), 183–8.

(20) K. GÖDEL, 'Die Vollständigkeit der Axiome des logischen Funktionenkalküls', *Mh. Math. Phys.*, xxxvii (1930), 349–60.

(21) K. GÖDEL, 'Einige metamathematische Resultate über Entscheidungsdefinitheit und Widerspruchsfreiheit', *Akad. der Wiss. in Wien, Mathematisch-naturwissenschaftliche Klasse, Akademischer Anzeiger*, lxvii (1930), 214–15.

(22) K. GÖDEL, 'Über formal unentscheidbare Sätze der *Principia mathematica* und verwandter Systeme I', *Mh. Math. Phys.*, xxxviii (1931), 173–98.

(23) K. GÖDEL, 'Eine Interpretation des intuitionistischen Aussagenkalküls', *Erg. math. Koll.*, iv (1933), 39–40.

(24) K. GRELLING and L. NELSON, 'Bemerkungen zu den Paradoxien von Russell und Burali-Forti', *Abhandlungen der Fries'schen Schule neue Folge*, ii (1908), 301–34.

(25) F. HAUSDORFF, *Mengenlehre* (Berlin und Leipzig, 1927).

(26) J. HERBRAND, 'Recherches sur la théorie de la démonstration', *Trav. Soc. Sci. Lett. Varsovie, Classe III*, Nr. 33 (1930).

(26a) H. HERMES, 'Zur Theorie der aussagenlogischen Matrizen', *Math. Z.*, liii (1951), 414–18.

(27) P. HERTZ, 'Über Axiomensysteme für beliebige Satzsysteme', *Math. Ann.*, ci (1929), 457–514.

(28) A. HEYTING, 'Die formalen Regeln der intuitionistischen Logik', *Sitzungsberichte der Preußischen Akad. der Wiss., Phys.-math. Klasse* (1930), 42–56.

(29) D. HILBERT, 'Die Grundlegung der elementaren Zahlenlehre', *Math. Ann.*, civ (1931), 485–94.

(29a) D. HILBERT, 'Die logischen Grundlagen der Mathematik', *Math. Ann.*, lxxxviii (1923), 151–65.

(30) D. HILBERT and W. ACKERMANN, *Grundzüge der theoretischen Logik* (Berlin, 1928).

(31) D. HILBERT and P. BERNAYS, *Grundlagen der Mathematik*, i (Berlin, 1934).

(31a) H. HIŻ, 'On primitive terms of logic', *J. Symbolic Logic*, xvii (1952), 156.

(32) E. V. HUNTINGTON, 'Sets of independent postulates for the algebra of logic', *Trans. Amer. math. Soc.*, v (1904), 288–309.

(33) E. V. HUNTINGTON, 'A set of postulates for abstract geometry exposed in terms of the simple relation of inclusion', *Math. Ann.*, lxxiii (1916), 522–59.

H h

458 BIBLIOGRAPHY

(34) E. V. HUNTINGTON, 'New sets of independent postulates for the algebra of logic', *Trans. Amer. math. Soc.*, xxxv (1933), 274–304, 551–8, and 971.

(35) E. V. HUNTINGTON, 'Independent postulates for the "informal" part of *Principia Mathematica*', *Bull. Amer. math. Soc.*, xl (1934), 127–36.

(36) S. JAŚKOWSKI, 'Recherches sur le système de la logique intuitioniste', *Actes du Congrès International de Philosophie Scientifique*, vi (Paris, 1936).

(36a) S. JAŚKOWSKI, 'Une modification des définitions fondamentales de la géometrie des corps de M. A. Tarski', *Ann. Soc. polon. Math.*, xxi (1948), 298–301.

(36b) S. JAŚKOWSKI, 'Trois contributions au calcul des propositions bivalent', *Studia Societatis Scientiarum Torunensis, Section A*, i (1948), 3–15.

(36c) B. JÓNSSON and A. TARSKI, 'Boolean algebras with operators', part I, *Amer. J. Math.*, lxxiii (1951), 891–939.

(37) T. KOTARBIŃSKI, *Elementy teorji poznania, logiki formalnej i metodologji nauk* (Elements of the theory of knowledge, formal logic, and the methodology of the sciences) (Lwów, 1929).

(38) C. KURATOWSKI, 'Sur la notion de l'ordre dans la théorie des ensembles', *Fund. Math.*, ii (1921), 161–71.

(39) C. KURATOWSKI, 'Évaluation de la classe borélienne ou projective d'un ensemble de points à l'aide des symboles logiques', *Fund. Math.*, xvii (1931), 249–72.

(40) C. KURATOWSKI, 'Über eine geometrische Auffassung der Logistik', *Ann. Soc. polon. Math.* ix (1931), 201.

(41) C. KURATOWSKI, *Topologie I* (Warszawa-Lwów, 1933).

(42) C. KURATOWSKI, 'Sur l'opération $\bar{A}$ de l'analysis situs', *Fund. Math.*, iii (1922), 192–5.

(43) C. KURATOWSKI, 'Les ensembles projectifs et l'induction transfinie', *Fund. Math.*, xxvii (1936), 269–76.

(44) C. H. LANGFORD, 'Analytical completeness of postulate sets', *Proc. London math. Soc.*, xxv (1926), 115–42.

(45) C. H. LANGFORD, 'Some theorems on deducibility', *Annals of Mathematics*, xxviii (1927), 16–40; 'Theorems on deducibility (Second paper)', ibid., 459–71.

(45a) S. LEŚNIEWSKI, *Podstawy ogólnej teorii mnogości I* (Foundations of general set theory I) (Moscow, 1916).

(46) S. LEŚNIEWSKI, 'Grundzüge eines neuen Systems der Grundlagen der Mathematik', §§ 1–11, *Fund. Math.*, xiv (1929), 1–81.

(47) S. LEŚNIEWSKI, 'Über die Grundlagen der Ontologie', *C. R. Soc. Sci. Lett. Varsovie, Classe III*, xxiii (1930), 111–32.

(47a) S. LEŚNIEWSKI, 'O podstawach matematyki' (On the foundations of mathematics, in Polish) *Przegląd Filozoficzny*, xxx (1927), 164–206; ibid., xxxi (1928), 261–91; ibid., xxxii (1929), 60–105; ibid., xxxiii (1930), 77–105; ibid., xxxiv (1931), 142–70.

(47b) S. LEŚNIEWSKI, 'Grundzüge eines neuen Systems der Grundlagen der Mathematik', § 12, *Collectanea Logica*, i (1938), 61–144.

(48) C. I. LEWIS, *A survey of symbolic logic* (Berkeley, Calif., 1918).

(48a) J. ŁOŚ, 'O matrycach logicznych' (On logical matrices) *Trav. Soc. Sci. Lett. Wrocław, Seria B*, Nr. 19 (1949).

(49) L. LÖWENHEIM, 'Über Möglichkeiten im Relativkalkül', *Math. Ann.*, lxxvi .(1915), 447–70.

(50) J. ŁUKASIEWICZ, 'O pojęciu wielkości' (On the concept of magnitude), *Przegląd Filozoficzny*, xix (1916), 35 ff.

(50a) J. ŁUKASIEWICZ, 'Logika dwuwartościowa' (Two-valued logic), *Przegląd Filozoficzny*, xxiii (1921), 189–205.

(51) J. ŁUKASIEWICZ, *Elementy logiki matematycznej* (Elements of mathematical logic) (1929).

(51a) J. ŁUKASIEWICZ, 'O znaczeniu i potrzebach logiki matematycznej' (On the significance and needs of mathematical logic), *Nauka Polska*, x (1929), 604–20.

(52) J. ŁUKASIEWICZ, 'Logistyka a filozofia' (Logistic and Philosophy), *Przegląd Filozoficzny*, xxxix (1936), 115–31.

(52a) J. ŁUKASIEWICZ, 'The shortest axiom of the implicational calculus of propositions', *Proc. Roy. Irish Acad.*, *Section A*, lii (1948), 25–33.

(53) S. MAZURKIEWICZ, 'Über die Grundlagen der Wahrscheinlichkeitsrechnung', *Mh. Math. Phys.*, xl (1934), 343–52.

(53a) J. C. C. MCKINSEY and A. TARSKI, 'On closed elements in closure algebra', *Annals of Mathematics*, xlvii (1946), 122–62.

(53b) J. C. C. MCKINSEY, 'A new definition of truth', *Synthèse*, vii (1948–9), 428–33.

(53c) A. MOSTOWSKI, 'Abzählbare Boolesche Körper und ihre Anwendung auf die allgemeine Metamathematik', *Fund. Math.*, xxix (1937), 34–53.

(53d) A. MOSTOWSKI, *Logika matematyczna* (Mathematical Logic) (Warszawa-Wrocław, 1948).

(53e) A. MOSTOWSKI, 'Some impredicative definitions in the axiomatic set-theory', *Fund. Math.*, xxxvii (1950), 111–24, and xxxviii (1951), 238.

(53f) A. MOSTOWSKI, *Sentences undecidable in formalized arithmetic* (Amsterdam, 1952).

(54) J. v. NEUMANN, 'Zur Hilbertschen Beweistheorie', *Math. Z.*, xxvi (1927), 1–46.

(55) J. NICOD, 'A reduction in the number of the primitive propositions of logic', *Proc. Cambridge phil. Soc.*, xix (1917), 32–41.

(56) J. NICOD, *La géometrie dans le monde sensible* (Paris, 1924).

(57) A. PADOA, 'Un nouveau système irréductible de postulats pour l'algèbre', *C. R. Deuxième Congrès International des Mathématiciens* (Paris, 1902), 249–56.

(58) A. PADOA, 'Le problème No. 2 de M. David Hilbert', *L'Enseignement Math.*, v (1903), 85–91.

(58a) C. S. PEIRCE, 'On the algebra of logic: a contribution to the philosophy of notation', *Amer. J. Math.*, vii (1885), 180–202.

(59) M. PIERI, 'La geometria elementare instituita sulle nozione di "punto" e "sfera"', *Memorie di Matematica e di Fisica della Società Italiana delle Scienze, Serie Terza*, xv (1908), 345–450.

(60) E. L. POST, 'Introduction to a general theory of elementary propositions', *Amer. J. Math.*, xliii (1921), 163–85.

(61) M. PRESBURGER, 'Über die Vollständigkeit eines gewissen Systems der Arithmetik ganzer Zahlen, in welchem die Addition als einzige Operation hervortritt', *C.R. Premier Congrès des Mathématiciens des Pays Slaves, Warszawa 1929* (Warszawa, 1930), 92–101.

(61a) A. ROSE, 'The degree of completeness of *M*-valued Łukasiewicz propositional calculus', *J. London math. Soc.*, xxvii (1952), 92–102.

(61b) J. B. ROSSER and A. R. TURQUETTE, 'Axiom schemes for *m*-valued propositional calculi', *J. Symbolic Logic*, x (1945), 61–82.

(61c) J. B. ROSSER and A. R. TURQUETTE, 'A note on the deductive completeness of *m*-valued propositional calculi', *J. Symbolic Logic*, xiv (1950), 219–25.

(61d) B. A. W. RUSSELL, *The principles of mathematics* (London, 1903).

(61e) B. A. W. RUSSELL, *Introduction to mathematical philosophy* (London, 1920).

(62) E. SCHRÖDER, *Vorlesungen über die Algebra der Logik (exakte Logik)* (Leipzig; i, 1890; ii, part 1, 1891; ii, part 2, 1905; iii, part 1, 1895).

(62a) K. SCHRÖTER, 'Deduktive abgeschlossene Mengen ohne Basis', *Mathematische Nachrichten*, vii (1952), 293–304.

(63) H. M. SHEFFER, *The general theory of notational relativity* (Cambridge, Mass., 1921).

(63a) H. M. SHEFFER, 'A set of five postulates for Boolean algebras with application to logical constants', *Trans. Amer. math. Soc.*, xv (1913), 481–8.

(64) TH. SKOLEM, 'Untersuchungen über die Axiome des Klassenkalküls und über Produktations- und Summationsprobleme, welche gewisse Klassen von Aussagen betreffen', *Skrift Videnskapsselskapet, I. Mat.-nat. klasse 1919*, No. 3 (1919).

(65) TH. SKOLEM, 'Logisch-kombinatorische Untersuchungen über die Erfüllbarkeit oder Beweisbarkeit mathematischer Sätze nebst einem Theorem über dichte Mengen', *Skrift Videnskapsselskapet, I. Mat.-nat. klasse 1920*, No. 4 (1920).

(66) TH. SKOLEM, 'Über einige Grundlagenfragen der Mathematik', *Skrifter utgitt av Det Norske Videnskaps-Akademi i Oslo, I. Mat.-nat. klasse 1929*, No. 4 (1929).

(66a) J. SŁUPECKI, 'Dowód aksjomatyzowalności pełnych systemów wielwartościowych rachunku zdań' (A proof of the axiomatizability of full systems of many-valued sentential calculus, in Polish), *C. R. Soc. Sci. Lett. Varsovie, Classe III*, xxxii (1939), 110–28.

BIBLIOGRAPHY 461

(66b) J. Słupecki, 'Pełny trójwartościowy rachunek zdań' (The full three-valued sentential calculus), *Annales Universitatis Mariae Curie-Skłodowska*, i (1946), 193–209.

(66c) B. Sobociński, 'Z badań nad teorją dedukcji' (Some investigations upon the theory of deduction) *Przegląd Filozoficzny*, xxxv (1932), 171–93.

(66d) B. Sobociński, 'Aksjomatyzacja pewnych wielowartościowych systemów teorji dedukcji' (Axiomatization of certain many-valued systems of the theory of deduction, in Polish), *Roczniki Prac Naukowych Zrzeszenia Asystentow Uniwersytetu Jozefa Pilsudskiego w Warszawie*, i (1936), 399–419.

(67) M. H. Stone, 'The theory of representations for Boolean algebras', *Trans. Amer. math. Soc.*, xi (1936), 37–111.

(68) M. H. Stone, 'Applications of the theory of Boolean rings to general topology', *Trans. Amer. math. Soc.*, xii (1937), 375–481.

(69) M. H. Stone, 'Applications of the theory of Boolean rings to general topology', *Fund. Math.*, xxix (1937), 223–303.

(70) M. H. Stone, 'Topological representations of distributive lattices and Brouwerian logics', *Časopis pro Pěstovánl Matematiky a Fysiky*, lxvii (1937–8), 1–25.

(71) A. Tarski, 'Sur les ensembles finis', *Fund. Math.*, vi (1925), 45–95.

(72) A. Tarski, 'Remarques sur les notions fondamentales de la méthodologie des mathématiques', *Ann. Soc. polon. Math.*, vii (1929) 270–2.

(73) A. Tarski, 'O pojęciu prawdy w odniesieniu do sformalizowanych nauk dedukcyjnych' (On the notion of truth in reference to formalized deductive sciences), *Ruch Filozoficzny*, xii (1930–1), 210–11.

(74) A. Tarski, 'Über definierbare Mengen reeller Zahlen', *Ann. Soc. polon. Math.*, ix (1931), 206–7.

(75) A. Tarski, 'Sur les classes d'ensembles closes par rapport à certaines opérations élémentaires', *Fund. Math.*, xvi (1930), 181–304.

(76) A. Tarski, 'Der Wahrheitsbegriff in den Sprachen der deduktiven Disziplinen', *Akad. der Wiss. in Wien, Mathematisch-natur-wissenschaftliche Klasse, Akademischer Anzeiger*, lxix (1932), 23–25.

(77) A. Tarski, *Cardinal algebras* (New York, 1949).

(78) A. Tarski, 'Einige Bemerkungen zur Axiomatik der Boole'schen Algebra', *C. R. Soc. Sci. Lett. Varsovie, Classe III*, xxxi (1938), 33–35.

(79) A. Tarski, 'On undecidable statements in enlarged systems of logic and the concept of truth', *J. Symbolic Logic*, iv (1939), 105–12.

(80) A. Tarski, 'Ideale in den Mengenkörpern', *Ann. Soc. polon. Math.*, xv (1937), 186–9.

(81) A. Tarski, 'Über additive und multiplikative Mengenkörper und

Mengenfunktionen', *C. R. Soc. Sci. Lett. Varsovie, Classe III*, xxx (1937), 151–81.

(82) A. TARSKI, 'The semantic conception of truth and the foundations of semantics', *Philosophy and Phenomenological Research*, iv (1944), 341–76.

(83) A. TARSKI, 'A problem concerning the notion of definability', *J. Symbolic Logic*, xiii (1948), 107–11.

(84) A. TARSKI, *A decision method for elementary algebra and geometry*, 2nd ed. (Berkeley, Calif., 1951).

(84a) A. TARSKI, 'Some notions and methods on the borderline of algebra and metamathematics', *Proc. International Congress of Mathematicians, Cambridge, Mass., 1950*, i (Providence, Rhode Island, 1952), 705–20.

(85) A. TARSKI and A. LINDENBAUM, 'Sur l' indépendance des notions primitives dans les systèmes mathématiques', *Ann. Soc. polon. math.*, v (1927), 111–13.

(86) O. VEBLEN, 'A system of axioms for geometry', *Trans. Amer. math. Soc.*, v (1904), 343–81.

(87) O. VEBLEN, 'The square root and the relations of order', *Trans. Amer. math. Soc.*, vii (1906), 197–9.

(87a) M. WAJSBERG, 'Aksjomatyzacja trówartościowego rachunku zdań' (Axiomatization of the three-valued propositional calculus)', *C. R. Soc. Sci. Lett. Varsovie, Classe III*, xxiv (1931).

(87b) M. WAJSBERG, 'Beiträge zum Metaaussagenkalkül I', *Mh. Math. Phys.*, xlii (1935), 221–42.

(87c) H. WANG, 'Remarks on the comparison of axiom systems', *Proc. Nat. Acad. Sci., Washington*, xxxvi (1950), 448–53.

(88) A. N. WHITEHEAD, *An enquiry concerning the principles of natural knowledge* (Cambridge, 1919).

(89) A. N. WHITEHEAD, *The concept of nature* (Cambridge, 1920).

(90) A. N. WHITEHEAD and B. A. W. RUSSELL, *Principia Mathematica* 2nd ed., i–iii (Cambridge, 1925–7).

(91) L. WITTGENSTEIN, *Tractatus logico-philosophicus* (London, 1922).

(92) E. ZERMELO, 'Über den Begriff der Definitheit in der Axiomatik', *Fund. Math.*, xiv (1929), 339–44.

SUBJECT INDEX

Abstraction, method of extensive, 29.
addition, logical, 114, 144, 175.
— — of systems ($\dotplus$), 351.
algebra, Boolean, atomistic, 334, 378.
— — extended, 323, 348.
— — ordinary, 320, 347.
— of logic, 168, 320, 347.
algorithm, sentential, 348, 374, 355.
analytical, 418.
antecedent, 39.
antinomy of heterological words (Grelling-Nelson), 165 n., 402.
— of Richard, 110, 119, 402.
— of the liar, 157.
atom (At), definition of, 334.
axiom, definition of, 179.
— logical, 284, 373.
— metamathematical, 31.
— of infinity, 243, 289.
— of intuitionistic calculus, 423.
— of sentential calculus, 423.
— system, finite, 35, 40, 76.
axiomatizability of system L_n, 49.
— — $L_{\aleph_0}$, 51.
axiomatizable, finitely, 35, 76.
— systems, 354.

Basis, of a set X of sentences, 35, 88.
— ordered, 76.

Calculus of classes, language of, 165, 168, 320.
— of systems, 342, 347, 350.
— — interpretations of, in set theory, 359.
— sentential, 38, 54, 348.
— — and topology, 421, 434.
— — interpretation of, in Boolean algebra, 449.
— — incomplete system of, 393.
— — interpretation of, in Boolean algebra, 449.
— — many valued, 47.
— — ordinary, 42.
— — of Heyting, 352, 400.
— — positive system of, 399.
— — restricted, 52.
— — topological interpretation of Two-valued and Intuitionistic, 434.

cardinal degree of capacity (of a system), 368.
— — of completeness, 368.
categorical, 390.
— (intrinsically and absolutely), 311.
— strictly, 313.
categoricity, 34, 100, 309, 313, 391.
category, semantical, 215, 218, 268.
— — first principle of, 216.
— — order of, 218.
— unifying, 230.
complementation, 123, 320, 347.
— double, law of, 124.
complete (defined), 93, 185, 287, 394.
— absolutely, 93.
— relatively, 94.
completeness, 34, 93.
— degree of, 35, 100.
— of concepts, 308.
— with respect to sets of terms, 313–14.
— with respect to specific terms, 311.
ω-completeness, 288.
concentric, 26.
concepts, completeness of, 308 f.
— semantical, definitions of, 252, 406.
conjunction (defined), 176, 422.
consequence, 30, 40, 63, 182, 252, 285, 345, 409 et seq.
— formal, 419.
— formalized concept of, 294.
— material, 419.
— of the nth degree (defined), 181.
consequent, 39.
consistency, 34, 90.
consistent (defined), 185, 287, 394.
ω-consistent, 288.
constants, 168.
contradiction, principle of, 197, 236, 257.
contradictory, 418.
convention T, 187, 236, 247.
convergence of an increasing sequence of sentences, 363.
— of systems, 363.
counter domain, 121.

Decision domain, 93.
deductive sciences, methodology of, 30.

INDEX OF NAMES OF PERSONS

INDEX OF SYMBOLS

$\overset{\circ}{\underset{k}{\sum}}$, 125.

$\overset{n}{\underset{k}{\sum}} t_k$, 175, 345.

$\underset{v \in X}{\sum} y$, 322, 325.

$\underset{x}{\sum} \phi(x)$, 144.

$\underset{\phi(x)}{\sum} f(x)$, 322.

$\underset{Y \in \mathfrak{R}}{\sum} Y$, 353.

$\mathfrak{S}$, 33, 70, 287, 350.

$\mathfrak{S}^X$, 359.

$\mathfrak{S}_X$, 359.

T_ω, 289.

T_∞, 292.

$T(\alpha)$, 380.

Tr, 9, 195, 250, 257, 275.

$tr(p)$, 3.

U_k, 176.

$\mathsf{U}_k^l \eta$, 283.

U_k, 123.

$\mathfrak{U}$, 35, 83.

V, 135, 283.

$\mathfrak{V}$, 34, 287, 365.

$\mathfrak{V}_\omega$, 288.

$\vee$, 2, 283, 422, 434, 451.

$\wedge$, 283, 422, 434, 451.

$\mathfrak{W}$, 34, 91, 287.

$\mathfrak{W}_\omega$, 288.

$\bar{X}$, 114, 347, 351, 387.

$\bar{x}$, 343.

ZK, 426.

$\mathfrak{ZR}$, 423.

ω, 61.

Ω, 61.

$\{\ \}$, 42 fn., 61.

$\sim$, (negation) 2, 424.

$\sim$, (equivalence) 34.

$\equiv$, 2, 348.

$\supset$, 2, 348.

$\curvearrowright$, 422, 434, 451.

$\underset{A}{\curvearrowright}$, 438.

$\underset{A}{\rightarrow}$, 438.

$\rightarrow$, 283, 422, 434, 451.

$.$, 2, 6, 114, 325, 347.

$\overset{\circ}{.}$, 123.

$+$, 114, 324, 344, 347.

$\dotplus$, 351.

$\overset{c}{+}$, 123.

$\nsubseteq$, 171.

$\subseteq$, 170, 321 fn.

$x \frown y$, 172, 281.

$=$, 171, 323, 347.

$\neq$, 171.

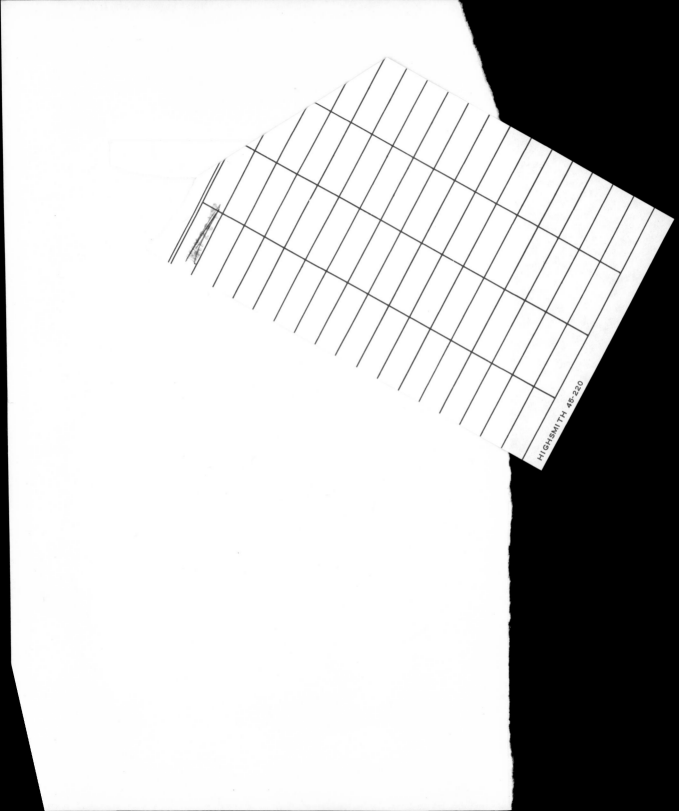

HIGHSMITH 45-220